Festschrift in Honor of the 75th Birthday
of Dr. Edwin P. Plueddemann

Silanes and other Coupling Agents

Editor: K.L. Mittal

Utrecht, The Netherlands
© 1992

VSP BV
P.O. Box 346
3700 AH Zeist
The Netherlands

© VSP BV 1992

First published in 1992

ISBN 90-6764-142-1

CIP-DATA KONINKLIJKE BIBLIOTHEEK, DEN HAAG

Silanes and other coupling agents / ed.: K.L. Mittal.
Utrecht: VSP. - Ill.
Festschrift in Honor of the 75th Birthday of Dr. Edwin P. Plueddemann.
Originally published in: Journal of Adhesion Science and Technology, ISSN 0169-4243;
Vol. 5 Nos. 4, 6 and 10, 1991; Vol. 6 No.1, 1992.
ISBN 90-6764-142-1 bound
NUGI 813
Subject heading: silanes

Printed in The Netherlands by Koninklijke Wöhrmann B.V.

CONTENTS

vi   *Contents*

Part 3: Applications of Silane Coupling Agents

Part 4: Non-Silane Coupling Agents

Silanes and Other Coupling Agents, pp. ix–x
Ed. K. L. Mittal
© VSP 1992

Preface

This book documents the proceedings of the Symposium on Silanes and Other Coupling Agents held in honor of Dr. Edwin P. Plueddemann, hosted by Dow Corning Corporation, in Midland, Michigan, April 3–5, 1991. Initially this event was planned to coincide with Ed's 75th birthday, but as you know he passed away on March 17, 1991, merely two weeks before this symposium. The symposium was a grand success and was very well attended. Ed might not be there in flesh and blood, but his spirit was with us. Naturally, he was fondly remembered and sorely missed throughout the symposium. Apropos, when I sent the program (containing 41 papers) to Ed in January 1991, he called me and said, 'Kash this is quite a program.' So we are pleased that he had the opportunity to see that hosts of researchers (both fundamental and applied) admired him and his work.

When the idea of organizing a symposium to honor Dr. Plueddemann was conceived, the choice of the topic for the symposium was quite patent. So without an iota of deliberation, we decided to hold the symposium under the rubric 'Silanes and Other Coupling Agents' as he had pioneered the development of silanes and had been working with this wonderful class of materials for many decades. The contributions to the symposium were by invitation only, and there was a consensus among those invited that this event was very opportune to honor this man who had done so much for the fields of adhesion and composites. I believe he had developed more silane coupling agents than any other human being. I had heard him being introduced as 'Mr. Silane', 'Pundit of Silane Chemistry' and with other very descriptive and apt appellations.

Here I will not detail the many and significant contributions of Dr. Plueddemann as these are nicely summarized by Prof. F. James Boerio in the following biography. However I would like to reinforce that Ed was a world-class applied scientist and his discoveries and developments of silanes have found tremendous usefulness in adhesively bonded joints and reinforced composites. Ed was really a man of multiplex tastes, as in addition to his scientific work, he had many other diversified interests. He was an excellent problem-solver and he could produce extremely valuable information with modest pieces of equipment and research outlay. He had told on many occasions that his universal substrate was a microscope slide and he screened his newly developed silanes first on this substrate. Also he could quantitatively measure peel strength by using different size apples. Ed was endowed with a megadose of creativity, and his creativity had brought him a legion of honors, awards and other accolades.

Now turning to the present volume, here the papers have been rearranged in a more logical manner *vis-à-vis* the order in which they were published in the special issues—Vol. 5, No. 4 (1991); Vol. 5, No. 6 (1991); Vol. 5, No. 10 (1991) and Vol. 6, No. 1 (1992)—of the *Journal of Adhesion Science and Technology* (JAST), which were dedicated to him as a Festschrift in his honor. As a large number of researchers and technologists evinced a great deal of interest in acquiring these four issues separately; concomitantly, we decided to bring out this commemorative volume.

The book opens with an excellent write-up summarizing the professional and personal life of Dr. Plueddemann by Dr. Boerio. The remainder of the text containing 37 papers is divided into four parts as follows: Part 1. General Papers; Part 2. Mechanisms of Silane Coupling Agents and Interfacial Studies; Part 3. Applications of Silane Coupling Agents; and Part 4. Non-Silane Coupling Agents. Among the topics covered include: Reminiscing on silane coupling agents (by Dr. Plueddemann himself); silanes as adhesion promoters; stability of silanes in aqueous solution, and the methods to improve the performance of silane coupling agents; kinetics of hydrolysis and condensation of silanes; adsorption of silanes studied by XPS, IETS, FT-IR; characterization of silanes by TOFSIMS; acid–base characteristics of silane treated glass; plasma polymerized organosilanes; applications of silanes in promoting adhesion, e.g. of polyimide coating and resist patterning layer, and in composites, adhesively bonded joints and bonding of dissimilar thermoset materials; and non-silane coupling agents: zirconium based, zircoaluminates, and metal alkoxides.

In essence this book documents the latest developments and offers commentary on current research activity in the area of silanes and other coupling agents. It should be of interest to both veterans and neophytes in the domain of coupling agents, and, hopefully, will provide a fillip for many rewarding developments in this class of materials. Also I would like to add that this book should serve as a constant reminder of the many and varied contributions of Ed and the high esteem he was held in by adhesionists and those engaged in composites research all over the globe.

Now I have the pleasant task of acknowledging the contribution, support and unflinching help of those involved in the organization of the symposium. First my thanks are due to the members of the Steering Committee (Drs. T. Lane, B. Arkles, J. Bell, M. Chaudhury and F. J. Boerio). My special appreciation goes to Drs. Thomas Lane, Manoj Chaudhury and Gary Wieber of Dow Corning Corporation for taking care of the myriad of details entailed in organizing a symposium. Last, but not least, the interest and cooperation of the appropriate management of Dow Corning Corporation in hosting this symposium is gratefully acknowledged. I must add that people at Dow Corning were very gracious hosts and we very much appreciated their hospitality.

K. L. Mittal
IBM U.S. Technical Education
500 Columbus Avenue
Thornwood, NY 10594
USA

Silanes and Other Coupling Agents, pp. xi–xviii
Ed. K. L. Mittal
© VSP 1992.

Edwin P. Plueddemann: Experimental laboratory chemist

F. JAMES BOERIO

Department of Materials Science and Engineering, University of Cincinnati, OH 45221, USA

Revised version received 4 January 1991

When Dr. Mittal invited me to write a personal account of Dr. Plueddemann for a symposium to be held in his honor and to commemorate his 75th birthday, I was pleased at the opportunity to present the story of his interesting life and productive career. Ed has strongly influenced my research and that of many others through his pioneering work in developing silane coupling agents for providing durability in polymer composites and in adhesive bonds. However, he has touched the lives of still others through his richly varied personal activities. I knew that there would be great interest in the stories of his career as a chemist and his life as a husband, father, preacher, musician, skier, stock market analyst, and golfer.

Edwin P. Plueddemann was born on April 1, 1916 in Galion, a small town in north central Ohio, about 50 miles north of Columbus. His grandfather, who emigrated to central Michigan from Germany in 1890, attended the University of Michigan for a time but left to enter a seminary affiliated with German Wallace College (now known as Baldwin Wallace College) in Berea, Ohio, where he studied to become a Methodist minister. His father was also a Methodist minister and graduate of the seminary at German Wallace College.

Since his family moved frequently, Ed attended school in many different communities. Several years of his early education were in schools in the small town of Pigeon, Michigan. He attended eighth grade in Detroit and high school in Akron, Ohio where his interest in chemistry began. During Ed's senior year in high school, his family moved to the area of Bay City, Michigan. Ed stayed in Akron to finish school and lived with the family of George Mayer, who was a chemist with a company providing analytical services to the rubber companies near Akron. Ed made frequent visits to Mayer's lab and soon developed an interest in the work which enabled him to 'demonstrate things'. However, Mayer advised him to wait until he was in college to begin studying chemistry.

After graduating from high school, Ed's intention was to follow his father and grandfather into the ministry. His first few years of higher education were at Bay City Junior College where he studied mostly liberal arts. However, in his second year, Ed took his first course in chemistry and knew immediately that was the career for him. He entered Baldwin Wallace College in 1936 and graduated with a B.S. Degree in Chemistry in 1938. He then entered Ohio State University and received the Ph.D. Degree in Chemistry in 1942. His dissertation research was concerned with organofluorine chemistry, especially the addition of hydrogen fluoride to acetylene.

After leaving Ohio State, Ed joined Westvaco Chlorine Products in Carteret, New Jersey where he conducted research on chlorine and phosphorous compounds and their application to insecticides, degreasing solvents, and plasticizers. While at Westvaco, Ed helped develop DDT and polyphosphate insecticides.

In 1947, Ed moved to the Plaskon Division of Libbey–Owens–Ford in Toledo, Ohio where he began what came to be a lifelong endeavor—research and development on organosilicon compounds and the application of these materials to problems related to adhesive bonding. Plaskon had been a successful producer of unsaturated polyesters for some time. However, the introduction of glass-fiber reinforced polymers created the need for innovative approaches for obtaining strong, durable bonds between the polymers and the fibers, especially during exposure to moisture at elevated temperatures. Ed theorized that making the glass surface hydrophobic would prevent water from accumulating at the interface and improve the durability of the composites but his initial experiments involving glass fibers treated with phenyltrimethoxysilane were disappointing. He then began investigating vinyltriethoxysilane, to which Plaskon had patent rights, and found that this compound, which could copolymerize with unsaturated polymers, produced an effective, water-soluble coupling agent when carefully hydrolyzed.

In 1955, Plaskon was sold to Allied Chemicals, which planned to halt production of silanes at Plaskon. Ed decided to leave and approached Dow Corning, which, at the time, had a policy of not hiring from competitors. However, Plaskon assured Dow Corning that it was no longer a competitor since it was discontinuing production of silanes and the way was cleared for Ed to join Dow Corning.

Upon joining Dow Corning, Ed Plueddemann began a systematic search for organofunctional silanes which would serve as coupling agents in glass-reinforced polyester and epoxy composites. What emerged was a class of compounds known as silane coupling agents which are extensively used for bonding dissimilar materials. Silane coupling agents have the general structure $X\mathrm{Si}Y_3$ where Y is an easily hydrolyzable functional group and X is a hydrolytically stable group usually chosen for reactivity with a particular polymer. Many theories have been suggested to explain the mechanisms by which silanes function but one of the most widely discussed is the chemical bonding theory. According to this theory, the trisilanol $X\mathrm{Si(OH)}_3$ is formed during hydrolysis of a silane in water. When inorganic substrates are treated with an aqueous solution of a silane, the trisilanols are adsorbed onto the substrates by hydrogen bonding with surface hydroxyl groups. During subsequent drying of the substrates, the silanols may condense to form a siloxane polymer and may also form metallo-siloxane bonds with the particles. The functional groups X are available for reaction with a matrix resin in a composite.

Ed recalls that the turning point in his work came about as the result of a meeting of the Society of the Plastics Industry (SPI) that he attended in 1961. At the meeting there was considerable discussion concerning the mechanisms by which coupling agents functioned but the results required to test the various theories were lacking. During the next year, Ed and his coworkers synthesized over a hundred new organosilicon compounds and tested them as coupling agents. They found that the compounds that performed best as coupling agents in reinforced composites were those that chemically reacted with the matrix resin. One of the key discoveries was that a crotonate-functional silane, which would not copolymerize with unsaturated polymers, was a poor coupling agent whereas the isomeric methacrylate-silane, which would readily copolymerize, was excellent. These results, which were presented at the SPI meeting in 1962, provided the first substantial support for the chemical bonding theory in which silane coupling agents are considered to form bonds between the filler and the matrix resin. Many inventions which eventually lead to significant improvements in reinforced polymers and adhesive bonding soon followed. In fact, γ-methacryloxypropyltrimethoxysilane, one of the compounds which was first prepared during this investigation, is still considered to be the standard for bonding unsaturated polymers to glass substrates.

Since joining Dow Corning, Ed has developed a wide range of materials that function as coupling agents. Included are epoxy functional silanes which are especially effective with urethanes and epoxies, unsaturated cationic amino-functional silanes which are used with a variety of thermoplastic and thermo-setting polymers, and anionic coupling agents which are widely used in vinylic resin composites. Although his initial goal was bonding polymers to inorganic substrates, Ed has also extended his work to important problems in bonding polymers to polymers. His current work is concerned with using silanes to obtain hydrothermally stable adhesive bonds to metal substrates. In his characteristic manner, Ed has set a goal of obtaining bonds that are impervious to the effects of moisture at elevated temperatures for approximately 1000 years.

During his 35 years with Dow Corning, Ed has developed numerous types of coupling agents, determined their effect on the properties of composites and adhesive bonds, and developed theories to explain his observations. He holds ninety-four US patents, has published thirty-six papers in journals, contributed chapters to nineteen books, lectured extensively in the US and abroad, and consulted with the US Government on applications of composites and adhesive bonding in space and the military. The book he authored entitled *Silane Coupling Agents* is considered indispensable reading for anyone interested in the strength and durability of composites and adhesive bonds.

Ed Plueddemann's work has brought him many awards. In 1971 he received the 'Man of the Year Award' from the Society of the Plastics Industry (SPI). He has received 'best paper' awards at SPI annual meetings six times, including the award for his paper in 1962. In 1984 he received the American Chemical Society Award for Creative Invention. That year he also received an award for Creative Development of Technical Innovation from the National Aeronautics and Space Administration. In 1984 he also received the Award for Outstanding Achievement and Promotion of the Chemical Sciences from the Midland, Michigan section of the American Chemical Society and was recognized as 'Inventor of the Year' by the Saginaw Valley Patent Law Association. In 1988, he was inducted into the Plastics Hall of Fame. That same year he received the Fred O. Conley Engineering Technology Award from the Society of Plastics Engineers, and the Adhesion Society Award for Excellence in Adhesion Science, sponsored by 3M.

The religion that was instilled in him by his father and grandfather has always played an important part in Ed's life. He preaches and leads the choir at Olson Community Church. His wife, Mary Margaret, whom he met while both were students singing in a choir at Ohio State University, is the church organist and Sunday school teacher. Three of Ed's children have been involved with the ministry. His son Jim, who is now a Professor at Wheaton College in Wheaton, Illinois, was a missionary in Nigeria for 14 years. His daughter Karen and her husband were associated with the Campus Crusade for Christ and were also missionaries in Nairobi, Kenya for 9 years. Daughter Beth and her husband now live in Midland where he works for Dow Chemical Co. but they spent several years in Nigeria while he taught the bible at a secondary school. Ed's third daughter, Margo, is married to a school teacher and lives in Ft. Wayne, Indiana where she helps him coach girls gymnastics. Ed and Mary Margaret made four trips to Africa while their children were missionaries there and three of their nine grandchildren were born in Africa.

Ed has many interests besides those in chemistry and religion. He is known as a musician, having played trombone with the Midland Symphony Orchestra, a skier, and stock market analyst. However, he is perhaps best known for his avid interest in golf. Ed is in great demand as a speaker and travels extensively but always takes his clubs with him. In fact, one of my first encounters with Ed was on a golf course in Kent, Ohio where we were both attending a meeting. I knew all about Ed and his work on silanes. In fact, it was partly my interest in his work that led me to begin my own work on silanes. However, no one had told me about Ed and golf. During a free afternoon, I joined Ed at a local course for what we agreed would be a quick game. Unfortunately, what was for him a game soon became a lesson for me. Ed hit virtually every shot onto the fairways and greens while I hit what seemed like a considerably higher number of shots into the woods. I have carefully avoided playing golf with Ed ever since.

Ed Plueddemann is now well past retirement age but shows few signs of slowing down. He was named Scientist Emeritus at Dow Corning in 1989 and continues research, teaching, inventing, publishing, and assisting his fellow scientists in industrial, government, and academic laboratories in solving practical and theoretical problems in composites and adhesive bonding. He says that the well of invention never runs dry and that he does not think there will ever be a natural conclusion to his work. I wish him many more years of extraordinary success in his research work and his other widespread interests.

PUBLICATIONS AND PATENTS OF EDWIN P. PLUEDDEMANN

Books

1. E. P. Plueddemann, *Silane Coupling Agents*, Plenum Press, New York, 1st edition (1982); 2nd edition (1990).
2. E. P. Plueddemann (Ed.), *Composite Materials*, Vol. 6, Academic Press, New York (1974).

Publications

1. A. L. Henne and E. P. Plueddemann, Addition of hydrogen fluoride to acetylenic compounds. *J. Am. Chem. Soc.* **65**, 587–589 (1943).
2. A. L. Henne and E. P. Plueddemann, Addition of hydrogen fluoride to haloolefins. *J. Am. Chem. Soc.* **65**, 1271–1272 (1943).

3. E. P. Plueddemann, Silicones join epoxies. *Chem. Eng. News* **36**(38), 72 (1958).
4. E. P. Plueddemann and G. Fanger, Epoxyorganosiloxanes. *J. Am. Chem. Soc.* **81**, 2632 (1959).
5. E. P. Plueddemann, Epoxysiloxane casting resins; a study in geometry and functionality. *J. Chem. Eng. Data* **5**, 59 (1960).
6. E. P. Plueddemann, H. A. Clark, L. E. Nelson and K. R. Hoffman, New silane coupling agents for reinforced plastics. *Mod. Plast.* **39**(12), 135 (1962).
7. H. A. Clark and E. P. Plueddemann, Bonding of silane coupling agents in glass-reinforced plastics. *Mod. Plast.* **40**(10), 133 (1963).
8. E. P. Plueddemann, Temperature effect in coupling thermoplastics to silane-treated glass. *Mod. Plast.* **43**(12), 131 (1966).
9. E. P. Plueddemann, Promoting adhesion of coatings through reactive silanes. *J. Coat. Technol.* **40**(516), 1 (1968).
10. E. P. Plueddemann, Silylating agents, in: *Kirk-Othmer Encyclopedia of Chemical Technology*, 2nd ed., Vol. 18, p. 260. Wiley and Sons, New York (1969).
11. E. P. Plueddemann, Silicone coupling agents in reinforced plastics, in: *Silicone Technology*, Chapter 2. Interscience Publishers, New York (1970).
12. E. P. Plueddemann, Reactive silanes as adhesion promoters to hydrophilic surfaces, in: *Treatise on Coatings, Film-Forming Compositions*, R. R. Myers and J. S. Long (Eds), Part III, Vol. I, Ch. 9, p. 381. Marcel-Dekker, New York (1970).
13. E. P. Plueddemann, Mechanism of adhesion of coatings through reactive silanes. *J. Paint Technol.* **42**, 600 (1970).
14. E. P. Plueddemann, Adhesion through silane coupling agents. *J. Adhesion* **2**, 184–201 (1970).
15. E. P. Plueddemann, Water is key to new theory on resin-to-fiber bonding. *Mod. Plast.* **47**(3), 92 (1970).
16. E. P. Plueddemann, Silanes in bonding thermoplastic polymers to mineral surfaces. *Appl. Polym. Symp.* **19**, 75 (1972).
17. E. P. Plueddemann, Cationic silane coupling agents for thermoplastics. *Polym.-Plast. Technol. Eng.* **2**, 89 (1973).
18. P. W. Erickson and E. P. Plueddemann, Historical background, in: *Composite Materials*, Vol. 6, Interfaces in Polymer Matrix Composites, E. P. Plueddemann (Ed.), Ch. 1. Academic Press, New York (1974).
19. E. P. Plueddemann, Mechanisms of adhesion, in: *Composite Materials*, Vol. 6, Interfaces in Polymer Matrix Composites, E. P. Plueddemann (Ed.), Ch. 6. Academic Press, New York (1974).
20. E. P. Plueddemann, Catalytic effects in bonding thermosetting resins to silane-treated fillers, in: *Fillers and Reinforcements for Plastics*, Advances in Chemistry Series, No. 134, Ch. 9, p. 86. American Chemical Society (1974).
21. E. P. Plueddemann and G. L. Stark, Catalytic and electrokinetic effects in bonding through silanes. *Mod. Plast.* **51**(3), 74 (1974).
22. E. P. Plueddemann, Mechanism of adhesion through silane coupling agents. *Composite Mater.* **6**, 173 (1974).
23. E. P. Plueddemann and W. T. Collins, Silane treated fillers in rubber, in: *Adhesion Science and Technology*, Part A, L. H. Lee (Ed.), p. 329. Plenum Press, New York (1975).
24. W. A. Korel, K. Hoefelmann and E. P. Plueddemann, Organofunctional silanes as coupling agents for reinforced plastics. *Kunststoffe* **65**, 760 (1975).
25. E. P. Plueddemann, Mechanism of adhesion of rubbers to minerals. *Adhesives Age* **18**(6), 36 (1975).
26. E. P. Plueddemann, Zwitterion silane-modified polymer latexes, in: *Polyelectrolytes and Their Application*, A. Heimbaum and E. Seligney (Eds), p. 119. D. Reidel Publishing, Dortrecht, The Netherlands (1975).
27. E. P. Plueddemann and W. T. Collins, The nature of rubber reinforcement by silane-treated mineral fillers. *Polym. Prepr. Am. Chem. Soc. Div. Polym. Chem.* **16**(1), 769 (1975).
28. E. P. Plueddemann and G. L. Stark, Optimizing rheological aspects of filler/coupling treatments. *Mod. Plast.* **54**(9), 102 (1977).
29. E. P. Plueddemann and G. L. Stark, Role of coupling agents in surface modification of fillers. *Mod. Plast.* **54**(8), 76 (1977).
30. E. P. Plueddemann, View from the inside. *Chemtech*, 217 (April 1977).

31. E. P. Plueddemann, Rheological aspects of coupling additives. *Rev. Plast. Mod.* **34**(259), 93 (1978).

32. E. P. Plueddemann, Silane coupling agents, in: *Additives for Plastics*, Vol. I, R. B. Seymour (Ed.), Ch. 6, p. 123. Academic Press, New York (1978).

33. E. P. Plueddemann, Effect of additives on viscosity of filled resins, in: *Additives for Plastics*, Vol. II, Ch. 6, p. 49. Academic Press, New York (1978).

34. E. P. Plueddemann and B. Thomas, Silane-treated mineral fillers in rubbers, in: *Developments in Rubber Technology*, **1**, A. Whelan and K. S. Lee (Eds), pp. 183–205. Applied Science Publishers, London (1979).

35. E. P. Plueddemann, Chemistry of silane coupling agents, in: *Silylated Surfaces*, D. E. Leyden and W. T. Collins (Eds), pp. 31–53. Gordon and Breach, New York (1980).

36. E. P. Plueddemann, Principles of interfacial coupling in fibre-reinforced plastics. *Int. J. Adhesion Adhesives*, 305 (October 1981).

37. E. P. Plueddemann, Silylating agents, in: *Kirk–Othmer Encyclopedia of Chemical Technology*, p. 962. Wiley and Sons, New York (1982).

38. E. P. Plueddemann, Silane adhesion promoters for polymeric coatings, in: *Adhesion Aspects of Polymeric Coatings*, K. L. Mittal (Ed.), p. 363. Plenum Press, New York (1983).

39. E. P. Plueddemann, Silane adhesion promoters in coatings. *Prog. Org. Coat.* **11**, 297–308 (1983).

40. E. P. Plueddemann, Bonding through coupling agents. *Polym. Prepr., Am. Chem. Soc., Div. Polym. Chem.* **24**, 196 (1983).

41. E. P. Plueddemann, Bonding through coupling agents, in: *Molecular Characterization of Composite Interfaces*, H. Ishida and G. Kumar (Eds), pp. 13–23. Plenum Press, New York (1985).

42. E. P. Plueddemann, Adhesion in mineral–organic composites, in: *Industrial Adhesion Problems*, D. M. Brewis and D. Briggs (Eds), Ch. 6. Orbital Press, Oxford (1985).

43. E. P. Plueddemann and P. G. Pape, Mixed silanes can give composites a performance boost. *Mod. Plast.* **62**(7), 78 (1985).

44. E. P. Plueddemann, Coupling and interfacial agents and their effects on mechanical properties of thermoplastics, in: *Mechanical Properties of Reinforced Thermoplastics*, D. W. Uegy and A. A. Collyer (Eds), Ch. 8. Elsevier Applied Science, New York (1986).

45. E. P. Plueddemann, P. G. Pape, and H. M. Bank, New coupling agents for improved corrosion-resistant composites. *Polym.-Plast. Technol. Eng.*, **25**, 223–232 (1986).

46. E. P. Plueddemann, Role of silanes in polymer–polymer adhesion, in: *Surface and Colloid Science in Computer Technology*, K. L. Mittal (Ed), p. 143. Plenum Press, New York (1987).

47. E. P. Plueddemann, History of silane coupling agents in polymer composites, in: *History of Polymeric Composites*, R. B. Seymour and R. D. Deanin (Eds). VNU Science Press, Utrecht, The Netherlands (1987).

48. M. K. Chaudhury, T. M. Gentle and E. P. Plueddemann, Adhesion mechanism of poly(vinyl chloride) to silane primed metal surfaces. *J. Adhesion Sci. Technol.* **1**, 29–38 (1987).

49. M. K. Chaudhury and E. P. Plueddemann, Bonding of vapor deposited gold to glass using organosilane primers. *J. Adhesion Sci. Technol.* **1**, 243–246 (1987).

50. E. P. Plueddemann, Silane primers for epoxy adhesives. *J. Adhesion Sci. Technol.* **2**, 179–188 (1988).

51. J. Jang, H. Ishida and E. P. Plueddemann, Hydrothermal stability and desorption behavior of a silane with a crosslinking additive on E-glass fibers, in: *Interfaces in Polymer, Ceramic, and Metal Matrix Composites*, H. Ishida (Ed.), p. 365. Elsevier Science Publishing, New York (1988).

52. J. Jang, H. Ishida and E. P. Plueddemann, The condensation and structure of a silane with a crosslinking additive in solution. *SAMPE Q.* **20**(4), 32 (1989).

53. J. Jang, H. Ishida and E. P. Plueddemann, Adsorption behavior of a silane with a crosslinking additive on a substrate. *44th Ann. Conf. Composites Inst. (SPI) 9-B* (February, 1989).

54. E. P. Plueddemann, Composites having ionomer bonds with silanes at the interface. *J. Adhesion Sci. Technol.*, **3**, 131–139 (1989).

55. E. P. Plueddemann, Coupling agents, in: *Concise Encyclopedia of Polymer Science and Engineering*, J. I. Kroschwitz (Ed.), p. 209. J. Wiley and Sons, New York (1990).

U.S. Patents

2,423,343	Degreasing Al and Mg Surfaces	1947
2,494,310	Purification of Alkyl and Substituted Alkyl Phosphates	1950
2,512,582	Mixed Alkyl Benzyl Phosphates and Their Production	1950
2,558,380	Producing Esters of Polyphosphoric Acids	1951
2,612,514	Esters of Strong Acids	1952
2,642,447	Preparation of Organosilicon Compounds	1953
2,717,900	Organopolysiloxane Fluids Stabilized with Organic Phosphites	1955
2,739,165	Stabilization of Aromatic Chlorosilanes	1956
2,888,475	Titanated Alkoxy Silanes	1959
2,946,701	A Method of Treating Glass and the Articles Made Thereby	1960
2,951,860	Organosilicon Hydroxy Phosphate Esters	1960
2,963,501	Organosilyl Peroxides	1960
3,046,250	Organosilicon Hydroxyalkylamine Polymers	1962
3,055,774	Epoxy-Functional Silicon Compounds on Textiles	1962
3,057,901	Hydroxyether Organosilicon Compounds	1962
3,079,361	A Treated Siliceous Article	1963
3,120,546	Epoxy Acyloxy Silanes	1964
3,137,599	Rocket Propellants	1964
3,146,248	Endblocked Polysilanes	1964
3,146,249	Polymethylhydrogensilanes	1964
3,179,612	α, β-Unsaturated Carboxylic-Ester-Substituted Organosilicon Compounds	1965
3,186,965	Vinyl Sulfide Organosilicon Compounds	1965
3,223,577	Silylated Polyepoxides	1965
3,240,754	Polymeric Silanes and Siloxanes Containing Oxetane Groups	1966
3,249,464	Siliceous Articles	1966
3,258,477	Acryloxyalkylsilanes, Compositions Thereof, and Methods of Fabricating Structures Therewith	1966
3,306,800	Bonding Thermoplastic Resins to Inorganic Materials	1967
3,317,369	Acryloxyalkylsilane Compositions and the Fabrication of Structures Therewith	1967
3,317,461	Organosilicon-Polysulfide Rubbers	1967
3,328,450	Silylalkyl Phenols	1967
3,338,867	Silanes and Siloxanes Containing Oxetane Groups	1967
3,388,144	Polymercaptoorgano and Polyhydroxyorgano Silanes and Siloxanes	1968
3,395,069	Bonding of Organic Resins to Siliceous Materials	1968
3,398,044	Nitroarylsilane Bonding Agents	1968
3,398,210	Acryloxyalkylsilanes, Compositions Thereof, and Methods of Fabricating Structures Therewith	1968
3,427,339	Alkoxyalkarylsilanes and Condensates Thereof	1969
3,427,340	Alkoxyalkarylsilanes and Condensates Thereof	1969
3,445,420	One-Component, Curable Organopolysiloxanes	1969
3,453,230	Room Temperature Curable Acrylate Rubbers	1969
3,455,877	Organosilicon Epoxides	1969
3,461,027	Organosilicon Primers for Siliceous and Metallic Materials	1969
3,461,095	Extracoordinate Silicone Complexes as Curing Agents for Epoxy Resins	1969
3,481,815	Silane Coupling Agents for Aryl-Containing, Thermally Stable Polymers II	1969
3,508,946	Pipe Coated with Epoxy Resin Composition Cured with Extracoordinate Silicon Complex and Process for Coating Said Pipe	1970
3,509,196	Aminoarylsilanolates and Siloxanols and Their Preparation	1970
3,554,952	Aqueous Dispersion of Aminoalkyl Silane-Aldehyde Reaction Products	1971
3,560,394	Curing Agent for Epoxy Resins	1971
3,560,543	Polyimino Organosilicon Compounds	1971
3,563,941	Silicone Modified Carnauba Wax	1971
3,567,497	Acryloxyalkylsilanes, Compositions Thereof, and Method of Fabricating Structures Therewith	1971

Part 1

General Papers

Silanes and Other Coupling Agents, pp. 3–19
Ed. K. L. Mittal
© VSP 1992.

Reminiscing on silane coupling agents

EDWIN P. PLUEDDEMANN

Dow Corning Corporation, Midland, MI 48686-0994, USA

Revised version received 30 December 1990.

Abstract—When K. L. Mittal asked me to provide a historical account of the applications of silane coupling agents in adhesion, I decided to write in the form of a personal account of my last 45 years in this line of study. No attempt is made to make the history comprehensive, or to recognize the host of other researchers who have contributed to our understanding of adhesion across an interface of dissimilar materials. It has been an immensely interesting area of study with many practical applications in composites and bonded structures.

Keywords: Chemical bonding; equilibrium bonding; hydrophobicity; interdiffusion; test methods.

1. INTRODUCTION

The first synthetic plastics were the phenol–formaldehyde resins introduced by Baekeland in 1907 [1]. Melamine and urea also react with formaldehyde to form intermediate methylol compounds which condense to cross-linked polymers much like phenol–formaldehyde resins. Paper, cotton fabric, wood flour or other forms of cellulose have long been used to reinforce these methylol-functional polymers. Methylol groups react with hydroxyl groups of cellulose to form stable ether linkages to bond filler to polymers. Cellulose is so compatible with these resins that no one thought of an interface between them, and the term 'reinforced composites' was not even used to describe these reinforced systems.

Glass fibers, introduced in the early 1940s, promised to bring in an era of improved composites resulting from their high strength and modulus. A glass surface, covered with silanol groups, in some ways resembles that of cellulose. The silanol groups are readily esterified by organic hydroxyl groups to form alkoxysilanes. It was soon found, however, that fiberglass was a very unsatisfactory reinforcement for phenolics, ureas, melamines, and the new unsaturated polyester resins. Although the resins bonded well to dry glass, the alkoxysilane bonds formed were easily hydrolyzed by water. Specific dry strength and modulus of fiberglass reinforced plastics exceeded that of aluminum or steel, but properties decreased dramatically under exposure to ambient humid conditions. Glass fibers and particulate minerals also have coefficients of thermal expansion much lower than plastics. Stresses across the interface resulting from this differential could exceed the strengths of the materials under extreme temperature cycling. Clearly, glass fibers and many mineral fillers were incompatible with organic polymers.

2. DEVELOPMENT OF THE COUPLING AGENT CONCEPT

When two materials are incompatible it is often possible to bring about compatibility by introducing a third material that has properties intermediate between those of the other two. Organosilicon compounds were obvious choices as potential coupling agents for glass-reinforced polymers since the silicon ends of the molecules are similar to glass, and organic groups on silicon could be synthesized for compatibility with organic polymers.

Early attempts at Plaskon Division of Libbey Owens Ford, and at Dow Corning in the late 1940s were to apply organosilanes to glass to make it hydrophobic. At Plaskon we treated glass and mineral fillers with phenyltriethoxysilane or ethyltriethoxysilane. The surfaces became hydrophobic, and we observed that treated fillers dispersed much more readily in liquid resins, but cured composites did not show much improvement in wet strength. Johannson at Dow Corning treated surfaces with methylchlorosilanes to obtain hydrophobic surfaces [2]. Treated ceramics maintained high electrical resistivity under humid conditions, but treated fiberglass or mineral fillers did not give improved polymer composites.

The Bjorksten organization was given a contract by the U.S. Air Force about 1949 (AFTR 6220) to explore the effect of glass fabric treatments on polyester laminate wet strength properties. A total of 2000 compounds were screened. The best of these, and still good by today's standards, was a nonaqueous solvent treatment (BJY) based on an equimolar adduct of vinyl trichlorosilane and β-chloroallyl alcohol [3].

Polyester laminate strength data obtained by the Bjorksten group using their BJY treatment are shown in Fig. 1. Fiberglass treatments with DuPont's hydrophobic chrome complex (Volan® 114) and no treatment (112) are also shown for comparison.

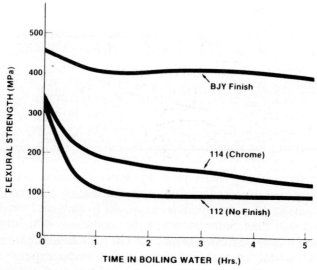

Figure 1. Wet-strength retention of fiberglass–polyester laminates with finishes of 1950 on the glass, from Bjorksten and Yaeger [3].

After 5 h in boiling water, the flexural strength of the BJY laminate was still above 400 MPa and was, in fact, substantially greater than the original dry strength of the other laminates. The BJY treatment never was a commercial success since it required organic solvents for application. Aqueous treatments based on a methacrylate-chrome [4] complex or vinyltriethoxy silane [5] soon became standard treatments for fiberglass in polyester composites. The two coupling agents were about equivalent in performance in composites, but the chrome complex provided glass cloth with very good hand and wettability. The industry generally associated the green color of chrome complex with coupled glass and were slower to accept colorless vinylsilane treatments. I was assigned a project of finding a green dye to put in vinyl silanes to mimic the appearance of chrome-treated glass. Somehow, I never got around to looking for green dyes. I felt that silanes should stand on their own merits and not try to mimic Volan®. Some representative commercial compounds that have been proposed as coupling agents are shown in Table 1.

2.1. Prior art

Someone had an early patent on the use of vinyl silanes in rubber compounding and was trying to extend this coverage to all plastic composites. At meetings dealing with interfaces, I found that the front row was usually occupied by lawyers from the reinforced plastics industry who were trying to find the extent of prior art in coupling agents. I commented that the coupling concept was not new and could not be patented. The Apostle Paul in 50 A.D. described coupling agents on a higher level when he wrote that "love is the bond of perfectness" [6]. Incompatible people can be made compatible through the bond of love. In 400 B.C. Plato used the coupling agent concept to explain how a universe comprising earth, air, fire and water could be a homogeneous whole when fire and water were incompatible [7]:

"It is not possible for two things to be fairly united without a third, for they need a bond between them which shall join them both, that as the first is to the middle so is the middle to the last, then since the middle become the first and the last, and last and the first both become middle, of necessity, all will come to be the same, and being the same with one another, all will be a unity."

At one meeting someone discussed prior art in glass-making in Egypt that went back to 2000 B.C. I then ad-libbed that prior art in coupling agents went back to "In the beginning God created the heavens and the earth" [8]. Before He created the world God knew that man would foul things up and make heaven and earth incompatible, so he purposed to send his Son to act as coupling agent between heaven and earth. The concept was before the creation of the world, while the reduction to practice was 2000 years ago.

2.2. Mechanisms

In 1960 there were three commercial coupling agents: a methacrylate–chrome complex (Volan®), vinyltriethoxysilane, and aminopropyltriethoxysilane. There was little evidence that Volan® or vinyl silanes on glass actually copolymerized with unsaturated resins in composites. It was even more difficult to explain how Volan® was almost as good as aminopropylsilanes in epoxy composites. There

Table 1.
Representative commercial coupling agents (1990)

Organofunctional group	Chemical structure
A. Vinyl	$CH_2=CHSi(OCH_3)_3$
B. Chloropropyl	$ClCH_2CH_2CH_2Si(OCH_3)_3$
C. Epoxy	$\overset{O}{\overset{\diagup\!\diagdown}{CH_2CHCH_2OCH_2CH_2CH_2Si(OCH_3)_3}}$
D. Methacrylate	$\overset{\quad CH_3}{\underset{\quad\vert}{CH_2=C-COOCH_2CH_2CH_2Si(OCH_3)_3}}$
E. Primary amine	$H_2NCH_2CH_2CH_2Si(OC_2H_5)_3$
F. Diamine	$H_2NCH_2CH_2NHCH_2CH_2CH_2Si(OCH_3)_3$
G. Mercapto	$HSCH_2CH_2CH_2Si(OCH_3)_3$
H. Cationic styryl	$CH_2=CHC_6H_4CH_2NHCH_2CH_2NH(CH_2)_3Si(OCH_3)_3 \cdot HCl$
I. Cationic methacrylate	$\overset{\quad CH_3 \qquad\qquad Cl^-}{\underset{\quad\vert \qquad\qquad\qquad \oplus}{CH_2=C-COOCH_2CH_2-N(Me_2)CH_2CH_2CH_2Si(OCH_3)_3}}$
J. Chrome complex	
K. Titanate	$(CH_2=\overset{CH_3}{\overset{\vert}{C}}-COO)_3TiOCH(CH_3)_2$
L. Crosslinker	$(CH_3O)_3SiCH_2CH_2Si(OH)_3$
M. Mixed silanes	$C_6H_5Si(OCH_3)_3 + F$
N. Formulated	Melamine resin + C

For J. Chrome complex:

$$CH_2=\overset{CH_3}{\underset{\vert}{C}}$$
$$\overset{\vert}{C}=$$

$$\begin{array}{ccc} R'OH & O \qquad\qquad O & R'OH \\ \diagdown\!\diagup & & \diagdown \\ Cl-Cr & & Cr-Cl \\ \diagup\quad\vert & & \vert\quad\diagdown \\ Cl\quad H_2O & O & H_2O\quad Cl \\ & \vert & \\ & H & \end{array}$$

was a rather heated discussion of mechanisms of coupling in a 1960 meeting of the Reinforced Plastics Division of the Society of Plastics Industry. Mechanisms proposed included:

- Weak boundary layers—coupling agents in some way eliminated weak boundary layers.
- Wettability—coupling agents gave improved wetting between polymers and glass. The ideal coupling agents would be a bromophenylsilane that would impart a high critical surface tension to treated glass.
- Acid–base effects—coupling agents were altering the acidity of a glass surface. The amine group of the aminopropyl silane probably bonded to glass.
- The button-down theory—coupling agents were sometimes seen to agglomerate into islands on a glass surface. The number and spacing of buttons would be important.

- A deformable layer theory—coupling agents produced a tough, flexible layer between glass and polymer.
- A restrained layer theory—coupling agents developed a highly cross-linked interphase region with a modulus intermediate between that of glass and the polymer.
- A chemical bonding theory—the coupling agent formed covalent bonds with both glass and polymer linking them together in their most stable manner.

During this rather heated discussion Dr. Fred McGarry of MIT whispered to me that at his school they taught that one should have adequate data before one proposed a general mechanism. I purposed in my mind that I would get that data. During the next year I cooperated with Lee Nelson, Hal Clark, and Ken Hoffmann at Dow Corning to prepare and test 120 new silanes as coupling agents in polyester composites and 22 different silanes in epoxy composites. We were able to report in 1961 that chemical reactivity with the polymer was of overwhelming importance in a silane to couple the polymer to glass [9]. The chemical bonding theory was rather firmly established (Fig. 2).

There were still two major problems with a simple chemical bonding theory that required later modifications.

(1) Oxane bonds between silane coupling agents and glass or metal oxides are rather easily hydrolyzed. For example, the Al–O–Si bond hydrolyzes rapidly in water, and yet very water resistant bonds between polymers and aluminum can be formed through silane primers.

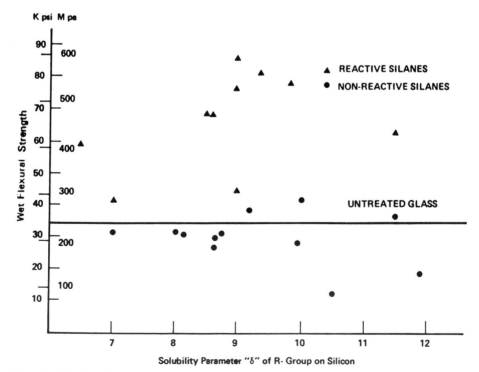

Figure 2. Polarity of non-reactive and reactive silanes related to their performance on fiberglass in polyester laminates after 2 h water-boil.

silica was catalyzed by aliphatic amines. Glass microbeads were treated with silanes with and without added amine catalysts and used to prepare polyester composites containing 22% by volume of filler. Tensile strengths of the composites were improved markedly by including an amine catalyst in the silane treatment (Table 3).

   2.3.2. A minimum penetration of water to the interface. Silane coupling agents, themselves, may contribute hydrophilic properties to the interface. This is especially true when amino-functional silanes are used as primers for reactive polymers such as epoxies and urethanes. The primer may supply much more amine functionality than can possibly react with the resin at the interface. Excess unreacted amine at the interface is hydrophilic and is responsible for the poor water-resistance of such bonds. The amount of excess amine at the interface may be minimized by using very dilute solutions of silane in the primer, or by washing the primed surface with water or organic solvent to remove all but a very thin layer of chemically adsorbed silane.

   A more effective way to use hydrophilic silanes in primers is to blend them with hydrophobic silanes such as phenyltrimethoxysilane (PS). Only a small proportion of an aminosilane (AS) is needed in the mixture to provide adequate reactivity with epoxies or urethanes (Table 4). Mixed siloxane primers containing

Table 3.
Effect of surface treatment on the tensile strength of glass beads-polyester composites

	Tensile strength of composites	
Surface treatment on glass	MPa	psi
None	40.6	5890
Methacryloxypropylsilane	54.9	7960
Methacryloxypropylsilane + amine	58.5	8490
Vinyltriethoxysilane	40.5	5880
Vinyltriethoxysilane + amine	50.9	7380

Table 4.
Adhesion of 1-component urethane to glass (cured 5 days at room temperature)

Ratio of silanes PS/AS in primer	Adhesion to glass (N/cm)		
	Dry	2 h boil H_2O	5 h boil H_2O
Unprimed control	3.0	nil	nil
0/100	C	nil	nil
50/50	C	nil	nil
80/20	C	C	0.1
90/10	C	C	C
95/5	C	C	C
99/1	C	C	1.1
99.8/02	C	C	0.7

   C = Cohesive failure in polymer at >20 N/cm; PS = phenyltrimethoxysilane; AS = aminosilane F of Table 1.

a major proportion of PS also have improved thermal stability typical of aromatic silicones. Note: PS, alone, is not a coupling agent for thermosetting polymers.

2.3.3. Polymer structures that hold silanols at the interface. Good examples of hydrolytically stable crosslinked structures are silica and silicate rocks. Although every oxane bond in these structures is hydrolyzable, a silicate rock is quite resistant to water. Each silicon is bonded to four oxygens under equilibrium conditions with a favorable equilibrium constant for bond retention. The probability that all four bonds to silicon can hydrolyze simultaneously to release soluble silicic acid is extremely remote. With sensitive enough analytical techniques it is possible to identify soluble silica as it leaches from rocks, but an individual rock will survive in water for thousands of years.

Silane coupling agents provide oxane bonds between organic polymer and metals or glass, but the interphase region is not highly crosslinked. Even though silane coupling agents are trifunctional in silanol groups there is a strong tendency for the silanols to condense to cyclic oligomers rather than to cross-linked structures. Addition of certain polyalkoxyfunctional silanes to standard coupling agents provides a high degree of siloxane crosslinking which gives water-resistant bonds that should be stable for a thousand years (Table 5).

Preferred additives are bis-trimethoxysilyl compounds of the general structure $(MeO)_3Si-R-Si(OMe)_3$. R may be a difunctional aliphatic radical $-(CH_2)_n-$, an aromatic ring (especially *meta* or *para*) or a substituted organic radical such as disiloxane, fumarate, or diamine. Bis-trimethoxysilylethane is being tested extensively as crosslinking additive for commercial silane coupling agents, and is available for limited sampling.

2.4. Silane bonds with polymers

2.4.1. General. Although simplified representations of coupling through organofunctional silanes often show a well aligned monolayer of silane forming a covalent bridge between polymer and filler (Fig. 5), the actual picture is much more complex. Coverage by hydrolyzed silane is generally many monolayer equivalents. The hydrolyzed silane condenses to oligomeric siloxanols that initially are soluble and fusible, but ultimately condense to rigid cross-linked structures. Contact of a treated surface with the polymer matrix is made while the siloxanols still have some degree of solubility. Bonding with the matrix resin can then take several forms. The oligomeric siloxanol layer may be compatible in the liquid matrix resin and form a true copolymer during resin cure. Alternatively, the oligomeric siloxanol may have partial solution compatibility with the matrix resin and form an interpenetrating polymer network (IPN) as the siloxanols and matrix resin cure separately with a limited amount of copolymerization. With thermoplastic resins, only the siloxane layer will crosslink, giving a pseudo-IPN (Fig. 6).

2.4.2. Acid–base reactions across the interface. This subject was well covered by a symposium on Acid–Base Interactions: Relevance to Adhesion Science and Technology in honor of Professor F. M. Fowkes on his 75th birthday. Most of these papers have been published in Vol. 4, No. 4, No. 5 and No. 8 (1990) and Vol. 5, No. 1 (1991) of the *Journal of Adhesion Science and Technology.*

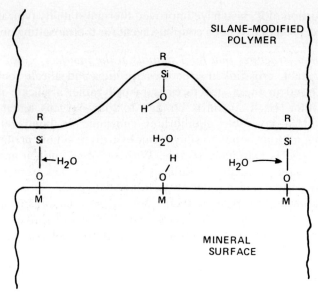

Figure 5. Idealized structure of a silane-modified interface.

Table 5.

Room-temperature-cured epoxy films on glass (hours in 70°C water to adhesion loss)

Primer on glass/curing agent	Polyamide	DMP-30	DEH-24
None	< 1	< 1	< 1
Z-6020	1	1	1
9/1 Z-6020/bis	2	3	1
8/2 Z-6020/bis	2	3	1
6/4 Z-6020/bis	3	3	2
1/1 Z-6020/bis	4	4	8
1/9 Z-6020/bis	C	C	C
Bis alone	C	6	C
Z-6040	C	8	C
9/1 Z-6040/bis	C	36	C
8/2 Z-6040/bis	C	C	C

bis $= (MeO)_3SiCH_2CH_2Si(OMe)_3$; C = No failure after 1 week in 70°C water.

The acid–base properties of polymers, fillers and silane additives, as described by Fowkes [14] can be used to predict the effect of silanes on the dispersion of fillers in polymer, and viscosity of the mix. In general, opposites attract (give good dispersion) while like materials repel (poor dispersion) [15]. The effect of cationic silane (Z-6032) on the dispersion of silica (acidic filler) in this particular unsaturated polyester resin (acidic polymer) is shown in Table 6. Addition of Z-6032 in increments to a silica-filled polyester resin lowered the viscosity of the mixture to a minimum at about 0.4% silane based on the filler.

Acid–base properties can also be used to orient amino-functional silanes on glass surfaces with strong effects on performance of the coupling agent in

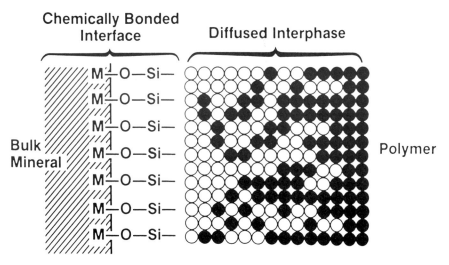

Figure 6. Bonding through silanes by interdiffusion. ○: regions of coupling agent, ●: regions of polymer.

Table 6.
Viscosities of silica-filled polyester (Z-6032 silane added in increments)

% Silane based on filler	Viscosities (cP) (Brookfield) at 45% filler (Minusil 5 μm)		
	30 rpm	12 rpm	6 rpm
None	12 200	23 000	40 000
0.1	11 000	21 000	33 000
0.2	8800	15 800	27 000
0.3	5400	9500	15 000
0.4	2600	2600	3000
0.5	2500	2600	2600
1.0	2200	2200	2200
2.0	3000	3000	3000

composites. A nonionic methacrylate ester-functional silane and a cationic vinyl-benzyl functional silane were compared as coupling agents on silica in polyester composites [16]. Silanes were applied to silica from aqueous solutions in a pH range of 2–12. Mechanical properties of composites were influenced by orientation of silanes on the silica surface as predicted by electrokinetic effects. Silica treated with the cationic silane is most sensitive to the pH of application, showing a steep improvement as the pH is reduced from 12–2. The nonionic silane is much less sensitive to pH of application, but also is most effective at a pH of 2 (Fig. 7).

I once mentioned in a lecture that acid–base properties across the interface were not important in developing water resistant bonds to metal surfaces and cited the data of Table 7 [17]. Professor Fowkes was present and commented that there are numerous examples where acid–base reactions across the interface

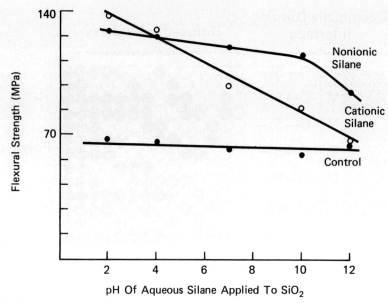

Figure 7. The pH of aqueous silane treament (1%) on SiO$_2$ vs. flexural strength of polyester castings (50% 5 μm minusil in resin).

Table 7.
Wet-adhesion of methylmethacrylate copolymers (4 mol% functional modification)

Functional group in polymer	Adhesion to surface (IEPS)							
	Si (2)	Sn (4)	Glass (5)	Ti (6)	Cr (7)	Al (9)	Steel (10)	Ag (12)
None	1	0	1	1	1	1	2	1
—OH	5	3	3	3	3	3	5	2
—COOH	2	3	2	5	2	3	5	3
—N(CH$_3$)$_2$	5	7	4	4	5	10	3	2
—Si(OCH$_3$)$_3$	10	9	10	9	6	9	9	3

Ratings from 0 = nil to 10 = best, soaking in water at room temperature.

were responsible for water resistant bonds, but he was too much of a gentleman to point out my obvious flaw in reasoning. Polymethylmethacrylate is a basic polymer, and with only 4 mol% acid modification it was both acid and basic, which explained why I saw no acid–base effect on different metals. When I repeated this work with styrene copolymers there was an obvious acid–base effect in bonding to different metals (Table 8).

Although acid–base effects can be important in adhesion our studies indicate that bonding to metals through silane coupling agents is not by an acid–base mechanism, but probably through Si–O–M oxane bonds. As with glass, the hydrolysis and formation of oxane bonds are true equilibria, but the individual equilibrium constants are not known.

Table 8.
Wet-adhesion of styrene copolymers (4 mol% functional modification)

Functional group in polymer	Adhesion to surface (IEPS)							
	Si (2)	Sn (4)	Glass (5)	Ti (6)	Cr (7)	Al (9)	Steel (10)	Ag (12)
None	1	0	1	1	2	2	2	1
—COOH	1	1	1	1	1	8	3	4
—N(CH$_3$)$_2$	2	3	3	2	2	8	1	5
—Si(OCH$_3$)$_3$	10	10	10	8	3	10	3	10

Ratings from 0 = nil to 10 = best, soaking in water at room temperature.

2.5. Test methods and test equipment

An important factor in evaluating hundreds of adhesion promoters in composites was to have very simple tests for initial screening. This was rather easily attained since the differences in performance were so great. A good coupling agent often improved the water resistance of a composite a thousand fold or more. An outstanding coupling agent may be ten times or a hundred times better than a good coupling agent.

2.5.1. Film tests. A favorite method of testing the water resistance of the bond of a polymer to glass or metals was to fuse or cure films (0.5 mm) of polymer on primed and unprimed glass microscope slides or metal coupons. Adhesion was observed by attempting to loosen the films with a razor blade. Samples were soaked in water at designated temperature until the films could be loosened. Films were rated according to the length of time they withstood the action of water (Table 9). Flexible polymers were rated according to the time they withstood water before their 90° peel strength dropped below a designated value, e.g.

Table 9.
Water resistance of epoxy bonds to metals (DER-331 + 10% DMP-30, 1 h, 80°C)

Surface	Adhesion in 70°C water, time to failure		
	Control (h)	2% Z-6020 added (days)	Primer[a] (days)
Glass	1	> 12	> 12
Tin	2	2	10
Titanium	2	1	5
Chromium	2	1	3
Solder	8	12	> 12
C.R. steel	2	3	7
St. steel 304	4	3	10
Brass	2	5	10
Copper	12	10	6
Nickel	2	3	7
Magnesium	4	1	2
Silver	8	3	7
Gold	4	3	10

[a] Primer = 20% Xl-6100 (M in Table 1) in methanol.

10 N/cm. Measuring 90° peel strength with an Instron® machine is rather complex and tedious, so we developed a simple Plutostron device shown in Fig. 8 that gave rapid, accurate results. A pail was clamped to the film and held at a 90° peel angle while adding water to the pail until initial peel was observed. The limit for the Plutostron on a one inch width film was 10.3 kg/in. or about 30 N/cm. To make the device more scientific we thought of pouring apples into the pail, counting the apples, and getting our answer directly in Newtons (as ~ 100 g apple makes a newton).

2.5.2. Impact tests. Notched Izod® tests do not differentiate well among different filled thermosetting polymer samples. We made 20 g castings in aluminum weighing dishes and measured impact strength by dropping them flat-side against a concrete floor. By giving them a spin they would drop nicely on the flat side. If they didn't break at one head height we tried two or three head heights, up to the ceiling beams or light fixtures, and got some good comparisons on impact strength. Gary Stark was working with me at that time, and since he was the taller, we settled on his head height as the standard unit of impact strength. This was very appropriate since 'stark' is German for 'strong' and a 4-Stark plastic was really a strong material.

2.5.3. Laminate tests. Silanes that gave favorable results in screening tests generally showed up well in fiberglass cloth-reinforced plastic composites. Results obtained with four different thermosetting resins are shown in Fig. 9. Although epoxy resins and phenolic resins, by nature, are better than polyesters and melamines in bonding to glass, with appropriate coupling agents they all give composites with comparable flexural strength. This suggests that interfacial failure has been eliminated as a variable in these composites and the full strength of glass was realized in reinforcement.

2.6. The free enterprise system

In 1961 on my way to an SPI Conference in Chicago, I stopped to give a talk on

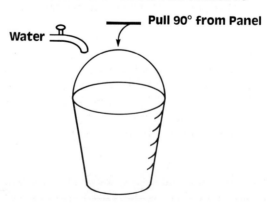

Static Peel Measurement

Water
Pull 90° from Panel

Figure 8. A simple laboratory device for measuring peel strength.

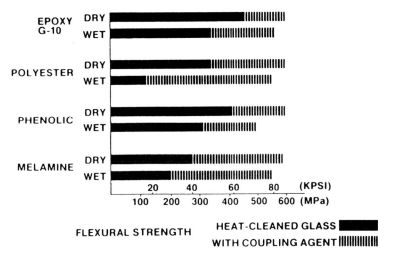

Figure 9. Effect of coupling agents on flexural strengths of glass cloth-reinforced thermosetting resins, dry and wet after 24 h boil.

adhesion at the Wheaton College Chemistry Department. At the SPI Conference I was going to introduce our new methacrylate-functional silane coupling agent that was a massive improvement over vinylsilanes or Volan® in polyester composites. I mentioned a Russian article I had read in which a methacrylate silane (undisclosed structure) was not as good as vinylsilanes and no improvement over mineral oil treatment on glass in polyester composites. But the Russians obviously failed to synthesize an organofunctional methacrylate silane. However, I predicted to the Wheaton students that our competitors would push us to action and that the methacrylate functional silane would soon be standard fiberglass treatment for polyester composites. In a totalitarian Russian system it would take much longer to introduce this new material. When I got to the SPI meeting, I found that our competitors had data sheets on the methacrylate-functional silane, and within a year it was standard treatment for fiberglass reinforced polyesters. Ten years later I met a chemist from a Czechoslovakian fiberglass plant. I asked him which of the Russian silanes he used on his glass. He said he used only Western silane coupling agents, the good ones were not available in the East.

I was fortunate as an industrial scientist for Plaskon and Dow Corning to be allowed to concentrate for over 40 years on organofunctional silanes and their applications in surface modification of minerals. I chose a scientific ladder rather than an administrative ladder, so I could stay in the laboratory with one or two assistants and develop a practical feel for polymer composites. Understanding of interfacial phenomena was helped immensely by academic workers such as Professors Koenig and Ishida at Case Western Reserve University and Professor Boerio at the University of Cincinnati. They and their students conducted extensive analytical studies of the interface to demonstrate the reality of some of the concepts I had proposed from indirect evidence of performance tests.

Applications were not limited to reinforced composites. After working with solid silicate surfaces for many years, I started playing with silane modification of

aqueous solutions of soluble silicates. I soon found ways to stabilize silicate
solutions over wide ranges of pH and temperature. One of these modifications
has become a commercial corrosion inhibitor for engine coolants.

3. RETIREMENT?

In 1985 I retired from Dow Corning Corporation, but stayed on as a full-time
private contractor. This conflicted with IRS guidelines on taxation, so within 6
months Dow Corning hired me back into regular employment. This rang a bell in
the company computer system and I was scheduled to attend a meeting for new
employees to hear about company policies and especially guidelines on sexual
harassment. I declined, and stated that after 30 years I was fairly familiar with
company policies, and that during this time none of the ladies had ever harassed
me sexually. I marvel at their self restraint.

In 1989 I again retired from full time employment into a new position of
Scientist Emeritus created by Dow Corning. In this position I still keep contact
with Dow Corning, have an office, a laboratory, a secretary, a telephone and
mailing address and am free to write, travel, lecture and consult with outside
companies. As long as my health allows, I hope to continue active participation
in the study of adhesion at interfaces.

4. EPILOGUE

Progress made in developing coupling agents for fiberglass-reinforced polyester
composites is illustrated in Fig. 10. The world standard silane for this application
was introduced in 1962. Someone asked me to bring this chart up-to-date, but
there is little to add. It is difficult to improve on perfection. Fiberglass reinforced
polyester boats have been in the water for over 20 years without any moisture-
induced debonding of polyester from glass. In certain primer applications on

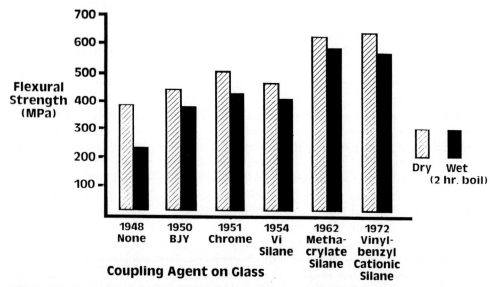

Figure 10. Progress in coupling agents for fiberglass–polyester composites.

metals, formulations with this silane can be improved by adding crosslinker (L of Table 1). This does not mean that there is no future for research in coupling agents. The science and technology of bonding through coupling agents still has many practical and theoretical challenges. Many of these will be met by people attending this symposium. One of my goals has been to use silanes to eliminate moisture induced failure across the interface in any application for 1000 years under conditions of use.

REFERENCES

1. L. H. Baekeland, *Ind. Eng. Chem.* **1**, 149–161 (1909), reprinted in *Chem-tech* **6**, 40–53 (1976).
2. O. K. Johannson and J. J. Torok, *Proc. Inst. Radio Engrs* **34**, 296 (1964).
3. J. Bjorksten and L. L. Yaeger, *Mod. Plast.* **29**, 124 (1952).
4. M. Gobel and R. Iler, U.S. Patent 2,455,666 to DuPont (1951).
5. R. Steinman, U.S. Patent 2,688,006 to Libby-Owens Ford (1954).
6. The Apostle Paul, Col. 3:14 (50 A.D.).
7. Plato in *Tinaeus*, English translation by R. D. Archer, Hinds, McMillan and Co., London (1888).
8. *The Bible*, Genesis 1:1.
9. E. P. Plueddemann, H. A. Clark, L. E. Nelson and K. R. Hoffmann, *Mod. Plast.* **39**, 136 (1962).
10. E. P. Plueddemann, *Silane Coupling Agents*, 2nd edn. Plenum Press, New York (1990).
11. H. Ishida and J. L. Koenig, *J. Polym. Sci., Polym. Phys. Ed.* **18**, 233 (1980).
12. E. R. Pohl and F. D. Osterholtz, in: *Silanes, Surfaces and Interfaces*, D. E. Leyden (Ed.), pp. 481–500. Gordon and Breach, New York (1986).
13. R. L. Kaas and J. L. Kardos, 32nd ANTEC meeting of Society of Plastics Engineers, Paper 22 (1976).
14. F. M. Fowkes, *J. Adhesion Sci. Technol.* **1**, 7–27 (1987).
15. E. P. Plueddemann, in: *Additives for Plastics*, Vol. 2, p. 49, R. B. Seymour (Ed.). Academic Press, New York (1978).
16. E. P. Plueddemann and G. L. Stark, *Mod. Plast.* **51**, 31 (1974).
17. E. P. Plueddemann and G. L. Stark, SPI 20th Ann. Tech. Conf. Reinforced Plastics 21-E (1973).

Silanes and Other Coupling Agents, pp. 21–47
Ed. K. L. Mittal
© VSP 1992.

Organosilanes as adhesion promoters

P. WALKER

AWE, Building SB43, Aldermaston, Reading, Berkshire RG7 4PR, UK

Revised version received 12 December 1990

Abstract—The use of organosilanes as adhesion promoters for surface coatings, adhesives and syntactic foams is described and reviewed in the light of published work. Data are presented on the beneficial effect of silanes, when used as pretreatment primers and additives, on the bond strength of two pack epoxide and polyurethane paints applied to aluminium and mild steel. It is shown that silanes when used as additives to structural epoxide and polyurethane adhesives are less effective than when used as pretreatment primers on metals but are highly effective on glass substrates. The compressive properties of glass microballoon/epoxide syntactic foams are shown to be markedly improved by the addition of silanes.

Keywords: Organosilane; adhesion; paint; adhesive; foam; epoxide; polyurethane; water.

1. INTRODUCTION

Silanes of the general structure $R-Si(OR^1)_3$, where R is an organo-functional group, e.g. vinyl, amino or mercapto, and R^1 may be a halide, acyloxy or alkoxide group, capable of hydrolysis, have been available commercially for many years. They have found extensive use in the composites field where the disastrous effects of water on the mechanical properties of glass reinforced phenolic and polyester resin based composites were recognized. More recently their value in improving the adhesion of surface coatings and adhesives has been investigated and in particular the improvement in the 'wet' adhesion of coatings and adhesives which results from their use [1–6].

From the 1960s considerable effort had been expended on both the theory and practice of silane chemistry and technology, and by the mid 1970s Erickson and Plueddemann were able to discuss the merits of several different theories of coupling [7], work reinforced and extended by Bascom [8] and Boerio and co-workers [9–10]. It is not clear from the literature to what extent the practical application of silanes and their technology has matched the interest and progress of the scientific studies, especially in the surface coatings and adhesives areas. It has been suggested that this may be explicable in terms of commercial secrecy, where manufacturers are unwilling to disclose a 'performance edge' in highly competitive industries [11], but it seems possible that their use is both widespread and expanding rapidly in both technologies.

The use of silanes as adhesion promoters in the polymer industries represents a relatively low-tech approach to obtaining material improvements in the initial and wet adhesion of adhesives and coatings and improved physical properties in some composites. However, it should be clearly understood that although relatively 'low-tech', the requirements, principles of use and technology must be understood, or the use of silanes may result in a marked deterioration in those properties which it is sought to improve.

The present paper will cover just some aspects of the use of silanes as adhesion promoters in the surface coating, adhesive and composite fields from a practical point of view, and review the results in the light of published work. The chemistry and reactions of silanes have already been covered in some detail and do not require further repetition here [12].

Throughout the paper the term 'bond strength' rather than adhesion has been used as it is the author's view that all adhesion values are in fact cohesion values [13], as paints and adhesives fail cohesively, while frequently appearing, especially under water-soaked conditions, to fail cleanly from the substrate. For all practical purposes these failures may be regarded as adhesional failures, and represent practical adhesion failures [14].

2. EXPERIMENTAL

2.1. Paints

The work reported was carried out on single coats of a proprietary two-pack polyamide cured epoxide and an aliphatic isocyanate adduct cured polyester paint both pigmented with rutile titanium dioxide and conforming to the government specifications DTD5555A, Exterior Glossy Finishing Schemes (Cold Curing Epoxide Type) and DTD5580A, Exterior and Interior Finishing Schemes (Cold Curing Polyurethane Type) respectively and known to be silane free. The oxidative cure paints were manufactured from commercial resins and pigmented with rutile titanium dioxide at a pigment volume concentration of 22%.

2.2. Silanes

The eight organosilanes used in the series of investigations are shown in Table 1.

2.3. Substrates

The work on surface coatings was carried out on mild steel panels to BS1449, Steel Plate, Part 1B, CR3/FF and aluminium, 99.9% pure to BS1470, Grade SIC-H8. The adhesive test specimens consisted of flat headed bolts, 12.5 mm diameter, manufactured from mild steel round bar stock to BS970 Part 1, stainless steel round bar stock to BS970 Part 1, 1/431/S29 and aluminium 99.9% pure.

2.4. Paint panel preparation

All panels, each $150 \times 100 \times 2$ mm, were cut from the same batch of sheet stock and individually identified. All panels were degreased with repeated washes in acetone (Analar). Panels selected for grit-blasting were lightly blasted using 3A Fused Blast Media, Grade WB180/220, supplied by Greyson Intl., brushed to remove particulate matter and re-degreased on completion of this operation. With minimum delay the prepared panels were coated using a conventional gravity feed spray gun, DeVilbiss Model GFG513 with an AV15-EX 1.8 spray tip and needle at 206 kPa to give a dry film thickness of 35–50 μm. Panels falling outside this range were rejected. All coated panels were allowed to cure at room temperature and humidity (21°C, 40% RH), for 14 days prior to testing.

2.5. Adhesive specimen preparation

All specimens were degreased by swabbing with methyl ethyl ketone (Analar) followed by a 20 min degrease in a Sohxlet apparatus using methyl ethyl ketone and allowed to cool before use. Grit-blasted specimens were treated in the same manner as the paint panels.

2.6. Adhesion tests

2.6.1. Direct pull-off-surface coatings. Four squares 31×31 mm ($1\frac{1}{4} \times 1\frac{1}{4}$ in) were cut from each test panel, avoiding the panel edges, and sandwiched between 25.4 mm diameter cylindrical test specimens using a two-pack cold curing epoxide adhesive (Versamid 115/Epikote 828 50/50 wt%. After 8 h in an alignment jig under a pressure of 45 N (10 psi) the resulting doublets were broken on an Instron Universal Testing Machine at a cross-head speed of 2 mm/min ($\frac{1}{16}$ in/min).

Where wet adhesion was to be measured, discs 31 mm in diameter were punched from the panels and bonded in the same manner, except that all discs were kept water wetted until bonded, and the alignment jigs stored at 100% RH. Specimens were removed and tested with minimum delay to avoid drying out.

2.6.2. Butt tensile—adhesives. Bolt shaped specimens were bonded and cured as for the surface coating specimens and tested in the same way.

2.6.3. Recording of data. In recording the results of the bond tests, the average failing load, standard deviation and approximate area of detachment, as judged by eye, have been recorded. Where the detached area differed widely between individual test specimens, the range has been recorded. It should be noted that the area of detachment value is a substitute for the site of failure in that it indicates where a failure occurred in the adhesive used to bond the specimens, i.e. a zero detachment, and serves as a useful method of indicating that a bond strength failure has occurred without having to describe the type of failure.

2.6.4. Application of silanes as pretreatments. Where silanes were used as pretreatment primers, they were applied from a 2 wt% solution in an 80/20 wt% ethyl alcohol/water mixture, by brush, and allowed to air dry for 4 h before coating or bonding. No hydrolysis catalyst was used and the solution was applied at the natural pH.

3. SILANES USED

The silanes used in a series of investigations are shown in Table 1.

3.1. Potential methods of use of silanes

Silanes may be used in four main ways, depending on both the application and

Table 1.
Silanes used in the investigations reported

Vinyltrimethoxysilane	(VTS)

$$CH_2=CH\ Si(OCH_3)_3$$

Vinyltris(2-methoxyethoxysilane)   (VTMS)

$$[CH_3O(CH_2)_2O]_3\ SiCH=CH_2$$

γ-methacryloxypropyltrimethoxysilane   (MAMS)

$$\underset{\underset{CH_2=C-CO\ (CH_2)_3\ Si(OCH_3)_3}{}}{\overset{CH_3\ \ \ O}{\overset{|\ \ \ \ ||}{}}}$$

β-(3,4,epoxycyclohexyl) ethyltrimethoxysilane   (ECMS)

$$O\underset{\overset{CH}{\underset{CH}{}}}{\overset{CH}{}}\overset{CH_2}{\underset{CH_2}{}}\overset{CH-(CH_2)_3-Si(OCH_3)_3}{\underset{CH_2}{}}$$

γ-glycidoxypropyltrimethoxysilane   (GPMS)

$$CH_2-CH-CH_2-O(CH_2)_3\ Si(OCH_3)_3$$
$$\diagdown\!\!\diagup$$
$$O$$

γ-mercaptopropyltrimethoxysilane   (MPS)

$$HS(CH_2)_3\ Si(OCH_3)_3$$

γ-aminopropyltriethoxysilane   (APES)

$$NH_2CH_2CH_2CH_2Si(OC_2H_5)_3$$

N-β-(aminoethyl)-γ-aminopropyltrimethoxysilane   (AAMS)

$$NH_2(CH_2)_2NH(CH_2)_3Si(OCH_3)_3$$

the polymer field of interest: (1) as pretreatment primers (surface coatings and adhesives); (2) as pretreatments for fillers (composites); (3) as formulated primers (surface coatings, adhesives and composites); and (4) as additives to the organic phase (surface coatings, adhesives and composites).

3.1.1. Pretreatment primers. In this method of use the silane may be applied from a solvent solution, by vapour phase deposition or by plasma deposition although solvent application is the more usual. The solution usually contains water and silane at a concentration of 1–2 wt%. The applied film may be water washed before subsequent coating/bonding and/or heat cured. The solvent(s) used may be important in both the stability of the solution and the performance, particularly in the wet adhesion. It has been shown that the presence of water either in the solution or as a final rinse is important, particularly in the case of AAMS and presumably other silanes [1]. Other factors which are important include: the concentration of silane; the pH of the solution; the thickness of the silane film deposited.

In general the concentration of silane should be below 4 wt% if excessively thick silane films are to be avoided. There are indications that freshly made solutions are preferable due to polymerization and precipitation of the silane on storage. In an experiment, not detailed here, in which films of AAMS were deposited on aluminium from an alcohol/water solution, the initial bond strength of a two-pack polyurethane paint was shown to fall markedly below the bond strength on a non-silane coated panel when the concentration of AAMS rose above 3 wt% [9]. This was attributed to cohesive failure within a weakly polymerized, excessively thick silane film. The pH of the solution has a profound effect on the chemistry and structure of the dried film as reported by Boerio and Williams [10] and on practical adhesion by Mittal and Suryanarayana [15].

3.1.2. Pretreatment for fillers. When used as a surface treatment for fillers or reinforcing materials, in which the silane is applied to the filler or fibre before incorporation into a resin matrix, the same factors as for pretreatment primers apply. In addition, the particle size and the absence/presence of water are important, and in a sense this application is only a variation on the former. It should be noted that silane treated fillers may have, or impart, different rheological properties to non-treated fillers, particularly particulates. A major disadvantage of this approach is that a general purpose silane may have to be used by a manufacturer rather than one specifically tailored to the use of a particular resin type and less than optimum properties are likely to be achieved in some cases.

3.1.3. Formulated primers. The use of formulated primers in which a silane (or silanes) is blended with a film former, a solvent, and perhaps a hydrolysis catalyst, has been described by Plueddemann [16] and in the Dow Corning trade literature [17]. Primers of this type which contain a film former should not be confused with the previously mentioned silane pretreatment primers, as they deposit a film of measureable thickness and strength. Such formulated primers are required to wet the substrate, be readily wet by the adhesive or coating and should provide a layer of intermediate modulus between substrate and resin system. As with the pretreatment primers a formulated primer can be tailored to both the substrate and the resin. Ageing of formulated primers may be a problem both in the storage container and on the substrate before subsequent coating.

3.1.4. Additives. The use of silanes as additives, incorporated into the organic resin either immediately prior to use or as a part of the formulation at the manufacturing stage, is probably the most effective method of utilization. The manufactured 'single pack' concept is almost universally desirable in surface coating and adhesive technology but there are several parameters which are critical if success is to be achieved. The first potential problem is reaction between the silane and the solvents or the resin and/or curing agent present. For example, amino-functional silanes such as APES and AAMS will react with oxygenated solvents present in many two pack systems to form a ketamine and the highly undesirable reaction product, water. As the silanes contain organo functional groups they may react with some resins and act as curing agents. AAMS, for example, will react with epoxide groups and GPMS will react with amine or isocyanate adduct curing agents. The result may be premature gelation, or the prevention of the silane moieties from concentrating at the substrate surface,

P. Walker

thus preventing reaction. Potential solvent reactions may be obviated by solvent substitution or alternative silanes, and resin reactions by incorporating the silane in the appropriate part of a two pack system.

The second problem is that of water, and most commercial solvents, pigments, and fillers contain water; it may be necessary to remove this before manufacture or incorporate molecular sieves, zeolites or proprietary drying agents [18]. Silane depletion by pigments or fillers may prove to be a further problem unless the surface is pre-passivated by a silane treatment. It should be noted that thixotropes are particularly prone to cause silane depletion and it may be necessary to change from an inorganic to an organic thixotropic agent in a particular formulation.

These practical formulation problems are most likely to arise in surface coatings technology, but some aspects cannot be discounted in both adhesive and composite technologies, particularly silane/resin reactions.

4. SILANES AND SURFACE COATINGS

4.1. Silanes and the surface—used as pretreatments

The effect of using different silanes as pretreatment primers on the initial bond strength of a two pack polyamide cured epoxy to mild steel and aluminium is shown in Table 2.

Reference to Table 2 will show that on degreased mild steel the use of all six silanes resulted in a higher bond strength, although in the case of MPS and GPMS the apparent increase was not statistically significant, AAMS provided the greatest improvement. Similarly on degreased aluminium AAMS provided

Table 2.
The effect of silanes on the bond strength of a two-pack epoxy paint: Direct pull-off, silane on surface

| | Substrate | | | |
| | Mild steel | | Aluminium | |
Silane/surface preparation	MPa	Area detached %	MPa	Area detached %
Degreased only	21.2 ± 1.14	30–40	22.6 ± 1.54	40–60
MAMS/degreased	28.4 ± 1.48	20	31.0 ± 1.09	20
ECMS/degreased	27.5 ± 1.38	0–20	25.8 ± 0.75	20–30
GPMS/degreased	25.1 ± 1.33	0–40	27.4 ± 1.29	10–40
MPS/degreased	23.6 ± 1.42	10–40	25.6 ± 1.74	10–30
APES/degreased	29.8 ± 1.91	10–20	28.7 ± 2.04	5–20
AAMS/degreased	31.9 ± 1.24	0	33.5 ± 1.98	0
Grit-blasted only	27.9 ± 1.76	20	33.3 ± 2.56	20
MAMS/grit-blasted	32.1 ± 1.86	10	35.9 ± 2.37	10
ECMS/grit-blasted	33.6 ± 1.34	0	38.0 ± 2.24	0
GPMS/grit-blasted	29.4 ± 2.32	10–20	37.8 ± 2.19	0
MPS/grit-blasted	33.7 ± 1.18	0	35.7 ± 2.21	0
APES/grit-blasted	35.0 ± 1.62	0	36.2 ± 0.98	0
AAMS/grit-blasted	36.2 ± 1.70	0	37.1 ± 1.44	0

the greatest improvement, and it should be noted that on both substrates no coating became detached, i.e. the values quoted are a minimum and the actual adhesion may have been much higher than that recorded. All the silanes produced an increase in bond strength on the grit-blasted surfaces, the improvement being the greatest on the surfaces treated with ECMS, MPS, APES and AAMS where no coating was detached. It is interesting to note that the use of AAMS on the degreased surfaces resulted in higher bond strengths than on the non-silane grit-blasted surfaces. The practical value of this approach is immediately obvious.

Reference to Table 3 will show the data for a two-pack polyurethane paint applied to mild steel and aluminium. It can be seen that all but one silane on mild steel and all but two on aluminium produced a significant improvement in bond strength, MAMS and AAMS giving the highest values on both degreased substrates. On the grit-blasted surfaces APES and AAMS give particularly high values with no coating detachment on mild steel, and on aluminium MPS also performed well. It is interesting to note that ECMS and GPMS performed relatively badly on the nongrit-blasted surfaces.

4.2. Silanes in the coating—used as an additive

Reference to Table 4 will show the bond strength data for epoxide paints containing MPS and AAMS at a range of concentrations from 0.1 to 1.0 wt% on the resin content. It can be seen that at a concentration of 0.2 wt% no paint was detached in the test, indicating a failure in the bonding adhesive. The recorded bond strength values are therefore a minimum and the actual bond strength of

Table 3.
The effect of silanes on the bond strength of a two-pack polyurethane paint: Direct pull-off, silane on surface

Silane/surface preparation	Substrate			
	Mild steel		Aluminium	
	MPa	Area detached %	MPa	Area detached %
Degreased only	18.3 ± 1.43	100	12.6 ± 1.27	100
MAMS/degreased	29.3 ± 1.62	10–20	31.4 ± 2.14	0–60
ECMS/degreased	14.0 ± 1.63	100	13.3 ± 2.11	100
GPMS/degreased	26.5 ± 1.54	40–100	11.2 ± 1.14	100
MPS/degreased	24.2 ± 1.14	20–40	18.7 ± 1.67	100
APES/degreased	27.5 ± 1.38	10	21.9 ± 1.12	10–30
AAMS/degreased	33.8 ± 2.26	0–5	24.8 ± 1.09	5–50
Grit-blasted only	34.2 ± 1.95	10–40	28.4 ± 1.62	10–30
MAMS/grit-blasted	35.6 ± 2.14	0–20	34.1 ± 1.74	10
ECMS/grit-blasted	27.3 ± 1.31	20–90	29.1 ± 1.43	10–50
GPMS/grit-blasted	31.0 ± 1.71	50–90	26.3 ± 1.66	10–50
MPS/grit-blasted	35.2 ± 2.50	0–30	35.9 ± 2.80	0
APES/grit-blasted	37.8 ± 2.68	0	37.1 ± 1.16	0
AAMS/grit-blasted	38.8 ± 2.13	0	33.2 ± 0.96	0

Table 4.
The effect of silanes on the bond strength of a two-pack epoxide paint: Direct pull-off, silane as an additive

| | Substrate | | | |
| | Mild steel | | Aluminium | |
Wt% silane	MPa	Area detached %	MPa	Area detached %
None	22.1 ± 1.97	60	21.1 ± 1.88	50
0.1% MPS	27.4 ± 1.59	60	30.0 ± 1.71	30
0.2% MPS	29.8 ± 1.43	0	39.9 ± 2.47	0
0.4% MPS	37.7 ± 2.87	0	43.0 ± 2.02	0
1.0% MPS	37.8 ± 2.68	0	43.8 ± 1.66	0
0.1% AAMS	23.8 ± 1.88	30	26.2 ± 1.78	40
0.2% AAMS	34.0 ± 2.72	0	36.8 ± 2.80	0
0.4% AAMS	33.8 ± 1.83	0	34.8 ± 1.74	0
1.0% AAMS	34.7 ± 2.08	0	32.3 ± 1.39	0

the paints to both degreased mild steel and aluminium were in excess of these. For all practical purposes there would seem to be little point in increasing the silane concentration above 0.2 wt%.

Similar data for the two-pack polyurethane paints are shown in Table 5. Although there is a slight apparent improvement in the bond strength of the paints containing MPS with increasing silane concentration on both mild steel and aluminium, the increases are not statistically significant. In the case of the paints containing AAMS no paint was detached at the 0.1 wt% level on mild steel and the 0.2 wt% level on aluminium. The 0.2 wt% level would appear to give almost optimum results for both silanes. Although not reported here, the silanes APES and GMPMS proved equally effective at the 0.2 wt% level.

5. THE EFFECT OF WATER IMMERSION ON BOND STRENGTH

5.1. Silanes on the surface

The deleterious effect of water on the bond strength of coatings has been previously described [19] and if an adhesion promoter is to be of maximum value it must improve not only the initial bond strength but also the wet and recovered bond strength.

To determine the wet bond strength coated panels were immersed in distilled water for 1500 h, removed and discs 25.4 mm in diameter stamped from them. The surfaces were wiped with a dry tissue and bonded between two cylindrical test pieces using a polyamide cured epoxide adhesive and immediately placed in a sealed container at 100% RH for the adhesive to cure. After 16 h the specimens were broken on an Instron Universal Test Machine with minimum delay. 'Recovered' values were measured after the panels had dried out at room temperature and humidity for 7 days. Clearly, it is unlikely that the values reported represent the minimum bond strengths, as some drying out is almost inevitable, but the values are directly comparable.

Reference to Table 6 shows the initial, wet and recovered bond strengths of the

Table 5.

Effect of silanes as additives on the bond strength of a two-pack polyurethane paint: Silane wt% – direct pull-off. All panels grit-blasted only

| | Substrate | | | | |
| | Mild steel | | Aluminium | | |
Wt% silane	MPa	Area detached %	MPa	Area detached %
None	23.8 ± 3.33	100	10.8 ± 1.32	100
0.1% MPS	31.2 ± 1.56	60	21.8 ± 1.96	30
0.2% MPS	32.9 ± 1.98	10	23.5 ± 1.46	30
0.4% MPS	32.5 ± 1.85	10	24.9 ± 1.44	20
1.0% MPS	33.3 ± 1.59	10	26.1 ± 1.12	20
0.1% AAMS	41.2 ± 2.81	0	31.4 ± 2.23	10
0.2% AAMS	42.6 ± 3.32	0	38.1 ± 2.21	0
0.4% AAMS	43.1 ± 3.44	0	41.6 ± 2.20	0
1.0% AAMS	41.9 ± 2.81	0	39.7 ± 1.87	0

two pack epoxide paint on mild steel and aluminium with silanes used as pre-treatments. It can be seen that in the absence of a silane treatment the initial bond strength on both the degreased and grit-blasted surfaces fell to below 10 MPa after water immersion and showed minimum recovery on drying out. On degreased mild steel all three silanes investigated produced a marked increase in the wet and recovered bond strengths over that of the non-silane control, AAMS being particularly effective with a loss of only 13% under water soaked conditions. Similar results were obtained on the grit-blasted substrates. On aluminium the values were generally lower than those on mild steel but still represented a major improvement in retention and recovery of bond strength.

Reference to Table 7 will show the data for the two-pack polyurethane paint, and indicate a similar improvement to that of the epoxide paint due to the use of silanes. Although there are differences in the values for particular silanes on the different substrates and surface treatments, the general picture is that AAMS and MPS are the most effective in epoxide and polyurethane paints.

Table 8 shows the data for two air drying (oxidative cure) paints based on a siliconized alkyd and a styrenated alkyd. It can be seen that when used as a pre-treatment only MAMS produced a significant increase in the initial bond strength, and then only on mild steel. The VTS and VTMS failed to effect the initial bond strength of either paint on both substrates. However, under water soaked conditions the use of VTMS resulted in an increase in the wet and recovered bond strength of both paints on both substrates as did the MAMS. Examination of the apparent site of failure of the specimens showed all the initial sites to be clearly cohesive within the paint films. This type of failure is common in coatings of the oxidative cure type where the adhesion to the substrate is much greater than the cohesive strength of the film. It is probable therefore that the degree of cure of these films varied as it was not possible to produce all the panels at the same time. However, the wet and recovered bond strength values of all the silane treated panels was greater than the non-silane controls. The effect of MAMS on the corrosion resistance of the mild steel panels should be noted.

Table 6.
Effect of water immersion on the bond strength of an epoxide paint: Direct pull-off, 1500 h, silane on surface

Silane/surface preparation	Substrate					
	Mild steel			Aluminium		
	Initial MPa/area of detachment	Wet MPa/area of detachment	Recovered MPa/area of detachment	Initial MPa/area of detachment	Wet MPa/area of detachment	Recovered MPa/area of detachment
None/degreased	19.9 ± 1.35/100	7.2 ± 0.81/100	11.0 ± 1.36/100	21.4 ± 1.75/90	5.7 ± 1.26/100	11.2 ± 1.67/100
MAMS/degreased	26.9 ± 1.41/20	19.3 ± 0.99/100	23.7 ± 1.91/100	30.2 ± 1.62/0	12.0 ± 1.03/30	19.7 ± 1.01/50
ECMS/degreased	27.4 ± 1.26/20	17.3 ± 1.07/100	21.8 ± 1.59/90	24.9 ± 1.80/30	13.3 ± 0.91/30	21.7 ± 1.53/50
AAMS/degreased	32.0 ± 1.74/0	28.1 ± 2.50/10	20.2 ± 2.25/10	31.2 ± 2.37/0	11.5 ± 0.85/30	20.5 ± 1.80/40
None/grit-blasted	25.0 ± 2.11/10	9.2 ± 1.16/100	21.0 ± 2.45/100	28.5 ± 2.91/30	8.5 ± 1.07/100	13.7 ± 1.90/80
MAMS/grit-blasted	33.4 ± 1.93/40	21.2 ± 1.54/50	30.9 ± 2.17/50	31.8 ± 2.38/10	13.3 ± 0.83/40	21.6 ± 1.22/50
ECMS/grit-blasted	27.7 ± 1.57/0	16.3 ± 1.33/30	31.7 ± 3.02/60	37.1 ± 3.05/0	15.7 ± 1.32/30	27.1 ± 1.37/30
AAMS/grit-blasted	33.6 ± 2.04/0	25.3 ± 2.19/50	27.9 ± 2.33/40	32.5 ± 2.98/0	13.0 ± 1.17/40	25.0 ± 2.00/30

Table 7.
Effect of water immersion on the bond strength of a polyurethane paint: Direct pull-off, 1500 h, silane on surface

Silane/surface preparation	Substrate					
	Mild steel			Aluminium		
	Initial MPa/area of detachment	Wet MPa/area of detachment	Recovered MPa/area of detachment	Initial MPa/area of detachment	Wet MPa/area of detachment	Recovered MPa/area of detachment
None/degreased	16.7 ± 1.54/100	5.7 ± 0.68/100	6.8 ± 1.48/100	12.6 ± 1.81/100	3.8 ± 0.46/100	9.9 ± 1.20/100
MAMS/degreased	30.3 ± 1.89/20	16.6 ± 1.57/100	18.4 ± 1.63/100	32.2 ± 2.45/30	10.1 ± 0.62/30	21.8 ± 2.15/30
MPS/degreased	25.2 ± 2.10/30	5.4 ± 0.71/100	12.1 ± 1.15/100	19.4 ± 2.13/100	9.8 ± 0.89/100	17.6 ± 1.69/20
AAMS/degreased	38.2 ± 1.85/0	7.4 ± 0.77/90	12.9 ± 1.36/90	26.3 ± 1.92/40	11.1 ± 0.72/30	22.8 ± 2.31/30
None/grit-blasted	25.7 ± 1.70/40	11.8 ± 1.45/90	20.8 ± 1.07/60	28.6 ± 2.44/10	8.5 ± 0.95/100	13.6 ± 2.60/80
MPS/grit-blasted	32.2 ± 1.85/5	23.7 ± 1.23/10	25.3 ± 1.32/0	34.2 ± 2.87/0	22.0 ± 1.67/100	19.9 ± 1.87/30
MAMS/grit-blasted	34.8 ± 1.92/10	26.1 ± 1.36/10	30.2 ± 1.47/0	33.7 ± 1.69/0	13.0 ± 0.90/20	22.4 ± 2.13/40
AAMS/grit-blasted	36.7 ± 2.13/0	22.8 ± 1.20/30	29.2 ± 1.91/0	34.0 ± 2.90/0	14.9 ± 1.13/20	22.0 ± 1.98/30

Table 8.

Effect of water immersion on the bond strength of air drying paints: 1500 h, direct pull-off, silane on surface

Paint/silane	Substrate					
	Mild steel			Aluminium		
	Initial MPa/area of detachment	Wet MPa/area of detachment	Recovered MPa/area of detachment	Initial MPa/area of detachment	Wet MPa/area of detachment	Recovered MPa/area of detachment
Siliconized alkyd						
None	18.4 ± 1.26/20	3.1 ± 0.59/100[a]	5.3 ± 0.82/100[a]	18.4 ± 1.15/90	5.2 ± 0.91/100	8.7 ± 0.99/100
VTS	18.5 ± 1.50/0–30	5.0 ± 0.61/100[a]	8.9 ± 0.91/100[a]	14.1 ± 1.22/100	7.8 ± 1.32/100	12.6 ± 1.29/100
VTMS	18.6 ± 1.49/0–30	7.8 ± 0.93/100[a]	14.3 ± 0.88/100[a]	20.3 ± 1.71/100	10.1 ± 1.39/100	17.4 ± 1.37/60
MAMS	23.5 ± 2.10/0	14.9 ± 1.53/100	21.7 ± 1.35/20	17.5 ± 1.83/100	9.5 ± 1.18/100	15.9 ± 1.40/20
Styrenated alkyd						
None	21.0 ± 1.92/0–50	6.1 ± 1.72/100[a]	10.3 ± 1.13/100[a]	13.2 ± 0.85/90	5.6 ± 0.92/100	8.9 ± 1.07/100
VTS	23.8 ± 1.85/0–60	7.2 ± 1.62/100[a]	14.4 ± 1.32/100[a]	15.7 ± 0.91/60	13.4 ± 1.18/100	13.1 ± 0.98/40
VTMS	21.1 ± 2.09/0–80	8.5 ± 1.53/100[a]	16.3 ± 0.98/100[a]	14.4 ± 1.23/90	14.9 ± 1.72/60	15.7 ± 1.27/30
MAMS	24.7 ± 2.16/0	14.7 ± 1.27/80	21.8 ± 2.04/30	17.4 ± 1.38/20	16.6 ± 1.83/60	16.8 ± 1.43/20

[a] Corrosion of substrate.

Although not reported here the silanes used in conjunction with the epoxide and polyurethane paints were largely ineffective with the oxidative cure systems.

5.2. Silanes in the paint

Reference to Table 9 will show the effect of silanes used as additives on the initial, wet and recovered bond strength of the epoxide paints containing silanes. It is interesting to compare the data for the control of non-silane containing paints in Table 9 with that for the controls in Table 6 when it will be seen that there is close agreement between the two sets of data suggesting some degree of reproducibility.

Without exception the initial bond strengths of the silane containing paints are higher than the non-silane control and closely reflect those shown in Tables 4 and 5, but it is the 'wet' bond strengths which are particularly impressive. In the case of those paints containing the higher levels of silanes MPS and AAMS, the retention of bond strength was increased by a factor of 3–4. Equally important are the recovered values which are all considerably higher than those for the controls. The relatively large number of specimens on which failure to detach the paint occurred, even under water soaked conditions, should also be noted.

Data for the polyurethane paint are shown in Table 10 where the total failure on drying out of the non-silane control on degreased aluminium is striking. It would seem possible that this panel had been inadequately degreased as failure of this magnitude is not characteristic of the paint. Here again the 'wet' and recovered bond strengths of all the silane containing paints were much higher, but the anomaly of the paint containing 0.4 wt% of AAMS having a considerably lower wet bond strength than that containing 0.2 wt% is inexplicable. In general the overall values for the paints containing silanes were greater than those obtained from the use of silanes as pretreatment primers.

5.3. Rate of loss of bond strength

The rate of loss of bond strength was measured by applying selected paints to the flat heads of bolts and after immersion, bonding a second unpainted bolt to the test surface to form a doublet and breaking this specimen, after cure at 100% RH.

Figure 1 shows the rate of loss of bond strength of an epoxide paint containing 0.2 wt% of MPS and that of the same paint without silane on a surface treated with AAMS applied as a 2 wt% solution, compared with a non-silane control. It can be seen that the bond strength of the non-silane control fell rapidly, losing over 50% of the original strength in the first 3 days and was still losing strength at 14 days. In contrast the paint containing MPS had equilibrated after 10 days, losing only 50% of its original strength in 15 days. The paint on the AAMS treated surface also reached equilibrium after 10 days although losing a greater part of its original bond strength than the MPS containing paint.

A similar set of curves is shown in Fig. 2 for the polyurethane paint. Here again the non-silane control lost most of its bond strength in the first 3 days and the loss was continuing after 15 days. The paint containing MPS and the AAMS treated surface equilibriated after 7 days.

Table 9.
Effect of water immersion on the bond strength of an epoxide paint: Direct pull-off, 1500 h, silane in paint

Wt% silane/surface preparation	Substrate					
	Mild steel			Aluminium		
	Initial MPa/area of detachment	Wet MPa/area of detachment	Recovered MPa/area of detachment	Initial MPa/area of detachment	Wet MPa/area of detachment	Recovered MPa/area of detachment
None/degreased	22.1 ± 1.87/40	9.7 ± 1.53/70	12.1 ± 1.37/50	18.2 ± 0.84/100	6.2 ± 0.94/100	11.6 ± 1.53/100
0.2 MPS/degreased	30.2 ± 2.63/0	21.9 ± 1.84/5	31.7 ± 2.69/0	27.6 ± 1.69/0	26.3 ± 2.70/0	27.3 ± 2.82/0
0.4 MPS/degreased	29.1 ± 1.99/0	24.1 ± 2.33/0	29.7 ± 2.81/0	42.5 ± 3.04/0	36.4 ± 3.41/10	39.9 ± 3.17/0
0.2 AAMS/degreased	31.4 ± 3.01/20	17.7 ± 1.82/10	26.6 ± 2.43/20	31.5 ± 2.63/0	25.3 ± 2.80/0	26.9 ± 2.55/20
None/grit-blasted	28.7 ± 2.51/10	9.1 ± 1.36/100	22.9 ± 1.90/100	21.0 ± 1.48/10	7.4 ± 1.39/100	16.2 ± 1.48/20
0.2 MPS/grit-blasted	33.7 ± 2.78/0	29.5 ± 2.84/0	31.4 ± 2.47/0	31.2 ± 2.34/0	25.1 ± 1.63/0	26.7 ± 2.53/5
0.1 AAMS/grit-blasted	35.7 ± 3.14/0	21.6 ± 1.94/0	27.5 ± 2.31/40	35.4 ± 2.88/0	33.6 ± 2.80/0	39.8 ± 3.17/0
0.2 AAMS/grit-blasted	34.8 ± 2.85/0	23.8 ± 2.20/0	25.3 ± 1.97/20	36.0 ± 2.43/0	27.8 ± 3.52/0	35.9 ± 3.06/0
0.4 AAMS/grit-blasted	43.6 ± 3.66/0	35.1 ± 2.93/0	37.3 ± 3.62/0	33.6 ± 3.21/0	28.2 ± 3.61/0	28.7 ± 1.89/0

Table 10.
Effect of water immersion on the bond strength of a polyurethane paint: Direct pull-off, 1500 h, silane in paint

Wt% silane/surface preparation	Substrate					
	Mild steel			Aluminium		
	Initial MPa/area of detachment	Wet MPa/area of detachment	Recovered MPa/area of detachment	Initial MPa/area of detachment	Wet MPa/area of detachment	Recovered MPa/area of detachment
None/degreased	15.6 ± 1.12/100	5.1 ± 0.70/100	3.2 ± 0.63/100	8.8 ± 1.09/100	1.3 ± 0.76/100	Peeled on drying out
0.4 MPS/degreased	20.8 ± 1.51/40	7.8 ± 0.92/100	12.8 ± 1.52/100	20.6 ± 2.32/100	11.7 ± 2.31/100	13.8 ± 1.38/30
0.2 AAMS/degreased	29.5 ± 2.01/0	19.8 ± 1.76/50	21.2 ± 1.94/10	21.0 ± 1.59/100	9.5 ± 1.08/100	14.5 ± 1.29/100
0.4 AAMS/degreased	27.8 ± 2.13/40	19.8 ± 1.84/60	20.2 ± 2.32/100	24.7 ± 2.07/0	15.1 ± 1.91/60	19.9 ± 1.74/10
None/grit-blasted	28.6 ± 2.54/20	12.8 ± 1.19/50	22.5 ± 2.09/40	29.3 ± 2.25/10	10.4 ± 2.15/70	15.2 ± 1.29/40
0.1 MPS/grit-blasted	38.5 ± 2.61/0	26.1 ± 2.33/10	35.2 ± 2.63/0	33.9 ± 2.81/0	19.1 ± 1.98/30	26.8 ± 2.05/20
0.4 MPS/grit-blasted	36.3 ± 3.12/0	20.9 ± 1.85/30	24.7 ± 2.14/20	31.5 ± 3.04/0	22.8 ± 2.34/10	25.4 ± 2.81/0
0.2 AAMS/grit-blasted	38.1 ± 2.55/0	29.0 ± 2.63/0	33.3 ± 2.88/0	33.6 ± 2.89/0	15.6 ± 2.31/70	25.2 ± 1.68/30
0.4 AAMS/grit-blasted	36.0 ± 1.97/0	21.3 ± 1.57/20	29.8 ± 2.62/0	39.2 ± 3.24/0	28.6 ± 1.84/0	35.1 ± 3.00/0

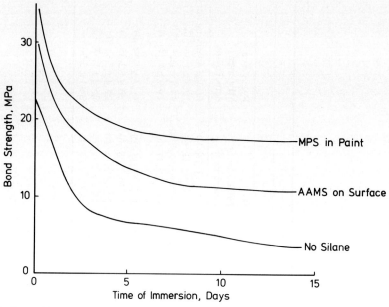

Figure 1. Rate of loss of bond strength of an epoxide paint on aluminium—under water immersion conditions.

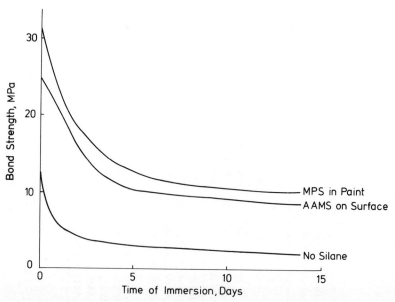

Figure 2. Rate of loss of bond strength of a polyurethane paint on aluminium—under water immersion conditions.

6. SILANES AND ADHESIVES

As it had been shown that silanes were effective as pretreatments for a variety of coatings and particularly so when used as additives, selected silanes were examined as pretreatments and additives in conjunction with a two pack polyamide cured epoxide adhesive (Epikote 828/Versamid 115, 1/1) and a structural polyurethane adhesive based on diphenylmethanediisocyanate and a polyester resin.

Four silanes, MAMS, GPMS, MPS and AAMS, and three substrates, mild steel, stainless steel and aluminium were investigated. Reference to Table 11 will show the data for the epoxide adhesive. It can be seen that when tested in the butt tensile mode MAMS, used as a primer, produced a considerable improvement in bond strength on all three substrates, being particularly effective on aluminium. When used as an additive on mild steel the bond strength was lower than the non-silane control. On stainless steel the values were almost identical and the adhesive failed cohesively. On aluminium the bond strength was lower when used as an additive but still considerably higher than the control. GPMS used as a primer produced a large increase in bond strength on all three substrates and only a minor improvement on mild steel and aluminium when used as an additive. MPS used as a pretreatment primer produced a marked improvement in bond strength on all three substrates, particularly on aluminium but little or no improvement when used as an additive. The use of AAMS as a pretreatment primer resulted in a marked increase on all three substrates but a smaller increase when used as an additive.

Reference to Table 12 will show the data for the structural polyurethane adhesive, from which it can be seen that the pattern for the epoxide adhesive is almost repeated, with all the silanes used as pretreatments producing marked

Table 11.
Comparison of silanes as pretreatments and additives for a structural epoxide adhesive: Butt tensile, grit-blasted metal, 2 wt% as pretreatment and additive

| | Silane | | | | | | | | |
| | None | MAMS | | GPMS | | MPS | | AAMS | |
		P	A	P	A	P	A	P	A
Mild steel									
Bond strength	38.3	48.8	21.8	52.0	39.5	48.2	42.0	52.3	41.8
(MPa)	±2.83	±3.07	±4.33	±2.34	±4.66	±3.66	±5.63	±1.31	±3.30
Failure	AF	CF	AF	CF	AF	CF	AF	CF	AF
Stainless steel									
Bond strength	53.0	59.4	59.9	65.0	63.0	66.0	54.6	66.3	61.6
(MPa)	±5.72	±6.36	±6.23	±3.25	±7.87	±3.17	±6.50	±4.24	±4.62
Failure	CF/AF	CF	CF	CF	CF	CF	CF	CF	CF
Aluminium									
Bond strength	30.2	63.2	44.7	60.6	36.7	60.3	36.0	49.7	41.7
(MPa)	±4.02	±5.69	±2.95	±5.94	±4.04	±2.40	±6.05	±2.88	±5.54
Failure	AF	CF	CF	CF	CF?	CF	CF?	CF	CF

P = primer; A = additive; AF = adhesional failure; CF = cohesive failure in adhesive.

Table 12.

Comparison of silanes as pretreatments and additives for a rigid polyurethane adhesive: Butt tensile, grit-blasted metal, 2 wt% as pretreatment and additive

	Silane								
		GPMS		MPS		APES		AAMS	
	None	P	A	P	A	P	A	P	A
Mild steel									
Bond strength	25.7	36.9	26.4	36.9	28.1	30.0	25.0	47.6	27.6
(MPa)	±2.29	±8.15	±2.67	±7.01	±3.23	±2.67	±2.30	±3.71	±2.24
Failure	AF	CF	AF	CF	AF	CF	AF	CF	AF
Stainless steel									
Bond strength	36.3	52.3	35.4	43.7	36.5	39.8	42.9	47.8	43.8
(MPa)	±3.23	±3.09	±4.04	±3.36	±4.96	±2.11	±5.71	±2.25	±2.85
Failure	CF/AF	CF	AF	CF	AF	AF	CF	CF	AF/CF
Aluminium									
Bond strength	45.9	52.9	38.7	49.4	37.0	51.7	46.1	57.2	43.0
(MPa)	±6.10	±3.91	±3.71	±4.25	±7.10	±9.67	±2.58	±5.55	±4.34
Failure	AF	CF	AF	CF?	AF	CF?	AF/CF	CF	CF

P = primer; A = additive; AF = adhesional failure; CF = cohesive failure in adhesive.

increases in bond strength, except for APES on stainless steel where there was less of an increase. The extent to which the silanes produced an improvement in bond strength is masked by the change from adhesional to cohesive failure. They were considerably less effective when used as additives, having little or no effect on the measured bond strength.

7. SILANES ON GLASS

The effect of silanes on the bond strength of two epoxide adhesives to glass is shown in Table 13. The test mode was that of 'Sandwich Butt Tensile' in which a glass disc was bonded between two aluminium specimens in the form of a sandwich, and the resulting composite specimen broken on an Instron Universal Test Machine. Clearly in such a composite test specimen, alignment is important and there are many potential sites and modes of failure.

The recorded bond strengths clearly show that MAMS and ECMS were totally ineffective as adhesion promoters on glass. MPS, APES and AAMS were all effective and their use resulted in a marked improvement in the bond strength of both adhesives. All three silanes resulted in a change of the site or mode of failure, the locus of failure transferring from the glass surface to the aluminium test specimen or within the adhesive.

8. EFFECT OF HUMIDITY ON SILANE/ADHESIVE JOINTS

8.1. Metal

The effect of storage at 100% RH at 21°C on stainless steel butt tensile/structural epoxide adhesive joints is shown in Fig. 3. AAMS, GPMS and MPS were examined as pretreatments and the choice of stainless steel as the substrate

Table 13.
Effect of silanes on bond strength to glass: Sandwich butt tensile, silane on surface

Adhesive/silane	Bond strength (MPa)	Type of failure
Amine cured epoxide		
(E815/TETA) None	29.3 ± 2.83	AFG
MAMS	13.5 ± 1.72	AFG
ECMS	19.8 ± 2.39	AFG
GPMS	30.5 ± 2.75	AFA
MPS	36.8 ± 3.09	AFA
APES	45.0 ± 2.98	CF
AAMS	43.4 ± 3.51	CF
Polyamide cured epoxide		
(E828/V125) None	26.9 ± 3.43	AFG
MAMS	7.9 ± 1.15	AFG
ECMS	13.9 ± 1.36	AFG
GPMS	36.1 ± 2.29	AFA
APES	38.5 ± 2.79	CF
AAMS	39.8 ± 3.61	CF

AFG = adhesional failure to glass; AFA = adhesional failure to aluminium; CF = cohesive failure in adhesive.

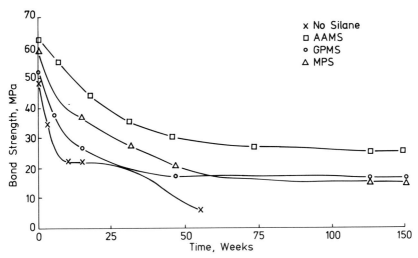

Figure 3. Effect of 100% RH at 21°C on the bond strength of a stainless steel joint—polyamide cured epoxide adhesive.

was dictated by the need to avoid the complication of corrosion. It can be seen that without a silane treatment the bond strength fell from over 50 to 22 MPa in the first 10 weeks of exposure and subsequently to 5 MPa in 52 weeks.

The best of the silane treatments, AAMS, equilibrated at 26 MPa after 70 weeks and all three silane treatments reached equilibrium at an acceptable level well above that of the non-silane control.

Figure 4 shows data for the structural polyurethane adhesive/stainless steel

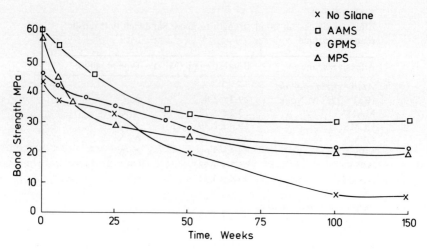

Figure 4. Effect of 100% RH at 21°C on the bond strength of a stainless steel joint—rigid poly-urethane adhesive.

joints stored at 100% RH at 21°C. Without silane the bond strength fell from 43 to 20 MPa in 50 weeks and to 6 MPa in 150 weeks. Again, the best of the silanes, AAMS, equilibrated at 31 MPa after 70 weeks. The GPMS and MPS joints reached equilibrium after 80 weeks at 23 and 21 MPa, respectively.

8.2. Glass

The effect of storage of 100% RH at 21°C on the bond strength of glass/ aluminium sandwich butt tensile joints, with and without silane, is shown in Fig. 5. It can be seen that without silane the bond strength effectively fell to zero in 150 days. With MPS on the surface, the bond strength equilibriated at 24 MPa at 220 days.

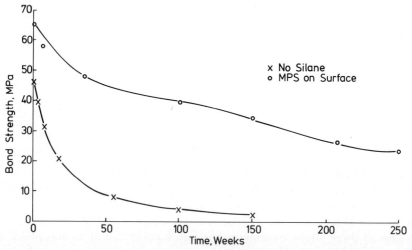

Figure 5. Effect of 100% RH at 21°C on the bond strength of a glass/aluminium joint—amine cured epoxide adhesive.

9. SILANES AND SYNTACTIC FOAMS

Syntactic foams manufactured from hollow glass or silica microspheres and an epoxide, phenolic or other matrix resin represent a class of lightweight structural materials used for buoyancy purposes, insulation and packaging. The effect of silanes on the mechanical properties of syntactic foams at a nominal density of 0.35 g/cm^3 is shown in Tables 14–16. The Proportional Limit is defined as the greatest stress which the foam is capable of sustaining without any deviation from proportionality of stress to strain (Hooke's Law).

Reference to Table 14 will show the effect of increasing levels of APES on the compressive properties of an anhydride cured epoxide/silica microballoon foam, the APES being added on the resin content. The notation w/r (wt% resin) has been used in the tables. Both the yield stress and strain to failure increased steadily with increased silane content, with a corresponding increase in compressive modulus. At the 5 wt% level there was no real increase in yield stress but a marked increase in strain to failure, resulting in a lower modulus. However, at the 4% level the compressive strength was more than double that of the non-silane control.

Table 15 shows the compressive data for anhydride cured foams containing 5 wt% of several silanes. In all cases the use of the silanes resulted in an increase in compressive strength and strain to failure, although in almost every case this was accompanied by a fall in the compressive modulus, i.e. the foams were less 'stiff'.

Table 14.
Effect of silane addition level on the compressive properties of an anhydride cured epoxide silica microballoon syntactic foam (APES, cured at 100°C for 4 h, nominal density 0.35 g/cm^3)

Addition level (w/r)	Density (g/cm^3)	Yield stress (MPa)	Yield strain (%)	Proportional limit (MPa/%)	Compressive modulus (MPa)
0	0.360	6.3	1.44	6.3/1.44	440
1	0.352	8.5	1.80	6.3/1.80	470
2	0.360	13.1	2.50	12.8/2.02	520
3	0.355	12.3	2.45	11.7/2.01	500
4	0.358	14.5	2.31	14.5/2.31	630
5	0.352	14.8	4.05	14.8/4.05	365

Table 15.
Effect of silanes on the compressive properties of an anhydride cured epoxide silica microballoon syntactic foam (cured at 100°C for 4 h, 5 w/r addition, nominal density 0.35 g/cm^3)

Silane	Density (g/cm^3)	Yield stress (MPa)	Yield strain (%)	Proportional limit (MPa/%)	Compressive modulus (MPa)
None	0.354	6.3	1.44	6.3/1.44	440
GPMS	0.360	8.0	1.96	7.2/1.76	410
ECMS	0.357	8.9	2.85	8.3/2.50	310
MPS	0.362	12.1	3.97	11.7/3.18	305
AAMS	0.347	12.3	2.70	12.3/2.70	455
APES	0.352	14.8	4.05	14.8/4.05	365

Table 16.
Effect of silanes on the compressive properties of a polyamide cured epoxide silica microballoon syntactic foam (cured at 21°C for 14 days, 5 w/r addition, nominal density 0.35 g/cm³)

Silane	Density (g/cm³)	Yield stress (MPa)	Yield strain (%)	Proportional limit (MPa/%)	Compressive modulus (MPa)
None	0.362	5.2	1.40	5.2/1.40	370
GPMS	0.362	7.5	2.15	6.2/1.62	350
AAMS	0.335	8.5	1.95	7.9/1.27	430
MPS	0.334	10.1	1.91	10.1/1.91	530
ECMS	0.336	10.8	2.14	9.1/1.27	510
APES	0.344	11.6	1.75	9.6/1.11	660

Table 16 shows similar data for a polyamide cured epoxide/silica foam at a nominal density of 0.35 g/cm³. All the silanes produced an increase in compressive strength and strain to failure, but showed less effect on the compressive modulus. In general this series of foams were 'stiffer' than the anhydride cured series and the silanes showed a different order of merit. Although it cannot be claimed that the level of addition chosen was necessarily optimum in all cases the value of silanes in increasing the compressive strength is obvious.

In packaging applications it is not only the compressive strength and modulus which are important but also the energy absorption characteristics. Comparison of these characteristics for individual foams with that of maximum possible at the same force and deflection gives the efficiency factor, where

$$\text{efficiency} = \frac{\text{area under curve}}{\text{maximum possible area}} \times 100.$$

In other words the area under the stress/strain curve is a measure of the energy absorbing characteristics. The effect of APES additions on the stress/strain properties of an anhydride cured epoxide syntactic foam at a nominal density of

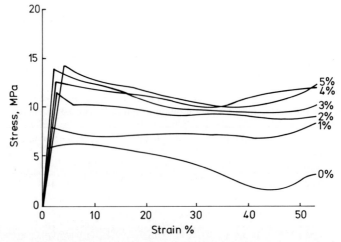

Figure 6. The effect of APES addition at increasing wt% on the stress/strain behaviour of an MNA/Epikote 828/glass microballoon syntactic foam.

0.35 g/cm^3 is shown in Fig. 6. At the 1 wt% level there is almost a flat response up to 50% strain. The efficiency increases up to 5 wt% but the 3–4 wt% would appear to be almost optimum.

Figure 7 shows the family of curves for anhydride cured foams containing 5 wt% of various silanes and it can be seen that although every silane improved the energy absorbing efficiency to a marked extent, the APES foam was the most efficient, followed closely by MPS and AAMS. Figure 8 shows similar data for the polyamide cured epoxide foams, from which it can be seen that the MPS foam was the most efficient. In these foams the flat response only held up to 25–30% strain.

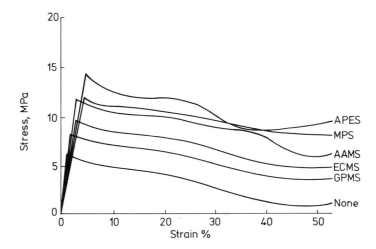

Figure 7. The effect of 5 wt% silane addition on the stress/strain behaviour of an MNA/Epikote 828/glass microballoon syntactic foam.

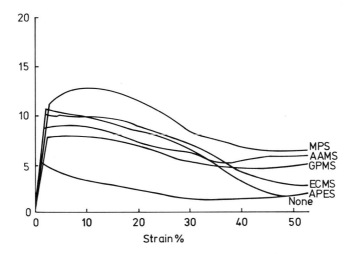

Figure 8. The effect of 2 wt% silane addition on the stress/strain behaviour of a Versamid 125/Epikote 828/glass microballoon syntactic foam.

10. DISCUSSION

If any sound conclusions are to be drawn from the large amount of bond strength data presented, it is necessary to consider the reproducibility of the test method and the scatter of results obtained. A meaningful statistical analysis of the data is impossible as the many specimens were prepared at different times and under different conditions and the practical nature of the investigation built in many uncontrollable variables. Further, the large number of cases where the actual bond strength was not measured due to non-detachment of the paint and therefore only a minimum value was recorded complicates the situation still further, as a precise value for the bond strength cannot be assigned. However, consideration of the data shows that where the same paint or adhesive has been applied to the same substrate in different experiments, the bond strength values are, broadly speaking, very similar. This indicates that the test methods are reasonably reproducible. The scatter of results as indicated by the standard deviation is generally less than 10% for the initial determinations and 15% for wet and recovered specimens, particularly where silanes are involved. The relatively low standard deviations and the large differences in bond strength due to the use of silanes allow sound conclusions to be drawn as to the effectiveness of silane additions and treatments.

The beneficial effects of silanes on the bond strength of coatings and adhesives to metallic and glass substrates have been amply demonstrated in the present paper. In some cases, particularly where a silane has been used as an additive in coatings, the initial bond strength has been doubled, but it is in mitigating the effect of water on bond strength where the silanes have proved the most effective. Silanes have also been shown to materially improve the recovered bond strength, in many cases the recovery has been almost to the original value, presumably due to the re-formation of the hydrogen bonds broken by water.

In determining how silanes improve wet bond strength it is necessary to know the locus of failure of the coating or adhesive, and the data on the failure of coatings are sparse. Work by the author, although not unequivocal, strongly suggests that most, if not all, apparent adhesion failures are cohesive failures [13] and Bikerman [20] strongly argues against the possibility of true adhesion failures. Boerio and Dillingham [21], in reporting the use of APES in titanium/ epoxide adhesive joints, discuss the possibility of a modified interface that is stronger for primed joints than for unprimed joints causing the locus of failure to shift from near the interface for unprimed joints to within the adhesive for primed joints. This strongly suggests that the failures in both cases were cohesive either in an oxide layer or in the adhesive. Reaction of the APES with the oxide layer increases the cohesive strength of the oxide and may also, in the case of aluminium, improve the wet strength by inhibiting hydration of the oxide [22]. A similar mechanism may be appropriate for any oxide coated metal. It may then be argued that the primary reason for improved bond strength and water resistance due to silanes is the reinforcement of the oxide layer or an increase in its water resistance. This has the effect of causing the failure to occur in the new weakest layer: the adhesive or coating. This may also explain, in part, the differences in bond strength achieved with different silanes and different metals as it is the nature of the oxide film and the degree of reinforcement that varies rather than any intrinsic property of a particular metal/silane combination.

Some credence must be attached to the view that the wet strength of an oxide layer is improved by silanes in the light of the findings reported by Walker [13]. XPS data showed that water immersion in the absence of silanes resulted in the presence of aluminium on both surfaces of a broken joint. Aluminium was neither detected in the presence of a silane, nor on joints broken without exposure to water. The presence of aluminium on both fracture surfaces indicates either a cohesive failure within an oxide layer, or detachment of the oxide layer, the former seems more likely.

It has been shown that with the surface coatings examined, the silanes gave the best performance when used as additives rather than pretreatments; but with adhesives, the reverse was true. As both coatings and adhesives were of the two pack type utilizing the same basic chemistry, this requires comment. The essential differences between the coatings and the adhesives are the presence or absence of solvents, the applied viscosity and the speed of cure, in so far as the latter factor affects the viscosity. Both adhesives were of the 100% solids, high viscosity type and in the case of the polyurethane had a short pot-life and rapid cure (20 and 120 min respectively). The surface coatings contained 30–40 wt% of solvent and were of low viscosity and relatively long cure time. It seems likely therefore that either the silanes in the adhesives were unable to migrate to the metal surface before being locked into the cross-linked resin matrix as the viscosity increased rapidly, or that the initial high viscosity of the silane/resin mixture was unable to penetrate the oxide layer. The latter would effectively prevent reinforcement and protection against water ingress.

The finding that the silanes were more effective as additives in coatings than as pretreatments may at first sight seem surprising as the reverse was true of adhesives. The explanation may be in the fact that in the case of pretreatments it is likely that the silanes were used under less than optimum conditions. All the silanes were used at their natural pH, no hydrolysis catalysts were added and no attempt was made to adjust the pH of the solutions before use. It has been shown that the structures of APES films on iron are dependent on solution pH, as is the availability of reactive groups [23].

The solution pH also has a profound effect on the structure of MPS films on copper [24] of MPS on aluminium [10] and APES on titanium [21]. More importantly for the present discussion it has been shown that aluminium joints primed with APES at pH = 8.5 retained 93% of their original strength after 20 days immersion in water at 60°C. Joints primed with APES at pH = 10.4 retained only 80% of their original strength under the same conditions [10]. Titanium joints primed with APES at pH = 5.5 retained less of their original strength than joints primed with APES at pH = 8.0 and 10.4 [21] and iron joints primed with APES at pH 8.0 showed greater strength retention after water immersion than joints primed at pH = 10.4 [9].

It is clear from the findings of Boerio and co-workers [9, 10, 21] that optimum water resistance is dependent on pH and that the optimum pH will vary with both the silane used and the metal to be bonded.

Further work on coatings will be required to establish if this in fact is the reason for the observed difference. However, in spite of the non-optimum conditions of use as pretreatments, the results detailed show the value and versatility of silanes as adhesion promoters. It might be considered that application of the acid–base theory in the field of surface coatings should prove effective. Since different metal oxide surfaces have different isoelectric points in water and may be regarded as acidic or basic, addition of material having acidic properties to coatings for use on basic substrates or basic materials for use on acidic substrates may improve adhesion [25]. The epoxide and polyurethane paints detailed in the paper may be regarded as basic in nature, certainly the pH of a surface from which a water soaked film has been stripped is known to be 8–10. The amino-functional silanes APES and AAMS are strongly basic and when applied to surfaces having isoelectric points in the range 9.1 (Al) to 12.0 (Fe^{2+}) may be expected to produce a basic surface. Neither silane could therefore be expected to enhance the adhesion of a basic polymer by an acid–base reaction but both have been shown to improve the initial and wet adhesion of epoxide and poly-urethane paints. It is Plueddemann's opinion [26] that although acid–base reactions may be important in filler dispersion, in silane orientation and provide a modest improvement in adhesion, the major factor in water resistant bonds is something other than acid–base reactions across the interface. It is perhaps unfortunate that pH is so important if maximum water strength is to be obtained, as although it is relatively simple to adjust the pH for a particular silane/solvent/substrate combination in the laboratory, general purpose solutions are required in the industrial field. On this basis the additive approach in coatings is extremely attractive provided that the coating/silane combinations are stable. It has been shown that the stability of epoxide and polyurethane paints containing MPS and AAMS and incorporating proprietary drying agents can be maintained for two years [2]. Further testing has established that this period is at least four years after which time there was no loss in bond strength.

It was considered in the past that an important factor in the mechanism by which silanes enhance bond strength is the reactivity of silane with some com-ponent of the coating or adhesive, the so-called 'chemical bonding theory' [7], now much modified [12]. Of the six silanes examined in detail in conjunction with epoxide and polyurethane resin systems, four, ECMS, GPMS, APES and AAMS will almost certainly react with either epoxide or isocyanate groups [7]. In the case of MPS it would seem theoretically possible for the mercapto group to react with an epoxide (1) or isocyanate group (2) according to:—

$$-\overset{|}{\underset{\underset{O}{\diagdown\diagup}}{C}}-\overset{|}{C} + HS-R \rightarrow HO-\overset{|}{\underset{|}{C}}-\overset{|}{\underset{|}{C}}-SR \tag{1}$$

$$R^1-NCO + HS-R \rightarrow R^1 NHCOSR \tag{2}$$

It seems less likely that MAMS is reactive towards either epoxide or iso-cyanate groups. In spite of this MAMS performed well, in some cases, notably the polyurethane paint on mild steel and aluminium, see Tables 3 and 7, better than ECMS and APES. Clearly chemical reaction is not a prerequisite for bond strength enhancement. Chain tangling, simple physical effects, may play an important role with both MPS and MAMS and possibly other silanes.

It has been pointed out by Plueddemann [26] that although a simplified representation of silane coupling indicates a well-aligned monolayer of silane, the coverage is far more likely to be equivalent to several monolayers. If contact is made between a polymer and uncross-linked siloxanols which still have some degree of solubility then bonding can take several forms including copolymer formation and interpenetrating polymer networks. Plueddemann concludes that the performance of the coupling agent may depend as much on physical properties resulting from the method of application as on the chemistry of the organo-functional silane used.

11. CONCLUSIONS

From the data reported in the paper it can be concluded that silanes have a wide potential application in polymer technologies where improved dry bond strength is desirable or bond strength under water soaked conditions is important.

Careful attention to detail is required if optimum effects on bond strength are to be obtained.

Acknowledgements

The author wishes to thank his colleagues C. J. Allen and C. Kerr for valued assistance in the work reported.

REFERENCES

1. P. Walker, *J. Oil Colour Chem. Assoc.* **65**, 415 (1982).
2. P. Walker, *J. Oil Colour Chem. Assoc.* **65**, 436 (1982).
3. P. Walker, *J. Oil Colour Chem. Assoc.* **66**, 188 (1983).
4. P. Walker, *J. Oil Colour Chem. Assoc.* **67**, 108 (1984).
5. P. Walker, *J. Oil Colour Chem. Assoc.* **67**, 128 (1984).
6. C. Kerr and P. Walker, in: *Adhesion 11*, K. W. Allen (Ed.). Elsevier Applied Science Publishers, Barking, UK (1987).
7. P. W. Erickson and E. P. Plueddemann, in: *Composite Materials*, Vol. 6, E. P. Plueddemann (Ed.). Academic Press, New York (1974).
8. W. D. Bascom, *ibid.*
9. F. J. Boerio and J. W. Williams, *Appl. Surf. Sci.* **7**, 19 (1981).
10. F. J. Boerio, C. A. Gosselin, R. G. Dillingham and H. W. Liu, *J. Adhesion* **13**, 159 (1981).
11. K. J. Sollman, Private communication, 1980.
12. M. R. Rosen, *J. Coating Technol.* **50**, (644), 70 (1978).
13. P. Walker. *Proceedings of the Loughborough International Conference on Adhesion*, April 1990 (in press).
14. K. L. Mittal, *Polymer Eng. Sci.* **17**, 467 (1977).
15. K. L. Mittal and D. Suryanarayana, *J. Appl. Polym. Sci.* **29**, 2039 (1984).
16. E. P. Plueddemann, *Prog. Organic Coatings* **11**, 297 (1983).
17. Dow Corning Trade Literature.
18. P. Walker, in: *Surface Coatings—1*, A. D. Wilson, J. W. Nicholson and H. J. Prosser (Eds). Elsevier Applied Science Publishers Ltd, Barking, UK (1987).
19. P. Walker, *J. Paint Technol.* **37**, 1561 (1965).
20. J. J. Bikerman, *The Science of Adhesive Joints*. Academic Press, New York (1961).
21. F. J. Boerio and R. G. Dillingham, in: *Adhesive Joints: Formation, Characteristics, and Testing*, K. L. Mittal (Ed.), pp. 541–553. Plenum Press, New York (1984).
22. A. Kaul and N. H. Sung, *Polym. Eng. Sci.* **26**, 768 (1986).
23. F. J. Boerio, L. Armogan and S. Y. Cheng, *J. Colloid Interface Sci.* **73**, 416 (1980).
24. F. J. Boerio and S. Y. Cheng, *J. Colloid Interface Sci.* **68**, 252 (1979).
25. J. C. Bolger and A. S. Michaels, in: *Interface Conversion for Polymer Coatings*, P. Weiss and G. D. Cheevers (Eds). American Elsevier Publishing Co. Inc., New York (1969).
26. E. P. Plueddemann, *Silane Coupling Agents*. Plenum Press, New York (1982).

Silanes and Other Coupling Agents, pp. 49–66
Ed. K. L. Mittal
© VSP 1992.

Controlling factors in chemical coupling of polymers to metals

J. P. BELL,[1,*] R. G. SCHMIDT,[2] A. MALOFSKY[1] and D. MANCINI[1]

[1] *Polymer Program, Institute of Materials Science, University of Connecticut, Storrs, CT 06269-3136, USA*
[2] *Dow Corning Corp., Research Center, Midland, MI 48686, USA*

Revised version received 21 June 1991

Abstract—An 'overview' discussion is presented on the effects of variables on the mechanical behavior and durability of the polymer–metal interfacial region when a polymeric coupling agent is employed to enhance bonding. Variables include locus of failure, effect of coupling agent thickness, effect of surface wetting, surface preparation, coupling agent functionality, durability in water, and corrosion vs. adhesion failure mechanisms. New data are presented on some of the subjects with previously published data included to complete the discussion. It is believed that most of the effects discussed are general in nature, although the data presented are largely for coupling agent systems investigated by the authors.

Keywords: Coupling agent; coupling agent thickness; durabilty; surface preparation; surface wetting.

1. INTRODUCTION

A polymer–metal bond is comprised of several regions or components, as shown in Fig. 1. The regions, and the interphases between regions, have their own unique set of properties and characteristics. When the bond is subjected to a tensile stress, the regions and interphases may be considered as links in a chain; failure will occur at the weakest link. If the weakest link is within the metal oxide, then improvement of the coupling agent or coupling agent–metal oxide interphase will be of no

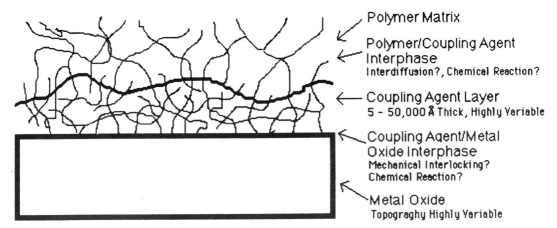

Figure 1. Schematic diagram of the polymer–coupling agent–metal oxide region.

---

*To whom correspondence should be addressed.

benefit until the oxide is regenerated or made stronger. The same considerations apply to joints subjected to shear stress, although the weak link could well be at a different location. These and other factors affecting epoxy–metal bonding have been discussed in detail [1].

Most polymer–metal oxide bonds that do not have mechanical interlocking will ultimately fail when subjected to prolonged moisture exposure and stress, even in the absence of corrosion. Polymer–metal oxide interactions are often dominated by hydrogen bond formation between a hydrated oxide and a partner, such as a hydroxyl group, on the polymer. A strong relationship between the concentration of hydroxyl groups in an epoxy and the bond strength to a metal (oxide) was shown many years ago by DeBruyne [2]. However, when water reaches the coupling agent–oxide interphase, the hydrogen bonds are interrupted and failure occurs.

Coupling agents having strong chemical interactions which are not rapidly disrupted by water are presently of great value, e.g. the organofunctional silanes, and are the hope for permanent bonds in the future. Reproducible, permanent bonds would open a myriad of new applications. At present, adhesives are not generally used for systems subjected to appreciable moisture and stress (without mechanical interlocking).

The class of materials known as 'primers' is distinct from coupling agents. While these are of benefit in providing better wetting of the surface, hindrance to water and O_2 transport, and provide greater compatibility with both oxide and adhesive, they do not (to our knowledge) provide the covalent bonding and strong, permanent interactions sought by coupling agents as a class. Also, primers are generally applied in much greater thickness.

This paper discusses some recent developments in the understanding of the various factors controlling the efficacy of coupling agents, and also emphasizes the importance of considering corrosion processes in addition to bond strengths.

2. RESULTS AND DISCUSSION

2.1. Locus of failure

Bond failure may occur at any of the locations indicated in Fig. 1. Visual determination of the locus of failure is possible only if failure has occurred in the relatively thick polymer layer, leaving continuous layers of material on both sides of the fracture. The appearance of a metallic-appearing fracture surface is not definite proof of interfacial failure since the coupling agent, polymer films, or oxide layers may be so thin that they are not detectable visually. Surface-sensitive techniques such as X-ray photoelectron spectroscopy (XPS) and contact angle measurements are appropriate to determine the nature of the failure surfaces; scanning electron microscopy (SEM) may also be helpful if the failed surface can be identified.

Metals such as iron, aluminum, and copper can have weak oxide or oxide/metal structures. In such a case, a coupling agent would be of little value until the weakness in the oxide layer had been remedied; failure will occur in the weakest region.

The presence of metal oxide on both sides of a fracture surface can be deceptive if the surface has microroughness (peaks less than 1 μm high), since the failure crack may propagate through the tops of the roughness peaks, carrying some oxide with the polymer fracture surface. In such an instance, the amount of oxide on the polymer side of the fracture, as measured by XPS, should be considerably less than the amount on the metal side, and the contact angle should also be intermediate between the polymer and the pure metal oxide. For scientific studies it is desirable to use surfaces which are as smooth as possible, such as a copper mirror carefully deposited on a silicon wafer.

Fracture cracks often propagate in an uncontrolled manner. Hence, adhesion tests should utilize cracks which are grown carefully and slowly, as is the case with the procedures commonly used for fracture toughness measurements (ASTM D3433-85).

2.2. Effect of the coupling agent layer thickness

Typically 'primers' are applied to metal surfaces by coating to a thickness of at least several micrometers. On the other hand, Schmidt and Bell [3a] have reported an optimum coupling agent layer thickness for polymer mercaptoester (EME, Fig. 2) containing coupling agents bonded to an epoxy–polyamide coating in the range of 150 Å, several orders of magnitude less. Additional data for EME coupling agents using other coating systems are shown in Figs 3 and 4; the procedure for the measurements will be described below. The results are in excellent agreement with the above earlier work [3a], showing an optimum concentration which is related to the optimum thickness by 0.07% coupling agent = 150 Å. We found failure to be near the coupling agent–polymer interface in earlier work [4]. The thickness effect data appear to be consistent with the results of Sung *et al.* [3b], who investigated the effect of the concentration of γ-aminopropyl silane (γ-APS) solutions on the 180° peel strength of polyethylene bonded to γ-APS-treated Al_2O_3. They also found that the peel strength went through a maximum as the coupling agent concentration increased.

Similarly, Plueddemann has reported [3c] an optimum silane coupling agent layer thickness of 50–200 Å for commercial glass fiber treatment for use in composite materials. On the contrary, optimum adhesion of polypropylene to aluminum was obtained with a silane primer layer between 0.5 and 10 μm thick. In

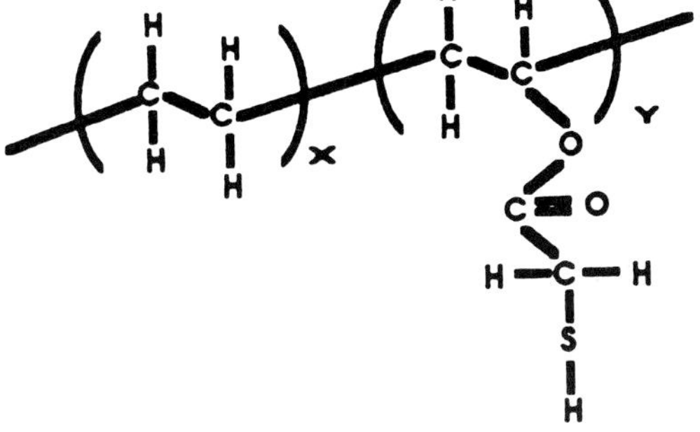

Figure 2. Ethylene mercaptoester polymeric coupling agent (EME); the reaction product of random ethylene/vinyl alcohol co-polymers with mercaptoacetic acid. The number after the EME (e.g. EME 47) refers to the weight percent of mercaptoester units along the chain. Essentially no vinyl alcohol units remain.

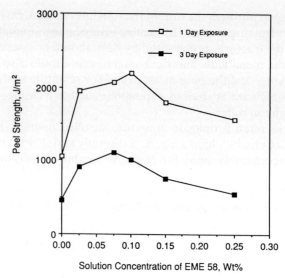

Figure 3. Effect of EME 58 (58 wt% mercaptoester units co-polymer) coupling agent concentration on the peel strength of flexible epoxy (amine-cured)/AD = acetone-degreased steel test panels following (a) 1 day and (b) 3 day exposure to 57°C condensing humidity. See Appendix 4 for epoxy resin and cure description.

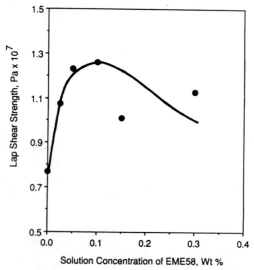

Figure 4. Effect of EME 58 (58 wt% mercaptoester unit co-polymer) coupling agent concentration on the lap shear strength of acrylic/steel joints following 7 day immersion in 57°C water. See Appendix 4 for acrylic resin formulation and Appendix 5 for shear strength procedure.

both cases, significant deviations from the optimum thickness resulted in substantial performance declines. Trivosono *et al.* [3d] discovered a similar thickness dependence when using silane coupling agents in fiberglass-reinforced polyester systems.

Measurement of such small thicknesses is difficult. Ellipsometry [3] and angle-dependent XPS [5] may be used with care to avoid complications caused by surface roughness.

A rather simple application technique to flat metal coupons which has been found to be helpful when the EME coupling agents are used at high dilution (0.07% in 50:50 acetone–hexane) is outlined below:

(1) Modify the metal oxide surface as desired (i.e. acid etch, sand-blasting, etc.)
(2) 'Degrease' the metal coupons in a high grade (99.9% +) solvent for 30 min.
(3) Air-dry or blow with a dried gas (air, nitrogen) for 3 min.
(4) Dip the coupons into the coupling agent solution, remove vertically, and let excess solvent drip off.
(5) Air-dry for 3 min.
(6) Apply the resin and assemble the joints. Cure under conditions appropriate to the resin.

For the 'effect of thickness' experiments, coupons were quickly removed from a 50:50 by volume acetone–hexane covered bath. The bath was mounted on a sensitive balance, such that the amount of liquid retained on the coupon when removed through a slit in the bath top could be measured. The sample was allowed to drip for a few seconds before removal from the bath. A correction was made for the rate of weight loss of the bath with no coupon. Calculation gave the amount of liquid retained on both sides of the coupon. Since the concentration of coupling agent in the liquid was known, the amount of coupling agent deposited (all that was in the liquid) was also known and its thickness could be calculated. Different thicknesses were obtained by using different concentrations of coupling agent in the bath. The method is not exact, but gives a qualitative estimate of thickness. This method has given results which are consistent with measurements using ellipsometry on very smooth surfaces [3a, 6].

There may be variations in the optimum coupling agent thickness from metal to metal and between various coupling agents. The existence of an optimum, however, appears significant.

2.3. Effect of surface wetting

For the purpose of minimizing corrosion, a coupling agent layer should be as hydrophobic as possible. The chemical groups participating in the coupling reactions are in general hydrophilic, but it is possible to space them such that the overall coupling agent is quite hydrophobic. An example of this is given in Table 1 [7].

The three EME coupling agents in Table 1 were analyzed using contact angle measurements to determine their polar and dispersion components of surface tension. From the surface tension data, wettability envelopes were constructed and compared with the surface tension properties of the epoxy coating [4]. These data predicted that EME 47 would be wet by the epoxy, but not EME 23. This is believed to be the reason for the very low peel strength when EME 23 was employed [4].

As noted, corrosion of the oxide by water ingression is a serious problem. Very hydrophobic polymers such as those containing fluorine are attractive in terms of protecting the metal oxide surface from water. However, wetting of the oxide by the coupling agent and of the coupling agent by the top coat must occur for the system to function effectively. Since chemical coupling reactions do not easily occur when there is insufficient wetting, a good match of surface energy at the

Table 1.
Effect of the coupling agent mercaptoester concentration on the peel adhesion and corrosion protection of epoxy/EME/steel test panels [7]

Wt% mercaptoester in EME	Peel strength[a] (g/cm) (after 1 h in 57°C water)	Relative corrosion protection (in water at 57°C). Time to form 3 corrosion pits (h)
No coupling agent	27	14
23	10	46
47	55	32
90	126	20

[a] Peel rate 1 cm/min, 90° peel test.

phase boundary between the coupling agent and top coat is necessary to obtain optimum coupling.

2.4. Surface preparation

Weak oxide layers must be removed and regenerated when they are the weakest link in the system. In some instances, sand-blasting and dust removal are sufficient. For aluminum, an etching process such as phosphoric acid anodization is normally used. The objective of surface treatments is to develop a thin, strong, and hydrolytically stable oxide layer which can bond to the coupling agent. Mechanical interlocking provided by structures generated during etching is an added bonus, especially for aluminum surfaces. Surfaces that are not mechanically or chemically treated must be, as a minimum treatment, degreased to remove invisible organic contaminants which are deposited from the air.

Not so obvious is the fact that some oxide usually enhances adhesion. Park and Bell [8] reduced a copper surface in an oxygen-free electrochemical bath, and then dried and coated the oxide-free copper with an epoxy composition. It was found (Table 2) that the peel strength was reduced relative to an almost identical experiment in which a small amount of air was bubbled through the cell after the copper oxide reduction process (Table 2). In early work with low molecular weight mercaptoesters [9] and later with EME coupling agents [6], a citric acid pretreatment was used to remove weak oxides from steel and copper. The need for this depends on the prior history of the metal surface. In recent experiments using clean, degreased steel panels of the type used for paint evaluation, the room-temperature citric acid treatment seemed to have little benefit. Data are shown in Fig. 5. In very recent experiments with higher temperature (80°C), shorter time (10 min) neutralized citric acid (pH 7.0) treatments and coupling agent application from refluxing solution, a much greater beneficial effect of the citric acid treatment has been observed. This subject is still under study.

2.5. Coupling agent functionality

Since the functional groups of the coupling agent that form specific chemical interactions with the metal are hydrophilic in nature (in all cases known to us), an

Table 2.
Effect of electrochemical reduction of copper and post-air exposure on epoxy/copper torsional joint strength [8]

Joint no.	Force to break (N)	
	Air bubbling	Nitrogen bubbling
1	1277	1157
2	1312	1090
3	1299	1068
4	1183	1112
Avg.	1268	1107
Standard deviation	50.2	32.9

Electrolyte: 0.3% KCl aq. solution prepared with twice-distilled water, containing 0.02 M benzotriazole. Polarization: Cathodic 5 V for 10 min and 10 min without potential (with air bubbling). Joints were tested after 24 h in boiling water. Adhesive: 100 parts diglycidyl ether of bisphenol A (Epon 828 from Shell Chemical Co.) cured with four parts dicyandiamide at 175°C for 2 h.

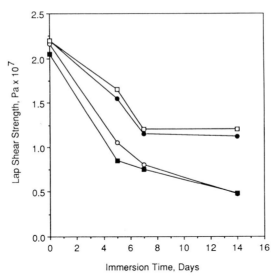

Figure 5. Epoxy/steel joint lap shear strength as a function of 57°C water immersion time for the following steel pretreatments: (—O—) citric acid treatment; (—●—) citric acid treatment followed by 0.08 wt% EME 58 dip; (—■—) acetone degrease; (—□—) acetone degrease followed by 0.08 wt% EME 58 dip.

increase in the number of such groups results in improved bonding due to increased interaction, but generally corrosion protection declines due to water more readily permeating the coupling agent–metal oxide interphase. In a dry environment high functionality is favorable, but for a steel substrate in a wet or wet/salty environment corrosion protection is usually limiting. Typical data are shown in Fig. 6. For many applications a balance between adhesion and corrosion protection is chosen; this compromise will be discussed in more detail in Section 2.8. Plueddemann and Pape [10] have also shown with hydrophobic/hydrophilic silane mixtures that bond durability in the presence of water can be improved by

2.7. Advantages of polymeric coupling agents

The usual schematic diagram of a coupling agent shows a relatively low molecular weight molecule with one end attached to the metal oxide and the other to the polymer above it. However, in practice it is difficult to obtain a uniform mono-molecular layer on a metal oxide surface. If, as usually occurs, the layer is multi-molecular, then the weak link in the strength chain is often between the layers in the coupling agent. This is a fundamental disadvantage of low molecular weight coupling agents. Silane coupling agents avoid this to a considerable degree by polymerizing on the surface and forming a network within the multilayer, resulting in a structure capable of load transmission. EME coupling agents, on the other hand, are high molecular weight polymers as originally synthesized.

Another advantage of polymeric coupling agents is that they have the ability to absorb a portion of the mechanical and thermal stresses generated in the inter-phase region which arise due to the mismatch between the moduli and thermal expansion coefficients of the metal and the polymer applied to it. If a small molecule is stretched between the metal and the polymer, it will have difficulty withstanding the stresses developed upon mechanical or thermal cycling (or H_2O absorption). A third advantage of polymeric coupling agents is that the ratio of hydrophobic to hydrophilic components can be controlled; the composition is not fixed, but can be adjusted to suit particular needs.

2.8. Corrosion vs. adhesion

An often overlooked factor in the failure of polymer–metal bonds is the balance between bond displacement and corrosion. Bond displacement involves breakage of the bonds at the polymer–metal oxide interface, whether by displacement of hydrogen bonds by water, fracture of the bonds because of thermal expansion or modulus differences between phases, or chemical degradation of the bonds because of acidity or basicity of the oxide. Corrosion is a quite different process which involves oxygen, water, and an electromotive driving force. Corrosion can occur in an oxide layer beginning at the edges of a bond, and have nothing to do with either the polymer or coupling agent layers. Alternatively, oxygen and water transport through the polymer top-coat and coupling agent, or through pinholes in these layers, provides the oxide with the ingredients for corrosion. As we have stated earlier, hydrophilicity of the coupling agent–metal interphase increases water solubility in this region and thus facilitates corrosion.

In oxygen-free hot water (57°C) or even in normal distilled water of the same temperature, the durability of epoxy–EME coupling agent–steel bonds was quite high relative to epoxy/steel joints without a coupling agent (Fig. 8). A substantial improvement was also observed with acrylic–EME–steel joints (Fig. 9). If the joints were suspended above the water (57°C), the improvement associated with the coupling agent was observed at short times, but disappeared at longer times of exposure (Fig. 10). Examination of the fracture surfaces revealed corrosion from the edges into the joint. Generation of corrosion products can place a tensile stress on the remainder of the joint, decreasing its strength even further. In the condensing humidity test (joint suspended over the water), the partial pressure or activity of the oxygen was much higher than when the joint was immersed; hence the corrosion rate was higher.

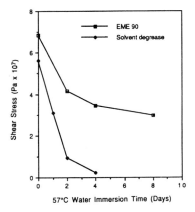

Figure 8. Shear strength durability in a 57°C water immersion of epoxy/steel torsional joints with and without EME 90 (90 wt% mercaptoester unit co-polymer) coupling agent pretreatment. From ref. 6. Adhesive: diglycidyl ether of bisphenol A (Epon 828) cured with a stoichiometric amount of methylene dianiline for 1 h at 120°C followed by 2 h at 150°C.

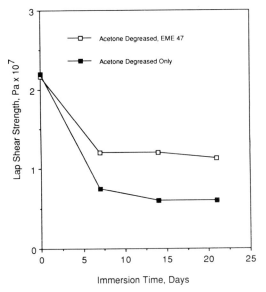

Figure 9. Lap shear strength durability in a 57°C water immersion of acrylic/steel joints with and without EME 47 (47 wt% mercaptoester unit co-polymer) coupling agent pretreatment. Resin composition is given in Appendix 4.

Corrosion may be hindered by slowing diffusion of oxygen and water into the coupling agent–oxide interphase, and by use of hydrophobic and low permeability coupling agents, and corrosion inhibitors. It is important to recognize in a specific problem whether disbonding or corrosion is the controlling mechanism.

2.9. Fundamental interactions across the interfaces

Of particular interest are the specific chemical interactions that occur between the coupling agent's active functionality and either the metal oxide or the polymeric top-coat. Recently, reflection angle infrared spectroscopy (RAIRS) [12, 13] and

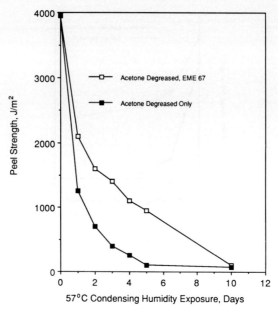

Figure 10. Peel strength durability in a 57°C condensing humidity of acetone-degreased and EME 67 (67 wt% mercaptoester unit copolymer) coupling agent treated 1010 carbon steel/epoxy peel test panels. The epoxy top-coat formulation is given in Appendix 4.

its Raman analog [14] have been utilized with increasing frequency in the study of such interfaces, for example, organized monomolecular assemblies [15, 16] and smooth fracture surfaces of model adhesive joints [17]. These two techniques are amenable to the study of the formation of chemical interactions at the two bond interfaces noted above.

As an example, both monofunctional and multifunctional polymeric mercapto-esters were deposited onto optically smooth silicon wafers coated with vapor-deposited copper. The copper had been oxidized to Cu_2O, as verified by XPS. Infrared reflectance (RAIRS) at 81° (4 cm^{-1} resolution, 2000 scans) using an MCT detector yielded information on both the nature and the durability of the mercaptoester bond to the metal oxide film. A 16 cm^{-1} shift (1740 → 1724 cm^{-1}) was observed in the carbonyl absorption of stearyl thioglycolate (STG) deposited onto the Cu_2O mirror. The absorption spectrum of the carbonyl region is illustrated in Fig. 11, both for the pure STG and the reacted monolayer.

The STG was cast from a 1×10^{-3} M solution into which the mirror had been immersed for from 10 to 20 min (see Appendix 1). The presence of a monolayer was confirmed by ellipsometry (18 Å). The spectral data agreed with data gathered on a similar system in a powder form. In this case, Cu_2O powder was immersed in a 0.01 M solution of isooctyl thioglycolate (OTG) in isopropanol for from 1 to 10 min, washed with pure isopropanol, dried in air, and analyzed via infrared transmission in a KBr dispersion pellet (see Appendix 2). A similar spectral shift of approximately 15 cm^{-1} (1739 → 1724 cm^{-1}) was observed and the lack of two distinct carbonyl absorbances suggested the formation of a monolayer. In both cases, the formation of a copper–mercaptoester salt may be responsible

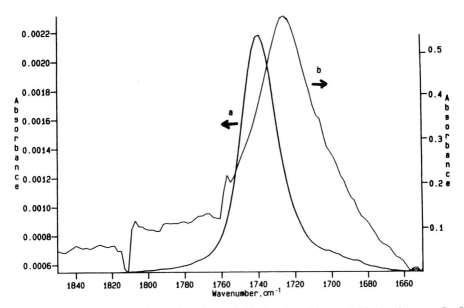

Figure 11. (a) Carbonyl absorption region of stearyl thioglycolate; (b) stearyl thioglycolate on a Cu_2O mirror, reflectance infrared spectrum (81° grazing angle, 4 cm^{-1} resolution, 2000 scans).

for the shift. The spectra of the original OTG and the coated powder are illustrated in Fig. 12.

Similar IR characterization was previously reported [18] using a trithiol glycolate compound and ferric oxide powder. In this work, the carbonyl stretch absorbance of the trithiol glycolate was found to exhibit a doublet centering at 1739 and 1725 cm^{-1} in a transmission spectrum of the neat material. These two carbonyl absorbances were attributed to the presence of both free and hydrogen-bonded components, respectively. However, upon reaction with ferric oxide the thiol was consumed, removing the opportunity for hydrogen bonding, and a single carbonyl absorbance centered at 1731 cm^{-1} was observed. It is believed that weak dipolar interactions with the iron resulted in a carbonyl spectral shift from 1739 to 1731 cm^{-1}, which is similar to, but less extensive than, that observed with Cu_2O.

Additional work involves probing the coupling agent–polymer interface. Utilizing spectral subtraction techniques, it can be determined whether reaction across the interface has occurred. Current experiments, using the RAIRS technique, involve the casting of thin layers < 100 Å) of a coupling agent (EME 68) onto an optically flat, reflective substrate followed by coating with a layer of an epoxy resin (Epon 828, an oligomeric bisphenol-A based epoxy). Subsequent heating coupled with infrared analysis of the detectable epoxide peaks (~ 1250 and 916 cm^{-1}) and thiol (2569 cm^{-1}) will provide information about the extent of reaction and inter-diffusion. Prior work [19] has shown that the thiol groups of the model compound trimethylol propane trithioglycolate (TTTG) when cast as a surface layer on an NaCl plate reacts readily at 80°C with an epoxy top-coat. The thiol absorbance at 2568 cm^{-1} and the epoxy absorption at 915 cm^{-1} disappeared over a few hours, while the corresponding hydroxyl absorption at 3460 cm^{-1} increased dramatically. A similar experiment using Epon 1001 on an EME 47-covered NaCl plate was in agreement with the model compound experiment. These experiments

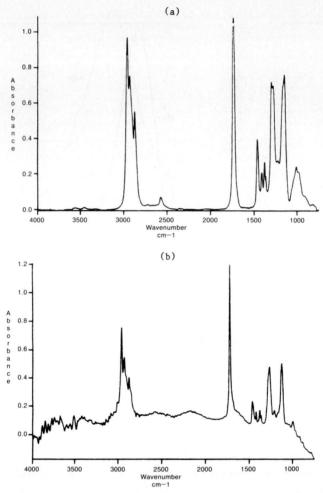

Figure 12. (a) Infrared spectrum of isooctyl thioglycolate; (b) infrared spectrum of isooctyl thioglycolate-treated Cu_2O powder dispersed in a KBr pellet.

showed the capability of EME coupling agents to react with epoxy groups; reaction with other species, such as vinyl groups, is also expected.

3. CONCLUSIONS

The factors controlling the mechanical behavior of polymer–coupling agent–metal oxide systems have been discussed in terms of the 'weakest link in a chain' concept. Determination of the locus of failure and thus the 'weak link' is not usually reliable by visual inspection, and surface roughness can cause misleading spectroscopic results if failure is near an interface.

There is an optimum coupling agent thickness of about 150 Å for poly(ethylene mercaptoester) (EME) co-polymer coupling agents, comparable to the 50–200 Å reported for silanes. An example is given of large initial strength loss when an EME coupling agent was altered such that effective wetting did not occur; wetting is vital to effective bond performance. Metal surface preparation to produce a stable, strong oxide is very important; the presence of some oxide is helpful relative to the pure metal. Hydrophobic coupling agents hinder water ingression, but lack the polar groups needed to bond to the metal oxide. Hydrophilic coupling agents may bond well chemically, but fail because of corrosion associated with high water and oxygen permeability.

Corrosion becomes much more important in 'condensing humidity' tests because of the higher partial pressure of oxygen that is present. Chemical bonding of top-coat polymers to coupling agents is enhanced if the top-coat is applied shortly after the coupling agent is applied. Reactive groups in the coupling agent will try to 'bury' themselves to minimize the surface free energy at the air interface, reducing their availability for reaction over time. Polymeric coupling agents have advantages in that a monolayer is not required for optimum performance. Other advantages include readily controllable hydrophilicity and viscoelastic character, providing mechanisms for thermal and chemical stress relief.

Fundamental studies by reflection angle infrared spectroscopy of the bonding of EME coupling agents to metal oxides reveal a significant shift in the carbonyl absorbance band when the coupling agent is applied as a very thin layer on a metal oxide. The shift is reproducible and the extent varies with the type of oxide. These results were obtained both by use of copper mirrors and from Cu_2O powder coated with very thin layers of model compounds. The compounds were not removable by isopropanol, a solvent for the bulk compound. The thiol absorbances of thin layers of model compounds were also found to decrease in relative intensity with time. This illustrates that a specific chemical interaction has occurred.

REFERENCES

1. R. G. Schmidt and J. P. Bell, in: *Advances in Polymer Science*, K. Dusek (Ed.), Vol. 75. Springer Verlag, New York (1986).
2. N. A. deBruyne, *J. Appl. Chem.* **6**, 303 (1956) (reprinted in ref. 1).
3. (a) R. G. Schmidt and J. P. Bell, *J. Adhesion* **27**, 135–142 (1989); (b) N. H. Sung, A. Kaul and S. Ni, in: *Adhesion Aspects of Polymeric Coatings*, K. L. Mittal (Ed.), p. 379, Plenum Press, New York (1983); (c) E. P. Plueddemann, *Silane Coupling Agents*, pp. 82–85. Plenum Press, New York (1982); (d) N. M. Trivosono, L. H. Lee and S. M. Shriner, *Ind. Eng. Chem.* **50**, 912 (1958).
4. R. G. Schmidt and J. P. Bell, *J. Adhesion Sci. Technol.* **3**, 515–527 (1989).
5. B. Tyler, D. Castner and B. Ratner, *J. Vac. Sci. Technol.* **A7**, 1646 (1989).
6. R. G. Schmidt, Ph.D. Thesis, University of Connecticut (1987).
7. R. G. Schmidt and J. P. Bell, *J. Adhesion* **25**, 85–107 (1988).
8. J. M. Park and J. P. Bell, in: *Adhesive Joints: Formation, Characteristics and Testing*, K. L. Mittal (Ed.), p. 532. Plenum Press, New York (1984).
9. A. DeNicola, Ph.D. Thesis, University of Connecticut (1981).
10. E. P. Plueddemann and P. O. Pape, 40th Annu. Tech. Conf. Reinforced Plastics, SPI Paper 17-F (1985).
11. S. Wu, *Polymer Interface and Adhesion*. Marcel Dekker, New York (1982).
12. R. Greenler, *J. Chem. Phys* **44**, 310 (1966).
13. R. Greenler, *J. Chem. Phys.* **50**, 1963 (1969).
14. F. J. Boerio, P. P. Hong, P. J. Clark and Y. Okamoto, *Langmuir* **6**, 721 (1990).
15. H. O. Finklea, L. R. Robinson, A. Blackburn, B. Richter, D. Allara and T. Bright, *Langmuir* **2**, 239 (1986).
16. D. Allara and R. Nuzzo, *Langmuir* **1**, 52 (1985).
17. D. J. Ondrus, F. J. Boerio and K. J. Grannen, *J. Adhesion* **29**, 22–42 (1989).
18. R. G. Schmidt, J. P. Bell and A. Garton, *J. Adhesion* **27**, 127 (1989).
19. R. G. Schmidt and J. P. Bell, in: *Adhesives, Sealants and Coatings for Space and Harsh Environments*, L. H. Lee (Ed.), p. 165. Plenum Press, New York (1988).

APPENDIX 1. PREPARATION AND ANALYSIS OF STEARYL THIOGLYCOLATE MONOMOLECULAR FILMS ON CUPROUS OXIDE MIRRORS

1. Materials

(A) Stearyl thioglycolate, 99.9 wt%, Pfaltz and Bauer.
(B) Isopropanol, 99.9 wt%, Aldrich Chemical.
(C) Copper, 99.9 wt%, 1 mm wire, Aldrich Chemical.
(D) Silicon wafers, 4″ diameter (optically polished, single crystal).

2. Instrumentation

(A) Edwards E306A Coating System Evaporator.
(B) Bio-Rad Digilab FTS-60 Spectrometer with a Bio-Rad Digilab 3200 Data Station.

3. Experimental procedure

A 2 cm length of isopropanol-washed copper wire was vapor-deposited onto 1″ × 3″ sections of silicon wafer at a pressure of about 10^{-5} Torr. The composition of the copper films was determined via XPS to be 99% cuprous oxide. The mirrors were then stored in a desiccator for up to 1 month without any change in composition. Next, a 0.01 M solution of stearyl thioglycolate in isopropanol was prepared. A single, isopropanol-rinsed copper mirror was then immersed into 200 ml of the stearyl thioglycolate solution for 10 min. The mirror was then removed from the solution, air-dried horizontally, and finally immersed into a swirling bath of pure isopropanol for about 30 s. After being air-dried, experimental analysis was then conducted on the prepared films.

4. Analytical procedure

All samples were analyzed at an incident angle of 80° using a variable angle accessory, model VRA-MX-1-RMA from Harrick Scientific. A nitrogen purge was used. From 256 to 1000 scans were taken at a frequency of 30 kHz with an open aperture using a low pass filter set at 4.5 kHz. A collector gain setting of 1 was used. A previously installed, liquid-nitrogen-cooled mercury–cadmium–telluride (MCT) detector was used to collect the signal. Reference spectra were taken from freshly evaporated copper mirrors, typically the same ones that were subsequently coated with stearyl thioglycolate.

APPENDIX 2. PREPARATION AND ANALYSIS OF ISOOCTYL THIOGLYCOLATE-COATED METAL OXIDE POWDERS

1. Materials

(A) Fe_2O_3 powder, 99.9%, Aldrich Chemical.
(B) Cu_2O powder, 99.9%, Aldrich Chemical.
(C) CuO powder, 99.9%, Aldrich Chemical.
(D) ZnO powder, 99.9%, Aldrich Chemical.
(E) Isooctyl thioglycolate, 99.9%, Pfaltz and Bauer.
(F) KBr, 99.9%, Aldrich Chemical.
(G) Isopropanol, 99.9%, Aldrich Chemical.

2. Instrumentation

(A) Bio-Rad Digilab FTS-60 Spectrometer with a Bio-Rad Digilab 3200 Data Station.

3. Experimental procedure

Each oxide powder was washed copiously with pure isopropanol, dispersed in a KBr pellet, and analyzed via infrared transmission for organic contaminants and to obtain reference spectra of the 'blank' oxides. The washed powders were then air-dried and placed aside in a desiccator. Next, 500 ml of 0.01 M solution of isooctyl thioglycolate (OTG) in isopropanol was prepared and 15 ml glass vials were then filled with approximately 0.50 g of each powder. This was followed by the addition of about 10 ml of the OTG solution to each vial. The vials were then sealed and agitated for about 10 min. This was followed by gravimetric filtration of each vial's contents using filter paper. Several filtration washes were made using pure isopropanol. The powders were then dried from 1 to 24 h at room temperature and stored in glass vials.

4. Analytical procedure

All samples were analyzed in transmission under a nitrogen purge. From 256 to 1000 scans were taken with the radiation normal to the KBr pellet. A scanning frequency of 5 kHz, an aperture set for 1 cm^{-1}, and a low pass filter setting for 1.12 kHz were used. A collector gain setting of 1 was used for all spectra. Reference spectra of black KBr pellets and pure oxide powder/KBr dispersions were used in the analysis of the reacted powders.

APPENDIX 3. EME SURFACE TENSION MEASUREMENT

1. Materials

(A) Water (triply distilled, deionized).
(B) Methylene iodide, 99.9%, Aldrich Chemical.
(C) EME 48, EME 58, EME 68.

2. Instrumentation

(A) Ramé-Hart Contact Angle Analyzer, Model No. 100-00-115.

3. Experimental procedure

First, the testing liquids, water and methylene iodide, were evaluated on polyethylene and polytetra-fluoroethylene plates to verify their suitability for contact angle measurements. When contact angle values $((\Theta_a - \Theta_r)/2)$ were found to be consistent with those established in the literature, the liquids were selected for surface tension measurements on the EME co-polymers. All contact angles used in the analysis were averages of the advancing and receding angles for the liquids. The polymeric films were cast onto the glass slides from dilute solution (0.06 wt%) and blown dry with nitrogen (~ 3 min).

To be sure that the observed drop in surface tension (Fig. 7) was not attributable to evaporation of residual solvent, several sample surfaces were exposed to vacuum during the aging process. Samples were removed from the chamber at periodic intervals and the contact angle was immediately measured. The results of these measurements were superimposable on the results of measurements in air (Fig. 7), showing that the observed surface tension drop was not related to the rate of residual solvent removal. To determine whether the decrease in surface tension was related to accumulating contamination from laboratory air, measurements of the surface tension of a cleaned (acetone-degreased, new) glass slide were made in laboratory air for times up to 1 h. The contact angles of fluid drops applied separately to the glass were essentially the same over the 1 h period, indicating that the glass was not rapidly becoming contaminated. A second method involved analyzing the EME surfaces with time by FTIR reflection after the final isopropanol wash and the nitrogen drying procedure. This was done for times up to 2 h, without any observed changes in the spectra; the appearance of contaminants was not observed.

APPENDIX 4. ADHESIVE AND COATING MATERIALS

1. Thermosetting acrylic adhesive formulation

	wt%
(A) Isopropyl methacrylate, Rohm Tech	56.0
(B) Butylene glycol dimethacrylates, Rohm Tech	2.0
(C) Styrene/butadiene/styrene triblock co-polymer (Stereon 840A, Firestone)	30.0
(D) Methacrylic acid, Rohm Tech	10.0
(E) Tert-butylperoxybenzoate	2.0
Total wt%	100.0

$T_{\text{soften}} \simeq 180°C$, heat-cured at 177°C for 24 h.

2. Thermosetting, flexible epoxy formulation

	wt%
(A) XB-4122 epoxy resin, Ciba Geigy	10.0
(B) Epon 828 epoxy resin, Shell Chemical	90.0
(C) Triethylenetetramine, Dow Chemical	(14 pph epoxy)
Total wt%	100.0

$T_{soften} \sim 15°C$, heat-cured at 75°C for 10 h.

APPENDIX 5. ACRYLIC ADHESIVE JOINT EVALUATION

1. Materials

(A) Acrylic adhesive (Appendix 4).
(B) $1'' \times 4''$ low carbon 1038 cold rolled steel, Q-panel.
(C) 0.06 wt% solutions of EME 48 and EME 57 in mixtures of acetone and hexanes (50 vol% acetone for EME 48 and 60 vol% acetone for EME 57).
(D) Acetone, 99.9 wt%, Aldrich Chemical.
(E) Water, triply distilled, deionized.
(F) High density polyethylene containers (3 l).

2. Instrumentation

(A) 4206 Model Instron (with internal data station).

3. Experimental procedure

All joints were evaluated by following the procedure outlined in ASTM D-1002. All joints were prepared with a one $\frac{1}{2}''$ overlap with no gap, clamped, and heat-cured at 177°C for 24 h. Water immersion tests were conducted by immersing the joints in the distilled, deionized water contained in the polyethylene containers. The containers were heated to 57°C in an air-circulating oven and exposed to the air via $\frac{1}{8}''$ holes in their covers. Water was replenished as necessary.

Actual joint preparation involved first acetone-degreasing the steel lap shear specimens for about 30 min (Independent evaluation via XPS found the metal surface to be 99% Fe_2O_3 along with some adventitious carbon.) These joints coated with coupling agent underwent the following procedure:

(A) Acetone-degreasing for 30 min.
(B) Air-dry 1 min.
(C) Immersion into coupling agent solution for 10 min.
(D) Air-dry 1 min.
(E) Immersion into pure, swirling acetone for 1 min.
(F) Air-dry 1–10 min (assemble joints as soon as possible).

Silanes and Other Coupling Agents, pp. 67–80
Ed. K. L. Mittal
© VSP 1992.

Surface modification by fluoroalkyl-functional silanes: A review

M. J. OWEN* and D. E. WILLIAMS

Analytical Research, Dow Corning Corporation, Midland, MI 48686-0994, USA

Revised version received 3 December 1990

Abstract—In the last decade there has been a resurgence of interest in fluorosilicone materials including fluoroalkyl-functional silanes. The various studies concerning surface modification with this type of silane are reviewed. The most commonly studied materials are self-assembled monolayers from trichloro- and trialkoxysilane solutions involving fundamental investigations of wettability, surface tension and surface composition. Adhesion, friction and wear behavior have also been studied. Likely applications are lubricating treatments, adhesion control, textile treatments and other low-soiling coatings. Monofunctional fluoroalkyl-containing silylating agents are also becoming important. Applications for such silanes are developing in the areas of chromatographic separation materials and as XPS surface derivatization reagents.

Keywords: Silanes; fluoroalkyl-functional groups; surface modification; surface tension; adhesion control; lubrication; silylating agents; release agents.

1. INTRODUCTION

Silane coupling agents to promote adhesion are the most familiar example of the R_nSiX_{4-n} class of organosilane materials having two types of substituent where R is a nonhydrolyzable organic group that may be relatively inert, such as a hydrocarbon radical, or may be reactive to particular organic systems. The X functionality is a hydrolyzable group, often an alkoxy group. In coupling agents $n = 1$ but n may also be 2 (linear monomer type) or 3 (simple silylating agent or end-blocker type). These organosilanes have been used in a multitude of ways to modify surfaces. This has been amply demonstrated in the book by Plueddemann [1], the symposium proceedings edited by Leyden and Collins [2–4], and the review by Arkles [5]. Applications as diverse as adhesion promoters, chromatography, immobilized catalysts, bound antimicrobial substances, surface energy control, ion concentration and removal by bound chelate-functional silanes, and coated metal oxide electrodes are discussed in these publications.

One type of R substituent that features only briefly in these reviews but which is growing in importance is the fluoroalkyl-containing group. There is renewed interest in fluorosilicone materials of all types including silanes. In line with the well-recognized low surface tension of aliphatic fluorocarbon-containing species, the prime interest is in the area of surface energy control. Examples of such silanes of all three types ($n = 1$, 2 and 3) have been reported and the purpose of this review is to summarize recent (last decade) developments in this topic.

---

*To whom correspondence should be addressed.

Emphasis is on direct application (i.e. polymerization *in situ*) of these silanes rather than their use as intermediates in pre-formed siloxane polymers.

2. LITERATURE REVIEW

2.1. Coupling agent type (n = 1)

This class is termed the coupling agent type for molecular structure reasons and not because of the applications. Since producing low surface tensions is the opposite (in terms of wettability) to what is often required in primers and coupling agents, most of the applications will be in non-traditional areas other than adhesion promotion. However, since compatibility is essential in the development of interpenetrating networks and such interaction has been shown to be one important mechanism of coupling agents [6], there is the potential for their use in bonding fluoropolymers to substrates such as glass and metal. Uses of this type are described in the patent literature, for example to coat reinforcing glass fibers in fluorine-containing elastomers and resins. Another related possibility is to use them in combination with conventional silane coupling agents to increase the hydrophobic character of the coupling region and diminish bond weakening by water ingress.

One approach to the fluoropolymer/glass adhesion problem is reported by Lee and McCarthy [7]. They start with hydroxylated polychlorotrifluoroethylene (PCTFE-OH) and produce the *in situ* fluoropolymer silane, PCTFE-OC(O)NH(CH$_2$)$_3$Si(OEt)$_3$, by reaction with 3-isocyanatopropyltriethoxysilane in the presence of dibutyltin dilaurate in THF at room temperature. This material adheres tenaciously to glass when bonded at 80°C for 12 h. The film could not be removed from the glass without tearing it. SEM and XPS analysis (Fig. 1) of the glass surface indicate cohesive failure in PCTFE. Covalent bonding between polymer surface Si(OEt)$_3$ groups and glass Si—OH is presumed to occur. The films of PCTFE-OH and PCTFE-OC(O)NH(CH$_2$)$_3$Si(O)OH show no tendency to adhere at all, falling from the glass slides when inverted.

The extent to which surface tension can be controlled by fluoroalkyl-containing coupling agent type treatments is summarized in Table 1. Its purpose is to simply illustrate the range of control possible; detailed comparisons are unwarranted because of differences in sample preparation and choice of substrate, data acquisition and treatment. Some of the critical surface tensions (σ_c) are obtained with *n*-alkanes, some with other liquids. Some of the dispersion force components (σ_s^d) and polar components (σ_s^p) of solid surface tension are derived by use of different equations. The reader is referred to the key references in Table 1 for full details.

The high values for the trifluoropropyl group are surprising but correspond with the liquid surface tension of polytrifluoropropylmethylsiloxane [13], although not in line with its critical surface tension and solid surface tension. The low values for the longer fluoroalkyl groups are comparable to the lowest surface tension fluoropolymers and correspond well with the values for related pre-formed fluorosilicone polymers [14–15].

New insights into solid/solid adhesion have been provided by Chaudhury and Whitesides [12] who have directly studied the adhesion forces between various silicone surfaces. They analyzed the deformations occurring on contact of small

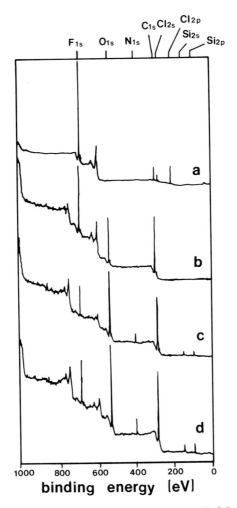

Figure 1. XPS survey spectra of PCTFE (a), PCTFE-OH (b), PCTFE-OC(O)NH(CH$_2$)$_3$Si(OEt)$_3$ (c), and PCTFE-OC(O)NH(CH$_2$)$_3$Si(O)OH (d). Reprinted with permission from K. W. Lee and T. J. McCarthy, *Macromolecules* **21**, 3353 (1988). Copyright (1988) American Chemical Society.

Table 1.
Surface tension of self-assembled layers from hydrolyzed silane solutions

R Group	X Group	σ_c (mN/m)	σ_s^p (mN/m)	σ_s^d (mN/m)	Ref.
CF$_3$(CH$_2$)$_2$	OMe	33.5	18.7	16.0	[8]
(CF$_3$)$_2$CFO(CH$_2$)$_3$	OMe	16–18	—	—	[9]
CF$_3$(CF$_2$)$_6$CH$_2$O(CH$_2$)$_3$	OEt	14–16	—	—	[9]
CF$_3$(CF$_2$)$_5$(CH$_2$)$_2$	Cl	8.1	1.8	13.5	[10]
CF$_3$(CF$_2$)$_7$(CH$_2$)$_2$	Cl	—	—	12.4	[11]
CF$_3$(CF$_2$)$_7$(CH$_2$)$_2$	Cl	—	—	8.7	[12]

Note: Me = methyl; Et = ethyl.

semispherical lenses of elastomeric polydimethylsiloxane (PDMS) and flat sheets of this material. The work of adhesion (W) can be obtained from such experiments using the theory of Johnson et al. [16].

Good agreement was found between values obtained from this approach and those derived from conventional contact angle studies. Their work also provides an interesting direct proof of Young's equation as it allows independent measurement of the solid surface tension of silicone surfaces and the interfacial tension between them. Most relevant to this review are the studies involving fluoro-alkyltrichlorosilane modification. The PDMS surface is oxidized by exposure to an oxygen plasma to produce a silica-like surface. There is currently great interest in close-packed monolayer films formed by spontaneous adsorption of amphiphilic molecules at liquid–solid interfaces [17]. Such self-assembled mono-layers are known to form on silica and should similarly form on the oxidized PDMS. $CF_3(CF_2)_7(CH_2)_2SiCl_3$ and $CH_3(CH_2)_9SiCl_3$ were used in this manner.

The determination of work of adhesion can be made while increasing or decreasing the contact load. For most systems studied by Chaudhury and Whitesides there was little hysteresis in these measurements but the fluoro-alkylsilane-treated case was an exception. Figure 2 [12] is a plot of a^3 against P for the surfaces of $CF_3(CF_2)_7(CH_2)_2Si\equiv$-treated PDMS where a is the radius of contact deformation and P is the external load. Values of W for loading and unloading were 14.2 mJ/m² and 42.0 mJ/m², respectively. Factors such as molecular-level roughness and partial interpenetration of fluoroalkyl chains are thought to be involved in this hysteresis. The interfacial interaction between alkylsiloxane and fluoroalkylsiloxane monolayers was also probed in this way.

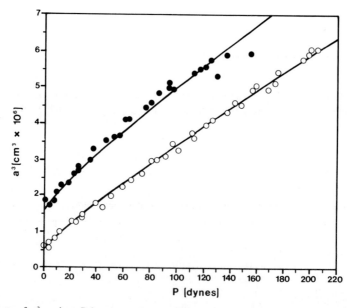

Figure 2. Plots of a^3 against P for the surfaces of $CF_3(CF_2)_7(CH_2)_2Si\equiv$-treated PDMS exhibit large hysteresis. ◯: represent data obtained from increasing loads. ●: represent data obtained from decreasing loads. Reprinted with permission from M. K. Chaudhury and G. M. Whitesides, *Langmuir*, in press. Copyright (1991) American Chemical Society.

The value of W is intermediate to the values obtained for the separate cases and fits a geometric combining rule.

A recent study by DePalma and Tillman [10] also demonstrates the potential of surface modification by self-assembled monolayers of low surface energy fluoroalkyl-containing silanes. Fatty acids, amines and alcohols have long been known to adsorb as monomolecular films on metals. Silane coupling agents have featured strongly in new studies to develop more robust films, covalently bound together and to the metal substrate.

DePalma and Tillman investigated self-assembled monolayer films from three silanes, tridecafluorooctyltrichlorosilane, undecyltrichlorosilane, and octadecyltrichlorosilane, on silicon, a popular model substrate for such studies with great relevance to potential semiconductor coating applications. They characterized the films by ellipsometry and contact angle measurements (data for tridecafluorooctyltrichlorosilane are included in Table 1), but more usefully from an applicational viewpoint, they carried out friction and wear measurements with a pin-on-disk device where the silicon wafer substrate, coated with monolayer, is moved under a spherical glass slider. Optical microscopy was used to assess wear. Table 2 summarizes DePalma and Tillman's data and their comparison with the classical self-assembled monolayer friction studies of Levine and Zisman [18].

Table 2.
Self-assembled monolayer friction behavior

Monolayer	Slider/Substrate	Coefficient of kinetic friction, μ_k	Ref.
None	Glass/Silicon	1.2 [a]	[10]
$CF_3(CF_2)_5(CH_2)_2SiCl_3$	Glass/Silicon	0.16	[10]
$CH_3(CH_2)_{10}SiCl_3$	Glass/Silicon	0.09	[10]
$CH_3(CH_2)_{17}SiCl_3$	Glass/Silicon	0.07	[10]
$CF_3(CF_2)_6CO_2H$	Steel/Glass	0.13	[18]
$CH_3(CH_2)_{10}CO_2H$	Steel/Glass	0.10	[18]
$CH_3(CH_2)_{16}CO_2H$	Steel/Glass	0.05	[18]

[a] Considerable stick-slip, wear, and material transfer to slider.

Evidently the effect of headgroup polymerization, which cross-links the silicon atoms via covalent bonding, on friction properties is minimal. However, they believe that the covalent bonding of the silyl group to the substrate surface hydroxyl groups will more rigorously exclude the possibility of mechanical transfer, thus opening up possible uses involving lubrication without employing adsorbed liquid films and for achieving boundary lubrication of surfaces by covalently bound ultra-thin films.

An interesting study involving fluoroalkyl-containing silanes is Cuddihy's [19] research on low soiling coatings for solar-energy devices such as solar thermal collectors and photovoltaic modules. He considered theoretical mechanisms and evaluated several possibilities. He suggested that the requirements for surfaces having resistance to the formation of these soil layers that are resistant to removal by rain are: (i) hard; (ii) smooth; (iii) hydrophobic; (iv) low surface energy; (v) chemically clean of sticky materials (surface and bulk); and (vi) chemically clean of water-soluble salts (surface and bulk). This list of requirements suggests that coatings based on fluorocarbon chemistry should be used. Non-reactive fluorinated acrylic polymers were shown to be readily dissipated from surfaces and could not be used. Two silanes with chemically reactive functional groups for chemical attachment to substrate surfaces were found to perform well on glass and plastic films after one year of outdoor exposure, retaining less soil compared to uncoated controls. The substrates used were outer cover materials being evaluated for photovoltaic modules: Sunadex® (ASG) soda-lime glass, and Acrylar® (3M) and Tedlar® (du Pont) UV-screening plastic films. The promising silanes were a 3M fluorinated silane L-1668 and Dow Corning E-3820-103B. This latter product is perfluorodecanoic acid chemically reacted with a conventional aminofunctional silane coupling agent, an Ed Plueddemann invention [20]. The chemical attachment of E-3820-103B to Acrylar and of L-1668 to Tedlar was weak and in these instances the surfaces of the plastic films were treated with ozone to generate polar groups for enhanced chemical reactivity to achieve the required degree of chemical attachment.

Another application area where low soiling is important is in fabric treatments. It is probably the oldest established use of these fluoroalkyl-containing silanes. The United States Department of Agriculture has a number of patents on this application. Bovenkamp and Lacroix have published some Canadian government reports on the topic. The latest [21] examines the durability of liquid repellency properties imparted to fabrics by solutions of the copolymers of fluoroalkyl-substituted silanes and non-fluorinated silanes. The fluorine-containing groups were $CF_3(CF_2)_nCONH(CH_2)_3$ where n is 6 and 8. The best available commercial finishes were examined at the same time and the results compared with those of the experimental finishes. The durability testing encompassed wear, washing, dry cleaning and weathering. Although the fluoroalkylsilane-containing copolymers wear well, they are somewhat inferior to the commercial finishes after a number of washes. Much better durability to washing was obtained when a small amount of resin was added to the finish, following the normal commercial practice. The properties found for the fluorosiloxane finishes indicate that they could be useful in some circumstances. Earlier aspects of this work have been published in the open literature [22].

Russian workers are also interested in this application. Makarskaya and coworkers [23] have reported on the use of acetoxy-functional silane coupling agents for the improvement of cotton fabrics. They studied a variety of compounds including $CF_3(CH_2)_2Si(OAc)_3$, and a variety of properties including water absorption, contact angle data, critical sliding angle, air permeability and abrasion resistance. Fabric treated with this trifluoropropyl-substituted silane exhibited a high water repellent effect. In a related paper, this group [24] also reported on the modification of glass surfaces with a similar set of acetoxy-functional silanes including the trifluoropropyl-substituted material.

A final potential for these fluoroalkyl-functional silanes that must be mentioned is as bound antimicrobial agents. The use of alkyl quaternary ammonium silanes in this application is well known; Plueddemann and Revis [25] have claimed fluoroalkyl-containing versions of these organosilicon quaternary ammonium antimicrobial compounds.

2.2. Linear monomer type (n = 2)

The class of silanes where $n = 2$ is typically the most important from a polymer intermediate perspective but the least important in regard to direct surface treatment, the subject of this review. This is the case for fluoroalkyl-containing silanes. We are aware of only one fundamental study of the latter sort, the molecular orientation investigation of Salaneck and co-workers [26] of $CF_3(CF_2)_5(CH_2)_2MeSiCl_2$, tridecafluorooctylmethyldichlorosilane (TDFS), on silicon. The popularity of this substrate has been noted previously. Salaneck *et al.* regard it as an excellent model surface for studying various biopolymer interaction phenomena such as protein exchange reactions, and conformational changes and the effect of surfactants on adsorbed proteins.

They studied the chemistry of oxidized Si(111) surfaces treated at two concentrations of the silane in trichloroethylene solution using angle-dependent X-ray photoelectron spectroscopy (XPS or ESCA). Although these are non-aqueous adsorption studies, sufficient surface silanol or adsorbed water is present for complete hydrolysis to occur because no trace of chlorine is seen in the XPS spectra. The two concentrations studied were 1% v/v, termed saturated, and <1/400% v/v, termed dilute. They lead to two distinct types of molecular bonding to the surface. $C(1s)$ XPS spectra of these two situations are shown in Fig. 3.

In the samples prepared from dilute solution, the silane appears to lie down flat on the surface of the oxidized silicon substrate. For samples prepared under saturation conditions, the molecules appear to be more tightly packed, with a tendency to stand up on the surface. The angle-resolved XPS data are consistent with a tilt of the tightly packed molecules at an angle with the surface determined by the expected sp^3 bonding configuration of the silane Si atom to the oxidized silicon surface. Detailed XPS core-level binding energy shifts are also consistent with these proposed different types of molecular interaction.

2.3. Silylating agent type (n = 3)

The class where $n = 3$ encompasses a variety of chromatographic substrate treatments and other surface silylation applications. For example, Berendsen and coworkers [27] have reacted $CF_3(CF_2)_7(CH_2)_2Me_2SiCl$ with an activated silica for use in chromatography. This perfluoro-bonded phase showed specific fluorine–fluorine interaction and enhanced retention for fluorine-containing compounds that increases with the number of fluorine atoms in the solute. The most promising feature noted was the separation of fluorine-containing compounds from their non-fluorine-containing analogues. Benzene and monofluorobenzene cannot be separated on alkyl phases but were readily separated on the fluoroalkyl substituted silane treated silica. The separation of fluorine-containing herbicides was also demonstrated. Subsequent workers found that such chromatographic materials show a strong reduction in retentivity for both small non-fluorine-containing compounds [28, 29] as well as for proteins [30, 31]. Carr and coworkers also noted that recovery of proteins from such materials was enhanced over that of the analogous alkylsilane-treated silica [30].

An example of surface energy control is given by Menawat, Henry and

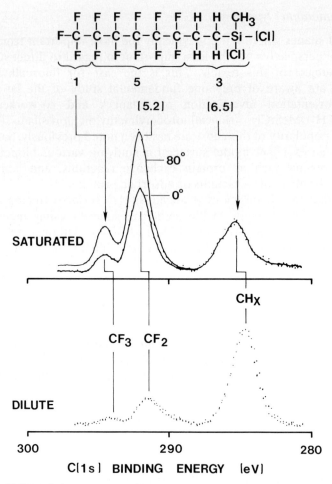

Figure 3. The XPS(θ) C($1s$) spectra of TDFS on silicon are shown for two types of samples, saturated at the top and dilute at the bottom: and for two values of θ, maximum bulk sensitivity for $\theta = 0$ deg, and maximum surface sensitivity for $\theta = 80$ deg. Reprinted from [26], with permission of Academic Press Inc.

Siriwardane [32] who modified the surface energy of glass plates by chemisorption of organochlorosilanes including an unusual aromatic fluorocarbon-containing one, pentafluorophenyldimethylchlorosilane. Contact angles of water in air and of xylene in water were measured and Auger electron spectroscopy was used to determine surface elemental composition. The authors' aim was to develop model surfaces for the study of particle transfer in liquid/liquid/solid systems. In the case of the fluorine-containing silane treatment, water in air contact angles increased with treatment level to a value of c. 77 deg (advancing) and 74 deg (receding). Xylene in water contact angles were in the range 115–134 deg (advancing) and 107–126 deg (receding). Large changes with time were generally not observed, indicating good stability of treated surfaces except at the lowest concentration where partial coverage occurred. Such temporal drift of the contact angle was attributed either to hydrolysis reactions or to desorption of weakly adsorbed molecules. Surfaces treated with low concentrations of triphenylchlorosilane or *t*-butyldiphenylchlorosilane did not show temporal drift but surfaces treated with low concentrations of the less sterically hindered *t*-butyldimethylchlorosilane did.

We have found several uses for fluoroalkylsilyl-*N*-methylacetamides such as trifluoropropyldimethylsilyl-*N*-methylacetamide (TFSA) and here we focus on two of these: Their use in DUAL ZONE™ silicas for biochemical and other chromatographic separation and the valuable but little known application to derivatize polymer and other surfaces to enhance the information provided by XPS. Williams and Tangney [33] have investigated the mechanism of surface silylation of porous silica. The rate of chemical reaction involving porous substrates can be limited either by low reactivity or by relatively slow diffusion in the pores. Both situations can prevail in silylation although reactivity rather than diffusion usually limits the reaction rate unless the silylating agent is too large to freely diffuse in some of the pores. Silylation by TFSA was the first example of pore-diffusion control caused by an ultrafast reaction rather than by pore-size restricted diffusion. This discovery is significant because its use makes possible DUAL ZONE™ materials in which the external surfaces bear one functional group and the internal surfaces another. Materials of this type may be useful in a variety of applications in fields such as catalysis, detoxification, and separatory materials. It has been shown that such DUAL ZONE™ materials in which the external zone is enriched in lipophobic fluoroalkylsilyl and polyether groups display enhanced resistance to adsorption and to column fouling by high loads of protein [34].

Figure 4 illustrates the effect on porous silica beads. The straight diagonal line shows the expectation for reactivity control where no preferential reaction at the external surface is encountered. The TFSA data show the reagent is immobilized on the outer surface before it can migrate to the porous internal surface. There is

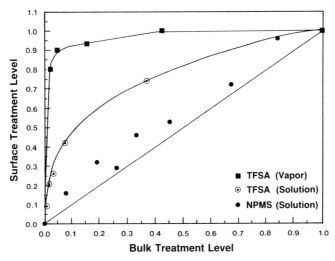

Figure 4. Silylation of porous silica beads. The fractional extent of completion of the reaction at the external surface and on the internal surface is shown on the ordinate and abscissa, respectively. The straight diagonal line shows expectation for reactivity control. The TFSA data show a high degree of pore diffusion control. (From [33], with permission from Gordon and Breach, Science Publishers S.A.)

a less strong degree of pore diffusion control from solution than from the vapor phase but it is still significant in comparison to silylation by *n*-butylamine catalyzed naphthylphenylmethylsilanol (NPMS). This is a much more sluggish silylating agent and closely follows the diagonal straight line expected for reactivity controlled silylation.

A recent example of the use of silylating agents in the derivatization of polymer surfaces is given by Davies and Munro's application of trimethyl-silylimidazole (TMSI) to poly(acrylic acid) (PAA) surface analysis [35]. Figure 5 shows the changes in core level XPS spectra for PAA after reaction with TMSI. Williams first proposed the use of TFSA for this purpose in 1981 [36] but this application was not subsequently published so we take this opportunity to provide more detail.

The chemical bonding information available from routine XPS studies of polymer surfaces can be rather limited and is often insufficient to distinguish between various functional group possibilities. A common example where binding energy information is not enough to make a clear choice between plausible options involves oxidized silicone polymer surfaces. Hydroxyl groups produced by the exposure of silicones to such environments as UV, plasma, corona discharge, flame etc., cannot readily be distinguished from backbone siloxane oxygen by binding energy chemical shifts [37]. Although such groups can be identified by infrared spectroscopy, the surface sensitivity is often insufficient for many surface applications, including the important topic of such hydrophilic treatments frequently used to enhance adhesion of silicone materials.

A suitable XPS derivatizing agent should have the following features: (1) A readily detectable labelling atom: (a) of high cross section for sensitive XPS

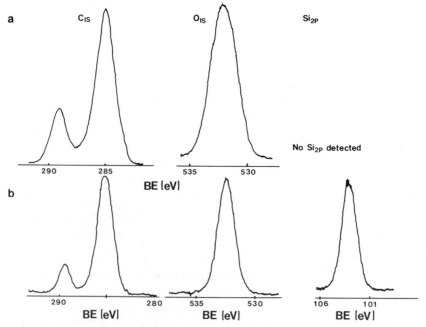

Figure 5. C(1*s*), O(1*s*) and Si(2*p*) XPS spectra for PAA before (a) and after (b) reaction with TMSI. Reprinted from [35], with permission of Butterworth–Heinemann Ltd.

detection; and (b) not already present in the surface region of interest. (2) Quantitatively reactive in the gas phase with the surface (no solvent swelling, reorientation effects or possible dissolution of low molecular weight species). (3) Other minimal perturbative effects: (a) room temperature derivatization is preferred; and (b) only innocuous, volatile, readily-removable by-products produced. (4) A stoichiometric enhancement factor consistent with as small a derivatizing group as possible to provide best chance of quantifying an extensively modified surface.

TFSA satisfies all these requirements for hydrophilic silicone surfaces. Using silica as a model for extensively oxidized silicone surfaces, we showed that silylation with vapor phase TFSA took place at room temperature under scrupulously dry conditions. In contrast, others have found for the more conventional silylating agents, the methylchlorosilanes and the methylmethoxy-silanes, that no reaction occurred at temperature below 200°C under similar dry conditions [38].

The reaction involved is:

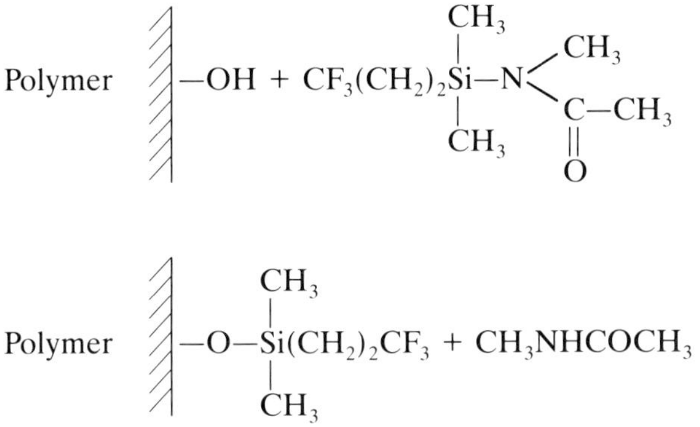

$$\text{Polymer}\!\!\left[\!-\text{OH} + \text{CF}_3(\text{CH}_2)_2\text{Si}\!-\!\text{N}\!\!\begin{array}{c}\text{CH}_3\\|\\\text{CH}_3\end{array}\!\!\overset{\displaystyle\text{CH}_3}{\underset{\displaystyle\underset{\displaystyle\text{O}}{\|}}{\diagup}\!\!\text{C}\!-\!\text{CH}_3}\right.$$

$$(1)$$

$$\text{Polymer}\!\!\left[\!-\text{O}\!-\!\overset{\displaystyle\text{CH}_3}{\underset{\displaystyle\text{CH}_3}{\text{Si}}}(\text{CH}_2)_2\text{CF}_3 + \text{CH}_3\text{NHCOCH}_3\right.$$

The *N*-methylacetamide is a readily removable, neutral species which causes no subsequent polymer attack. The fluorine atom is easily detectable by XPS and is not present in most silicone surfaces. The trifluoropropyl group is the smallest fluorine-containing nonreactive group on Si that is chemically stable and provides a stoichiometric enhancement factor of three.

The only complication is a gradual loss of fluorine in prolonged XPS measurements due to evolution of HF. In practice XPS surface analysis can be carried out before appreciable X-ray damage of this sort has occurred. The vapor pressure of TFSA is very low at room temperature, on the order of 0.05 Torr. For this reason, very wet samples are unsuitable since the reagent will be destroyed before it reaches the sample. We have used TFSA to derivatize plasma-treated PDMS elastomers (Fig. 6). The introduction of the fluorine atom is clearly evident in the well-defined F ($1s$) peak at 688 eV and the F KLL Auger peak in the 600–650 eV range. The silicon region is magnified in the figure. Actual atomic composition for the main elements present in the treated sample (b) is 6.4% F, 24.3% O, 51.3% C, and 15.8% Si.

2.4. Note on patent situation

A review of the surface treatment patent literature was made to identify the

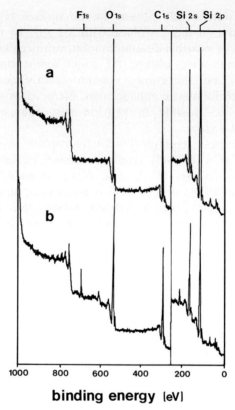

Figure 6. XPS survey spectra for plasma-treated PDMS before (a) and after (b) reaction with TFSA. Note the 0–240 eV binding energy region is magnified.

silanes of greatest commercial interest, the companies involved, and the types of uses being claimed. Much of our information comes from abstracts (for example, from Derwent Publications Ltd), not the original patents.

The coupling agent, $n = 1$ $RSiX_3$, structure is the most common structure involved. The $n = 2$ and 3 structures are hardly considered. Although a variety of linking groups between the fluoroalkyl group and the silicon atom are claimed including ether, thio and ester linkages, by far the most common organo-functional R group on silicon in the patent literature is $CF_3(CF_2)_n(CH_2)_2-$. Methoxy, ethoxy, acetoxy and chlorohydrolyzable X groups all feature in these patents.

These fluoroalkyl-containing surface modification patents are overwhelmingly assigned to Japanese companies in recent years. The bulk of the patents claim coatings with a variety of benefits such as release, lubrication, solvent resistance, abrasion and scratch resistance, low reflectivity, gloss retention and anti-contamination (e.g. reduced staining by toner of electrostatic image-carrying particles in photocopiers and lower soiling of optical glass in cameras). The two most common uses are as mold release agents for such items as plastic lenses and lubricating/protecting coatings for magnetic recording media films. Filler treatments for fluorine-containing elastomers and resins are also claimed as well as treatments for the dispersed solid phase in antifoam compositions.

We have chosen not to reference patents in this review but as this symposium is in honor of Ed Plueddemann we have made exceptions in his case.

3. SUMMARY

Fluoroalkyl-containing silanes are featuring in the resurgence of interest in fluorosilicone materials in the last decade and Ed Plueddemann's inventions are playing their part in this revived development. A number of significant fundamental surface characterizations and property determinations have been carried out involving contact angle, XPS, ellipsometry, optical microscopy, and coefficient of friction. The most promising application areas seem to be self-assembled lubricating monolayers, adhesion control, textile treatments and low-soiling coatings, materials for chromatographic separations, and XPS surface derivatization reagents.

Acknowledgements

We much appreciate the involvement of Jim Malek and Carl Voigt in literature and patent searches.

REFERENCES

1. E. P. Plueddemann, *Silane Coupling Agents*. Plenum Press, New York (1982).
2. D. E. Leyden (Ed.), *Silanes, Surfaces and Interfaces*. Gordon and Breach, New York (1986).
3. D. E. Leyden and W. T. Collins (Eds.), *Silylated Surfaces*, Midland Macromolecular Monographs, Vol. 7. Gordon and Breach, New York (1980).
4. D. E. Leyden and W. T. Collins (Eds.), *Chemically Modified Surfaces*. Gordon and Breach, New York (1988).
5. B. Arkles, *Chemtech.* **7**, 766 (1977).
6. M. K. Chaudhury, T. M. Gentle and E. P. Plueddemann, *J. Adhesion Sci. Technol.* **1**, 29 (1987).
7. K-W. Lee and T. J. McCarthy, *Macromolecules* **21**, 3353 (1988).
8. L.-H. Lee, in: *Adhesion Science and Technology*, Part B, L.-H. Lee (Ed.), p. 647. Plenum Press, New York (1975).
9. W. D. Bascom, *J. Colloid Sci.* **27**, 789 (1968).
10. V. DePalma and N. Tillman, *Langmuir* **5**, 868 (1989).
11. R. Maoz, L. Netzer, J. Gun and J. Sagiv, *J. Chim. Phys. Phys-Chim. Biol.* **85**, 1059 (1988).
12. M. K. Chaudhury and G. M. Whitesides, *Langmuir*, **7**, 1013 (1991).
13. M. J. Owen, *J. Appl. Polym. Sci.* **35**, 895 (1988).
14. M. M. Doeff and E. Lindner, *Macromolecules* **22**, 2951 (1989).
15. H. Kobayashi and M. J. Owen, *Macromolecules* **23**, 4929 (1990).
16. K. L. Johnson, K. Kendall and A. D. Roberts, *Proc. R. Soc. Lond.* **A324**, 301 (1971).
17. G. M. Whitesides and G. S. Ferguson, *Chemtracts: Org. Chem.* **1**, 171 (1988); G. M. Whitesides and P. E. Laibinis, *Langmuir* **6**, 87 (1990).
18. O. Levine and W. A. Zisman, *J. Phys. Chem.* **61**, 1068, (1957).
19. E. F. Cuddihy, in: *Particles on Surfaces 1: Detection, Adhesion and Removal*, K. L. Mittal (Ed.), p. 91. Plenum Press, New York (1988).
20. E. P. Plueddemann, U.S. Patent No. 4,617,057 (1986) (assigned to Dow Corning Corporation). We have chosen not to reference patents in this review but make an exception in this case and in reference 25.
21. J. W. Bovenkamp and B. V. Lacroix, *Fluoroalkyl Siloxanes as Liquid-Repellent Fabric Finishes, Part 3*. Report DREO-TN-83-4 (1983). Available from NTIS (CA *101*: 56428S).
22. J. W. Bovenkamp and B. V. Lacroix, *Ind. Eng. Chem. Prod. Res. Dev.* **20**, 130 (1981).
23. V. M. Makarskaya, M. G. Voronkov, L. P. Ignat'eva, A. A. Stotskii and A. A. Kharkharov, *Zh. Prikl. Khim. Leningrad* **51**, 661 (1978).

24. M. G. Voronkov, L. P. Ignat'eva, E. V. Kukharskaya and V. M. Makarskaya, *Zh. Prikl. Khim. Leningrad* **54**, 1392 (1981).
25. E. P. Plueddemann and A. Revis, U.S. Patent No. 4,866,192 (1989) (assigned to Dow Corning Corporation).
26. E. W. Salaneck, K. Udval, H. Elwing, A. Askendal, S. Welin-Klintstrom, I. Lundstrom and W. R. Salaneck, *J. Colloid Interface Sci.* **136**, 440 (1990).
27. G. E. Berendsen, K. A. Pikaart, L. de Galan and C. Olieman, *Anal. Chem.* **52**, 1990 (1980).
28. H. A. Billiet, P. J. Schoenmakers and L. de Galan, *J. Chromatography* **218**, 443 (1981).
29. P. C. Sadek and P. W. Carr, *J. Chromatography* **288**, 25 (1984).
30. G. Xindu and P. W. Carr, *J. Chromatography* **269**, 96 (1983).
31. D. E. Williams, in: *Chemically Modified Surfaces*, D. E. Leyden and W. T. Collins (Eds.), p. 515. Gordon and Breach, New York (1988).
32. A Menawat, J. Henry, Jr. and R. Siriwardane, *J. Colloid Interface Sci.* **101**, 110 (1984).
33. D. E. Williams and T. J. Tangney, in: *Silanes, Surfaces and Interfaces*, D. E. Leyden (Ed.), p. 471. Gordon and Breach, New York (1986).
34. D. E. Williams and P. M. Kabra, *Anal. Chem.* **62**, 807 (1990).
35. C. Davies and H. S. Munro, *Polymer Communications* **29**, 47 (1988).
36. D. E. Williams and J. P. Cannady, *Analysis of polymer surface functional groups*, Abstracts, 182nd ACS National Meeting, p. INDE-23 (1981).
37. N. H. Sung, A. Kaul, I. Chin and C. S. P. Sung, *Polym. Eng. Sci.* **22**, 637 (1982).
38. M. L. Hair and W. Hertl, *J. Phys. Chem.* **75**, 2181 (1975).

Silanes and Other Coupling Agents, pp. 81–90
Ed. K. L. Mittal
© VSP 1992.

Silanes as the interphase in adhesive bonds

K. W. ALLEN

Adhesion Science Group, City University, London EC1V 0HB, UK

Revised version received 19 July 1991

Abstract—Substantial films of polymerized γ-glycidoxytrimethyl silane were prepared and found to be mechanically very weak. It is suggested, by analogy with natural rubber, that their structure is a spiral with pendant hydroxyl groups along the core which are thus inaccessible. Reflectance spectroscopy of much thinner films of this silane indicated a loss of freedom of the hydroxyl groups as the thickness diminished. It is suggested that this is due to destruction of the spiral structure with consequent availability of the hydroxyl groups for hydrogen bonding to the oxide surface of the substrate.

Keywords: Coupling agents; interphase; silanes.

1. INTRODUCTION

With the development of adhesion science and technology, it has become relatively straightforward to produce adhesive bonds of very reasonable strength between most materials which are in general use. An informed choice of adhesive together with appropriate preparation of the surface will produce a strong bond.

It is far more difficult to produce bonds which are durable and which will retain their strength over an extended period, particularly in a moist environment. For practical purposes, the quality of durability is of enormous significance and a great deal of effort has been devoted to developing techniques to reduce or overcome the problems. The most satisfactory solution, or at least alleviation, has been the development of 'coupling agents'. While much of this work has been directed towards conventional adhesive bonding, similar requirements arise in the production of a variety of composite and filled materials, as well as in a range of coating products (e.g. paints and inks). So, techniques and materials have been developed for use in all these situations as well as in adhesive bonding. However, our concern here is entirely within the context of adhesive bonding.

The coupling agents which have been most extensively used are various organo-silicon compounds, although several other types of compound of quite different chemistry, including zirconium and titanium comlexes [1], have also been used. It is with silane materials that we are concerned here.

One of the earliest successes arose from the development of glass-fibre reinforcement of organic polymer resins. The earliest of these composites had higher strength-to-weight ratios than aluminium or steel, but very rapidly lost a large proportion of their strength on exposure to moisture. By 1950 a pretreatment of the fibre with a silane had been developed which revolutionized the situation, with improvements of the extent indicated in Table 1, and led, for example, to the production of highly serviceable glass-reinforced plastic boat

Table 1.
Flexural strength of the polyester/glass fibre laminate

Fibres	Initial strength (MPa)	Strength after 5 h in boiling water (MPa)
Untreated	320	100
Silane-treated	450	400

From ref. 14.

hulls. As pragmatic experience accumulated, so more theoretical explanations were sought and, to some extent, found. One may now recognize several distinct and separate approaches. The oldest is the chemical bond theory, which postulates the formation of recognizable covalent bonds between the substrate, the coupling agent, and the adhesive [2]. This approach has been supplemented, if not displaced, by suggestions of enhanced wettability of surfaces, of the involvement of a region of intermediate mechanical properties, and of reversible bond formation. All of these have been reviewed by Rosen [3] and more recently by Walker [4].

Whether one supports one of these theories or prefers a combination of them, there is one incontrovertible fact. Between the surface of the adherend and that of the finally cured adhesive there is a layer of silane. At one extreme this may be no more than a single molecule in thickness, but undoubtedly in most situations it is thicker. Frequently it is up to ten times thicker, according to Tutas *et al.* [5] and Bascom [6], and perhaps even thicker in some areas, depending on the exact technique used during its application. Thus, between an adherend and the bulk of the adhesive there is an interphase which is essentially the silane. It was the recognition of this fact that led us to investigate some of the properties of the silane as a material in its own right, in the state in which it was believed to exist in the interphase. This investigation was in two quite separate and distinct parts.

2. PART 1

2.1. Experimental

2.1.1. Materials and hydrolysis. For all this work γ-glycidoxypropyltrimethoxy-silane (Dow Corning Z6040 or Union Carbide A187) was used. The silane was hydrolysed and polymerized in solution to give a polysiloxane solid material. The original monomer was mixed with a 3 M excess of water containing a trace of hydrochloric acid. After 1 h, the pH was adjusted to 3.85 with more hydrochloric acid, a considerable excess of methanol was added, and the mixture was boiled under reflux for 30 min. Then aqueous methanol was distilled off at 65–70°C to leave a viscous, colourless liquid of refractive index 1.446. This liquid was poured into a mould and kept at 40°C for 24 h, by which time it had gelled to a material with sufficient strength to allow it to be removed from the mould and handled carefully. This gelled material was used as the initial product for further curing and examination.

2.1.2. Infra-red spectra and extent of condensation. Comparison of the IR spectra of the original monomer, the liquid before addition of the methanol, and the final viscous liquid revealed two factors. Firstly, during the hydrolysis a strong peak at 1650 cm^{-1} appeared, suggesting that some of the epoxide groups had been hydrolysed to a glycol, which had then been dehydrated to give terminal vinyl groups. Secondly, the peaks at 2840 and 1190 cm^{-1}, characteristic of the methoxy groups, remained in the final product [7]. Thus, only a proportion of the methoxy groups had been replaced by hydroxyl groups and some of them remained in the final product, which is consistent with the results of Boerio and Greivenkamp [8].

The extent of condensation of contiguous silanol groups was followed through the IR spectra of specimens which had been further cured by heating in air to 70°C. The ratio of peak heights at 3330 cm^{-1} (due to OH stretching) to the mean of the heights of the two peaks at 840 and 770 cm^{-1} (due to Si—C) was eventually chosen as the best way to follow the dehydroxylation.

The results of this are shown in Fig. 1, which clearly shows that while accurate quantitative conclusions are not possible, the hydroxyl content decreases by about a third during curing and a considerable amount remains even after prolonged curing at 70°C. If we assume that the freshly gelled material is essentially linear, then only one in three of the pendant hydroxyl groups is eliminated in later curing.

An attempt was made to exchange with deuterium the hydrogen of the remaining hydroxyl groups in the fully cured material. The IR transmission spectrum was unchanged by this, indicating that only negligible exchange could have occurred. However, nine reflection internal reflectance spectroscopy of the surface showed a broad peak at 2500 cm^{-1} due to O—D and a corresponding reduction in the O—H absorption at 3400 cm^{-1}. This suggested that exchange occurred only on the exposed surface of a specimen and did not penetrate into the bulk to any significant extent.

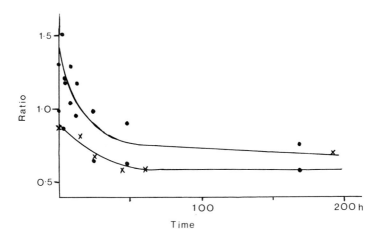

Figure 1. Ratio of the peak height at 3330 cm^{-1} (due to O—H stretching) to the mean of the heights of the two peaks at 840 and 770 cm^{-1} (due to Si—C) plotted against the time of curing (in h) after initial gelling at 70°C for films of polysiloxane. Used for quantitative estimation of the hydroxyl concentrations. (●) Results from KBr disc specimens; (×) results from films cast on NaCl plates.

2.1.3. Mechanical properties. The mechanical properties of films of the fully cured polysiloxane were quite unexpected because they were very flexible, easily torn, of low extensibility, and quite friable. A film 5 mm thick broke immediately when twisted, and disintegrated into crumbs when rubbed gently between the fingers.

It seemed that an analogy might be found between this behaviour and that of natural rubber [9], as follows. As the extent of crosslinking of a natural rubber increases, the strength first increases, passes through a maximum, and then decreases sharply. A vulcanizate containing about 7% sulphur is weak and friable, and quite similar in behaviour to these polysiloxanes. So, it is quite probable that they should be regarded as essentially rubbery in structure, rather than very weak and brittle solids. Hence, it was reasonable to explore the degree of crosslinking by measuring the swelling and the compression modulus of material which had been swollen in various solvents.

Small cylinders of uncured but gelled polysiloxane were immersed for 6 days in various organic liquids. The extent of swelling was determined by weighing each cylinder before and after immersion and swelling. These trials were all carried out in duplicate. The data on swelling, together with the solubility parameters and hydrogen bonding characteristics of the solvents, are given in Table 2.

These results are plotted in Fig. 2, from which it can be seen that they give two curves, one from liquids with strong and the other from liquids with weak or moderate hydrogen-bonding properties. The curve with liquids of weaker hydrogen-bonding properties indicates a solubility parameter for the gelled siloxane of about 19 $(mJ\ m^{-3})^{1/2}$.

Table 2.
Swelling of polysiloxane film in liquids

Liquids	Solubility parameter $(mJ\ m^{-3})^{1/2}$	Hydrogen bonding character	% Increase in weight—mean of duplicate runs
n-Hexane	14.9	Poor	0.8
Diethyl ether	15.1	Moderate	13.5
Petroleum ether (100-120)	~15.5	Poor	7.2
Cyclohexane	16.7	Poor	2.0
Carbon tetrachloride	17.6	Poor	45.0
n-Butyl acetate	17.7	Moderate	52.0
Ethyl acetate	18.6	Moderate	57.5
iso-Propan-2-ol	20.4	Strong	22.5
n-Butan-1-ol	23.7	Strong	16.0
Ethanol	26.1	Strong	24.5
Methanol	29.6	Strong	33.0
Water	47.9	Strong	51.5
Xylene	18.0	Poor	
Toluene	18.2	Poor	
Benzene	18.7	Poor	Samples disintegrated
Chloroform	19.0	Poor	
Dichloromethane	19.8	Poor	
Dichloroethane	20.0	Poor	
Acetone	20.4	Moderate	

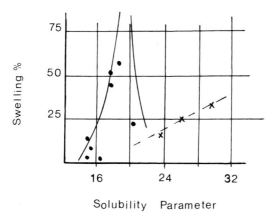

Figure 2. Estimation of the solubility parameter of gelled polysiloxane polymer. Made by plotting the percentage by weight swelling against the solubility parameter [(in mJ m^{-3})$^{1/2}$] of the liquid concerned. (●) Results from liquids of weak hydrogen-bonding character; (×) results from liquids of strong hydrogen-bonding character.

This puts this polymer remote from the usual range of silicone rubbers, which have solubility parameters of about 15 (mJ m^{-3})$^{1/2}$, and close to polycarbonates, which have a value of 19.4 (mJ m^{-3})$^{1/2}$, and nitrile rubbers and polyvinyl acetates, which are similar.

An attempt was made to derive the extent of crosslinking by measuring the compression modulus of specimens which had been allowed to swell in ethanol. While this technique has been successfully used for natural rubbers [10], no adequately constant results were obtained to make this calculation possible for this material.

The torsional pendulum principle, via a braid sample, was used to measure the variation of shear modulus with temperature, and the loss angle, and hence to derive the glass transition temperature. This gave for the gelled but uncured material $T_g \sim 80°C$, and for the material which had been kept at 70°C for 120 h, and could therefore be regarded as fully crosslinked, $T_g \sim 60°C$.

The only other mechanical property which could practically be measured was the compression modulus. Load–displacement curves were obtained for a series of samples after increasing times of curing. All these curves were similar, with an initial linear section corresponding to a low modulus, followed by a second approximately linear section corresponding to a considerably higher modulus, before 'barrelling' of the test specimen became appreciable. Moduli were calculated from the slopes of both parts of these curves and are given in Table 3. In every case, the change from the initial part of the curve to the other occurred at a stress of about 27 kN m^{-2}. This appears to represent the completion of a consolidation of the specimen, after which the true compression modulus is apparent.

2.1.4. Molecular refractivity. Molecular refractivity is an additive quantity which is often used to assist in determining molecular structures via the Lorentz–Lorentz equation. This generally gives good agreement between values derived from experimental measurements of density and refractive index and values

Table 3.
Compression moduli of polysiloxane films

Time of cure at 70°C (h)	Compression modulus (mN m^{-2})	
	Initial	Over major part of curve
Uncured	1.72	2.60
2	2.19	3.70
5	2.88	4.75
10	3.45	11.61
15	3.23	10.3
25	4.54	12.5 [a]
50	5.06	14.0 [a]

[a] Based on the apparent slope of a very irregular set of data.

calculated from the sum of contributions from the various atoms and groups. However, for siloxane compounds it is less satisfactory because of uncertainty in the proper values for the contribution of silicon-containing groups. The atomic refractivity of silicon depends very much more on the radicals to which it is bonded than is the case for other atoms [11].

Nevertheless, it was decided to make the measurements of density and of refractive index for specimens after different times of curing. Since the molecular weight of the polymerized silane was not known, the molecular refractivity could not be calculated but only the ratio of this to the molecular weight. Nevertheless, interesting results were obtained with a linear relationship between the cure time and refractive index up to 18 h. The relationship between the cure time and the ratio of molecular refractivity to molecular weight is shown in Fig. 3 and indicates a steady increase in molecular complexity up to 24 h, when it approaches a constant value suggesting a final state of crosslinking.

2.2. Discussion

It is clear from the IR spectra that in spite of the vigorous conditions, hydrolysis is never complete and that some methoxy groups persist, even after full cure. Thus, condensation of these methoxy groups with hydroxyl groups does not occur during the final stages of curing.

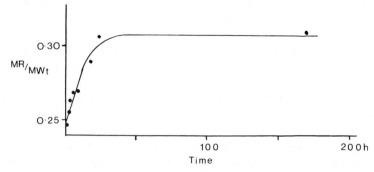

Figure 3. Relationship of the time of cure (in h) of the gelled polysiloxane to the ratio between molecular refractivity and molecular weight.

The hydroxyl content of the condensed polymer decreases during curing but remains substantial even after curing to the fullest extent possible.

Attempts to determine quantitatively the hydroxyl content were unsuccessful and one can only speculate on its value. However, if it is assumed

(1) that the remaining methoxy groups are few in number (i.e. that the initial hydrolysis was nearly complete) and so can be ignored,

(2) that the viscous liquid obtained by hydrolysis and removal of methanol consists of linear oligomers, and

(3) that when this liquid gels only a small extent of crosslinking occurs,

then for the gelled material, on average, one hydroxyl group per silicon atom would remain (two having been consumed in the formation of the linear oligomers). Then if about a third of these remaining hydroxyl groups are consumed during full curing of the material (as has been inferred from Fig. 1), there is left approximately one quarter of all the hydroxyl groups which resulted from hydrolysis of the methoxy groups in the original starting material.

The fact that these remaining hydroxyl groups are not generally accessible to deuteration suggests that they are held within fixed structures, perhaps rings or spirals, protected from reaction but free to vibrate.

The behaviour of the fully crosslinked material in liquids strongly suggests that it is an assembly of particles. This would lead to the very low strength and crumbling behaviour observed. No fully extended network of bonds could be so weak unless it were exceedingly swollen by liquid. Further, the fact that the compression stress–strain relationship is divided into two sections strongly suggests packing of particles.

Torsional braid analysis indicates that the —Si—O—Si— flexible chain is present and persists, although its flexibility is somewhat impaired, even when the material is fully cured. This indicates that there is a substantial proportion of the R—Si—O— units which are neither crosslinked nor involved in ring formation. However, a spiral structure with the hydroxyl groups enclosed along the core meets the requirements. Such a structure has been suggested by Birchall *et al.* [12] for silanol-terminated poly(oxymethylenes).

It is entirely clear that the structure of these free films must be totally different from that which exists in the films used as coupling agents. Films like those considered so far would be a source of extreme weakness in a bond rather than improving its strength. To that extent this investigation was not very helpful in considering the interphase in adhesive bonds including silane coupling agents. Hence further studies, using thinner films, were necessary.

3. PART 2

3.1. Experimental

Again the silane coupling agent used was γ-glycidoxypropyltrimethoxysilane (Union Carbide A187).

The techniques used in this stage of the work were all IR spectroscopy using a Perkin-Elmer spectrometer 599 and appropriate attachments. Conventional transmission through a sodium chloride cell of path length 0.1 mm, and multiple specular reflectance from aluminium films with a mirror finish were both used.

The aluminium mirror surfaces were prepared by evaporation from high purity aluminium wire onto clean glass surfaces at approximately 10^{-6} Torr. Films of about 0.3 μm were prepared which are well known [13] to develop an oxide film very rapidly on exposure to air, although retaining a mirror surface. Immediately after they had been prepared, they were immersed for about 45 min in a 1% solution of silane Union Carbide A187. This solution was made with twice-distilled water of pH 5, which was sufficiently acidic to dissolve and hydrolyse the silane, provided the solution was allowed to stand for 0.5–1 h before being used.

To validate the whole technique, a transmission spectrum of the neat liquid silane was compared with one obtained by specular reflectance from an aluminium mirror surface on which a film of silane had been cast from 1% solution in anhydrous methanol. These two spectra were virtually indistinguishable, as regards both position and intensity of all peaks. They contained a considerable number of peaks, most of which could be assigned with complete satisfaction in terms of the known structure of γ-glycidoxypropyltrimethoxy-silane.

Films produced by the deposition of silanes from aqueous solutions were investigated and the effects of thickness were explored using specular reflectance.

The thickest films were obtained by the evaporation of small volumes of solution, thinner ones by allowing the solution to drain away from the aluminium, and the thinnest by following drainage by washing with distilled water.

The thickness of the thickest films was estimated from the area covered, the weight of silane, and its density, which gave a value of about 5 μm. Comparison of the intensities of some of the more prominent IR peaks enabled some estimate to be made of the thickness of the thinner films.

3.2. Discussion

In comparing the spectra of the liquid silane and those from the aqueous solutions, the most obvious difference was the appearance of a large, broad absorption band in the 3400 cm^{-1} region and, less marked and weaker, a group of peaks around 1650 cm^{-1} due to the hydroxyl groups and to the water molecule, respectively. The broad band at 770–860 cm^{-1}, which has been attributed to alkoxy groups, had almost disappeared from the spectra of the hydrolysed films.

The spectra from the hydrolysed films showed a significant alteration in the band attributed to hydroxyl groups around 3400 cm^{-1} as their thickness varied. As the thickness decreased, the peak of this band moved to a lower wavenumber, going from 3400 cm^{-1} for the thickest to 3200 cm^{-1} for the thinnest. Additionally, the intensity was considerably lower for the thinnest film, all as shown in Fig. 4.

A shift of this sort in this band normally corresponds to a loss of freedom of the hydroxyl group and increased hydrogen bonding, especially to chelation, where this group is most tightly bound. Hence, it is suggested that as a silane film is made thinner, the hydroxyl groups are more tightly bound to the substrate, probably through hydrogen bonding.

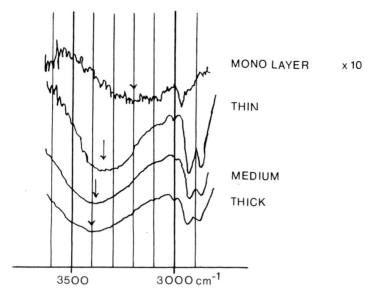

MONO LAYER   x 10

THIN

MEDIUM

THICK

3500   3000 cm^{-1}

Figure 4. Variation in the specular reflectance IR spectra of films of different thicknesses of silane A187 on aluminium mirror surfaces. The sensitivity used for the monolayer is ten times that used for the other spectra.

Nothing in these results is inconsistent with the theory that the silane film is bound to the metal oxide surface through hydroxyl groups which originate from hydrolysis of the methoxy groups in the original silane.

4. CONCLUSIONS

It has been shown in Part 1 of this paper that thick films of polymerized silanes bear no resemblance to the films which are of practical utility as adhesion promoters or primers. Their structure and proportions are such that they would constitute an exceedingly weak interphase between the substrate and adhesive in a bonded joint. It is suggested following Birchall *et al.* [12] that these thick films comprise spiral —Si—O—Si—O— structures with hydroxyl groups which are largely within the spiral and hence inaccessible. When the films become sufficiently thin, then this spiral structure is distorted and destroyed, and the hydroxyl groups become available. Then bonding between these hydroxyl groups and the oxide of the adherend becomes dominant and a firmly attached film of silane is produced.

Thus, a sufficiently thin layer or film of silane is an essential requirement for successful use as an adhesion promoter or coupling agent.

Acknowledgements

I must make free acknowledgement of the contributions of A. K. Hansrani and M. G. Stevens, who did the experimental work in Parts 1 and 2 respectively, as well as of the invaluable help and support in all this, as in so much else, of the late Dr William C. Wake.

REFERENCES

1. P. Walker, in: *Surface Coatings 1*, A. D. Wilson, J. W. Nicholson and H. J. Prosser (Eds), pp. 212–223. Elsevier, Barking (1987).
2. P. W. Erickson and E. P. Plueddemann, in: *Composite Materials*, E. P. Plueddemann (Ed.), Vol. 6, Ch. 1. Academic Press, New York (1974).
3. M. R. Rosen, *J. Coatings Technol.* **50**, 70–82 (1978).
4. P. Walker, in: ref. 1, pp. 193–195.
5. D. S. Tutas, R. R. Stromberg and E. Passaglia, *SPE Trans.* **4**, 256 (1964).
6. W. D. Bascom, *J. Colloid Interface Sci.* **27**, 789 (1968); *Macromolecules* **5**, 792 (1972).
7. Interpretation of all IR spectra involved: G. Socrates, *Infrared Characteristic Group Frequencies.* John Wiley, Chichester (1980).
8. F. J. Boerio and J. E. Greivenkamp, SPE 32nd Annu. Tech. Conf. Reinforced Plastics, Section 4-A (1977).
9. N. C. H. Humphreys and W. C. Wake, *Trans. Inst. Rubber Ind.* **25**, 334 (1950).
10. L. D. Loan, *Monograph No. 17*, p. 24. Soc. Chemical Industry, London (1963).
11. A. D. Petrov, B. F. Mironov, V. A. Ponomarenko and E. A. Charrystov, *Synthesis of Organosilicon Monomers* p. 186. English translation, Heywood, London (1964).
12. J. D. Birchall, J. G. Carey and A. J. Howard, *Nature* **266**, 154 (1977).
13. V. F. Henley, *Anodic oxidation of Aluminium and its Alloys.* Pergamon Press, Oxford (1982).
14. E. P. Plueddemann, *Silane Coupling Agents*, p. 2. Plenum Press, New York (1982).

Silanes and Other Coupling Agents, pp. 91–104
Ed. K. L. Mittal
© VSP 1992.

Factors contributing to the stability of alkoxysilanes in aqueous solution

B. ARKLES,* J. R. STEINMETZ,† J. ZAZYCZNY and P. MEHTA

Huls America, Inc., 2731 Bartram Road, Bristol, PA 19007, USA

Revised version received 4 November 1991

Abstract—Parameters controlling intrinsic stability and reactivity of organosilanols generated from alkoxysilanes in aqueous environments have been elucidated in several experiments. Data involving kinetics, equilibrium, phase separation, and bonding studies of alkyl and organofunctional alkoxysilanes are presented. The studies indicate that the rates of hydrolysis of alkoxysilanes are generally related to their steric bulk, but demonstrate that after rate-limiting hydrolysis of the first alkoxy group steric effects are much less important. Aqueous hydrolysis of alkylalkoxysilanes was studied to determine equilibrium constants and the extent of oligomerization up to phase separation. In the case of propyltrialkoxysilane, phase separation is coincident with the formation of tetramer. The equilibrium constant for esterification of silanols is

$$K_{eq} = \frac{[R_3SiOCH_3][H_2O]}{[R_3SiOH][CH_3OH]} = 2.5 \pm 0.3 \times 10^{-2}.$$

Also, the performance properties of new water-borne silanes were evaluated and in most cases, their performance equalled or exceeded their traditional silane counterparts.

Keywords: Aqueous solution; hydrolysis; alkoxysilanes.

1. INTRODUCTION

The silanes employed in improving adhesion and surface modification are usually alkoxysilanes. Before or during application and bonding processes, alkoxysilanes are hydrolyzed, initiating a complex cascade of reactions. A simplified view of the reaction cascade is depicted in Fig. 1. In most application protocols, a catalyst which initiates alkoxysilane hydrolysis but which also affects silanol condensation and substrate interaction dictates reaction paths. In order to achieve inherent control and reproducibility of silane modification of surfaces, it would appear necessary to control the rate and path of reactions leading to the formation and consumption of silanol-rich species, particularly silanol-rich oligomers. Maximizing the availability of silanol-rich species, in terms of both time and concentration, should lead to improved reproducibility and greater surface interaction.

The objective of the studies presented in this paper was to develop fundamental information for silanol intermediates concerning their kinetics of formation, equilibrium with other silicon-containing species, and phase separation behavior. Based on these results, it was the further objective of the studies

---

*Present address: Gelest, Inc., 612 William Leigh Drive, Tullytown, PA 19007, USA.
†To whom correspondence should be addressed.

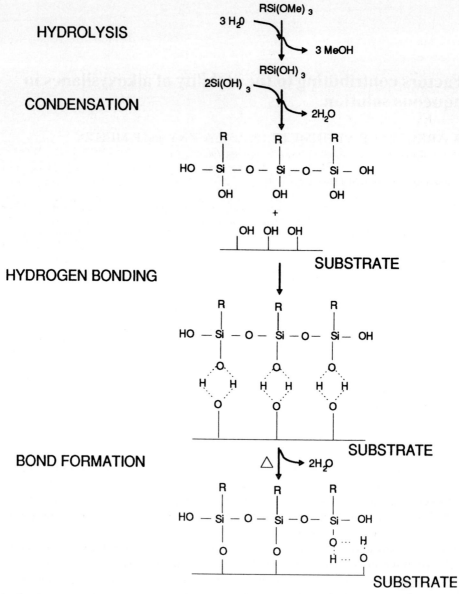

Figure 1. Reaction and bonding mechanism of alkoxysilanes.

to develop a series of stable silanol-rich oligomers and to evaluate their performance in sealant and composite systems.

2. SILANE HYDROLYSIS CHEMISTRY

In the vast majority of silane surface treatment applications, the alkoxy groups of trialkoxysilanes are hydrolyzed to form silanol-containing species. The silanol-containing species are highly reactive intermediates which are responsible for bond formation with the substrate. Hydrolysis of trialkoxysilane alkoxy groups may occur during the formal preparation of aqueous solutions or the reaction of the silane with adsorbed moisture on substrate surfaces. In principle, if silanol materials were stable, they would be preferred for surface treatments. Most silanes employed in surface treatments do not form stable monomeric silanols. Silanols usually condense with themselves or with alkoxysilanes to form siloxanes. Trialkoxysilanes are stable sources for silanols, but they have low intrinsic reactivity, low solubility in aqueous solutions, and by-products of hydrolysis which may be undesirable from a flammability and toxicity perspective.

The general mechanism for silane bond formation, depicted in Fig. 1, has been reviewed in detail [1, 2]. Alkoxysilanes undergo hydrolysis by both base- and acid-catalyzed mechanisms. In contact with high-purity (18×10^6 Ω cm) water under neutral, low ionic conditions in non-glass containers, alkoxysilanes bearing no autocatalytic functionality are stable for weeks or months. In contrast to chlorosilanes and acetoxysilanes, the products of alkoxysilane hydrolysis do not propagate the hydrolysis reaction. In contact with 'tap' water, hydrolysis of alkoxysilanes is substantial, if not complete, within hours. The same factors which accelerate the hydrolysis of alkoxysilanes also accelerate the condensation of silanols with other silanols and their alkoxy precursors. The overall pathway for the hydrolysis and full condensation of an alkoxysilane is complicated. If effects beyond the next nearest neighbor during condensation are ignored, there are six possible hydrolysis paths, 21 possible water-producing condensations, and 36 possible alcohol-producing reactions. Kay and Assink have presented a model for tetraalkoxysilane hydrolysis and condensation [3]. Presented in Fig. 2 is a scheme derived for trialkoxysilanes from that model. The numeric presentation indicates the number of alkoxy (OR) substitution in the first digit, the number of hydroxyl (OH) substitutes in the second digit, and the number of siloxane substitutions (OSi) in the third digit. Thus, a trialkoxysilane is 300 and a silanetriol is 030. Like the model for tetraalkoxysilanes, extended condensation or polymerization which results in phase separation and its kinetic consequences are not considered. Figure 3 shows the gross changes in silane condensation and polymerization in an acid vs. base condition. It can be related to the Kay and Assink model by observing that base hydrolysis favors a 300 to 003, where acid hydrolysis favors a 300 to 030 pathway.

It is important to note that catalysts for alkoxysilane hydrolysis are usually catalysts for condensation. In typical silane surface treatment applications, alkoxysilane reaction products are removed from equilibrium by phase separation and deposition of condensation products. The overall complexity of hydrolysis and condensation has not allowed simultaneous determination of the kinetics of silanol formation and reaction. Eqiulibrium data for silanol formation and condensation, until now, have not been reported.

2.1. Silane hydrolysis—background

2.1.1. Alkoxysilane hydrolysis—effects of substituents.
A series of hydrolysis studies [4–9] have elucidated general trends. Under basic conditions, the hydrolysis of alkoxy groups usually takes place in a stepwise manner. Carbon-bonded substituents can have profound effects on the rate of hydrolysis. With the

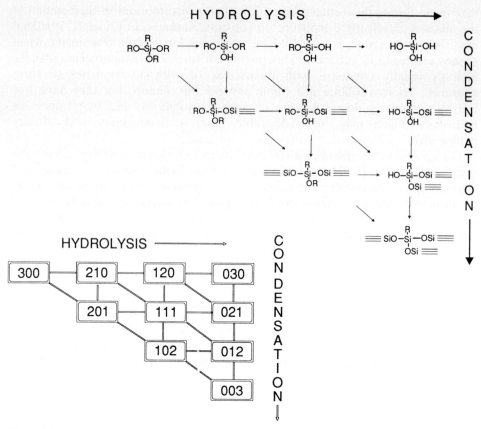

Figure 2. Numeric equivalents for the alkoxysilane hydrolysis cascade reaction.

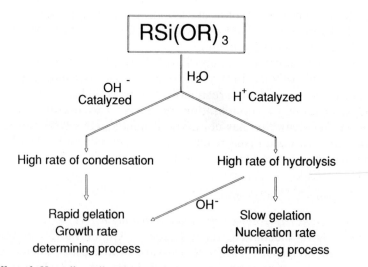

Figure 3. Effect of pH on alkoxysilane hydrolysis, adapted from Prassas [20].

exception of aminosilanes, most silanes are employed in surface treatment applications under acid-catalyzed conditions. The rate of acid hydrolysis is significantly greater than base hydrolysis and is minimally affected by other carbon-bonded substituents. The hydrolysis is preceded by protonation of the OR group. The rates of hydrolysis of the alkoxy groups are generally related to their steric bulk: $CH_3O > C_2H_5O > t\text{-}C_4H_9O$. A methoxysilane, for example, hydrolyzes at 6–10 times the rate of an ethoxysilane. As for base condensation, the hydrolysis of alkoxy groups is stepwise; however, the differential in hydrolysis rate between the first and second alkoxy group is greater. Finally, increased organic substitution enhances the hydrolysis rate, i.e. $Me_3SiOMe > Me_2Si(OMe)_2 > MeSi(OMe)_3$ [7]. The difficulty in straightforward application of the conclusions of the various hydrolysis studies to alkoxysilanes employed in surface treatments is that most of the studies employ non-functional silanes and provide data at pH extremes, i.e. outside the range of normal surface treatment regimens. Moreover, many of the studies incorporate solvents. In most normal surface treatment regimens, there are consequences of the insolubility of alkoxysilanes.

2.1.2. Condensation and persistence of silanols. In terms of devising a stable water-borne silane system, the silanol form of silanes is desirable since the silanols have greater solubility and reactivity than their alkoxy precursors and much greater solubility and reactivity than their siloxane products. One avenue for enhancing silanol persistence is to devise a system in which the rate of hydrolysis is substantially greater than that of condensation. A second avenue is to allow condensation of the polyhydroxylic silanol only to the extent that water solubility is maintained, allowing steric factors to preclude further condensation. Another consideration is that while silanols undergo S_N2 reactions rapidly with other silanols and methoxysilanes, their reaction with ethoxysilanes and higher alkoxysilanes is slow.

Individual examples of monomeric silanols such as triethylsilanol [10] and phenylsilanetriol [11] have been prepared under regimens different from those used for surface treatments and they exhibit extended stability. To date, no monomeric silanetriols have been isolated from aqueous hydrolysates of alkoxysilanes. The kinetics of silanetriol condensation have been studied [12]. The conditions which promote the hydrolysis of alkoxysilanes also promote condensation of silanols; the persistence of monomeric silanetriols for more than a few hours in typical solutions is unlikely. However, the persistence of silanols in reaction mixtures containing condensed structures have been observed empirically [13] and by ^{29}Si-NMR [14, 15].

The fact that silanol persistence can be favored by equilibrium conditions rather than control of condensation kinetics by steric or electronic factors is usually not considered. The phase separation which results from highly condensed systems continuously removes material from deposition solutions, depleting soluble silane species. While condensed silanols or siloxanes are typically not regarded as participating in a reversible reaction with water or alcohol, they do indeed participate in an equilibrium reaction. Iler [16] has shown that even hydrated amorphous silicon dioxide has an equilibrium solubility in methanol, which implies the formation of soluble low molecular weight species. The equilibrium concentration of SiO-containing species in methanol is 7 ppm, compared with 70–120 ppm in water.

The equilibrium reactions for alkoxysilane hydrolysis mixtures are given in equations (1)–(3). Data on equilibrium constants for these reactions have not been reported.

$$R_3SiOR + H_2O \rightleftharpoons R_3SiOH + ROH \tag{1}$$

$$2 R_3SiOH \rightleftharpoons R_3SiOSiR_3 + H_2O \tag{2}$$

$$R_3SiOSiR_3 + ROH \rightleftharpoons R_3SiOR + R_3SiOH. \tag{3}$$

The equilibrium as well as the kinetics of hydrolysis and condensation of trialkoxysilanes is influenced by the organic substitution. A special case exists for many aminosilanes such as aminopropyltriethoxysilane, where a stable zwitterionic silanolate forms that is stable in solution [17–19]. It has been reported that at concentrations lower than 0.2% the aminosilane exists as a monomeric silanetriol [19].

2.2. Silane hydrolysis—experimental

All silanes were obtained from commercial production at Huls America. Ultra-pure water was prepared in a Millipore purification unit. FTIR studies were conducted with a Nicolet Model 740 Spectrometer using ATR (attenuated total reflectance) of samples in a zinc selenide cell. NMR spectra were obtained using a General Electric QE-plus 300 MHz NMR Spectrometer.

2.3. Silane hydrolysis—results and discussion

A series of studies were conducted to define parameters related to reactivity, solubility, and stability of alkoxysilane hydrolysis mixtures for the purpose of generating model compounds stable in water solution and maintaining coupling agent activity.

2.3.1 Mixed alkoxysilane hydrolysis kinetics. Since the displacement of an alkoxy group with a hydroxyl group enhances water solubility, the possibility of generating mixed trialkoxysilanes which would react with water to form stable dialkoxysilanols was considered. If the kinetic differential of stepwise hydrolysis of alkoxy groups could be extended 'practically' stable silane–water solutions might be possible. The salient characteristics of acid-catalyzed hydrolysis of alkoxysilanes are illustrated by a series of FTIR experiments, the results of which are given in Figs 4–6. Isobutyltriethoxy, trimethoxy, and mixed ethoxymethoxy-silanes were hydrolyzed by adjusting mixtures with 0.1 M HCl to 6 mM H^+ concentration. Isobutyltrimethoxysilane hydrolyzes 7.7 times faster than its ethoxy analog. In the case of the mixed ethoxymethoxysilane, the overall rate of hydrolysis is 87% of that of the trimethoxysilane group. The nearly equivalent hydrolysis rates are a consequence of the fact that while the hydrolysis of the methoxy group is slowed down by 30% by the steric effects of the ethoxy groups, the hydrolysis of the methoxy group promotes the hydrolysis of the slower ethoxy group by 40%. Thus, under acidic conditions the hydrolysis of the first

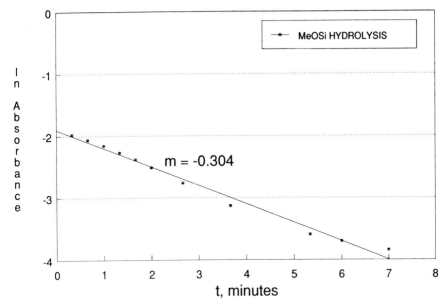

Figure 4. Hydrolysis of *i*-butyltrimethoxysilane.

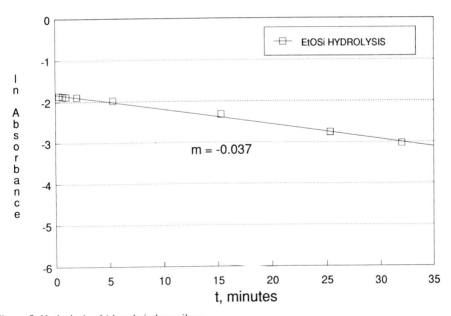

Figure 5. Hydrolysis of *i*-butyltriethoxysilane.

alkoxy group is rate-limiting. Preparing mixed alkoxysilanes would have no substantial benefit to extending silane aqueous solution stability.

2.3.2. Extent of silane condensation (degree of polymerization). In order to provide insight into the extent of condensation of silanols and their phase separation as a function of time, the hydrolysis of propyltrimethoxysilane was

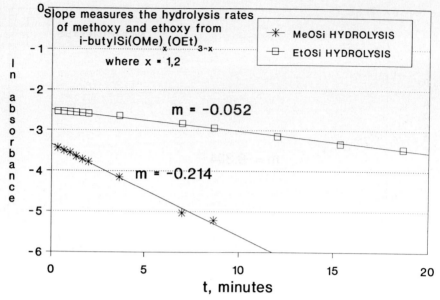

Figure 6. Hydrolysis of *i*-butyl mixed alkoxysilanes.

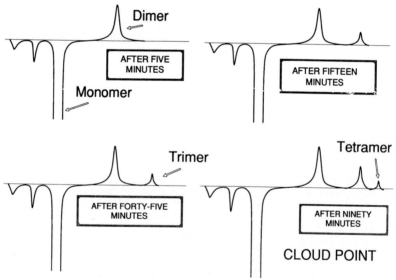

Figure 7. Hydrolysis of propyltrimethoxysilane.

studied. The aqueous hydrolysis of propyltrimethoxysilane was catalyzed with HCl and followed by HPLC, as shown in Fig. 7. It can be observed that the formation of dimeric species proceeds much faster than that of trimeric species. By the time tetramer was observed by HPLC, the solution had become hazy, indicating phase separation. The generalized model of propyltrimethoxysilane condensation is shown in Fig. 8.

It is interesting to note that the appearance of haze or onset of phase separation is coincident with the first opportunity for branching or cyclic

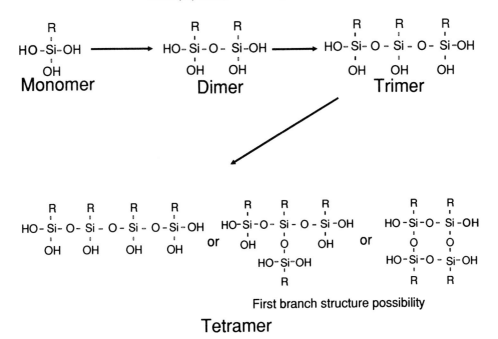

Figure 8. Oligomerization of silanes.

formation. More rapid molecular weight growth is possible in branched vs. linear systems due to the fact that silanediol chain ends are preferred for polymer growth. It may be argued, therefore, that the formation of the branched structure presages phase separation. In terms of preparing aqueous silane-treating solutions, it is clear that the maximum degree of polymerization for alkyltri-alkoxysilanes should be limited to 3.

2.3.3. Equilibrium constant determination for alkoxysilane hydrolysis. Triethyl-silanol was selected as a model compound for determination of the equilibrium constant for equation (1), since under neutral conditions the condensation to disiloxane was observed to take place only over an extended period of time (i.e. years), eliminating equilibria (2) and (3) as interfering factors.

A series of triethylsilanol/methanol/water mixtures were prepared as shown in Table 1 and stored in clean high-density polyethylene containers at room temperature. They were monitored for changes in component concentration and it was determined that after 100 days they had reached equilibrium.

2.3.4. NMR—results. All spectra were taken using a GE QE-plus 300 MHz NMR spectrometer. ^{13}C-NMR spectra (H decouple) were relatively easy; however, to eliminate the nuclear Overhauser effect, spectra were obtained in a reverse gated decouple mode. The T_1 spin lattice relaxation times for the samples were between 3 and 5 s. Therefore, all spectra were taken with a 30 s delay sequence. A typical spectrum with assignments is given in Fig. 9. The average equilibrium constant [Et$_3$SiOMe][H$_2$O]/[Et$_3$SiOH][MeOH] for the series of mixtures was $2.7 \pm 0.4 \times 10^{-2}$.

Table 1.

Initial reactant concentrations in the equilibrium study of the triethylmethoxysilane–H_2O system

Sample no.	Moles methanol	Moles triethylsilanol
1	0.000	0.101
2	0.006	0.045
3	0.010	0.041
4	0.021	0.031
5	0.062	0.044
6	0.089	0.022
7	0.095	0.010
8	0.112	0.000

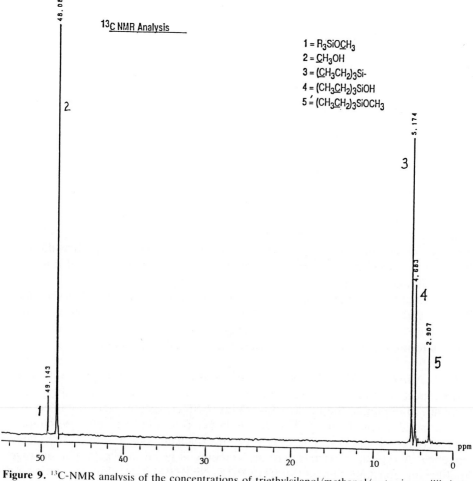

Figure 9. ^{13}C-NMR analysis of the concentrations of triethylsilanol/methanol/water in equilibrium mixture.

2.3.5. FTIR—results. All spectra were obtained using a Nicolet Model 740 FTIR spectrophotometer. A typical spectrum is given in Fig. 10. The spectra were deconvoluted using Lab Calc software, from Galactic Industries. The concentrations were derived by applying Beer's Law. The average equilibrium constant $[Et_3SiOMe][H_2O]/[Et_3SiOH][MeOH]$ for the series of mixtures was $2.3 \pm 0.3 \times 10^{-2}$.

Together, the NMR and IR results indicate that the equilibrium constant for $[R_3SiOR'][H_2O]/[R_3SiOH][R'OH]$ can be taken as $2.5 \pm 0.3 \times 10^{-2}$. There is, of course, some question as to how the results of monoalkoxysilanes can be translated to trialkoxysilanes. It is clear, however, that the equilibrium, while favoring hydrolysis, is not so large that stabilization by alcoholic species is precluded.

2.4. Silane hydrolysis—conclusions

New data have been presented in the context of a review of the aqueous behavior of silanes which elucidate their behavior, including mixed alkoxysilane hydrolysis kinetics, silane solubility, and the determination of the equilibrium constant for the alkoxy hydrolysis reaction.

The following factors enhance the stability of silanols:

(1) neutral conditions;
(2) limited condensation;

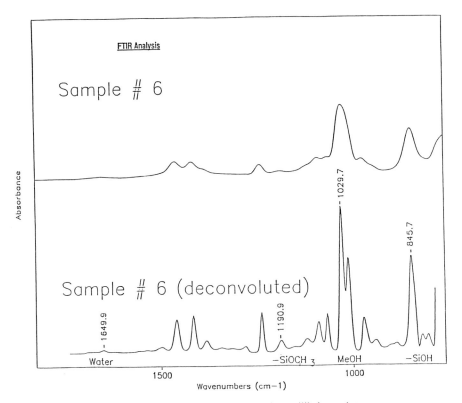

Figure 10. FTIR analysis of triethylsilanol/methanol/water in equilibrium mixture.

(3) the presence of hydroxylic species; and
(4) silanes forming stable zwitterions.

On the other hand, simply substituting different alkoxy groups on a silane or preparing mixed alkoxy-substituted silanes appears to offer little advantage in preparing stable water-borne silanes.

3. WATER-BORNE SILANES WITH STABLE SILANOLS

3.1. Water-borne silanes—background

A number of schemes were developed which incorporated the conclusions of the alkoxysilane hydrolysis studies in the first part of this paper. A series of water-borne silanes were developed having high active silanol contents which are stable in water for periods of more than 6 months. Low molecular weight alcohols were not incorporated in the solutions since even at concentrations as low as 1% they contribute to flammability.

Hydrosil® is the name used to designate a stable water-borne silane. The properties of various Hydrosil silanes are described in Table 2.

Table 2.
Physical properties of water-borne silanes

Silane type	Designation	Specific gravity	pH	Viscosity (mPa s)
Amine	Hydrosil 2627	1.1	10–10.5	5–10
Amine	Hydrosil 2628	1.1	10.5–11	5–10
Diamine	Hydrosil 2776	1.1	10–11	10–20
Diamine	Hydrosil 2774	1.1	10–11	65–70
Triamine	Hydrosil 2775	1.1	10–11	35–50
Epoxy	Hydrosil 2759	1.0	6–8	3–6
Mercapto	Hydrosil 2788	1.2	6–8	10–15
Vinyl	Hydrosil 2810	1.1	3.5–4.5	6–12
Vinyl-amino	Hydrosil 2781	1.1	10.5–11	4–9

3.2. Water-borne silanes—experimental

3.2.1. Acrylic latex sealants. Basic acrylic latex sealant formulations were compounded in a sigma blade mixer, as shown in Table 3. The water-borne silanes were then incorporated into a base. Standard wet and dry peel adhesion determinations from various substrates were performed periodically in accordance with ASTM C794-8.

3.3. Water-borne silanes—results and discussion

3.3.1. Acrylic latex sealants. Adhesion strengths on three substrates after 3 months sealant storage (0.5% silane added) were evaluated and the results are presented in Table 4.

The data substantiate that the silanes significantly enhance acrylic sealant bonding to various substrates. The significant results are the wet bond strength. Without silane addition there is virtually no wet bond strength. All silanes impart a substantial wet bond strength. The water-borne silanes, however, give wet bond strengths up to 30% greater than conventional silane coupling agents.

Table 3.
Latex sealant formulation

Latex (Rhoplex 1785 Rohm & Haas)	1915.0 parts
Non-ionic surfactant (TDET N407, Thompson Hayward)	59.4 parts
Dispersant (Calgon T)	49.9 parts
Plasticizer (Santicizer 160—Monsanto)	54.9 parts
Biocide (Fungitrol 234—Huls)	9.1 parts
Mineral spirits	59.2 parts
Ethylene glycol	63.2 parts
Calcium carbonate	3401.9 parts
Mix 30 min	
Latex (Rhoplex 1785)	534.4 parts
Ammonium hydroxide (28%)	4.5 parts
Silane	3.1 parts
Mix 15 min	

Table 4.
Effect of silanes on adhesion strength of latex formulations

(A) Adhesion to carbon steel Peel strength (N/m)	Control	AMEO[a]	Hydrosil 2627	DAMO[b]	Hydrosil 2774	Hydrosil 2776
Dry	5254	7000	6130	5254	5254	5954
Wet	<100	1750	2277[c]	525	876	2100[c]

(B) Adhesion to glass Peel strength (N/m)	Control	AMEO[a]	Hydrosil 2627	Hydrosil 2628	Hydrosil 2775
Dry	1750	2627	3152	4378	4378
Wet	<100	2627	3152[c]	2977[c]	2100

(C) Adhesion to aluminum Peel strength (N/m)	Control	AMEO[a]	Hydrosil 2628	Hydrosil 2775	Hydrosil 2776
Dry	2627	3327	5954	2627	2977
Wet	<100	525	875[c]	1400[c]	700

[a] Aminopropyltriethoxysilane.
[b] Aminoethylaminopropyltrimethoxysilane.
[c] Highest values observed.

3.3.2. Mineral-filled nylon 6/6 composites. Kaolin (Burgess) was treated with the appropriate silane in a Patterson Kelly 8-qt. twin shell blender with intensifier bar. Treatment levels were 1% active silane. The kaolin was compounded with nylon 6,6 (Vydyne 21x—Monsanto) in a Leistritz twin screw extruder. The compounds were tested in a 'dry as molded' condition in accordance with ASTM D-638 (see Table 5).

In virtually all the cases tested, the water-soluble silane hydrosil gave values better than that from a conventional control silane in solution.

3.4. Water-borne silanes—conclusions

Based on observations of silanol stability, a new series of silanes were developed

Table 5.
Tensile strength of 40% clay-filled nylon 6/6 as a function of treatment

Surface treatment	Tensile strength (MPa)	% of control
Control	64.8	100
AMEO-40[a]	71.7	111
HS2627[b]	75.8	117
HS2628	77.90	118
HS2781	80.66	124
HS2774	73.76	114
HS2776	77.2	119
HS2775	77.2	119

[a] 40% solution of aminopropyltriethoxysilane in water.
[b] Hydrosil.

with long-term stability in aqueous systems. These materials offer intrinsic process flexibility due to the elimination of volatile by-products and reduced flammability.

The performance of the new water-borne silanes was studied. The water-borne silanes were comparable to conventional silanes in most cases. In several cases, notably with acrylic latices, the new silanes excelled. In particular, the data obtained suggest that the new amine functional materials may exhibit superior performance to conventional silanes in moist environments.

REFERENCES

1. B. Arkles, *CHEMTECH* **7**, 766 (1977).
2. E. P. Plueddemann, *Silane Coupling Agents.* Plenum Press, New York (1982).
3. B. D. Kay and R. A. Assink, *J. Non-Cryst. Solids* **104**, 112 (1988).
4. K. McNeil, J. DiCarpio, D. Walsh and R. Pratt, *J. Am. Chem. Soc.* **102**, 1859 (1980).
5. E. Pohl, *Proc. 38th SPI Reinforced Plastics Composites Conf.*, Section 4-B (1983).
6. E. Pohl and F. Osterholtz, in: *Silane Surfaces and Interfaces*, D. Leyden (Ed.). Gordon and Breach, New York (1986).
7. K. Smith, *J. Org. Chem.* **51**, 3827 (1986).
8. M. Voronkov, V. P. Mileshkevich and Y. A. Yuzhelevskii, *The Siloxane Bond.* Consultants Bureau, New York (1978).
9. J. R. Steinmetz and B. Arkles, *22nd Organosilicon Proc.*, Petrarch, Bristol, PA (1989).
10. L. Sommer, E. Pietrusza and F. Whitmore, *J. Am. Chem. Soc.* **68**, 2282 (1946).
11. T. Takeguchi, *J. Am. Chem. Soc.* **81**, 2359 (1959).
12. E. Pohl and F. Osterholtz, in: *Chemically Modified Surfaces in Science and Industry*, D. Leyden and W. Collins (Eds). Gordon & Breach, New York (1988).
13. J. R. Steinmetz, *Mod. Paint Coatings*, **73**, 45 (1983).
14. R. A. Assink and B. D. Kay, *J. Non-Cryst. Solids* **107**, 35 (1988).
15. R. A. Assink, in: *Sol–Gel Science*, C. J. Brinker and G. W. Scherer (Eds), p. 111. Academic Press, New York (1990).
16. R. K. Iler, *The Chemistry of Silica*, p. 61. John Wiley, New York (1979).
17. E. Plueddemann, *Proc. 24th SPI/Reinforced Plastics Composites Conf.*, Section 19-A (1969).
18. H. Ishida, S. Naviroj, S. K. Tripathy, J. J. Fitzgerald and J. L. Koenig, *J. Polym. Sci. Polym. Phys. Ed.* **20**, 701 (1982).
19. H. Ishida and Y. Suzuki, in: *Composite Interfaces*, H. Ishida (Ed.), p. 317. Elsevier, New York (1986).
20. M. Prassas and L. L. Hench, in: *Ultrastructure Processing of Ceramics*, L. Hench and D. Ulrich (Eds), p. 100. John Wiley, New York (1984).

Silanes and Other Coupling Agents, pp. 105–116
Ed. K. L. Mittal
© VSP 1992.

Methods for improving the performance of silane coupling agents

PETER G. PAPE* and EDWIN P. PLUEDDEMANN

Dow Corning Corporation, Midland, MI 48686-0994, USA

Revised version received 1 May 1991

Abstract—The use of silane coupling agents in mineral- and glass-reinforced composites is well known. They impart improved initial mechanical properties, but, more importantly, they cause mechanical properties to be retained during the use of the composite. The main cause of loss of mechanical properties is attack of water at the interface. Recent research has focused on imparting more durable bonding of the silane coupling agent to both the polymer and the reinforcement. Improved silane coupling agent systems have been developed by utilizing several techniques: blends of hydrophobic silanes with hydrophilic silanes to give greater hydrophobic character; use of 1,2-bis-(trimethoxysilyl)ethane as an additive to give increased siloxane crosslinking; use of more thermally stable silanes such as phenyltrimethoxysilane and *N*-[2-(vinylbenzylamino)-ethyl]-3-aminopropyltri-methoxysilane to give increased thermal stability; and the use of a carboxy-functional silane with a carboxy-functional polymer and zinc salt to give ionomer bonds at the interface. The effectiveness of these new coupling agent systems was tested by measuring the flexural strength of composites and the adhesion strength of coatings on inorganic substrates. The results show that composites have increased flexural strength and better strength retention during thermal aging; coatings have greater adhesion strength; there is greater resistance of interfacial bonding to degradation by moisture; and thermoplastic composites have better properties after high shear processing.

Keywords: Adhesion; silane; composites; hydrophobic silane; crosslinked silane; ionomer bonding.

1. INTRODUCTION

In polymer composites reinforced with glass fiber or minerals, the interphase region involves a complex interplay of physical and chemical factors [1]. Traditional coupling agents act in the interphase to improve adhesion, give increased mechanical strength, and impart resistance to degradation of properties during the life of the composite [2]. They are applied to the inorganic substrate from water or an organic solvent, or, sometimes, added to the polymer during compounding. Research has been directed at improving the performance of silane coupling agents with particular attention towards improving water resistance, thermal stability, and the degradation of adhesive bonds by high shear forces during processing. Some of these modification techniques will be described.

2. CHEMISTRY OF SILANE COUPLING AGENTS

A silicon compound that contains both inorganic and organic reactivities in the

---

*To whom correspondence should be addressed.

same molecule, e.g. $(RO)_3SiCH_2CH_2CH_2{-}X$, where X is an organofunctional group, can function as a coupling agent. Coupling agents act at the interface between an inorganic substrate (such as glass, metal, or mineral) and an organic substrate (such as an organic polymer, coating, or adhesive) to bond the two dissimilar materials together. A simplified picture of the bonding mechanism is shown in Fig. 1.

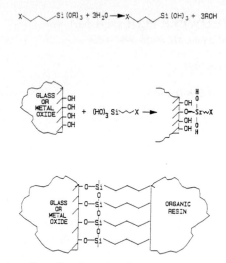

Figure 1. Simplified depiction of bonding through a silane coupling agent.

3. THE SILANE BOND TO THE POLYMER

The bonding of a coupling agent to an organic polymer is complex. The simplest way to choose an effective silane coupling agent for thermosetting polymers is to match its reactivity with the reactivity of the organic polymer, i.e. an epoxysilane will bond to an epoxy resin and a methacrylate silane will bond to an unsaturated polyester resin. Typical silane coupling agents and their chemical formulas are shown in Table 1. In a given composite system, significant improvements in composite properties can be obtained by varying the organic structure and reactivity of the coupling agent (see Fig. 2).

The multi-functional silane coupling agent with a cationic vinylbenzylamine organic group (coupling agent H in Table 1) is a very effective coupling agent for a wide variety of thermoset and thermoplastic polymers. As shown in Fig. 2, it can impart improved properties to an unsaturated polyester composite compared with the state-of-the-art methacrylate silane D. It is also preferred over an epoxysilane C for epoxy printed circuit-board laminates. Heat-cleaned fiberglass cloth was treated with 1% aqueous dispersions of epoxysilane C or vinylbenzylaminosilane H, and laminates with liquid epoxy resin were cured with *meta*-phenylenediamine. The superior flexural strength retention imparted by silane H is shown in Table 2.

The silanes of Table 1 are functional alkyltrialkoxysilanes (G and I are modifiers, and not coupling agents) with the thermal stability of typical aliphatic

Table 1.
Representative silane coupling agents

Organofunctional group	Chemical structure	Dow Corning designation
A. Vinyl	$CH_2=CHSi(OCH_3)_3$	Q9-6300
B. Chloropropyl	$ClCH_2CH_2CH_2Si(OCH_3)_3$	Z-6076
C. Epoxy	$CH_2CHCH_2OCH_2CH_2CH_2Si(OCH_3)_3$ (with epoxide O)	Z-6040
D. Methacrylate	$CH_2=C(CH_3)\text{—}C(=O)\text{—}OCH_2CH_2CH_2Si(OCH_3)_3$	Z-6030
E. Primary amine	$H_2NCH_2CH_2CH_2Si(OC_2H_5)_3$	Z-6011
F. Diamine	$H_2NCH_2CH_2NHCH_2CH_2CH_2Si(OCH_3)_3$	Z-6020
G. Methyl	$CH_3Si(OCH_3)_3$	Z-6070
H. Vinylbenzylamino	$CH_2=CHC_6H_4CH_2NHCH_2CH_2CH_2NH(CH_2)_3Si(OCH_3)_3$ · HCl	Z-6032
I. Phenyl	$C_6H_5Si(OCH_3)_3$	Q1-6124
J. Carboxy	Condensation product—0.9 mol silane F/1.0 mol isophthalic acid	7169-45B

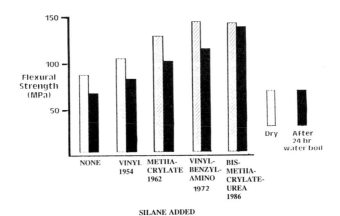

Figure 2. Performance of silane coupling agents added (0.12% on mix) to 50:50 silica–polyester castings.

chemicals. Aminosilanes E and F are among the most versatile adhesion promoters of Table 1, but in addition to suffering from poor thermal stability, they produce very hydrophilic layers on primed surfaces.

4. THE SEARCH FOR BETTER COUPLING AGENTS

Much effort has been expended in synthesizing and testing new potential coupling agents with improved thermal stability and hydrophobicity. A study

Table 2.

Effect of coupling agents on the flexural strength of a structural epoxy–fiberglass laminate

Coupling agent on glass (Table 1)	Flexural strength (MPa)	
	Dry	72 h H_2O boil
None	394	221
Epoxysilane C	605	476
Vinylbenzylaminosilane H	670	518

sponsored by the Air Force Materials Laboratory under Contract AF 33(615)-1163 was conducted for the purpose of developing coupling agents, sizings, and finishes having greater thermal stability than those currently available for AF-994 glass. Such materials are required to realize the elevated temperature strength properties of polybenzimidazole (PBI), polyimide (PI), and other high-temperature resins in glass fiber laminates. Preliminary results of the study have been described in the Summary Technical Report, AFML-TR-65-316, July 1965 and reported by E. P. Plueddemann to the Society of Plastics Industry (SPI) [3].

Aliphatic organofunctional silanes recommended as coupling agents for glass-reinforced polyesters, epoxies, or phenolics have heat stabilities of typical aliphatic organic chemicals and are not at all comparable to the methyl or phenyl silicones found in silicone polymers and resins.

The thermal stabilities of silane coupling agents were estimated by thermo-gravimetric analysis (TGA) in helium and in air, and by isothermal weight loss in air at 300°C. In each case, the silane was hydrolyzed and condensed to a silicone resin by drying briefly at 200°C. The resinous materials were cooled and ground to powders for thermal testing. It is recognized that variable amounts of hydroxyl groups remain on silicone resins dried at this temperature. Small initial changes with thermal aging were therefore attributed to silanol condensation.

TGA curves were run on 100–200 mg samples of the resins heated at 3°C/min in air or helium at a flow rate of 22.0 m^3/s. Weight losses before 200°C were not shown on the curves since they were attributed to loss of water from silanol condensation.

Typical aliphatic and aromatic organofunctional silanes are compared (Figs 3 and 4) in the TGA curves of polysiloxanes of (γ-cyanopropyl)- and (γ-cyanophenyl)-trimethoxysilane. Not only is the aromatic silane more stable than the aliphatic structure, but also the onset of weight loss is less sensitive to a change from helium to oxygen atmosphere. This suggests that aromatic structures would be most desirable for resistance to high-temperature oxidation.

Isothermal weight losses in air were measured on 3–4 g samples of powdered silicone resin spread on the bottom of 2-inch diameter aluminum dishes and exposed in a circulating air oven held at 300°C. The silicones generally fused to thin films in the dishes. The samples were weighed after 3, 24, and 100 h. Weight loss was used to calculate the percent of R-group remaining as $RSiO_{1.5}$ (Fig. 5) assuming ultimate burning to SiO_2.

A new generation of high-temperature plastics containing stable aromatic and heterocyclic rings in the polymer chains have use-temperature ceilings that tax

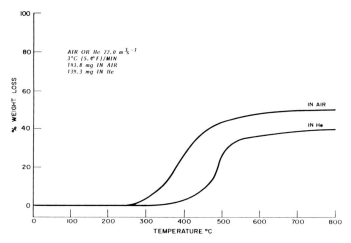

Figure 3. TGA curve of the polysiloxane of (γ-cyanopropyl)-trimethoxysilane.

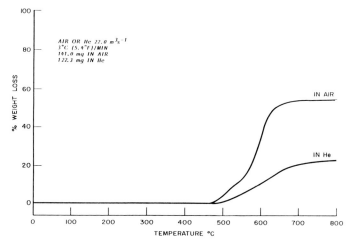

Figure 4. TGA curve of the polysiloxane of (γ-cyanophenyl)-trimethoxysilane.

the heat stability of even the best aromatic silicon compounds. Thermal stabilities of a few typical polymers are compared with typical silane coupling agent hydrolyzates by TGA as shown in Fig. 6.

Twenty-eight aromatic silane coupling agents were examined with somewhat disappointing results. The most heat-stable aromatic silanes were not reactive enough with polymers to act as coupling agents. Carboxyphenylsilanes were very heat-stable, and were effective coupling agents for high-temperature epoxies and polybenzimidazoles, but not for polyimides. Aminophenylsilanes were reactive with polyimide resins but had very poor thermal stability and were not much better than aliphatic aminosilane coupling agents. More recently, it was shown that when aminophenylsilanes are imidized with aromatic dianhydrides, they provide wholly aromatic imide silanes that give good adhesion retention to

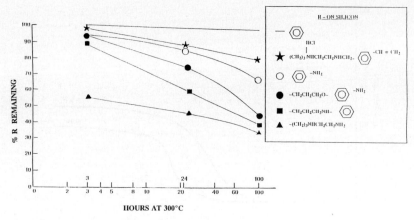

Figure 5. Isothermal weight loss of RSiO$_{1.5}$ in air at 300°C.

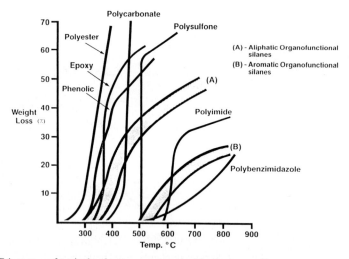

Figure 6. TGA curves of typical polymers compared with silane coupling agents.

polyimide adhesives after prolonged exposure to elevated temperature [4]. Introduction of new coupling agents by this route is tedious and expensive. It would be much simpler to try to upgrade the silanes of Table 1.

5. IMPROVED BONDING AT THE INTERFACE

Several modifications of commercial silane coupling agents have been evaluated to search for improved bonding at the interface. These approaches attempt to use modifiers to counteract basic faults of the individual silanes. Some of these modifications include increased hydrophobic character, increased crosslinking of the siloxane structure, increased thermal stability, and ionomer bond formation to reduce shear degradation at the interface.

5.1. Hydrophobic character

In order to counteract the effect of water absorption in the interphase region, hydrophobic substituents were added to the coupling agent [5]. Methyltrimethoxysilane G, vinyltrimethoxysilane A, and 3-chloropropyltrimethoxysilane B are good sources of hydrophobic character, but phenyltrimethoxysilane I gives a good combination of hydrophobic character, thermal stability, and polymer compatibility.

Blends of 10% aminosilane F and 90% hydrophobic silanes, i.e. vinylsilane A, chloropropylsilane B, methylsilane G, and phenylsilane I, gave superior adhesion of three types of polyurethane (RIM, thermoplastic, and one-component rigid) to glass compared with aminosilane F alone. Table 3 shows that the blend with phenylsilane I gave the best adhesion overall to all three polyurethanes after 5 h in boiling water. This improved performance is attributed to the enhanced hydrophobicity of the interphase region which is conferred by the replacement of most of the hydrophilic aminosilane with hydrophobic silane.

Table 3.
Bonding of urethanes to glass with a hydrophobic silane primer (peel strength, N/cm)

9:1 Blend	Rim			Thermoplastic			1-Component		
RSi(OMe)$_3$/aminosilane F	Dry	A	B	Dry	A	B	Dry	A	B
Unprimed	1.0	Nil	—	11.6	Nil	Nil	3.0	Nil	Nil
Aminosilane F alone	—	—	—	5.8	Nil	Nil	C	Nil	Nil
MeSi(OMe)$_3$	C	0.5	—	C	0.8	Nil	0.5	6.0	—
ViSi(OMe)$_3$	C	C	1.5	C	1.2	Nil	C	C	—
Cl(CH$_2$)$_3$Si(OMe)$_3$	C	C	2.5	C	25.0	C	0.8	0.7	—
PhSi(OMe)$_3$	C	C	C	C	19.0	17.0	C	C	C

RIM = Reaction injection-molded urethane, cure 15 min, 80°C. Thermoplastic = BF Goodrich Estane® 5701, fuse at 175°C. 1-Component = Dow Corning Urethane Bond®, cure 3 days, room temperature. A = Age in boiling water, 2 h; B = age in boiling water, 5 h; C = cohesive failure in polymer, >30 N/cm.

5.2. Crosslinking of the siloxane structure

Increased crosslinking of the siloxane structure in the interphase region can give increased bond strength and superior resistance to moisture. A series of potential crosslinking agents were evaluated as additives to conventional silane coupling agents [6]. The crosslinking agents are designed to react into the siloxane structure of the coupling agent to increase the number of siloxane bonds and make a tighter siloxane network. Significant improvement in the physical properties of composites and adhesion of polymeric coatings to glass and metals can be obtained. A hexamethoxysilane additive, 1,2-bis-(trimethoxysilyl)ethane, $(MeO)_3SiCH_2CH_2Si(OMe)_3$, was tested with aminosilane F and epoxysilane C in primers for bonding epoxy resins (three different curing agents) to glass (Table 4). It was possible to formulate primers that provided bonds that resisted 70°C water for more than 1 week while bonds without crosslinker failed in less than 1 h. Similar improvement was obtained with the crosslinker and

Table 4.

Effect of the crosslinkable adhesion enhancer (MeO)$_3$SiCH$_2$CH$_2$Si(OMe)$_3$ on the adhesion of epoxy films to glass

	DER 331 (Dow) liquid epoxy resin cured as indicated, h, 70°C water, to lose adhesion		
	Curing agent		
Primer on glass	Polyamide	DMP-30	DEH-24
None	< 1	< 1	< 1
Aminosilane F	1	1	1
1:9 Silane F–enhancer	C	C	C
Epoxysilane C	C	8	C
9:1 Silane C–enhancer	C	36	C
8:2 Silane C–enhancer	C	C	C

C = Cohesive failure in polymer after 1 week.

methacrylate silane D in bonding a crosslinkable ethylene vinyl acetate to titanium and steel (Table 5). Most experiments have shown that crosslinker levels of 10–50% with a conventional silane give a significant increase in adhesion compared with the silane alone.

The deposition of a mixture of the hexamethoxysilane crosslinker and methacrylate silane D onto glass from an aqueous solution was studied [7]. The crosslinker shows great selectivity for adsorption on the glass surface. A tight siloxane network of the crosslinker concentrates close to the glass surface and a more diffuse methacrylate structure is present away from the glass surface. This highly crosslinked siloxane structure close to the inorganic substrate supports the proposed mechanism for enhanced adhesion with this additive.

Table 5.

Effect of the crosslinkable adhesion enhancer (MeO)$_3$SiCH$_2$CH$_2$Si(OMe)$_3$ on the bond strength of crosslinkable ethylene vinyl acetate to titanium and steel

Primer on metal 10% in *i*-PrOH	Wet adhesion to metals (N/cm)	
	Titanium	Cold-rolled steel
None	Nil	Nil
Methacrylate silane D	0.25	7.0
Silane D + 10% enhancer	10.75	28.0 (cohesive failure)

90° peel strength after 2 h in 80°C water.

5.3. Thermal stability

Methods of upgrading the thermal stability of standard silanes were studied. Standard aliphatic silane coupling agents have adequate thermal stability for fabrication temperatures up to 250°C, but may decompose at higher temperatures. Surprisingly, tests for high-temperature stability showed the vinylbenzylaminosilane H of Table 1 to have remarkable thermal stability in isothermal weight loss tests in air at 300°C (see Fig. 5). Part of this thermal stability may be due to extra crosslinking obtained by polymerization of the vinylbenzyl groups at elevated temperatures. The most thermally stable silane, phenylsilane I of Table 1, gives the isothermal weight loss shown as the top line in Fig. 5. Thermogravimetric studies of phenylsilane blended with other less stable silanes showed that the thermal stability of properly blended silanes approaches the thermal stability of the phenylsilane [5].

The improved thermal oxidative stability of phenyltrimethoxysilane mixtures, as shown by TGA studies, translates to improved performance of fiberglass-reinforced high-temperature polymer composites. Polyimide laminates were prepared with S-glass fiber that was treated with a blend of 90% phenylsilane I and 10% aminosilane F and compared with treatment with aminosilane E alone (D. Vaughan, private communication, 1985). Table 6 shows superior retention of the flexural strength of the laminate containing the phenyl/aminosilane blend after aging for 2000 h at 260°C. Proprietary silane sizes for silicon carbide fiber have been formulated using these concepts to give polyimide composites with as much as a 50% increase in the initial flexural strength and with improved long-term flexural strength at 315°C (G. Stark, private communication, 1988) (see Table 7). Recommended commercial silane coupling agents for high-temperature polymers are either the vinylbenzylaminosilane H or a product based on combinations of phenylsilane I and one of the other silanes of Table 1.

Table 6.
Performance of thermally stable coupling agents in S-glass/polyimide laminates

Properties of laminate (MPa)	Coupling agents on glass	
	9:1 Phenyl I/amino F	Aminosilane E
Flexural strength, initial	544	476
1000 h at 260°C	409	258
2000 h at 260°C	306	134

Table 7.
Strength retention of polyimide (PMR-15)–silicon carbide fiber composites at 315°C

Silicon carbide fiber size	Flexural strength (MPa)			
	Initial	100 h	500 h	1000 h
Epoxy	1480	1320	1000	610
DCC-1	2350	2320	1660	890
DCC-2	2120	2060	1670	980

DCC-1 and DCC-2, proprietary silane sizes, Dow Corning Corporation.

5.4. Ionomer bond formation

The improvement in flexural strength imparted by silanes in thermoplastic composites is generally greater when specimens are compression-molded (low shear) than when they are injection-molded (high shear) (Table 8) [8]. Silanes that are used to treat fillers for injection-molded thermoplastics can give improvement over the control, but their performances are rather alike even though they differ markedly as adhesion promoters for polymers pressed against primed glass microscope slides [9]. Shear rates during injection molding can be up to 1000 times greater than those in compression molding [10]. Any structure building up in the interphase region is torn from the filler during this period of high shear.

An 'ionomer-bonded' silane is suggested for high shear applications. Commercial ionomers are thermoplastic polymers modified with up to 10% acrylic (or methacrylic) acid and neutralized to about 30% with sodium or zinc ion [11]. Compared with the parent polymer, ionomers have lower coefficients of thermal expansion, greater toughness, and greater tear resistance. At high temperatures and under high shear, the ionomer clusters are fluid, but they act as crosslinks when cooled to room temperature. In the 'ionomer bonding' system, an acid-functional silane is bonded to the reinforcement through stable siloxane bonds. The acid-functional silane bonds to an acid-functional polymer through an ionomer bond that is mobile at conditions of molding, but sets to a tough water-resistant bond at room temperature. When the matrix polymer contains no acid group, an acid-functional polymer that is compatible with the matrix may be included in formulating the ion-modified silane primer. A compatible acid-functional polymer may also be added in small proportion to the matrix polymer and allowed to react with the treated reinforcement during the molding operation [12].

Heat-cleaned fiberglass cloth was treated with 0.5% carboxysilane J (Table 1) and compression-molded into a laminate with nylon 6,6 polymer. Laminates were compared to state-of-the-art silane H. Table 9 shows that silane H provided a significant improvement in flexural strength over the control, expecially after a 2 h water boil, but a carboxysilane/zinc ion ionomer system gave an even better strength improvement.

Table 8.
Improvement imparted by silanes in fiberglass–thermoplastic composites

Polymer–silane system	% Flexural strength improvement			
	Compression-molded		Injection-molded	
	Dry	Wet[a]	Dry	Wet[a]
Nylon–aminosilane E	55	115	40	36
Nylon–vinylbenzylaminosilane H	85	133	40	45
PBT–aminosilane E	21	–	23	24
PBT–vinylbenzylaminosilane H	60	47	28	11
Polypropylene–silane E	8	18	7	10
Polypropylene–silane H	86	89	16	16

[a] Wet = after 2 h in boiling water.

Table 9.

Strengths of compression-molded fiberglass–nylon composites

Coupling agent on glass	Flexural strength (MPa)	
	Dry	After 2 h H$_2$O boil
None	179	73
Silane H	236	141
Carboxysilane J	173	122
Carboxysilane J + Zn^{2+} [a]	241	186

Six layers heat-cleaned fiberglass, 0.5% silane; nylon 6,6.
[a] Carboxysilane, 0.25 Zn^{2+} per free carboxy.

This concept was tested in injection-molded polypropylene composites. E-glass microbeads were treated with 1% of three silanes: carboxysilane J, vinyl-benzylsilane H, and aminosilane F [13]. Carboxylated polypropylene (10%) was added to virgin polypropylene to give an ionomer bond to the carboxysilane in the presence of zinc ions. Table 10 shows that the ionomer bond gave an improvement over the non-ionomer silane H and similar to an aminosilane/carboxypolypropylene system. More testing is being done with the ionomer bonding concept to optimize the ratio of Zn^{2+} to carboxyl groups in the primer for systems were an unknown concentration of carboxyl is supplied by the polymer.

Table 10.

Effect of silane treatment on the physical properties of polypropylene/E-glass bead composites

Silane system	% Difference vs. control			
	Tensile strength	Flexural strength	Izod	Charpy
No silane/GPPP	Control	Control	Control	Control
Silane H/GPPP	29.9	7.3	13.1	15.1
Carboxysilane/Zn^{2+}/carboxy PP	40.8	9.7	− 5.6	21.1
Carboxysilane/carboxy PP	20.7	1.5	13.8	29.3
Silane D/carboxy PP	44.7	16.7	− 1.3	29.3

1% silane on Potters E3000 glass beads. GPPP = General Purpose Polypropylene Himont Profax 6523; Carboxy PP = 10% BP Polybond 1001/90% Profax 6523; Notched Izod and Charpy impact strengths.

6. CONCLUSIONS

Factors which affect the performance of a silane coupling agent in the composite interphase region include chemical reactivity, hydrophobic character, siloxane crosslink network, ionomer bonding, and thermal stability. Proper choice of a single silane coupling agent, combinations of coupling agents, or modification of the coupling agent network with additives can give significant improvements in strength properties. These concepts can result in greater resistance of the interphase region to attack of moisture and greater resistance to thermal degradation. Improvements are possible by simple modification of standard commercial silane coupling agents. Some of the coupling agent systems described here are finding commercial acceptance.

REFERENCES

1. P. G. Pape and E. P. Plueddemann, in: *History of Polymeric Composites*, R. B. Seymour and R. D. Deanin (Eds), pp. 103–139. VNU Science Press, Utrecht, The Netherlands (1987).
2. E. P. Plueddemann, *Silane Coupling Agents*. Plenum Press, New York (1982).
3. E. P. Plueddemann, 22nd Reinforced Plastics/Composites Conf., SPI, Paper 9-A (1967).
4. G. C. Tesoro, G. P. Rajendran, C. Park and D. R. Uhlmann, *J. Adhesion Sci. Technol.* **1**, 39–51 (1987).
5. E. P. Plueddemann and P. G. Pape, 40th Reinforced Plast./Composites Conf., SPI, Paper 17-F (1985).
6. E. P. Plueddemann and P. G. Pape, 42nd Reinforced Plast./Composites Conf., SPI, Paper 21-E (1987).
7. J. Jang, H. Ishida and E. P. Plueddemann, 44th Reinforced Plast./Composites Conf., SPI, Paper 9-B (1989).
8. E. P. Plueddemann, in: *Additives for Plastics*, R. B. Seymour (Ed.), Vol. 1, p. 140. Academic Press, New York (1978).
9. W. T. Collins and J. L. Kludt, 30th Reinforced Plast./Composites Conf., SPI, Paper 7-D (1975).
10. O. J. Onufer and E. C. Staley, *Plast. Des. Proc.* **7**, 9 (1976).
11. R. Longworth, in: *Ionic Polymers*, L. Holliday (Ed.), Ch. 2. John Wiley, New York (1987).
12. E. P. Plueddemann, *J. Adhesion Sci. Technol.* **3**, 131 (1989).
13. G. Smith, *Proc. Soc. Plast. Eng., ANTEC '90 Meet.*, pp. 1946–1948 (1990).

Part 2

Mechanisms of Silane Coupling Agents
and Interfacial Studies

Silanes and Other Coupling Agents, pp. 119–141
Ed. K. L. Mittal
© VSP 1992.

Kinetics of the hydrolysis and condensation of organofunctional alkoxysilanes: a review

F. D. OSTERHOLTZ and E. R. POHL*

Union Carbide Chemicals and Plastics Company Inc., Old Saw Mill River Road, Tarrytown, NY 10591, USA

Revised version received 6 November 1991

Abstract—A review of the literature is presented for the hydrolysis of alkoxysilane esters and for the condensation of silanols in solution or with surfaces. Studies using mono-, di-, and trifunctional silane esters and silanols with different alkyl substituents are used to discuss the steric and electronic effects of alkyl substitution on the reaction rates and kinetics. The influences of acids, bases, pH, solvent, and temperature on the reaction kinetics are examined. Using these rate data, Taft equations and Brønsted plots are constructed and then used to discuss the mechanisms for acid and base-catalyzed hydrolysis of silane esters and condensation of silanols. Practical implications for using organofunctional silane esters and silanols in industrial applications are presented.

Keywords: Silane; ester; silanol; hydrolysis; condensation; surfaces; kinetics; mechanisms; catalysis; steric; polar; functionality; Taft; Brønsted.

1. INTRODUCTION

The commercial use of organic silane derivatives, functionalized on the organic group and bearing hydrolyzable groups (e.g. alkoxy groups) on silicon, has developed steadily since the 1950's [1]. The major uses may be categorized as 'coupling agents' and 'crosslinkers'.

To define a molecule as a 'coupling agent' provides an operational definition, not a structural definition. Coupling agents are chemicals which promote adhesion between mineral phases and organic phases, typically in mineral-filled or glass-reinforced polymer composites. They are also commonly used in adhesives, sealants, and adhesion-promoting primers for coatings, adhesives, and sealants. Their most prominent benefit is to maintain bond strength or composite strength during or after exposure to wet environments. Organofunctional alkyltrialkoxysilanes are the chemicals most often successfully used for this function.

'Crosslinkers' are chemicals used to form chemical bonds (usually covalent) between polymer chains. The macroscopic results include reduced flow of polymers at high temperature and the creation of insoluble networks. There are many chemistries used for polymer crosslinking, selected according to polymer type and end-use needs. The large-scale commercial availability of organofunctional silanes with a variety of reactive functional groups has led to the development of new crosslinking technologies for organic materials [2, 3].

Understanding of the best ways to use alkoxysilanes as coupling agents, adhesion promoters, or crosslinkers has developed empirically. As the scale of

---

*To whom correspondence should be addressed.

use expands, quantitative understanding of the characteristic hydrolysis and condensation reactions becomes (at least) useful and, very often, essential. Quantitative understanding of reaction rates allows rational design of equipment and processes for use of the materials.

Several authors have contributed to this understanding by exploring the most significant reaction processes [1, 4–17]. Often, the lack of suitable analytical techniques made these earlier studies both tedious and less informative than desired. Many of these studies have relied on the isolation of stable products to help infer intermediate mechanisms [18–24].

In the last decade, instrumental techniques have advanced to the point where many of these limitations can be removed. Quantitative solution-phase studies, making use of ^{13}C- and ^{29}Si-NMR on high-field instruments, can now include direct measurement of low concentrations of multifunctional intermediates on a time scale appropriate for kinetic studies [25–31]. Studies of reactions with surfaces at submonolayer concentrations are now becoming feasible [32]. Both solid-state and liquid-state reactions can be measured simultaneously [32].

In this review, we will briefly discuss selected literature on hydrolysis and condensation reactions at silicon which are relevant to organofunctional silanes. Some studies will involve simpler (lower functionality) models; some may involve related chemical reaction classes illustrating a mechanistic point. We will then discuss selected references focused on silanes of the types used as coupling agents and crosslinkers. Finally, some reactions at the interface will be mentioned.

1.1. Key reactions

Organic silane derivatives with hydrolyzable groups on silicon (—OR, —Cl, —OC(O)R, etc.) are usually derived from a chlorosilane. Before or after formation of a carbon–silicon bond by hydrosilylation, the chlorosilanes are commonly converted to alkoxysilanes:

$$R'SiCl_3 + excess\ ROH \rightleftharpoons R'Si(OR)_3 + 3\ HCl. \qquad (1)$$

This reaction is called 'esterification' and the alkoxysilanes produced are called 'esters', due to the close resemblance to silicic acid esters.

The two most commercially important reactions of the trialkoxysilanes are *hydrolysis* to silanetriols:

$$R'Si(OR)_3 + excess\ H_2O \rightleftharpoons R'Si(OH)_3 + 3\ ROH, \qquad (2)$$

and *condensation* of the silanetriols to siloxanes (reaction 3):

$$R'Si(OH)_3 + R'Si(OH)_3 \rightleftharpoons R'Si(OH)_2OSi(OH)_2R' + H_2O. \qquad (3)$$

Alcoholysis or reactions with other nucleophiles may also occur in use, and condensation may take place between two silanol molecules [reaction (3)] or between silanol and the solid substrate [reaction (4)].

$$R'Si(OH)_3 + surface-SiOH \rightleftharpoons R'Si(OH)_2OSi-surface + H_2O. \qquad (4)$$

All of these reactions are reversible, and many are equilibria with substantial concentrations of both products and reactants present under typical conditions of use [19, 27].

1.2. Reaction mechanism studies

Hydrolysis and condensation reactions of silanes may be considered in the broad category of nucleophilic substitutions at silicon. The common nomenclature for these reactions is $S_N X$-Si, where 'X' represents the kinetic order or molecularity, 'Si' indicates that silicon is the reaction center, and 'S_N' indicates that the reaction is a nucleophilic substitution. Nucleophilic reactions at silicon have been reviewed thoroughly and have been the subject of fundamental studies by several laboratories over the last three decades [33]. The literature is not as voluminous as the literature on the corresponding reactions at carbon. A general mechanistic view of these reactions has, however, emerged. There are many parallels to carbon-centered reaction mechanisms. One distinction from carbon-centered reactions is clearly apparent. Silicon is able to form relatively stable higher coordinated (pentavalent) intermediates; carbon is not [33].

2. FUNDAMENTAL STUDIES OF MECHANISM—HYDROLYSIS

Studies of the mechanisms of hydrolysis have explored the possible reaction pathways, including the possibility of higher coordinated intermediates. Substituent effect studies and other physical–organic chemical probes have been employed to better define the transition states and intermediates in these reactions [29, 34–43]. Often, to simplify the kinetic analyses, monofunctional reagents are employed and co-solvents are used to increase the solubility of the silanes in order to form homogeneous solutions. Good mechanistic insights can be obtained from studies of monofunctional materials because polycondensation products do not complicate the studies [34–40]. Many older studies of hydrolysis at relatively high concentrations of di- or trifunctional material routinely gave condensation products as the first isolable products [15, 16, 20].

One approach to this problem has been to characterize the practical consequences of silane hydrolysis. Visual observation of the hydrolysis behavior of typical organofunctional silanes, supplemented by some spectroscopic data, and trapping of silanols with trimethylsilanol were reported by Plueddemann [1, 14]. Comparative data give some measures of the ease of hydrolysis and the solution stability. The data are quite helpful in the practical use of hydrolyzed silane solutions. They are not presented in a way that allows quantitative kinetic conclusions.

One of the first homogeneous phase hydrolysis studies of a trifunctional silane under kinetically 'clean' conditions was carried out in Pratt's laboratory [43]. Using dilute aqueous solutions (0.1–0.5 mM or 0.03–0.17 wt%) of tris-(2-methoxyethoxy)phenylsilane, Pratt's group was able to measure the extent of hydrolysis by the change in optical absorption at 220 nm when the alkoxysilane hydrolyzed. Over a wide pH range (approximately pH 1–pH 13), the observed rate constants fit a V-shaped profile with a minimum at pH between 6 and 7 (Fig. 1). Specific acid and general base catalysis were observed. At pH greater than approximately 10, hydrolysis of the isolable first intermediate, $C_6H_5Si(OR)_2OH$, was inhibited, likely due to ionization of the SiOH group. At lower pH, the hydrolysis occurred in steps, with each succeeding step (alkoxy removal) faster than the first. Pratt and co-workers attributed these observations to relief of steric hindrance at silicon [43].

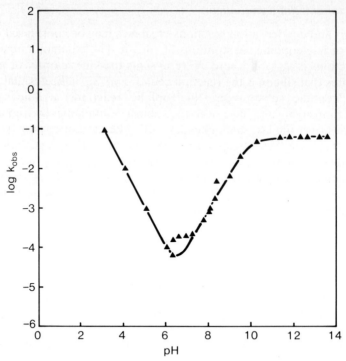

Figure 1. A pH profile for the hydrolysis of phenyl bis-(2-methoxyethoxy)silanol in aqueous solution at 25°C.

From the work of Pratt and co-workers and analysis of the available literature, a reasonable model for hydrolysis mechanisms can be developed. The studies of the hydrolysis of mono-, di-, and trifunctional silane esters in aqueous and aqueous–organic solutions lead to a kinetic expression describing the rate of silane ester disappearance (hydrolysis):

$$\frac{-d[S]}{dt} = k_{spon}[H_2O]^n[S] + k_H[H^+][H_2O]^m[S]$$
$$+ k_{HO}[HO^-]^o[H_2O]^p[S] + k_B[B][H_2O]^q[S].$$

The symbol S represents the silane ester and B represents any basic species other than hydroxide anion. There appears to be no need for a term containing HA, the non-ionized acid species in solution. The following sections discuss what is known about each of the terms in the kinetic equation.

2.1. Order in water—hydrolysis

The kinetic order in water for the spontaneous hydrolysis reaction, n, and the hydronium ion catalyzed reaction, m, varies depending on the structure of the silane ester and the solvent conditions [36, 40]. The difficulty in determining the kinetic order of water in aqueous–organic solvents arises from the observation that as the concentration of water is varied, the polarity of the solvent and the activity of the acid change [40]. A plot of the logarithm of the rate constant vs. the logarithm of water concentration often does not yield a straight line. These types of plots are often used to determine the kinetic order. The slope of the line is the kinetic order of water in the reaction. For example, in one study the slope of the plot of the logarithm of the rate of hydronium ion catalyzed hydrolysis vs. the logarithm of water concentration was negative at low water concentrations (below 10% aqueous dioxane) and positive at high concentrations (above 20% aqueous dioxane) [40].

The solvent also affects the order of the reaction. For example, the order in water for the spontaneous hydrolysis varies from $n = 0.8$ in aqueous propan-2-ol to $n = 4.4$ in 30–40% aqueous dioxane. For the hydronium ion catalyzed hydrolysis, the order of water varies from $m = 0$ in aqueous propan-2-ol to $m = 2.6$ in 30–40% aqueous dioxane [36, 40]. More water molecules are required for the spontaneous hydrolysis reaction than for the hydronium ion catalyzed reaction. The order in water for the hydroxide anion catalyzed hydrolysis term p and for the general base catalyzed hydrolysis term q has not been determined [36–39, 41–43].

2.2. Orders in hydroxide anion and hydronium ion—hydrolysis

The orders in hydroxide anion and hydronium ion are generally found to be equal to 1 [36–39, 41–43]. However, Swain *et al.* reported that hydroxide anion catalyzed hydrolysis of triphenylmethoxysilane in 40% aqueous acetonitrile is second order in hydroxide anion below 3×10^{-3} M [44]. It becomes first order above that concentration. They attributed the change in order to a change in the rate-determining step of the reaction. Eaborn has suggested that the apparent change in order of the hydroxide anion is due to acidic impurities in the acetonitrile, and that the reactions are first order in hydroxide [33].

2.3. Effects of leaving groups and substituents—hydrolysis

A critical concern in the use of organofunctional silanes is the effects that the organofunctional groups and the silane ester groups have on the hydrolysis reactions occurring at silicon.

A common probe of reaction mechanisms used to infer charge distribution in the transition state involves variation of substituent groups near the reaction center. From the variation in reaction rate produced by electron-donating and electron-withdrawing groups or by the steric hindrance of various sized groups, transition state characteristics can be inferred. Two empirical correlations have been proposed and refined which provide a common framework for this process. The Hammett equation is applied to aromatic systems [45]. The Taft correlation is applied to aliphatic systems [45]. Definitions of terms, collections of substituent constants (steric and electronic effects for various substituents), and listings of observed reaction response parameters (for typical reaction types) have been collected [45].

Several studies using these tools have examined the effects of the silane ester structure on the hydrolysis rates for reactions catalyzed by hydronium ion, hydroxide anion, and 'general' bases [35, 36, 38, 41–43]. In the following sections, we will discuss how the structure of the alkoxysilane impacted on the hydrolysis rate for each type of catalysis and, using these data, speculate about the possible reaction mechanisms.

2.3.1. Spontaneous reaction rates—hydrolysis. Although the spontaneous hydrolysis of alkoxysilanes has been reported, there are insufficient data to develop an understanding of how structure impacts on rate or to give insights into the reaction mechanisms [35–37, 40]. The spontaneous hydrolysis usually contributes only a small amount to the observed rate of hydrolysis under most conditions. In many instances, the contribution of the spontaneous hydrolysis is so small that it is not detectable [25, 39, 42, 43]. The impact of the silane structure on the spontaneous hydrolysis of alkoxysilanes will therefore not be discussed.

One study of related silanes addressing structural factors involves displacement of hydride rather than alkoxide. A series of papers by Saratov *et al.* have examined the effect of *omega* amino substitution on SiH displacement by nucleophiles [46–48]. These authors proposed that there is an effect of structure due to intramolecular interaction (coordination) between the silicon reaction site and nitrogen located along the chain. They have discussed reactions with alkyl or aralkyl alcohols [46]. The most notable rate enhancement was for structures with nitrogen on the carbon attached to silicon. With three or five methylene links between nitrogen and silicon, only a modest rate difference was observed [46]. Interestingly, in this alcoholysis reaction, the first substitution of the RO— for H— at silicon was 3–37 times faster than the second displacement. The work was extended to phenols and acids [47]. Hammett and Taft parameter correlations were constructed and the effect of solvents was studied [48]. The observation that electron-donating solvents inhibit the reaction was explained by postulating solvent interference with intramolecular coordination at silicon by nitrogen. An $S_N i$-Si mechanism was proposed [48].

2.3.2. Base catalysis—hydrolysis. Pohl studied the hydrolysis in aqueous solutions of a series of trialkoxysilanes of $R'Si(OCH_2CH_2OCH_3)_3$ structures in which R' was an alkyl or a substituted alkyl group [42]. The reactions were followed using an extraction/quenching technique. Silanes were studied at concentrations ranging from 0.001 to 0.03 M and pHs adjusted from 7 to 9. The hydrolysis was found to be first order in silane. The order in water was not determined because the reactions were carried out in a large excess of water (water was the solvent). The rate constants for the hydroxide anion catalyzed hydrolysis reactions and reaction half-lives are reported in Table 1.

Table 1.
Rate constants for the hydroxide anion catalyzed hydrolysis of alkyl tris-(2-methoxyethoxy)silanes in aqueous solution at 25°C and ionic strength = 1.0 [42]

R' in $R'Si(OCH_2CH_2OCH_3)_3$	pH range	k_{HO} (M^{-1} s^{-1})	$t_{1/2}$ at pH 9.0 (min)
ClCH$_2$	6.99–7.51	3.2×10^4	0.04
CH$_2$=CH	7.00, 7.50	6.5×10^2	1.8
CH$_3$	7.99–8.47	1.7×10^2	6.8
C$_6$H$_5$[a]	8.00–12.0	8.0×10^1	14
CH$_3$CH$_2$	8.41–8.46	4.6×10^1	25
CH$_3$CH$_2$CH$_2$	7.47, 8.36	2.0×10^1	58
Cyclo-C$_6$H$_{11}$	8.96–9.14	3.4	340

[a] Value estimated from ref. 43.

The hydroxide anion catalyzed rate constants for the series of alkyl tris-(2-methoxyethoxy)silanes obtained by Pohl were used to define a modified Taft equation, $\log(k_{HO}/k_{HO_0}) = 2.48\sigma^* + 1.67 E_s$ [42]. A good correlation was obtained, except for vinyl and phenyl substituents. The anomalous behavior observed for phenyl and vinyl tris-(2-methoxyethoxy)silanes may have resulted because the steric parameter or the polar parameter may be influenced by the carbon–carbon double bond. The steric parameter for α,β-unsaturated substituents may include an appreciable resonance effect. The polar parameter values may be influenced by the ability of silicon to back-bond through d orbitals with the α,β-unsaturated system [37, 49].

In a similar fashion, Shirai *et al.* constructed a modified Taft correlation for the hydrolysis of a series of trialkylalkoxysilanes of R_3SiOR' structures (in which R was varied to include different alkyl groups) in 55% aqueous acetone using a similar extraction/quenching technique [35]. The resulting modified Taft equation was $\log(k_{HO}/k_{HO_0}) = 1.50\sum\sigma^* + 2.72\sum E_s$. Although the structures of the silane esters and the reaction conditions were very different for the Pohl and Shirai studies, the values of ρ^* and s were quite similar. (ρ^* and σ^* are empirically derived coefficients for the polar parameter and the steric parameter, respectively.)

The values of ρ^* and s are often used to qualitatively describe the changes in structure and charge distribution as the reaction proceeds along the reaction coordinate from reactants to transition state [45]. The moderately large ρ^* ($\rho^* = 2.48$ for the alkyltrialkoxysilane series and $\rho^* = 1.50$ for the trialkylalkoxy-silane series) suggests that the transition state is stabilized more than the reactants by electron-withdrawing groups attached to silicon. This implies considerable negative charge development on or in the vicinity of the silicon at the transition state. The moderately large s ($s = 1.67$ for the alkyltrialkoxysilane series and $s = 2.72$ for the trialkylalkoxysilane series) suggests that the transition state is more sterically crowded than the reactants [50].

These data are consistent with a bimolecular nucleophilic displacement reaction (S_N2^{**}-Si or S_N2^*-Si) with a pentacoordinate intermediate; S_N2^{**}-Si is rate-determining formation ($k_{-1} < k_2$) and S_N2^*-Si is rate-determining breakdown ($k_1 > k_2$) of the pentacoordinate intermediate, as shown in Fig. 2 [42, 43]. In the proposed mechanism, negative charge development on silicon in the transition state (TS 1 or TS 2) is considerable. Electron-withdrawing substitution on the alkyl groups will stabilize the developing negative charge and should lower the energy of the transition state. A significant increase in the observed rate of the hydrolysis should then occur as the alkyl group becomes more electron-withdrawing. Experimentally, the hydrolysis of chloromethyl tris-(2-methoxyethoxy)silane is 1600 times faster than the hydrolysis of the n-propyl substituted silane. Chloromethyl and n-propyl have similar steric bulk as measured by the Taft steric parameter [45]. A large, positive ρ^* is consistent with the mechanism proposed in Fig. 2.

The value of the steric coefficient s also suggests a bimolecular displacement reaction with a pentacoordinate intermediate. In this mechanism, the sp^3 hybridized silicon is rehybridized to an sp^3d-like transition state. If the transition state TS 1 is the highest energy point along the reaction coordinate, there should be considerable bond formation between the incoming hydroxide anion and the

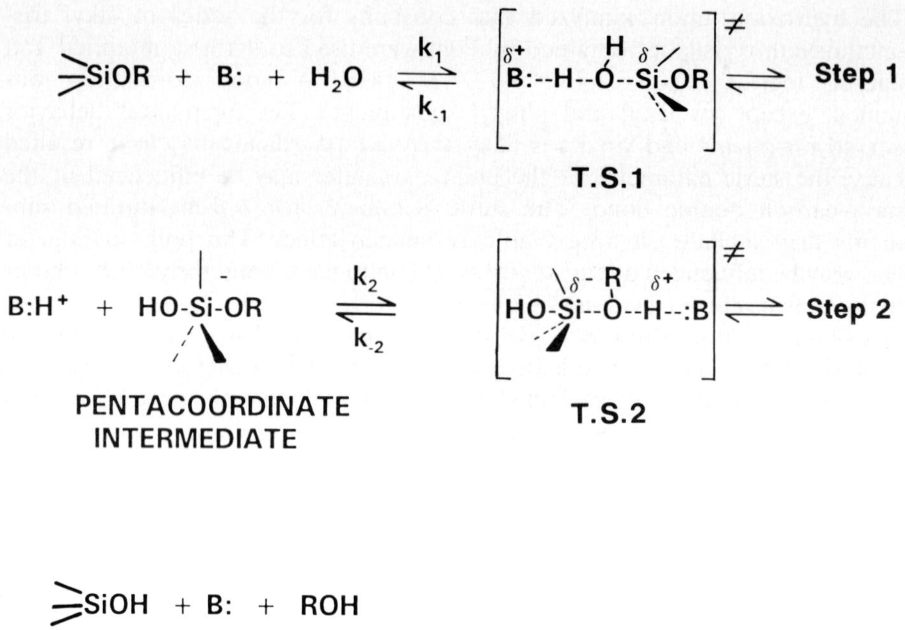

$$\equiv\!\!\text{SiOR} + \text{B:} + \text{H}_2\text{O} \underset{k_{-1}}{\overset{k_1}{\rightleftharpoons}} \left[\overset{\delta+}{\text{B:--H--}}\overset{H}{\underset{}{\text{O}}}\overset{\delta-}{\text{--}}\underset{}{\text{Si-OR}} \right]^{\neq} \rightleftharpoons \quad \textbf{Step 1}$$

T.S.1

$$\text{B:H}^+ + \overset{|}{\underset{\diagdown}{\text{HO-Si-OR}}}^- \underset{k_{-2}}{\overset{k_2}{\rightleftharpoons}} \left[\text{HO-}\overset{\diagdown\delta-}{\underset{\diagup}{\text{Si}}}\text{--}\overset{R}{\underset{}{\text{O}}}\text{--H--}\overset{\delta+}{:}\text{B} \right]^{\neq} \rightleftharpoons \quad \textbf{Step 2}$$

PENTACOORDINATE
INTERMEDIATE   **T.S.2**

$$\equiv\!\!\text{SiOH} + \text{B:} + \text{ROH}$$

Figure 2. Proposed mechanism for the base catalyzed hydrolysis of silane esters.

silicon. Also, the bond order between the silicon and oxygen of the leaving group remains essentially unchanged at 1. The high negative charge density on silicon and on the oxygen of the nucleophile should produce a tight transition state structure [51]. As the number of substituents surrounding the central atom increases, so should the steric crowding. The mechanism described in Fig. 2 requires that the crowding around the silicon increase significantly as the reaction proceeds from reactants to transition state and the data support this mechanism. Similar arguments can be made if the transition state TS 2 is the highest energy point along the reaction coordinate.

An alternative S_N2-Si mechanism may be considered, without a penta-coordinate intermediate. In this alternative, the silicon is rehybridized from sp^3 to sp^2. If bond order is conserved, then it is reasonable to ascribe a bond order of 0.5 to the Si—O bond of both the entering nucleophile and the leaving group [52]. It has been shown that the bond order is related to the bond distance [45]. Even though there are five substituents in the vicinity of the silicon in an S_N2-Si type mechanism, two of the substituents are significantly further away from the central atom (silicon) than the other three substituents. A looser transition state structure than that for an S_N2^{**}-Si or an S_N2^*-Si process results. The steric effects of the alkyl group bonded to silicon should, therefore, be considerably less in an S_N2-Si type mechanism.

The relatively large value of 1.67 for s seems to fit an S_N2^{**}-Si or S_N2^*-Si mechanism better rather than an S_N2-Si type mechanism. In addition, S_N2-type mechanisms generally give a small negative ρ^* value. For example, solvolysis of primary alkyl p-toluenesulfonates in ethanol has a p^* value of -0.746 [53]. The large positive ρ^* for base catalyzed hydrolysis is inconsistent with a simple S_N2-type mechanism.

Further insights into the reaction mechanism can be obtained by studying the effects of the silane ester (leaving) group on the hydroxide catalyzed hydrolysis. Akerman has studied the hydroxide anion catalyzed hydrolysis of substituted triethylphenoxysilanes in 48.6% aqueous ethanol [38]. Humffray and Ryan have investigated a similar set of silane esters in 40% aqueous dioxane [36].

An approach to analyzing the effects of the silane ester group on the reaction rate is to plot the logarithm of the rate constant of the hydroxide anion catalyzed reaction vs. the logarithm of the dissociation constant of the conjugate acid of the alkoxide or phenoxide leaving group. This type of plot is referred to as a Brønsted plot. Brønsted plots have been used as a measure of the amount of bond breaking or bond formation that is occurring in the transition state [54].

The Brønsted plots for the hydroxide catalyzed hydrolysis of the two series of triethylphenoxysilanes studied by Akerman [38] and Humffray and Ryan [36] in aqueous ethanol and aqueous dioxane are shown in Fig. 3. The impact of changing the substituted phenoxy (leaving) groups of the silane ester on the hydrolysis is measured by the slope of the Brønsted plot. The slopes, β_{1g}, for the hydrolysis in aqueous ethanol and aqueous dioxane are -0.68 and -0.65, respectively. The negative slopes indicate that there is considerable negative charge formation occurring in the transition state. If the reaction is proceeding by rate-determining formation (S_N2^{**}-Si mechanism) of a pentacoordinate intermediate, then the effective charge that is 'seen' by the polar substituent on the

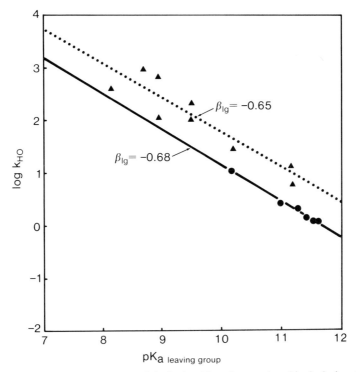

Figure 3. Brønsted plot of the logarithm of the hydroxide anion catalyzed hydrolysis rate constants vs. the pK_a of the conjugate acid of the leaving group for a series of substituted aryloxytriethylsilanes in 40% water/dioxane at 30°C (▲) and in 48.6% water/ethanol at 25°C (●).

leaving group is relatively large. Because the hydroxide anion is a stronger base (and therefore a poorer leaving group) than the substituted phenoxy anion leaving groups, it is reasonable to favor the rate-determining formation rather than rate-determining breakdown of the pentacoordinate intermediate. The large effective charge that is seen by the leaving groups appears to be consistent with the large positive ρ^* found for the modified Taft equations. In the analogous carbon system, the slope β_{1g} of the Brønsted plot for the acyl transfer between thiol anion and a series of acetate esters is equal to -0.33 for rate-determining attack and approximately -0.9 for rate-determining breakdown [54].

The hydrolysis of silane esters is subject to general base catalysis. In 'general base' catalysis, any basic species accelerates the reaction by assisting the removal of a proton from water in the transition state. In 'specific' base catalysis, only the hydroxide anion accelerates the reaction rate by directly attacking the substrate. Slebocka-Tilk and Brown found that the hydrolysis of t-butyldimethyl-3-nitrophenoxysilane in 70% aqueous dioxane is accelerated by oxyanions which are 'general' bases [41]. The Brønsted plot of the logarithm of the general base rate constant vs. the logarithm of the dissociation constant of the conjugate acid of the oxyanions yielded a slope of $\beta_B = 0.45$. Pratt and co-workers observed a similar enhancement in the rate of hydrolysis of phenyl tris-(2-methoxyethoxy)silane and phenyl bis-(2-methoxyethoxy)silanol in the presence of a series of oxygen and nitrogen bases [43]. The rates from oxygen and nitrogen bases fell on the same Brønsted line, as shown in Fig. 4. The β_B slopes of the line for phenyl tris-(2-methoxyethoxy)silane and phenyl bis-(2-methoxy-ethoxy)silanol are 0.67 and 0.65, respectively. The point for hydroxide anion, in both cases, is below the line. The larger β_B slopes and the poorer correlation of hydroxide anion for the phenylsilanes than for the sterically hindered alkylsilane, t-butyldimethyl-3-nitrophenoxysilane, may be attributed to a solvation effect [55]. The observation that rates from the oxygen and nitrogen bases fell on the same Brønsted line suggests that the catalysis is the general base type, as shown in Fig. 2, rather than the kinetically equivalent nucleophilic catalysis [43]. Pratt has argued that the mechanism for hydroxide anion catalysis proceeds through a pentacoordinate intermediate rather than an S_N2-Si type mechanism. If the hydrolysis proceeds by an S_N2-Si type mechanism, then the reaction should be both general acid and general base catalyzed. General acid catalysis has not been observed [43].

2.3.3. Acid catalysis—hydrolysis. Several series of alkylsilane esters were studied to determine the effect of silane structure on the hydronium ion catalyzed hydrolysis reaction. The hydronium ion catalyzed hydrolysis rate constants for a series of alkyl tris-(2-methoxyethoxy)silanes in aqueous solution were used to define the modified Taft equation $\log(k_H/k_{Ho}) = 0.39\sigma^* + 1.06E_s$, where k_{Ho} is the rate of hydrolysis for methyl tris-2-(methoxyethoxy)silane [42]. The hydronium ion catalyzed hydrolysis rate constants and the reaction half-lives are reported in Table 2. In a similar manner, the hydronium ion catalyzed hydrolysis rate constants for a series of trialkylalkoxysilanes in 55% aqueous acetone were used to obtain the modified Taft equation $\log(k_H/k_{Ho}) = -0.37\sum\sigma^* + 2.48\sum E_s$, where k_{Ho} is the rate of hydrolysis for trimethylalkoxy-silane.

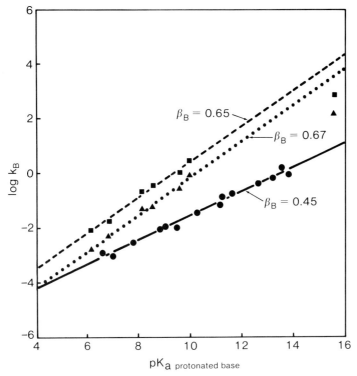

Figure 4. Brønsted plot of the logarithm of the general base catalyzed hydrolysis rate constants vs. the pK_a of the conjugate acids of the oxygen and nitrogen bases (buffers) for phenyl tris-(2-methoxyethoxy)silane (▲) and phenyl bis-(2-methoxyethoxy)silanol (■) in aqueous solution at 30°C, and for tert-butyldimethyl-3-nitrophenoxysilane (●) in 70% water/dioxane at 37°C.

Table 2.
Rate constants for the hydronium ion catalyzed hydrolysis of alkyltrialkoxysilanes in aqueous solution at 25°C and ionic strength = 1.0 [42]

R' in R'Si(OCH$_2$CH$_2$OCH$_3$)$_3$	pH range	k_H (M^{-1} s^{-1})	$t_{1/2}$ at pH 5.0 (min)
CH$_2$=CH	4.48–6.07	8.8 × 10^1	13
ClCH$_2$	4.98	7.8 × 10^1	15
CH$_3$	4.30–5.04	3.7 × 10^1	21
CH$_3$CH$_2$	4.30–5.04	3.7 × 10^1	31
C$_6$H$_5$[a]	2.00–6.00	3.5 × 10^1	33
ClCH$_2$CH$_2$	4.30–4.96	2.5 × 10^1	46
CH$_3$CH$_2$CH$_2$	4.30–4.90	2.1 × 10^1	55
Cyclo-C$_6$H$_{11}$	3.99–5.01	7.0	165
R in CH$_3$O(OCH$_2$CH$_2$O)$_3$CH$_2$CH$_2$CH$_2$Si(OR)$_3$	pH range	k_H (M^{-1} s^{-1})	$t_{1/2}$ at pH 5.0 (min)
CH$_3$	4.92	2.7 × 10^2	4.3
CH$_3$CH$_2$	5.01	5.0 × 10^1	23
CH$_3$CH$_2$CH$_2$	4.00–4.98	2.8 × 10^1	41

[a] Value estimated from ref. 43.

The values of ρ^* and s suggest that the mechanism for acid catalyzed hydrolysis is a rapid equilibrium protonation of the substrate, followed by a bimolecular S_N2-type displacement of the leaving group by water, as shown in Fig. 5 [5, 43]. The small positive value of ρ^* for the series of alkyl tris-(2-methoxyethoxy)silanes suggests a small increase in the negative charge on silicon in the transition state [45]. The small negative value of ρ^* for the series of trialkylalkoxysilanes suggests a small increase in the positive charge on silicon in the transition state [45]. In S_N2-type displacements at carbon, there is, generally, a small positive charge development on carbon [53]. The small negative ρ^* value for the series of trialkylalkoxysilanes is inconsistent with the formation of a siliconium ion in an A-1 type mechanism. Acid catalyzed hydrolysis of formals, which proceeds through a carbocation, has a ρ^* of -4.173 [56]. It is therefore reasonable that an S_N2-type displacement at silicon proceeds with a small negative or positive charge development on silicon in the transition state.

The smaller s values for the acid catalyzed reaction imply a less crowded transition state than that for the base catalyzed reaction. Several factors may influence the sensitivity of the steric parameters, such as solvation. The smaller values of s are at least consistent with a change in mechanism from a bimolecular displacement with a pentacoordinate intermediate to a mechanism that has more S_N2-type character [43]. Although there are still five substituents around silicon in an S_N2-Si type mechanism, the incoming nucleophile and outgoing leaving group are further from the sp^2 hybridized silicon atom than the nucleophile and leaving group in an sp^3d hybridized silicon atom of a pentacoordinate inter-mediate.

The effect of the leaving group on the acid catalyzed hydrolysis was also measured. The relative rates of hydrolysis of the alkyltrimethoxy-, triethoxy-, and

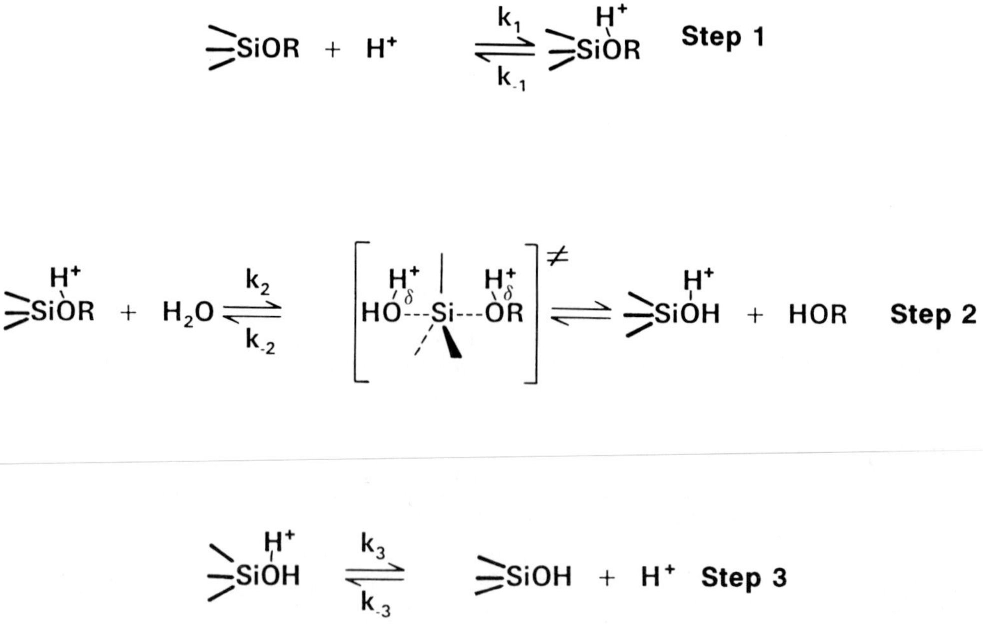

Figure 5. Proposed mechanism for the hydronium ion catalyzed hydrolysis of silane esters.

tris-(2-methoxyethoxy)silanes are 1.0, 0.19, and 0.10, respectively [42]. Chipperfield and Gould studied a series of substituted phenoxytriethylsilane esters in 48.6% aqueous ethanol [40]. The Brønsted plot, shown in Fig. 6, of the logarithm of the hydronium ion catalyzed hydrolysis rate constants vs. the logarithm of the dissociation constants of the conjugate acid of the leaving group suggests that the hydrolysis rate is not strongly dependent on the leaving group. The slope of the Brønsted line, β_{1g}, equals 0.18. Akerman found a similar Brønsted slope ($\beta_{1g} = 0.24$) for a series of substituted phenoxytributylsilane esters in 40% aqueous dioxane [38]. Chipperfield and Gould argued that the substituents affected both the equilibrium protonation of the silane ester and the rate at which water can displace the protonated leaving group [40]. For example, electron-withdrawing groups will make the oxygen of the Si—O—Ar group less basic. The equilibrium protonation of the silane ester shown in step 1 of Fig. 5 will therefore be shifted to the left. There will be fewer protonated species in solution. The rate of reaction should be slower due to concentration effects. However, the electron-withdrawing groups will make the protonated leaving group more reactive. The second step of the reaction mechanism in Fig. 5 should therefore be faster. These two effects may be of similar magnitude; the hydrolysis reaction rate would then appear to be insensitive to changes in the leaving groups.

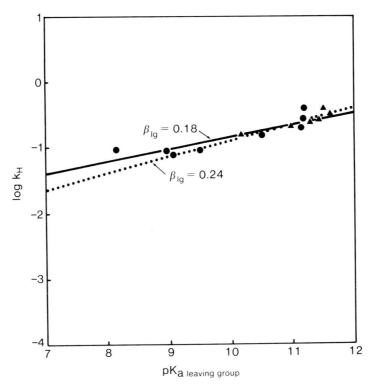

Figure 6. Brønsted plot of the logarithm of the hydronium ion catalyzed hydrolysis rate constants vs. the pK_a of the conjugate acid of the leaving group for a series of substituted aryloxytriethylsilanes in 48.6% water/ethanol at 25°C (●) and for a series of substituted aryloxytributylsilanes in 40% water/dioxane at 30°C (▲).

Thus, to summarize the hydrolysis data, the alkyl substitutents on silicon were found to have a significant effect on the rates of aqueous hydrolysis of alkyl-trialkoxysilanes under basic and acidic conditions. The base- and acid-catalyzed rate constants were found to correlate well with modified Taft equations. The Taft coefficients, ρ^* and s, suggest a change in mechanism as the aqueous solution is varied from basic to acidic pH. Large positive s and ρ^* under basic conditions suggest a two-step mechanism with a pentacoordinate intermediate. The smaller, positive ρ^* and s under acidic condition are consistent with an S_N2-type mechanism.

3. FUNDAMENTAL MECHANISTIC STUDIES—CONDENSATION

Condensation of silanol species is of major industrial importance as the mechanism by which siloxane backbone polymers are formed. Mono-, di-, tri-, and tetrafunctional silanol species are all used extensively in these processes and products. As a result, many studies of condensation mechanisms have been published. Most infer primary chemistry from the condensed products obtained [8, 10, 11, 17, 23, 24, 28, 31, 57, 58]. Accurate determination of rate data and mechanistic insights can provide fundamental support for these processes and others, such as sol–gel processes and treatment of mineral fillers and glass fibers with solutions of reacting silanols.

The difficulty in obtaining accurate rate data on the condensation of silanols is due to the numerous consecutive or parallel competing reactions. In practice, hydrolysis and condensation occur concurrently unless special efforts are made to separate the steps [28]. Devreux *et al.* studied the sol–gel condensation of rapidly hydrolyzed vinyltriethoxysilane, methyltriethoxysilane, and tetraethyl-orthosilicate in acid medium and excess water [30]. ^{29}Si-NMR and small-angle X-ray scattering techniques were used to identify the numerous structures of the condensed species and to monitor their formation with time. Smith studied the polycondensation of methyltrimethoxysilane by a similar NMR technique (59). The reaction process was modelled from the rates of a set of propagation and termination reactions. Molecular weight distribution and other macromolecular properties were calculated. A particularly interesting study of alkyltrialkoxy-silane hydrolysis and condensation was undertaken by Crandall and Morel-Fourier (personal communications). Isopropyltrimethoxysilane was hydrolyzed with one equivalent of water. Dimers, trimers, and low molecular weight oligomers were isolated, and detailed characterization was carried out. Linear and cyclic products were obtained. Good insights into the early and intermediate condensation steps can be obtained from the product mixtures. Nishiyama *et al.* [31, 60, 61] reported several studies on the hydrolysis and condensation of γ-methacryloxypropyltrimethoxysilane in solution or in the presence of colloidal silica. Solution-phase hydrolysis and condensation were carried out with no acid, base, or buffer in ethanol/water mixtures of varying proportions. ^{13}C- and ^{29}Si-NMR techniques were used to track the behavior of these solutions for up to 60 days. Detailed spectral assignments were made for all possible degrees of hydrolysis products (monomers) and condensation products (dimeric, trimeric, and higher oligomers). Polymeric species containing SiOH bonds were observed. Prud'homme and co-workers also studied the condensation of γ-methacryloxypropyltrimethoxysilane in aqueous solution [29]. They found that the main species resulting from the condensation of a concentrated silane ester solution was a cyclic trimer or tetramer.

Several studies have been carried out under conditions which greatly reduced the number of competing reactions. One method is to use monofunctional silanols. Rutz *et al.* studied the condensation of trimethylsilanol to hexamethyl-disiloxane in aqueous dioxane and in the presence of acids and bases [62]. The trimethylsilanol can only condense to the disiloxane. Lasocki extended the study to a series of dialkyl substituted silanediols in aqueous dioxane [63–67]. During the early stages of the reaction, the dialkylsilanediols condensed primarily to the disiloxane. Because these disiloxanes contained reactive silanols, further conden-sation to cyclic and linear oligomers was observed. The condensation of the disiloxanes to oligomers was considerably slower than the initial condensation of the dialkylsilanediol to the disiloxane. For instance, further condensation of tetramethyldisiloxanediol was 0.03 times as fast as the initial condensation of the dimethylsilanediol to the disiloxane [63]. Kinetic isolation of the single reaction, dialkylsilanediol to disiloxane, was achieved by measuring only the initial reaction and carrying out the reaction under dilute conditions. The reactions of trialkylsilanols, dialkylsilanediols, and alkylsilanetriols to the disiloxanes were studied by Pohl and Osterholtz in dilute aqueous solutions [25–27]. Under dilute conditions and short reaction times, condensation of the disiloxane to oligomers was not observed. The formation of oligomers, however, was observed at long reaction times.

From these studies of condensation of mono-, di-, and trifunctional silanols in aqueous and aqueous–organic solutions, a kinetic expression describing the rate of silanol disappearance (condensation of silanol to disiloxane) is obtained:

$$\frac{-d[S']}{dt} = k_H[H^+][S']^2 + k_{HO}[HO^-][S']^2 + k_B[B][S']^2.$$

The symbol S' represents the silanol and B represents any basic species other than hydroxide anion.

The condensation of the silanols to siloxanes in non-protic organic solvents proceeds to completion. Lasocki has observed that the rate of silanol conden-sation is affected by small amounts of water added to the non-protic (dioxane) solvent [68]. The maximum effect is observed at approximately 0.75 M water. The impact of water on the rate may be due to solvation of ionic intermediates, its effect on the activity of the acids or bases, and its effect on the polarity of the solvent.

The condensation reactions of the silanols to siloxanes in protic solvents, such as water, methanol, ethanol, etc., approach equilibrium rather than completion. The competing reactions of silanols in protic solvents, such as methanol, are [19]:

$$(CH_3)_3SiOH + CH_3OH \underset{k_{-1}}{\overset{k_1}{\rightleftharpoons}} (CH_3)_3SiOCH_3 + H_2O \qquad (5)$$

$$(CH_3)_3SiOH + (CH_3)_3SiOCH_3 \underset{k_{-2}}{\overset{k_2}{\rightleftharpoons}} (CH_3)_3SiOSi(CH_3)_3 + CH_3OH \qquad (6)$$

$$2\,(CH_3)_3SiOH \underset{k_{-3}}{\overset{k_3}{\rightleftharpoons}} (CH_3)_3SiOSi(CH_3)_3 + H_2O. \qquad (7)$$

The equilibrium constant K may be represented as

$$K = \frac{[(CH_3)_3SiOCH_3]^2[H_2O]}{[(CH_3)_3SiOSi(CH_3)_3]}.$$

The concentration of trimethylsilanol at equilibrium was not detectable. In this study, the rate of condensation of trimethylsilanol appears to be faster by reaction pathways (5) and (6) rather than the direct pathway (7) [19]. The equilibrium constants, K, for the reaction of silanols in various alcohols are given in Table 3.

Table 3.
Equilibrium constants, K, for the reaction of silanols
in various alcohols at 25°C [19]

Silanol	Alcohol	K
$(CH_3)_3SiOH$	CH_3OH	0.13
$(CH_3)_3SiOH$	CH_3CH_2OH	0.030
$(CH_3)_3SiOH$	$CH_3CH_2CH_2OH$	0.0072
$(CH_3CH_2)_3SiOH$	CH_3OH	0.055

Pohl and Osterholtz determined the equilibrium (pre-equilibrium) constants for the reaction in aqueous solution:

$$2\,R_aSi(OH)_{4-a} \underset{k_{-1}}{\overset{k_1}{\rightleftharpoons}} R_aSi(OH)_{3-a}-O-Si(OH)_{3-a}R_a + H_2O \qquad (8)$$

where a is 1, 2 or 3. The equilibrium constants were not determined for higher degrees of polymerization. The equilibrium constant, K, is defined as

$$K = [disiloxane][H_2O]/[silanol]^2.$$

The equilibrium constants are given in Table 4. They decrease significantly as functionality decreases.

Table 4.
Equilibrium constants for the first step of condensation of alkylsilanols to disiloxanes in aqueous solution [27]

Silanol	Temperature (°C)	Solvent	K
$D_2NCH_2CH_2CH_2Si(CH_3)_2OD$	35	D_2O	10.2
$OCH_2CHCH_2OCH_2CH_2CH_2Si(CH_3)(OD)_2$	35	D_2O	210
$CH_2=C(CH_3)CO_2CH_2CH_2CH_2Si(OH)_3$	25	H_2O	485

3.1. Base catalysis—condensation

The condensation reactions of silanols are catalyzed by bases [19, 25–27, 66, 69]. Grubb found that the condensation of trimethylsilanol in methanol was catalyzed by alkali metal salts of hydroxides [19]. The rate of condensation was dependent on the concentration of hydroxide anion. The type of alkali metal cation did not affect the rate. He postulated that the hydroxide anion deprotonated the silanol to form a silanolate anion. The silanolate anion could attack the trimethylmethoxysilane molecules in a second step of the reaction to form the hexamethyldisiloxane and methoxide anion [19]. Chojnowoski and Chrzczonowicz found that secondary and tertiary amines (piperidine, triethylamine, tri-*n*-butylamine) catalyzed the condensation of a series of dialkyl- or diarylsilanediols in aqueous dioxane [71]. They proposed that primary and secondary amines catalyzed the reactions by nucleophilic attack on silicon which was followed by the rapid attack of dialkylsilanediol, as shown below:

$$R_2Si(OH)_2 + R'_2NH \rightleftharpoons R_2Si(OH)NR'_2 + H_2O \qquad (9)$$

$$R_2Si(OH)NR'_2 + R_2Si(OH)_2 \rightleftharpoons R_2Si(OH)-O-Si(OH)R_2 + HNR'_2. \qquad (10)$$

Tertiary amine catalyzed reactions proceeded by a general base catalyzed mechanism.

Pohl and Osterholtz studied the acetate anion catalyzed condensation of aqueous solutions of a series of alkyl substituted silanetriols [26]. The modified Taft equation for a limited number of alkyl groups (methyl, ethyl, *n*-propyl, *n*-butyl, and *i*-butyl) indicated a large dependence on polar effects [26]. The modified Taft equation was $\log(k_{Ac}/k_{Aco}) = 6.7\sigma^* - 0.07E_s$.

It is worth noting that a modified Taft equation based on such a limited number of data points is very susceptible to large errors in the ρ^* and s coefficients. DeTar has suggested that the σ^* values for simple alkyl groups may be an artifact [70].

In earlier papers, Pohl and Osterholtz [25–27] reported the hydroxide catalyzed condensation of mono-, di-, and trisilanols. The reported rate constants did not take into account the influence of the acetate anion in the solution. Acetate buffers were used to adjust the pH of the solutions.

The activation energies for the condensation of chloromethylmethylsilanediol and dichloromethylmethylsilanediol were determined in aqueous dioxane in the presence of amine catalysts [69]. Pohl and Osterholtz found an activation energy of 16.7 kcal/mol for the condensation of methylsilanetriol in aqueous solution in the presence of the acetate buffer [71].

3.2. Acid catalysis—condensation

The condensation reactions of silanols are catalyzed by acids [19, 25–27, 63–68, 72]. Grubb measured the hydrogen chloride catalyzed silanol condensation reaction of trimethylsilanol in methanol [19]. Lasocki and Michalska studied the effect of acid type on the condensation of dialkylsilanediols in dioxane [68]. Under anhydrous conditions, the rate of acid catalysis by strong acids (such as hydrogen bromide and perchloric acid) was directly related to the acid concentrations. The catalytic effects of weaker acids, such as hydrogen chloride, were not linearly related to the concentration. They postulated that in anhydrous dioxane, the strong acids were completely ionized while the weaker acids were not [68]. When small amounts of water were added to the solvent, all the acids behaved in a similar manner. Lasocki [64–67] extended the studies to examine the effects of alkyl or aryl substitution of silanediols on the condensation rate in aqueous dioxane [64–67]. The rate constants for acid catalyzed condensation of substituted alkylmethylsilanediol in aqueous dioxane are presented in Table 5. They were surprisingly well correlated to the Taft equation, $\log(k_H/k_{H_0}) = 5.3\sigma^*$.

The above Taft equation is very poor at predicting the rate constants for reactions of substrates bearing strongly electron-withdrawing substituents. For example, the Taft equation would predict the rate constant for the condensation of dichloromethylmethylsilanediol to be $8 \times 10^9\,M^{-2}\,s^{-1}$. The rate constant was measured and found to be $5.5 \times 10^{-4}\,M^{-2}\,s^{-1}$. Pohl and Osterholtz also found a very large dependence on the rate constants for the condensation of a few alkylsilanetriols in aqueous solution on polar effects [26]. If the rate constants for the strongly electron-withdrawing substituents are included in defining the modified Taft equation, the polar effects become a much smaller influence and the steric effects are more important. Figure 7 is a plot of $\log(k_H/k_{H_0})$ vs. the modified Taft equation, $0.5\sigma^* + 2.0E_s$. Although the correlation is only good, it is better at predicting the effects of the strongly electron-withdrawing groups than the Taft equation proposed by Lasocki [66] and Pohl and Osterholtz [26]. A contributing factor in the poor correlation may be the fact that substituting an alkyl group on silicon affects both the nucleophile and the electrophile in such a condensation reaction. For example, a strongly electron-withdrawing group may make the silanol a better electrophile (more positive charge on silicon in the ground state), but a poorer nucleophile (the oxygen of SiOH may have less electron density because it has been delocalized to the more positive silicon). The small positive ρ^* value and the large s value are similar, though, to values found for the acid catalyzed hydrolysis of silane esters. If the modified Taft correlation shown in Fig. 7 better describes the true effect of alkyl substitution, then the reaction mechanism would most likely proceed by protonation of the silanol group followed by attack of non-protonated silanol in an S_N2-Si type process.

Lasocki determined the activation energies for acid catalyzed condensation of dialkylsilanediols in aqueous dioxane. The activation energies were generally very small, between 1.25 and 2.8 kcal/mol [64]. Pohl and Osterholtz measured an activation energy of 10.9 kcal/mol for the condensation of methylsilanetriol in acid aqueous solution [71].

Thus, to summarize the condensation data, the condensation of silanols in

Table 5.

Rate constants for the hydrogen chloride catalyzed condensation of silanols in aqueous dioxane at 25°C

R in $R(CH_3)Si(OH)_2$	Solvent	$k_H\,(M^{-2}\,s^{-1})$	Reference
CH_3	0.2% H_2O/dioxane	0.42	63
CH_3CH_2	0.2% H_2O/dioxane	0.11	63
$CH_3CH_2CH_2$	0.2% H_2O/dioxane	0.08	63
$CH_3CH_2CH_2CH_2$	0.2% H_2O/dioxane	0.085	63
$(CH_3)_2CHCH_2$	0.2% H_2O/dioxane	0.068	64
$CH_2=CH$	0.2% H_2O/dioxane	0.059	63
$ClCH_2$	0.2% H_2O/dioxane	0.016	69
$(CH_3)_2CH$	0.2% H_2O/dioxane	0.016	64
C_6H_{11}	0.2% H_2O/dioxane	0.015	64
Cl_2CH	0.2% H_2O/dioxane	0.00055	69

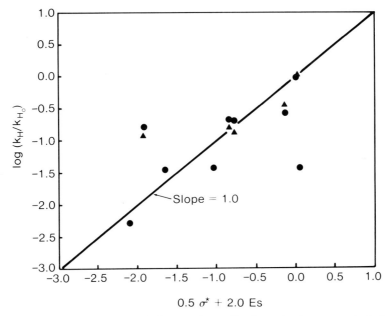

Figure 7. A plot of the logarithm of the experimentally determined rate constants for silanol condensation divided by the rate constant for the methyl substituted silanol vs. the modified Taft equation for a series of alkyl substituted methylsilanediols in dioxane at 25°C (●) and for a series of alkyl substituted silanetriols in deuterium oxide at 35°C (▲).

non-protic solvents goes to completion, while in protic solvents the reactions approach equilibrium. The reactions are catalyzed by acids and bases. For the acid catalyzed reaction, a modified Taft equation was found that correlates only fairly well with the experimental values. The small polar and large steric effects suggest that the acid catalyzed condensation proceeds through an S_N2-Si type mechanism. Base catalyzed condensation is less well understood.

4. REACTIONS OF SILANOLS WITH SURFACES

Chemical modification of siliceous surfaces with silylating agents has found important applications in many areas [73–76]. For many of these applications, the silylation reactions are carried out by treating these siliceous surfaces with aqueous or aqueous–organic solutions of the silylating agent [5, 75]. The structures of these silylating agents on siliceous surfaces have also been characterized using many physical methods [76–79]. {1H–29Si} CP-MAS NMR techniques have been particularly useful in structure determinations [80, 81]. For example, Sindorf and Maciel have reported the structures of a homologous series of methylchlorosilanes that have been reacted with silica gel [83]. Pohl and Blackwell used a similar {1H–29Si} CP-MAS NMR technique to follow the condensation reactions of dilute aqueous solutions of methylsilanetriol with silica gel [32]. They were able to monitor the disappearance of the methylsilanetriol and the formation of siloxane bonds between the silylating agent (methylsilanetriol) and the silica surface with time. Although no detailed kinetic information was obtained, a qualitative picture of the rates of reactions of silanetriols with siliceous surfaces was presented. There appears to be an acceleration of the condensation of the solution-phase silanols in the presence of the surface.

5. PRACTICAL IMPLICATIONS—HYDROLYSIS AND CONDENSATION

The common uses of alkoxysilanes as coupling agents or crosslinkers depend on hydrolysis and condensation reactions.

Most commercial alkoxysilanes (except aminoalkylsilanes) have limited water solubility until the alkoxysilane groups are converted to hydrophilic silanol groups by hydrolysis. Acids or bases may be used to catalyze hydrolysis. The slowest rate is at approximately neutral pH (pH 7). Each change of pH by 1 pH unit in either the acid or the basic direction produces a ten-fold acceleration in hydrolysis rate, assuming an excess of available water. Thus, at pH 4, the hydrolysis is about 1000 times faster than at pH 7. The anion of any acid (e.g. acetate) may also further accelerate the hydrolysis.

Condensation is also pH-dependent. The rate minimum is at approximately pH 4. (Substitution patterns at silicon can shift this minimum.) If fast hydrolysis followed by slow condensation is desired—to maximize the solution life for silanol species as in primers—acidic catalysis is preferred. If base catalysis is used, condensation to siloxane may occur even before all the alkoxy groups are hydrolyzed, promoting gelation. Large alkoxy groups hydrolyze more slowly than small ones; if the rate for β-methoxyethoxy equals 1, then the ethoxy rate equals 2 and the methoxy rate equals 10, under acid catalyzed conditions.

Both hydrolysis and condensation are reversible. Alcohols will 'reverse' the silane hydrolysis stabilizing solution of silanols for a period of time. Solution stability measured in days or weeks can be achieved.

Condensation of silanes is second order in silane concentration. Doubling the concentration of silanol species multiplies the rate of formation of condensed products by four-fold. To minimize condensation, aqueous solution concentrations of silanols below 1% by weight of typical organofunctional silanes should be used.

The equilibrium constant for condensation of two silanetriols to the simple dimer siloxane—forming one siloxane link—is in the range of 400–500. If the equilibrium constant for the second and third condensation is of comparable size, siloxane bond formation is highly favored over reversal. This observation is probably the basis for the extreme effectiveness of silane coupling agents in protecting composites against wet environments.

Difunctional and monofunctional silanes condense as well, but the equilibrium constant is much smaller. The position of the equilibrium may favor solution stability relative to trifunctional materials but implies less effective coupling in wet environments.

6. SUMMARY

The hydrolysis of alkoxysilane esters has been extensively studied. The influences of the silane ester structure and acids, bases, solvent, and temperature on the reaction rates and kinetics have been investigated. Linear free energy relationships, such as Taft equations and Brønsted plots, were used to gain insights into the reaction mechanisms and transition state structure. From the available data, it was argued that the base catalyzed hydrolysis of mono-, di-, and trifunctional alkoxysilane esters proceeded by an attack of hydroxide anion (or base assisted attack of water) in an S_N2^{**}-Si or S_N2^*-Si type mechanism with a pentacoordinate intermediate. Acid catalyzed hydrolysis of these same silane esters more likely proceeded by a rapid protonation of the leaving alkoxy group followed by attack of water in an S_N2-Si type mechanism.

The condensation of silanols in solution or with surfaces has not been as extensively studied and therefore is less well understood. The limitation until recently has been the lack of suitable analytical methods necessary to monitor in real time the many condensation products that form when di- or trifunctional silanols are used as substrates. With the advent of high-field ^{29}Si-NMR techniques, this limitation has been overcome and recent studies have provided insights into the effects of silanol structure, catalysts, solvent, pH, and temperature on the reaction rates and mechanisms. Analysis of the available data has indicated that the base catalyzed condensation of silanols proceeds by a rapid deprotonation of the silanol, followed by slow attack of the resulting silanolate on another silanol molecule. By analogy with the base catalyzed hydrolysis mechanism, this probably occurs by an S_N2^{**}-Si or S_N2^*-Si type mechanism with a pentavalent intermediate. The acid catalyzed condensation of silanols most likely proceeds by rapid protonation of the silanol followed by slow attack on a neutral molecule by an S_N2-Si type mechanism.

REFERENCES

1. E. P. Plueddemann, *Silane Coupling Agents.* Plenum Press, New York (1982 and 1991).
2. K. Isayama and I. Hatano, U.S. Patent No. 3,971,751, assigned to Kanegafuchi Kagaku Kogyo (1976).
3. G. L. Brode and L. B. Conte, Jr., U.S. Patent No. 3,632,557, assigned to Union Carbide Corporation (1972).
4. S. Jänichen, K. Rühlmann, R. Lehnert, A. Perzel and U. Scheim, *Plaste Kautsch.* **36**, 145 (1989).
5. M. G. Voronkov, V. P. Meleshkevisk and Y. A. Yuzkelvski, *The Siloxane Bond*, pp. 375–380. Plenum Press, New York (1978).
6. L. H. Sommer, *Steriochemistry, Mechanism and Silicon*, Ch. V. McGraw-Hill, New York (1977).
7. T. W. Zerda and G. Hoang, *J. Non-Cryst. Solids* **109**, 9 (1989).
8. C. J. Brinker, *J. Non-Cryst. Solids* **100**, 31 (1988).
9. D. E. Leyden, R. S. Shreedhara Murthy, J. B. Atwater and J. P. Blitz, *Anal. Chim. Acta* **200**, 459 (1987).
10. K. Shimada and T. Tarutani, *Bull. Chem. Soc. Jpn* **53**, 3488 (1980).
11. B. W. Peace, K. G. Mayhan and J. F. Montle, *Polymer* **14**, 418 (1973).
12. R. J. P. Corriu, G. Dabosi and M. Martineau, *J. Organomet. Chem.* **186**, 25 (1980).
13. S. Kozuka and T. Kitamura, *Bull. Chem. Soc. Jpn* **52**, 3384 (1979).
14. E. P. Plueddemann, *Proc. 25th Annu. Tech. Conf., Reinforced Plastics/Composites Inst.* Section 19-A (1969).
15. L. H. Shaffer and E. M. Flanigan, *J. Am. Chem. Soc.* **61**, 1591 (1957).
16. L. H. Shaffer and E. M. Flanigan, *J. Am. Chem. Soc.* **61**, 1595 (1957).
17. G. Xue, *Angew. Makromol. Chem.* **151**, 85 (1987).
18. R. Campostrini, G. Carturan, B. Pelli and P. Traldi, *J. Non-Cryst. Solids* **108**, 143 (1989).
19. W. T. Grubb, *J. Am. Chem. Soc.* **76**, 3408 (1954).
20. M. M. Sprung and F. O. Guenther, *J. Polym. Sci.* **28**, 17 (1958).
21. J. F. Brown, Jr. and L. H. Vogt, Jr., *J. Am. Chem. Soc.* **87**, 4313 (1965).
22. J. F. Brown, Jr., *J. Am. Chem. Soc.* **87**, 4317 (1965).
23. J. D. Miller and H. Ishida, *Polym. Mater. Sci. Eng.* **50**, 435 (1984).

24. K. D. Keefer, *Mater. Res. Soc. Symp. Proc.* **32**, 15 (1984).
25. E. R. Pohl and F. D. Osterholtz, in: *Molecular Characterization of Composite Interfaces*, H. Ishida and G. Kumar (Eds), pp. 157–170. Plenum Press, New York (1985).
26. E. R. Pohl and F. D. Osterholtz, in: *Chemically Modified Surfaces in Science and Industry*, D. E. Leyden and W. T. Collins (Eds), Vol. 2, pp. 545–558. Gordon and Breach, New York (1987).
27. E. R. Pohl and F. D. Osterholtz, in: *Silanes, Surface and Interfaces*, D. E. Leyden (Ed.), pp. 481–485. Gordon and Breach, New York (1986).
28. C.-C. Lin and J. D. Basil, *Mater. Res. Soc. Symp. Proc.* **73**, 585 (1986).
29. S. Savard, L. P. Blanchard, J. Leonard and R. E. Prud'homme, *Polym. Composites* **5**, 242 (1984).
30. F. Devreux, J. P. Boilot and F. Chaput, *Phys. Rev. A* **41**, 6901 (1990).
31. N. Nishiyama, K. Horie and T. Asakura, *J. Appl. Polym. Sci.* **34**, 1619 (1987).
32. E. R. Pohl and C. S. Blackwell, in: *Controlled Interphases in Composite Materials*, H. Ishida (Ed.), pp. 37–50. Elsevier, New York (1990).
33. A. R. Bassindall and P. G. Taylor, in: *The Chemistry of Organic Silicon Compounds*, S. Patai and Z. Rappoport (Eds), pp. 839–892. John Wiley, New York (1989); C. Eaborn, R. Eidenschink and D. R. M. Walton, *J. Chem. Soc. Chem. Commun.* 388 (1975).
34. J. A. Deiters and R. R. Holmes, *J. Am. Chem. Soc.* **109**, 1686 (1987).
35. N. Shirai, K. Moriya and Y. Kawazoe, *Tetrahedron* **42**, 2211 (1986).
36. A. A. Humffray and J. J. Ryan, *J. Chem. Soc. B* 1138 (1969).
37. E. Akerman, *Acta Chem. Scand.* **10**, 298 (1956).
38. E. Akerman, *Acta Chem. Scand.* **11**, 373 (1957).
39. K. A. Smith, *J. Org. Chem.* **51**, 3827 (1986).
40. J. R. Chipperfield and G. E. Gould, *J. Chem. Soc. Perkin Trans.* II 1324 (1974).
41. H. Slebocka-Tilk and R. S. Brown, *J. Org. Chem.* **50**, 4638 (1985).
42. E. R. Pohl, *Proc. 38th Annu. Tech. Conf., Reinforced Plastics/Composites Inst.* Section 4-B (1983).
43. K. J. McNeil, J. A. DiCapri, D. A. Walsh and R. F. Pratt, *J. Am. Chem. Soc.* **102**, 1859 (1980).
44. C. G. Swain, K. R. Pröschke, W. Ahmed and R. L. Schowen, *J. Am. Chem. Soc.* **96**, 4700 (1974).
45. R. W. Taft, Jr., in: *Steric Effects in Organic Chemistry*, M. S. Newman (Ed.), Ch. 13. John Wiley, New York (1956).
46. I. E. Saratov, I. V. Shpak, S. A. Markov, S. Y. Lazareu and V. O. Reikhsfel'd, *Zh. Obshch. Khim.* **51**, 2030 (1981).
47. I. E. Saratov, I. V. Shpak, S. A. Markov, S. Y. Lazarev and V. O. Reikhsfel'd, *Zh. Obshch. Khim.* **51**, 2034 (1981).
48. I. E. Saratov, I. V. Shpak and V. O. Reikhsfel'd, *Zh. Obshch. Khim.* **51**, 2038 (1981).
49. H. Kwart and K. G. King, *d-Orbitals in the Chemistry of Silicon, Phosphorus and Sulfur*, Ch. V. Springer-Verlag, New York (1977).
50. D. F. DeTar, *J. Org. Chem.* **45**, 5166 (1980).
51. W. P. Jencks and K. Salveson, *J. Am. Chem. Soc.* **93**, 4423 (1971).
52. W. J. Alberg, *Pure Appl. Chem.* **51**, 949 (1979).
53. A. Streitweiser, Jr., *J. Am. Chem. Soc.* **79**, 4935 (1956).
54. D. J. Hupe and W. P. Jencks, *J. Am. Chem. Soc.* **99**, 451 (1977).
55. E. R. Pohl, D. Wu and D. J. Hupe, *J. Am. Chem. Soc.* **102**, 2759 (1980).
56. A. Skrabal and H. H. Eger, *Z. Phys. Chem.* **122**, 349 (1926).
57. R. Aelion, A. Loebel and F. Eirich, *Recl Trav. Chim. Pay-Bas* **69**, 61 (1950).
58. F. Devreux, J. P. Boilot and F. Chaput, *Phys. Rev. A* **41**, 6901 (1990).
59. K. A. Smith, *Macromolecules* **20**, 2514 (1987).
60. N. Nishiyama, K. Horie and T. Asakura, *J. Colloid Interface Sci.* **129**, 113 (1989).
61. N. Nishiyama, K. Horie and T. Asakura, *J. Colloid Interface Sci.* **124**, 14 (1988).
62. W. Rutz, D. Longe and H. Kelling, *Z. Anorg. Allg. Chem.* **528**, 98 (1985).
63. Z. Lasocki and S. Chrzcznowicz, *J. Polym. Sci.* **59**, 259 (1962).
64. Z. Lasocki, *Bull. Pol. Acad. Sci.* **11**, 637 (1963).
65. Z. Lasocki, *Bull. Pol. Acad. Sci.* **12**, 223 (1964).
66. Z. Lasocki, *Bull. Pol. Acad. Sci.* **12**, 227 (1964).
67. Z. Lasocki, *Bull. Pol. Acad. Sci.* **12**, 281 (1964).
68. Z. Lasocki and Z. Michalska, *Bull. Pol. Acad. Sci.* **13**, 34 (1965).
69. J. Chojnowski and S. Chrzczonowicz, *Bull. Pol. Acad. Sci.* **13**, 41 (1965).

70. D. F. DeTar, *J. Am. Chem. Soc.* **102**, 7988 (1980).
71. E. R. Pohl and F. D. Osterholtz, unpublished data (1989).
72. S. Chrzczonowicz and Z. Lasocki, *Rocz. Chem.* **36**, 275 (1962).
73. C. Chian, H. Ishida and J. L. Koenig, *J. Colloid Interface Sci.* **74**, 396 (1980).
74. E. Gushka, *Bonded Stationary Phases in Chromatography.* Ann Arbor Publications, Ann Arbor (1974).
75. D. E. Leyden and G. H. Luttrell, *Anal. Chem.* **47**, 1612 (1975).
76. K. L. Loewenstein, *The Manufacturing Technology of Continuous Glass Fibers*, Ch. VI. Elsevier, New York (1983).
77. S. Shinoda and Y. J. Saito, *J. Colloid Interface Sci.* **103**, 554 (1985).
78. H. Ishida and J. L. Koenig, *J. Colloid Interface Sci.* **64**, 555 (1978).
79. K. Tanaka, S. Shinoda, N. Takai, H. Takashi and Y. Saita, *Bull. Chem. Soc. Jpn* **53**, 1242 (1980).
80. G. S. Caravajal, D. E. Leyden, G. R. Quinting and G. E. Maciel, *Anal. Chem.* **60**, 1776 (1988).
81. G. E. Maciel, D. W. Sindorf and O. J. Bartuska, *J. Am. Chem. Soc.* **102**, 7607 (1980).
82. D. Sindorf and G. E. Maciel, *J. Am. Chem. Soc.* **103**, 4263 (1981).

Silanes and Other Coupling Agents, pp. 143–157
Ed. K. L. Mittal
© VSP 1992.

Hydrolysis and condensation of alkoxysilanes investigated by internal reflection FTIR spectroscopy

DONALD E. LEYDEN* and JOHN B. ATWATER†

Department of Chemistry, Colorado State University, Fort Collins, CO 80523, USA

Revised version received 25 April 1991

Abstract—The hydrolysis of a series of alkyl-substituted alkoxysilanes and an antimicrobial quaternary ammonium silane was investigated in water–acetone solvents by Fourier transform infrared (FTIR) spectroscopy. An internal reflectance cell was employed for the investigation. Rates of sequential hydrolysis of the first, second, and in one case the third alkoxy groups were extracted by nonlinear regression of curves of methanol concentration plotted versus time using data obtained from the FTIR spectra of the reaction mixture. Acid catalysis of the hydrolysis was observed. Condensation of the silanol groups produced by the hydrolysis was investigated using similar techniques and rate data are presented.

Keywords: Alkoxysilanes; hydrolysis; Fourier transform infrared spectroscopy.

1. INTRODUCTION

Alkoxysilanes with the general formula $(RO)_n SiR'_{(4-n)}$, where $n = 1$, 2 or 3, are an important class of organosilicon compounds used to chemically modify a variety of surfaces. Such modifications may be for the purpose of enhancement of adhesion between two different materials, in which case the silanes are frequently called coupling agents. The alkoxy groups are used to stabilize the compounds against condensation reactions to form siloxane bonds. These alkoxy groups are hydrolyzed prior to or in concert with a reaction to the surface. Use of mono-, di- or trialkoxysilanes will provide different possibilities for bonding to substrates via siloxane bonds [1, 2]. The organofunctional group, R', determines the ultimate properties of the modified surface. Whereas the R' group $—C_{18}H_{37}$ bound to silica will render the surface highly hydrophobic, suitable for reversed-phase chromatography [3], the group $—(CH_2)_3N^+(CH_3)_2C_{18}H_{37}$ Cl$^-$ imparts an antimicrobial property to a substrate [4].

Quaternary ammonium compounds have been used in several types of applications as germicides, disinfectants, surgical antiseptics, and sanitizing agents. Silane compounds, such as (3-trimethoxysilyl)propyloctadecyldimethylammonium chloride, impart antimicrobial properties to the substrates to which they are bound. Knowledge of the species in solutions used to react alkoxysilane compounds containing the active functional group is an important step in understanding the nature of the modified surface. Therefore, time-dependent information on the hydrolysis of the alkoxy groups, and the condensation of the

---

*To whom correspondence should be addressed. Present address: Philip Morris USA R&D, P.O. Box 26583, Richmond, VA 23261, USA.

†Present address: Sweetwater Guest Ranch, Gypsum, CO 81637, USA.

resulting silanol groups, is of value. Although the compound investigated is not a traditional coupling agent, it is reported here as an example of a case in which all three rates of hydrolysis of alkoxy groups could be discerned with the methods used in this investigation.

Several methods have been applied to investigate the aspect of the hydrolysis of alkoxysilanes and condensation of the silanol groups. Plueddemann devised a simple test to estimate the relative rates of hydrolysis of alkoxy groups, as well as a measure of the ultimate solubility of an alkoxysilane in water [5]. FTIR and laser Raman spectroscopy have been used to obtain band assignments of aryl- and vinylsilanetriols in aqueous solution [6, 7]. However, these were not extended to obtain quantitative information from which the kinetics of hydrolysis or condensation to form siloxane bonds could be determined. Pohl and Osterholtz [8] used extraction of alkoxysilanes into hexane and quantitative infrared spectroscopy to determine the rates of hydrolysis as a function of the pH of the aqueous media. Pohl [9] investigated the steric and inductive effects of the alkyl group on the rate of the first step of hydrolysis of trialkoxysilanes. McNeil *et al.* [10] used an extraction method with UV spectrometry to investigate the dynamics of hydrolysis of a silicate triester. In each of these cases, the extent to which the hydrolysis was allowed to approach completion is not clearly reported. Millimolar concentrations were used to avoid exceeding the solubility of the silane in the aqueous medium used.

Miller and Ishida [11] combined size-exclusion chromatography with diffuse reflectance FTIR spectroscopy to identify the hydrolysis and condensation products of phenyltriethoxysilane in mixed organic–aqueous media. Pohl and Osterholtz [8, 12] followed the condensation of alkylsilanols using ^{3}C- and ^{29}Si-NMR spectroscopy and found condensation to be catalyzed by both deuteroxide anion and deuterium ion. Savard *et al.* [13] used ^{1}H-NMR spectroscopy to follow the rate of hydrolysis and condensation of 3-methacryloxypropyltrimethoxy-silane as a function of the pH. These authors used the Lentz technique of derivatization to determine the structures of some of the condensation products.

Transmission infrared spectroscopy has not been used extensively to investigate aqueous solutions of silanes because of the practical difficulties in using water-insoluble cells of sufficiently short pathlengths to combat the strong absorbance of water in the mid-infrared region. The cylindrical internal reflection (CIR) method has proved valuable to these types of investigations. A commercial CIR device known as the CIRCLE™ (Spectra-Tech) cell has been described elsewhere [14, 15]. This cell combines high energy throughput with short effective pathlengths of 5–15 μm. The CIRCLE™ device has an optical geometry well suited for FTIR spectroscopy because the circular aperture of the internal reflection element (IRE) matches the circular cross-section of an FTIR spectrometer beam as compared with the rectangular slit images of dispersive instruments. The design of the CIRCLE™ device also overcomes liquid sample handling problems by placement of the liquid in contact with the IRE. Detailed theory of internal reflection is given by Harrick [16].

This paper contains results of the investigation of the dynamics of the hydrolysis of several alkoxysilane compounds and the condensation of the silanols resulting from that hydrolysis. FTIR spectroscopy used in conjunction with the CIRCLE™ cell has proved to be well suited for these investigations.

2. EXPERIMENTAL

2.1. Materials

Standard solutions of methanol were prepared in distilled water using GR methanol (EM Science, 99.8%). Trimethylmonomethoxysilane (TMMS), methyltrimethoxysilane (MTMS), and ethyltrimethoxysilane (ETMS) were obtained from Petrarch Systems Inc., Bristol, PA. All the silanes were distilled under vacuum prior to use and stored at 5°C. A liquid-phase ^{13}C-NMR spectrum of each silane showed no presence of methanol, which indicates little or no hydrolysis of the methoxy groups.

(3-Trimethoxysilyl)propyloctadecyldimethylammonium chloride (SiQAC) was obtained from Dow Corning Corporation, Midland, MI, as a 40% solution in methanol. A suitable aliquot of this solution was rotary-evaporated to remove most of the methanol and then diluted to 10 ml with an appropriate solvent. An infrared spectrum was recorded immediately after this procedure to obtain the initial methanol concentration. It was observed that complete evaporation of the methanol to solid SiQAC results in hydrolysis of the methoxy groups and condensation of the silanol groups to form siloxane bonds. Scheme 1 shows the structures and abbreviations of all the compounds used in this investigation.

2.2. Instrumentation

Fourier transform infrared spectra were acquired using a Nicolet 60SX FTIR spectrometer (Nicolet Analytical Instruments, Madison, WI) purged with dry air and equipped with a liquid nitrogen-cooled wide-band mercury cadmium telluride detector, a germanium-on-KBr beam splitter, and a water-cooled Globar source. The optical retardation velocity was 1.626 cm/s. Data collection parameters were adjusted to obtain a single scan in 0.18 s at a nominal resolution of 8 cm^{-1}. The data processing time permitted the acquisition and storage of a single spectrum every 1.52 s. The number of co-added spectra for each time interval in a rate study was a compromise between acceptable signal-to-noise and the necessary time resolution.

All spectra were recorded using the micro-CIRCLE™ accessory (Spectra-Tech Inc., Stamford, CT), fitted with the 'open-boat' sampling cell and a zinc selenide internal reflection element. The temperature of the solutions was maintained at 24 ± 5°C during the course of the experiments. No measureable loss of methanol at the concentrations used in water or water–acetone mixtures was detected even at the longest reaction times investigated. The buffer solutions used for pH adjustment were made by mixing appropriate proportions of 0.02 M monochloroacetic acid and sodium monochloroacetate, or 0.01 M acetic acid and sodium acetate. Buffer was added by syringe directly to the cell or mixed with the sample which was transferred to the cell depending on the expected rate of hydrolysis reaction. All spectral data treatment was performed with the FTIR software supplied with the Nicolet instrument.

2.3. Data analysis

Quantitation of methanol was attempted by several means. The one which gave the most consistent results, and therefore the one used, involved fitting the

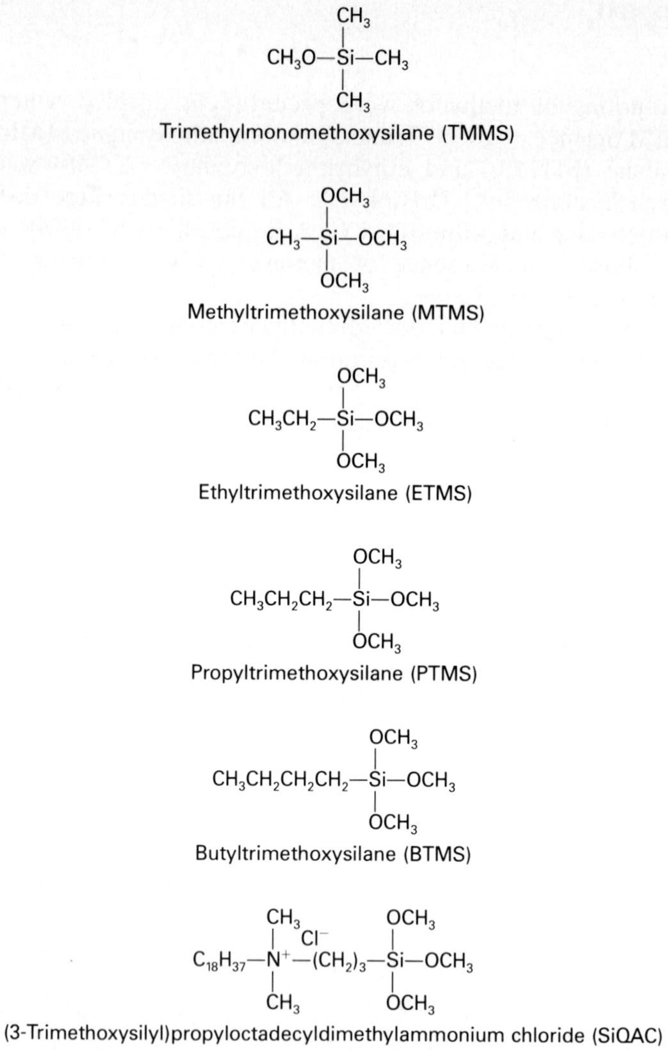

$$CH_3O-\underset{\underset{CH_3}{|}}{\overset{\overset{CH_3}{|}}{Si}}-CH_3$$

Trimethylmonomethoxysilane (TMMS)

$$CH_3-\underset{\underset{OCH_3}{|}}{\overset{\overset{OCH_3}{|}}{Si}}-OCH_3$$

Methyltrimethoxysilane (MTMS)

$$CH_3CH_2-\underset{\underset{OCH_3}{|}}{\overset{\overset{OCH_3}{|}}{Si}}-OCH_3$$

Ethyltrimethoxysilane (ETMS)

$$CH_3CH_2CH_2-\underset{\underset{OCH_3}{|}}{\overset{\overset{OCH_3}{|}}{Si}}-OCH_3$$

Propyltrimethoxysilane (PTMS)

$$CH_3CH_2CH_2CH_2-\underset{\underset{OCH_3}{|}}{\overset{\overset{OCH_3}{|}}{Si}}-OCH_3$$

Butyltrimethoxysilane (BTMS)

$$C_{18}H_{37}-\underset{\underset{CH_3}{|}}{\overset{\overset{CH_3}{|}}{N^+}}-(CH_2)_3-\underset{\underset{OCH_3}{|}}{\overset{\overset{OCH_3}{|}}{Si}}-OCH_3 \quad Cl^-$$

(3-Trimethoxysilyl)propyloctadecyldimethylammonium chloride (SiQAC)

Scheme 1. The structure and abbreviation of each methoxysilane used in the present study.

spectrum of a standard methanol solution to the methanol band in the sample. This was accomplished by use of a factor to adjust the standard spectrum so that the adjusted spectrum is subtracted from the sample spectrum to obtain as close as possible a null difference spectrum for methanol. Because the methanol standard spectrum has a factor of unity, the factor selected to give the null difference spectrum multiplied by the concentration of methanol in the standard gives the concentration of methanol in the sample. This method provided more consistently accurate results than the use of a working curve of integrated absorption band area vs. concentration of methanol in a set of standards. However, it is important that the spectrum of the standard has a similar methanol absorbance to that of the sample spectrum.

The initial concentration of the silanes was calculated from the weight of silane used and dilution factors. TMMS was measured volumetrically because of its high volatility. In all cases, the concentration of methanol produced as measured by FTIR spectroscopy after exhaustive hydrolysis was within ±6% of the predicted stoichiometric amount. In spite of the methanol present in the SiQAC samples and the onset of hydrolysis when the methanol is removed, the observed methanol concentration after exhaustive hydrolysis was within ±3% of that predicted.

Methanol concentration vs. time data in the investigations of hydrolysis dynamics were fitted sequentially to a variety of assumed rate laws. The Derivative Free Nonlinear regressions package (PAR) in the BMDP [17] statistical software system executed on a CDC Cyber 835 computer was used to fit the data. PAR estimates the parameters of a nonlinear function by least squares and can be used to compute maximum likelihood estimates. The function is fitted by an iterative pseudo-Gauss–Newton algorithm. When linear fits were appropriate, these were performed using Statgraphics (STSC Inc., Rockville, MD) executed on an NCR PC-8 computer equipped with an 80287 co-processor.

3. RESULTS AND DISCUSSION

3.1. Quantitation of methanol with the CIRCLE™ cell

An investigation of the quantitation of methanol with the CIRCLE™ cell was performed to determine the validity of the method and that Beer's law relationships hold [18]. Water has two very strong infrared absorption bands in the 4000–400 cm^{-1} region: a broad H—O—H stretching band between 3750 and 3000 cm^{-1}, and an H—O—H bending mode between 1850 and 1450 cm^{-1}. Useful analytical data in these regions are unobtainable because the strong bands lead to residual spectral artifacts when subtracted. This is a result in part of the high absorbance, and small band shifts as a result of solute–solvent interactions, pH, and other factors. It is possible to work with the methanol band found between 1055 and 965 cm^{-1}.

Table 1 shows the results of an investigation of the reproducibility of the quantitation of methanol by use of integrated peak intensity for three different sets of determinations. Each run consisted of three measurements. Run 1 consisted of triplicate measurements in which the solution in the CIRCLE™ cell was not removed, but the cell was removed and replaced in the spectrometer. Run 2 involved removing the solution after each run and cleaning and refilling the cell. Finally, run 3 was made by separating each measurement in the set by a period of several days during which the cell accessory was removed from the spectrometer. The data in Table 1 show that day-to-day reproducibility in methanol determination is approximately 1–2% relative standard deviation (RSD) except at the lowest concentrations investigated. To test the quantitative reliability further, the data shown in Table 1 were analyzed by linear regressions. In the concentration range of 0.14–57.7% (w/w) methanol in water, a linear fit was obtained with a slope of 1.039 ± 0.004, an intercept of 0.014 ± 0.090, and $R^2 = 0.9999$. Therefore, although the method of matching a spectrum of a methanol standard to that of the sample offers some advantages, the use of integrated absorbance and a working curve confirms the validity of the quantitation. Similar confirmations were made for the solvent mixtures employed.

Table 3.
Estimated pseudo-first-order rate constants for MTMS, ETMS, PTMS, and BTMS in 40% water–acetone

$-\log [H^+]$	Silane	$k_1 \, (s^{-1})$	$k_2 \, (s^{-1})$
1.69	MTMS	—[a]	—[a]
2.22	MTMS	$3.56 \, (\pm 0.80) \times 10^{-1}$	$9.47 \, (\pm 1.40) \times 10^{-2}$
2.61	MTMS	$1.94 \, (\pm 0.22) \times 10^{-1}$	$3.24 \, (\pm 0.20) \times 10^{-2}$
3.24	MTMS	$3.53 \, (\pm 0.11) \times 10^{-2}$	$1.87 \, (\pm 0.08) \times 10^{-2}$
1.69	ETMS	$4.58 \, (\pm 1.70) \times 10^{-1}$	$2.34 \, (\pm 0.21) \times 10^{-2}$
2.22	ETMS	$1.70 \, (\pm 0.40) \times 10^{-1}$	$1.97 \, (\pm 0.10) \times 10^{-2}$
2.62	ETMS	$7.63 \, (\pm 0.52) \times 10^{-2}$	$3.30 \, (\pm 0.20) \times 10^{-2}$
3.24	ETMS	$1.58 \, (\pm 0.07) \times 10^{-2}$	$9.66 \, (\pm 0.73) \times 10^{-3}$
1.69	PTMS	$2.62 \, (\pm 1.90) \times 10^{-1}$	$4.00 \, (\pm 0.30) \times 10^{-3}$
2.22	PTMS	$1.06 \, (\pm 0.06) \times 10^{-1}$	$3.04 \, (\pm 0.11) \times 10^{-2}$
2.61	PTMS	$4.63 \, (\pm 0.33) \times 10^{-2}$	$4.19 \, (\pm 0.47) \times 10^{-2}$
3.24	PTMS	$1.08 \, (\pm 0.03) \times 10^{-2}$	$5.25 \, (\pm 0.57) \times 10^{-3}$
1.69	BTMS	$1.90 \, (\pm 0.17) \times 10^{-1}$	$6.43 \, (\pm 0.77) \times 10^{-3}$
2.22	BTMS	$7.02 \, (\pm 0.72) \times 10^{-2}$	$4.06 \, (\pm 0.50) \times 10^{-2}$
2.61	BTMS	$2.90 \, (\pm 0.30) \times 10^{-2}$	$2.50 \, (\pm 0.15) \times 10^{-2}$
3.24	BTMS	$8.36 \, (\pm 1.50) \times 10^{-3}$	$1.35 \, (\pm 0.54) \times 10^{-2}$

[a] Hydrolysis too fast to follow under the experimental conditions used.

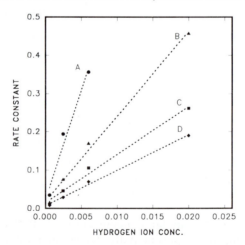

Figure 6. Plot of rate constant k_1 vs. $[H^+]$. (A) MTMS; (B) ETMS; (C) PTMS; (D) BTMS.

as the alkyl group becomes larger. MTMS behaves somewhat differently than the other silanes shown. This suggests that changes in steric effects may be significant upon substituting an ethyl group for a methyl group. Whatever the effect, it becomes less impacted by an additional methylene group on going in the series ethyl, propyl, and butyl.

3.4. Investigations of (3-trimethoxysilyl)propyloctadecyldimethylammonium chloride

Although this compound is not a traditional coupling agent, it does provide for a biologically effective interface between microbes and a variety of surfaces, and provides an opportunity to demonstrate the value of FTIR spectroscopy and the CIRCLE™ cell sampling technique. Also, the speciation distribution as a function of time of the commercially important antimicrobial agent is of interest. Therefore, this material was investigated in water and water–acetone mixtures. The methodologies employed are essentially those described for the alkylsilanes.

Figure 7 shows the time dependence of the Si—O—C, Si—OH, Si—O—Si, and methanol bands for a 4% (w/w) solution of SiQAC under three sequentially adjusted pH conditions. A solution of SiQAC has a pH of approximately 5.5. At this pH, little hydrolysis of the methoxy groups occurs even in several hours, as confirmed by the presence of the CH_3O- rock band at 1193 cm^{-1} and the Si—O—C band at 1082 cm^{-1}. In a matter of minutes after adjusting the pH to 2.7, extensive hydrolysis has occurred, as shown by the decrease in the intensity of the Si—O—C band, and the appearance of the methanol band at 1030 cm^{-1} and the Si—OH band at 920 cm^{-1}. There is no evidence of significant formation of siloxane bands even if the solution is permitted to stand for several hours. Adjustment of the pH to 7.0 initiates rapid and extensive condensation of silanol groups to siloxane bonds, as shown by the decrease in the Si—OH band and the appearance of a Si—O—Si band at 1100 cm^{-1}.

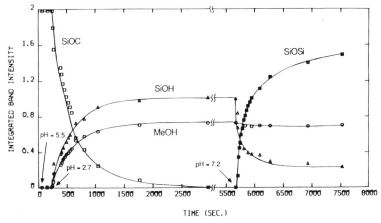

Figure 7. Time dependence of the concentration of species of (3-trimethoxysilyl)propyloctadecyl-dimethylammonium chloride.

Figure 8 shows a plot of the concentration of methanol produced by the hydrolysis of SiQAC at pH 4.07 in water and the nonlinear regression curve of equation (18) assuming three consecutive, irreversible first-order reactions. A summary of the observed rate constants at each pH studied is shown in Table 4. Regression fits produced R^2 values of better than 0.99 for all the pH values investigated. Plots of the observed values of k_1, k_2, and k_3 vs. pH are linear in all cases, with R^2 values greater than 0.99, and with slopes of -0.997, -0.992 and -0.999, respectively. The ratio of $k_1 : k_2 : k_3$ is approximately 20:3:1.

4. CONCLUSIONS

The combination of FTIR spectroscopy and the CIRCLE™ cell sampling technique proved useful in the investigation of the hydrolysis and condensation

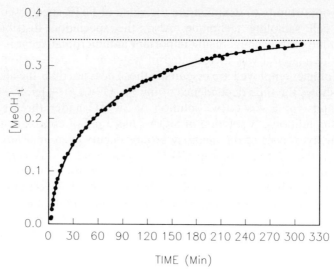

Figure 8. Plot of the concentration of methanol produced vs. time during the hydrolysis of an 8×10^{-2} M solution of (3-trimethoxysilyl)propyloctadecyldimethylammonium chloride in water at pH 4.07.

Table 4.
Average observed pseudo-first-order rate constants for the hydrolysis of the three methoxy groups of SiQAC in aqueous solution

pH	$k_{1,obs}$ (s^{-1})	$k_{2,obs}$ (s^{-1})	$k_{3,obs}$ (s^{-1})
4.07	$1.67 \, (\pm 0.83) \times 10^{-3}$	$2.47 \, (\pm 0.50) \times 10^{-4}$	$1.95 \, (\pm 1.3) \times 10^{-5}$
3.05	$2.43 \, (\pm 0.27) \times 10^{-2}$	$4.80 \, (\pm 1.20) \times 10^{-3}$	$1.68 \, (\pm 2.8) \times 10^{-3}$
2.39	$1.11 \, (\pm 0.27) \times 10^{-1}$	$1.53 \, (\pm 0.67) \times 10^{-2}$	$3.67 \, (\pm 2.8) \times 10^{-3}$
2.18	$1.44 \, (\pm 0.98) \times 10^{-1}$	$2.08 \, (\pm 1.20) \times 10^{-2}$	$9.67 \, (\pm 3.5) \times 10^{-3}$
2.13	$1.63 \, (\pm 0.18) \times 10^{-1}$	$2.52 \, (\pm 0.83) \times 10^{-2}$	$6.10 \, (\pm 8.4) \times 10^{-4}$

of alkoxysilanes in water and acetone–water solutions. Monoalkoxysilanes provide a simple model for the development of the methods and the data analysis because there is only one hydrolysis product and one siloxane bond condensation product.

Trimethoxyalkylsilanes were readily investigated using nonlinear regression methods to extract rate constants from time-dependent concentrations of methanol generated by the hydrolysis. For these compounds, only two consecutive first-order rate constants could be reliably extracted by the regression methods. This result indicates that the second and third rate constants may be similar in magnitude and are not resolved by regression methods. Furthermore, the use of curve-fitting techniques may be subject to considerable error for the third rate constant because of the decreased sensitivity of the rate of change of the concentration of methanol with time in the latter portions of the run. This is, in part, reflected by the large error of the values of k_3 shown in Table 4.

(3-Trimethoxysilyl)propyloctadecyldimethylammonium chloride is a silane with antimicrobial properties of some considerable commercial importance. Three consecutive first-order rate constants are found by regression of the concentration of methanol produced versus time during hydrolysis. Furthermore, reaction to form siloxane bonds is minimal provided the pH is kept in the acidic range suitable for catalysis of hydrolysis.

The relative rates of the consecutive hydrolysis reactions and the rate of condensation to form siloxane bonds depend on the structure of the silane molecule. Data such as those presented here provide the knowledge necessary to permit one to tailor the composition of species of alkoxysilane compounds in solution to obtain the desired rate and degree of hydrolysis of alkoxy groups, and the extent of siloxane bond formation. This is accomplished by the use of several easily observed bands in the mid-infrared spectrum.

Acknowledgements

This research was supported in part by Research Grants CHE-8513247 and CHE-8712457 from the National Science Foundation, and funds from the Dow Corning Corporation. The Nicolet 60SX FTIR spectrometer was purchased in part by Grant CHE-8317079 from the National Science Foundation.

REFERENCES

1. T. G. Waddell, D. E. Leyden and M. T. DeBello, *J. Am. Chem. Soc.* **103**, 5303 (1981).
2. J. P. Blitz, R. S. S. Murthy and D. E. Leyden, *J. Colloid Interface Sci.* **126**, 387 (1988).
3. R. K. Gilpin and M. E. Gangoda, *J. Chromatogr. Sci.* **21**, 352 (1983).
4. J. R. Malek and J. L. Speier, *J. Colloid Interface Sci.* **89**, 68 (1982).
5. E. P. Plueddemann, *Proc. SPI 24th Annu. Conf. Reinforced Plastics Composites Div.* Section 19-A (1969).
6. H. Ishida and J. L. Koenig, *Appl. Spectrosc.* **32**, 462 (1978).
7. H. Ishida and J. L. Koenig, *Appl. Spectrosc.* **32**, 469 (1978).
8. E. R. Pohl and F. D. Osterholtz, in: *Molecular Characterization of Composite Interfaces*, H. Ishida and G. Kumar (Eds), pp. 157–170. Plenum Press, New York (1985).
9. E. R. Pohl, *Proc. 38th Annu. Conf. Reinforced Plastics, SPI*, Section 4-B (1983).
10. K. J. McNeil, J. A. DiCaprio, D. A. Walsh and R. F. Pratt, *J. Am. Chem. Soc.* **102**, 1859 (1980).
11. J. D. Miller and H. Ishida, *Anal. Chem.* **57**, 283 (1985).
12. E. R. Pohl and F. D. Osterholtz, in: *Chemically Modified Surfaces*, D. E. Leyden (Ed.), Vol. 1, pp. 481–500. Gordon and Breach, New York (1986).
13. S. Savard, L. P. Blanchard, J. Leonare and R. E. Prud'homme, *Polym. Composites* **5**, 242 (1984).
14. A. Reing and P. Wilks, Jr., *Am. Lab.* 152 (Oct. 1982).
15. E. G. Bartick and R. G. Messerschmidt, *Am. Lab.* 56 (Nov. 1984).
16. N. J. Harrick, *Internal Reflection Spectroscopy*. John Wiley, New York (1967).
17. W. J. Dixon (Ed.), *BMDP Statistical Software*. University of California Press, Berkeley (1981).
18. E. H. Braue, Jr. and M. G. Pannella, *Appl. Spectrosc.* **41**, 1057 (1987).
19. P. W. Atkins, *Physical Chemistry*, p. 858. W. H. Freeman, San Francisco (1978).
20. C. Capellos and B. H. J. Bielski, *Kinetic Systems*, p. 46. Wiley Interscience, New York (1972).

Silanes and Other Coupling Agents, pp. 159–179
Ed. K. L. Mittal
© VSP 1992.

Kinetics and mechanism of the hydrolysis and alcoholysis of alkoxysilanes

DAVID J. OOSTENDORP, GARY L. BERTRAND and JAMES O. STOFFER*

Department of Chemistry, Materials Research Center, University of Missouri-Rolla, Rolla, MO 65401, USA

Revised version received 29 August 1991

Abstract—As part of our work with the use of silane coupling agents, we have been investigating the transformations of alkoxysilanes into siloxanes. The influence of a para-substituted phenyl group attached to the silicon was investigated and the rates of acid catalyzed hydrolysis and alcoholysis of these para-substituted phenylalkoxysilanes have been determined. The kinetics and mechanism of the reactions are presented. These reactions are of interest because of their role in the use of silane coupling agents as adhesion promoters, in the preparation of zinc-rich silicate coatings, in the sol-gel process and in the preparation of silicones in general.

The hydrolysis reaction was found to be first order in acid, zero order in water and to have a Hammett ρ of -1.42 when catalyzed by sulfuric acid. These results are consistent with current opinion that the reaction mechanism is SN_1 and involves a positive intermediate, possibly a siliconium intermediate.

The alcoholysis reaction was found to be first order in both the silane and the catalysts, and of intermediate order in the alcohol, when catalyzed by carboxylic acids. When catalyzed by dichloroacetic acid, the reaction has a Hammet ρ value of $+0.43$. This is consistent with a concerted displacement reaction between the alkoxysilane and the complex involving the alcohol and the carboxylate anion. The intermediate is a negatively charged intermediate, probably a penta-substituted silicon species.

Keywords: Alkoxysilanes; hydrolysis; alcoholysis; siliconium; coupling agents.

1. INTRODUCTION

The mechanism of the conversion of one alkoxysilane into another alkoxy-silane—alcoholysis—was widely studied in the 1960s and early 1970s. Since then, little work has been done in the area. The importance of this reaction has again increased because of three commercial developments. The first is the widespread use of silanes as adhesion promoters in various applications [1–3]. The second is the use of partially hydrolyzed tetraethoxysilane (TEOS) as the binder in zinc-rich coatings [4]. These coatings are used as a substitute for hot-dip galvaniz-ation when the hot-dip method is not practical due to the size or location of the object to be protected. This was further enhanced by the discovery of a new method to produce tetramethoxysilane $Si(OCH_3)_4$ (TMOS). This new method involves a direct reaction between elemental silicon and methanol [5]. The third development was the emergence of the sol-gel process, a low-temperature technique for making inorganic glasses and ceramics. This process involves the hydrolysis and condensation of TEOS and other alkoxides in solution [6]. The

---

*To whom correspondence should be addressed.

resulting products are similar to those made by traditional melt procedures, but require much less energy to produce.

While it is possible to use TMOS in the sol-gel process or in coating binders, the by-product of the hydrolysis is methanol. The toxicity of methanol makes TMOS unsuitable, so it must be converted into TEOS before it can be used. The conversion is carried out by reacting TMOS with ethanol. The reaction is catalyzed by both acids and bases. Since acid catalysis is used in the industrial preparation, this is the reaction that we have concentrated on.

$$Si(OCH_3)_4 + 4 C_2H_5OH \underset{}{\overset{H^+}{\rightleftharpoons}} Si(OC_2H_5)_4 + 4 CH_3OH. \quad (1)$$

An excellent review of the literature dealing with the alcoholysis can be found in 'The Siloxane Bond' by Voronkov, Mileshkevich, and Yuzhelevskii [7]. In 1967 and 1973, Schowen [8, 9] studied the base catalyzed methanolysis of para-substituted aryloxytriphenylsilanes. The substituent effect yielded a Hammett ρ value of 1.2 which indicated the development of a negative charge in the transition state. He suggested that this could be explained by a concerted-displacement mechanism:

$$B + CH_3OH + (C_6H_5)_3SiOAr \longrightarrow \left[\begin{array}{c} CH_3 \\ \diagdown \\ O\text{-}\text{-}\text{-}\underset{|}{Si}\text{-}\text{-}\text{-}O \\ H \qquad\qquad Ar \\ B \end{array} \right] \longrightarrow$$

$$B + HOAr + (C_6H_5)_3SiOCH_3. \quad (2)$$

This hypothesis was supported in 1974 when he and others [10] reported that the reaction can be simultaneously catalyzed by acids and bases. The methanolysis of aryloxysilanes was carried out in sodium formate-formic acid buffers. The rate law developed showed a first-order dependence on both the base and the acid. This led to the proposal of a concerted-displacement mechanism:

$$
\begin{array}{c}
B \\
CH_3 \qquad\qquad H \\
\diagdown \qquad | \qquad \\
O\text{-}\text{-}\text{-}\underset{|}{Si}\text{-}\text{-}\text{-}O \\
H \qquad\qquad Ar \\
B
\end{array}
\qquad (3)
$$

A pair of papers by Boe in 1972 and 1973 [11, 12] have the most relevance to the work conducted here. Boe also carried out both the acid and base catalyzed alcoholysis of para-substituted phenylalkoxysilanes. Reactions studied were reactions of $(p\text{-}XC_6H_4)(CH_3)_2Si[OCH(CH_3)_2]$ with n-propanol and sulfuric acid, and $(p\text{-}XC_6H_4)(CH_3)_2SiOPh$ with ethanol and sulfuric acid or butylamine/butylamine hydrochloride. He found a Hammett ρ value of -0.41 in the first acid reaction and a ρ of -0.57 in the second. In the base catalyzed reaction, he found a Hammett ρ value of +0.19. He also found that the acid catalyzed reaction was first order in acid, but he did not determine the order of the alcohol. The small negative values for ρ in the acid catalyzed reaction indicated to him that there was a small positive charge developed on the silicon during the rate determining step. He proposed that protonation of the alkoxy group was followed by a rate-determining attack by the alcohol. He could not determine if the reaction was a simple SN_2 or one involving a penta-coordinate complex.

$$Me_2Si(OR)(OR') + H^+ \rightleftharpoons [Me_2Si(OR)(OHR')]^+ \text{ (fast)}$$

$$[Me_2Si(OR)(OHR')]^+ + ROH \rightarrow \begin{array}{c} \backslash \ / \\ [RO\text{---}Si\text{---}OR']^+ \\ | \quad | \quad | \\ H \quad \quad H \end{array} \rightarrow$$

$$Me_2Si(OR)_2 + R'OH_2^+ \text{ (slow)}. \tag{4}$$

In his base catalyzed work, Boe agreed with Schowen and proposed the following transition state:

$$\begin{array}{c} \backslash \ / \\ [X\text{—}Si\text{—}OPh] \\ | \end{array}$$

$$X = EtO - HB \text{ or } X = B. \tag{5}$$

He felt that the smaller value of Hammett ρ compared to that found by Schowen indicated that there was less development of charge on the silicon site than on the phenoxy site.

We will next deal with the hydrolysis reaction. Traditionally, most industrial processes used to make products containing the siloxane bond, such as silicones and silicates, involve the hydrolysis of chlorosilanes. This causes hydrochloric acid to be produced as a by-product. Hydrochloric acid is, however, an un-acceptable by-product in products such as zinc-rich silicate coatings and glasses and glass ceramics made by the sol-gel process. The alternative is to use alkoxysilanes, which when hydrolyzed, have alcohols as a by-product.

Until 1982, most alkoxysilanes had been produced from chlorosilanes and alcohols. Hydrochloric acid was therefore still a problem. In 1982, a process was developed in which TMOS could be made directly from elemental silicon and methanol [5]. In the production of silicate coatings, TMOS is first converted to TEOS by an alcoholysis reaction with ethanol. This prevents toxic methanol vapors from escaping from the curing coating. The TEOS is partially hydrolyzed with the rest of the hydrolysis occurring at the time of application. This is therefore a way to produce silicates without chlorine. (If a practical method for converting alkoxysilanes to alkylsilanes could be found, there would also be a nonchlorine method of production of silicones.)

Interest in this method of preparation of silicates has led us to study the mechanisms involved in the various reactions. In a previous paper we investigated the kinetics and mechanism of the alcoholysis of TMOS [13]. Because alkoxysilanes are important in the formation of silicones and silicates, their hydrolysis and subsequent condensation have been widely studied. Excellent reviews of the field can be found in 'The Siloxane Bond' by Voronkov,

Mileshkevich and Yuzhelevskii [7]. The generally accepted sequence of the acid-catalyzed reaction involves first the replacement of the alkoxy group with a hydroxyl group:

$$R_3SiOR + H_2O \underset{}{\overset{H^+}{\rightleftharpoons}} R_3SiOH + ROH. \tag{6}$$

This is then followed by the condensation to a siloxane:

$$R_3SiOH + R_3SiOR \underset{}{\overset{H^+}{\rightleftharpoons}} R_3SiOSiR_3 + ROH. \tag{7}$$

While reaction (6) is generally considered to be the rate-determining step, the exact nature of the transition state has yet to be determined conclusively.

In 1952, Khaskin [14] showed that under acidic conditions, Equation (1) occurs with breakage of the silicon oxygen bond. This was proven using ^{18}O labeled water. The reaction proceeded with the inclusion of the ^{18}O in the resulting silanol as follows:

$$R_3Si-OR + H^{18}OH \underset{}{\overset{H^+}{\rightleftharpoons}} R_3Si-^{18}OH + ROH. \tag{8}$$

About this time, it was postulated that the transition state was a trivalent siliconium ion followed by the rapid addition of water. In 1982, this view was supported by Schmidt et al. [15]. They studied the hydrolysis of tetra-alkoxy-silanes and various alkoxyalkylsilanes. It was noted that the alkyl-containing silanes reacted significantly faster than the analogous alkoxysilanes. They attributed this result to the stabilizing effect that the electron donating character of the alkyl group (+I effect) would have on a siliconium ion.

The accepted sequence for the base catalyzed reaction is given by the work of Schowen [8, 16]. He reported the $\sigma-\rho$ relationships for the methanolysis of a series of phenoxy substituted silanes. The base-catalyzed reaction is presented as having a penta-sustituted silicon intermediate.

The kinetics of the hydrolysis and condensation of organic functional trialkoxy silanes has been reported by Pohl and Osterholtz [17–19]. The silane coupling agents used as adhesion promoters [1–3] usually have a trialkoxy silane as one of the functional groups, i.e. $(MeO)_3Si-(CH_2)_3-O_2CC(Me)=CH_2$. If this attaches to a glass substrate, it will form Si—O—Si bonds; or if it attaches to metal sub-strates, it can form M—O—Si bonds. Thus, the work described here can be applicable to providing additional understanding for those processes.

Our interest in silane coupling agents and in the preparation of silicates has led us to study the mechanisms involved in the various reactions. In a previous paper, we investigated the kinetics and mechanism of the alcoholysis of TMOS [13]. This present work studies the effects of changing the substituents on a para-substituted phenyl attached directly to the alkoxysilane. The alkoxysilane silanes used have only one alkoxy group present to eliminate complications from a competing second or third alcoholysis reaction. The resulting Hammett plot yields additional information on the mechanism of this reaction.

Since our earlier work gave little insight into the nature of the transition state of our proposed mechanism, the work reported in this paper intends to fill that gap. By attaching a phenyl group to the silicon and varying its electronic character with substituents, we can observe the influence of different partial charges on the silicon. This was done for both the alcoholysis and the hydrolysis reactions.

2. EXPERIMENTAL SECTION

2.1. Materials

The following materials were used in the preparation of the various silanes. The magnesium turnings and anhydrous diethylether were supplied by Fisher Scientific. Bromobenzene and the substituted bromobenzenes, p-bromochlorobenzene, p-bromotrifluoromethylbenzene, p-bromotoluene and p-bromoanisole, were supplied by Aldrich Chemical Co. and used as provided after the purity was checked by GC analysis. The diethoxydimethylsilane was supplied by Petrarch Systems.

The following materials were used for the kinetic reactions. Distilled and dried reagent grade methanol was used. Reagent grade dichloroacetic acid was used as the catalyst. Mesitylene was used as the gas chromatography (GC) internal standard after it was distilled and its purity verified by GC. Reagents were obtained from Fisher Scientific and Aldrich Chemical Co. The various silanes were made by the following methods.

2.1.1. Synthesis of phenyldimethylethoxysilane. The phenyldimethylethoxysilane (A) was prepared by reacting diethoxydimethylsilane with an aryl Grignard reagent. Alkoxy groups on silanes are subjected to displacement by the R group of the Grignard. Crushed magnesium turnings (24 g) were placed in a three-necked, 1-liter, round-bottomed flask equipped with a reflux condenser and a pressure-equalized dropping funnel. A crystal of iodine was added and the system was flushed with argon. A mixture of 15 ml of diethylether and 10 g of bromobenzene was placed in the dropping funnel and then introduced into the flask. The reaction was started with gentle heating from a steam bath. A solution consisting of 300 ml of diethylether and 147 g of bromobenzene was then placed in the dropping funnel and slowly added to the reaction mixture. After addition, the mixture was heated on the steam bath for 30 min to drive the reaction to completion. A 1-liter round-bottomed flask was charged with 264 g of distilled reagent grade diethoxydimethylsilane. A two-fold excess of diethoxydimethylsilane was used to reduce the chances of both ethoxy groups reacting with the Grignard reagent. A stir bar was added to the flask. Once the heating of the Grignard reagent was complete, the contents of the flask were added to the silane with vigorous stirring. The resulting salts were filtered from the solution and the phenyldimethylethoxysilane product was isolated by vacuum distillation using a silver jacketed distillation column. Yield: 53.6 g, 29.8%. Bp: 33°C at 1.0 mm Hg. H NMR (60 MHz) d −0.22 (s, 6 H), 0.57 (t, 3 H), 3.03 (q, 2 H), 6.88 (m, 5 H) Anal. Calcd [20] for $C_{10}H_{16}OSi$: C, 66.61; H, 8.94; Si, 15.58. Found: C, 65.82; H, 8.85; Si, 16.11.

2.1.2. Synthesis of p-chlorophenyldimethylethoxysilane. A 2-liter three-necked round-bottomed flask equipped with a pressure-equalized dropping funnel and a thermometer was flushed with argon and charged with 287 g (1.5 moles) of p-bromochlorobenzene dissolved in 600 ml anhydrous diethylether. The flask was chilled to −10°C and the solution was stirred with a magnetic stir bar. 400 ml of chilled 2.5 M butyllithium in ether (1 mole) was transferred into the dropping funnel. Drops of butyllithium were added over a period of 3 h, allowing a temperature of −10 to −5°C to be maintained. This was necessary to prevent the formation of biphenyls. Standard anaerobic transferring procedures were used. During addition of the butyllithium, a 2-liter round-bottom flask was flushed with argon and charged with 300 g (2.02 moles) of dimethyldiethoxysilane and cooled in an ice bath. In order to remove the lithium salt by-products, the product mixture was centrifuged in a large centrifuge and the supernatant was decanted away from the lithium salts. The product trapped in the salt was extracted with ether, centrifuged again and added to the rest of the product. The ether and other low boiling materials were removed by simple distillation. The p-chlorophenyldimethylethoxysilane (2.1.2) was then isolated by vacuum distillation using a vacuum-jacketed spinning band column. Since the product was the highest boiling material in the reaction mixture, tetraglyme (a higher boiling liquid) was added to aid the distillation. Yield: 62.9 g, 25.4%. Bp: 51°C at 0.2 mm Hg. H NMR (60 MHz) d −0.02 (s, 6 H), 0.82 (t, 3 H), 3.28 (q, 2 H), 7.02 (dd, 4 H). Anal. Calcd for $C_{10}H_{15}OClSi$: C, 55.93; H, 7.04; Si, 13.08. Found: C, 55.09; H, 6.78; Si, 12.77.

2.1.3. Synthesis of p-trifluoromethylphenyldimethylethoxysilane.

The general procedure for the synthesis of p-trifluoromethylphenyldimethylethoxysilane (2.1.3) was the same as for (2.1.2). A 1-liter three-necked round-bottomed flask equipped as in Section 2.1.2 was flushed with argon and charged with 101 g (0.45 moles) of p-bromotrifluoromethylbenzene dissolved in 300 ml anhydrous diethylether. The flask was chilled to −10°C and the solution was stirred. The dropping funnel was charged with 170 ml of chilled 2.5 M (0.43 moles) butyllithium in ether. Drops of butyllithium were added over a period of 1.5 h, allowing a temperature of −5°C to be maintained. During addition of the butyllithium, a 2-liter round-bottom flask was flushed with argon and charged with 178 g (1.2 moles) of dimethyldiethoxysilane and cooled in an ice bath. After all the butyllithium was added, the reaction mixture was slowly transferred into the diethoxydimethylsilane. The product mixture was centrifuged and the supernatant was decanted away from the lithium salt. The product trapped in the salt was extracted with ether, centrifuged again and added to the rest of the product. The p-trifluoromethylphenyldimethylethoxysilane was then isolated by simple distillation and then by vacuum distillation using a vacuum-jacketed spinning band column. Yield: 74.5 g, 66.9%. Bp: 42°C at 0.7 mm Hg. H NMR (60 MHz) d −0.23 (s, 6 H), 0.57 (t, 3 H), 3.08 (q, 2 H), 7.00 (dd, 4 H). Anal. Calcd for $C_{11}H_{15}OF_3Si$: C, 53.21; H, 6.09; Si, 11.31. Found: C, 52.77; H, 5.98; Si, 11.86.

2.1.4. Synthesis of p-tolyldimethylethoxysilane.

The general procedure for the synthesis of p-tolyldimethylethoxysilane (2.1.4) was the same for (2.1.2). A 2-liter three-necked round-bottomed flask equipped as in Section 2.1.2 was flushed with argon and charged with 180 g (1.05 moles) of p-bromotoluene dissolved in 600 ml anhydrous diethylether. The flask was chilled to −10°C and the solution was stirred. The dropping funnel was charged with 400 ml of chilled 2.5 M (1.0 moles) butyllithium in ether. Drops of butyllithium were added over a period of 1.5 h, allowing a temperature of −5°C to be maintained. During addition of the butyllithium, a 2-liter round-bottomed flask was flushed with argon and charged with 222 g (1.5 moles) of dimethyldiethoxysilane and cooled in an ice bath. After the butyllithium was added, the reaction mixture was slowly transferred into the diethoxydimethylsilane. The product mixture was centrifuged and the supernatant was decanted away from the lithium salt. The product trapped in the salt was extracted with ether, centrifuged again and added to the rest of the product. The *p*-tolyldimethylethoxysilane was then isolated by simple distillation and then by vacuum distillation using a vacuum-jacketed spinning band column. Yield: 22.3 g, 11.5%. Bp: 53°C at 1.0 mm Hg. H NMR (60 MHz) d −0.23 (s, 6 H), 0.55 (t, 3 H), 1.65 (s, 3 H), 3.00 (q, 2 H), 6.62 (dd, 4 H). Anal. Calcd for $C_{11}H_{18}OSi$: C, 67.98; H, 9.34; Si, 14.45. Found: C, 66.64; H, 9.21; Si, 15.38.

2.1.5. Synthesis of p-*methoxyphenyldimethylethoxysilane.* The general procedure for the synthesis of *p*-methoxyphenyldimethylethoxysilane (2.1.5) was the same as for (2.1.2). A 1-liter three-necked round-bottomed flask equipped as in Section 2.1.2 was flushed with argon and charged with 101 g (0.45 moles) of *p*-bromoanisole dissolved in 600 ml anhydrous diethylether. The flask was chilled to −10°C and the solution was stirred. The dropping funnel was charged with 400 ml of chilled 2.5 M (1.0 moles) butyllithium in ether. Drops of butyllithium were added over 1.5 h, allowing a temperature of −5°C to be maintained. During addition of the butyllithium, a 2-liter round-bottom flask was flushed with argon and charged with 235 g (1.59 moles) of dimethyldiethoxysilane and cooled in an ice bath. After the butyllithium was added, the reaction mixture was slowly transferred into the diethoxydimethylsilane. The product mixture was centrifuged and the supernatant was decanted away from the lithium salt. The product trapped in the salt was extracted with ether, centrifuged again and added to the rest of the product. The *p*-methoxyphenyldimethylethoxysilane was then isolated by simple distillation and then by vacuum distillation using a vacuum-jacketed spinning band column. Yield: 33.4 g, 15.9%. Bp: 42°C at 0.7 mm Hg. H NMR (60 MHz) d −0.24 (s, 6 H), 0.58 (t, 3 H), 3.10 (s, 3 H), 6.35 (d, 2 H), 6.98 (d, 2 H). Anal. Calcd [20] for $C_{11}H_{18}O_2Si$: C, 62.81; H, 8.63; Si, 13.35. Found: C, 61.10; H, 8.91; Si, 11.26.

3. RATE MEASUREMENTS

The reactions were carried out at 25°C in a constant temperature bath in Kimble 20 ml polyethylene scintillation vials with polyethylene lined caps. Polyethylene was used rather than glass to prevent the silanol-rich surface of the glass from catalyzing the reaction. Mixing was accomplished using a Fisher Vortex-Genie. The reaction mixtures were kept at a constant temperature in a Sargent-Welch precision thermostatic water bath. The mixtures were analyzed by gas chromatography. The chromatograph used was a Varian 3700 containing a 30-meter borosilicate glass capillary column coated with SE-30 dimethylsilicone and a flame ionization detector (FID). Integration was with an HP 3390A integrator.

The alcoholysis reaction involves the acid-catalyzed exchange of an ethoxy group for a methoxy group on various ethoxysilanes. The progress of the reaction was followed by measuring the disappearance of the ethoxysilane vs. an internal standard (mesitylene). Response factors for each of the ethoxysilanes were determined by measuring the response of a mixture containing known masses of the silane and mesitylene.

In the alcoholysis reactions, four different components make up the reaction mixture: the ethoxysilane, methanol, dichloroacetic acid and mesitylene. Because of the small amount of acid required to catalyze this reaction and to eliminate the catalyst–alcohol equilibration problem, large quantities of methanol and acid were weighed and premixed. This also allowed the ratio of catalyst to methanol to be constant in all of the reactions. To start a reaction, the ethoxysilane was weighed into a vial, sealed and placed in the constant temperature bath. Next, the mesitylene and the acid–alcohol mixture were weighed into a second vial, sealed and placed in the bath. All materials were weighed to 0.001 g. Both vials were allowed to equilibrate for at least 30 min. The reaction was initiated by removing the vials from the bath, quickly opening the vials, pouring the silane from one vial into the other and resealing. A timer was started, the solution was mixed momentarily on the Vortex-Genie and then returned to the bath.

For the hydrolysis reactions, five different components make up the reaction mixture: the silane, water, acid, ethanol (to solubilize the water) and mesitylene (the internal standard). The water, acid, ethanol and mesitylene were weighed (to 0.001 g) into the vial, which was then sealed with the polypropylene cap. The vial was placed in a constant temperature bath and allowed to reach equilibrium (approximately 30 min). The silane to be used was kept in the constant temperature bath as well. The reaction was initiated by removing the reaction vial from the bath, quickly weighing in the desired amount of the silane, mixing the solution momentarily with a Vortex-Genie and then returning it to the bath.

The changes in concentrations of the reactants and products were followed by GC. Prior to running the reactions, response factors of the pure silanes were determined vs. mesitylene. At time intervals, the reaction vial would be quickly opened, 0.5 μl of solution would be removed by syringe and injected into the chromatograph. Once the reaction was complete, the concentrations of all species were calculated.

4. RESULTS

4.1. Synthesis

While attempts were made to prepare the ethoxysilanes via the classic Grignard reaction, only phenyldimethylethyoxysilane was prepared in this manner. The Grignard reaction proved to be unsuitable for the preparation of the four para-substituted phenyldimethylethoxysilanes either because of sensitivity to Grignard formation or difficulty in isolating the product from the reaction mixture since the by-product salts were unfilterable. Therefore, all four silanes (p-chlorophenyl, p-methoxyphenyl, p-tolyl and p-trifluoromethylphenyldimethyl-ethoxysilane) were prepared via an organolithium reaction.

4.2. Kinetics

The substituent effects for the alcoholysis and hydrolysis reactions of para-substituted phenyldimethylethoxysilanes have been investigated. Compounds with the general formula $p\text{-}XC_6H_4(CH_3)_2SiOC_2H_5$ were used, where X = H, CH$_3$,

OCH_3, Cl and CF_3. This allows for the analysis of the electronic effect on the silicon atom while minimizing the steric effects. Mono-alkoxysilanes were used to avoid needless complication of the rate expression.

The kinetic reactions for the alcoholysis study were carried out by mixing one of the para-substituted ethoxysilanes with a solution containing methanol, dichloroacetic acid and an internal standard, keeping the reaction mixture at 25°C in a constant temperature bath:

$$(p\text{-}XC_6H_5)(CH_3)_2SiOEt + CH_3OH \underset{}{\overset{CHCl_2COOH}{\rightleftharpoons}} (p\text{-}XC_6H_5)(CH_3)_2SiOMe + C_2H_5OH$$

$$X = H, CH_3, OCH_3, Cl, CF_3.$$

A plot of the ln[silane] vs. time (Fig. 1) exhibits the similar phenomenon found in our earlier work [13] with the tetraalkoxysilanes. The reverse reaction quickly becomes important and one must correct for it, though the earlier work had an additional complication of successive substitution reactions. As in the earlier paper, this could take the form:

$$C + E \rightleftharpoons CE$$

$$C + M \rightleftharpoons CM$$

$$ES + CM \underset{k_{-1}}{\overset{k_1}{\rightleftharpoons}} MS + CE$$

$$\frac{d[ES]}{dt} = -k_1[ES][CM] + k_{-1}[MS][CE]$$

$$-\frac{d\ln[ES]}{dt} = \frac{k_1 K C_C^0 C_M^0}{1 + K C_M^0}\left[1 - \left(\frac{C_M^0 - C_M}{C_M^0}\right)\left(1 + \frac{[MS]}{[ES]}\right)\right]. \qquad (9)$$

$$\downarrow \qquad\qquad\qquad \downarrow$$
$$k' \qquad\qquad\qquad Z$$

CE and CM are complexes for a previous equilibrium step involving a catalyst and solvent. K is the equilibrium constant for the interaction of the catalyst with

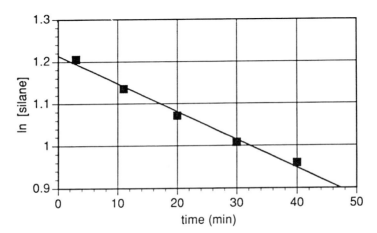

Figure 1. ln[silane] vs. time for phenyldimethylethoxysilane.

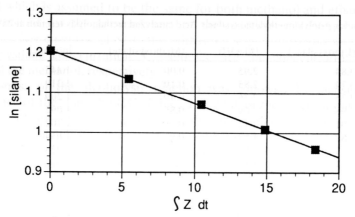

Figure 2. ln[silane] vs. ∫Zdt for phenyldimethylethoxysilane.

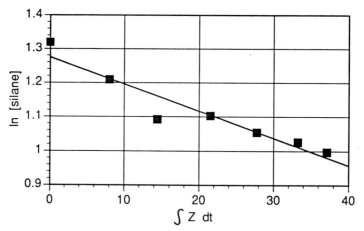

Figure 3. ln[silane] vs. ∫Zdt for *p*-toluenedimethylethoxysilane.

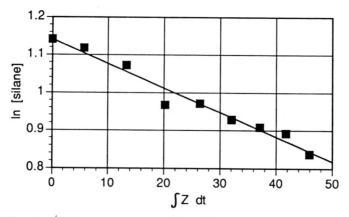

Figure 4. ln[silane] vs. ∫Zdt for *p*-methoxyphenyldimethylethoxysilane.

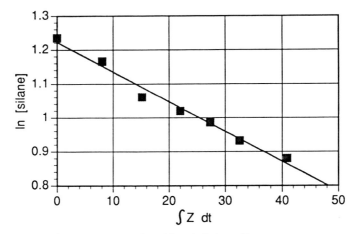

Figure 5. ln[silane] vs. ∫Zdt for *p*-chlorophenyldimethylethoxysilane.

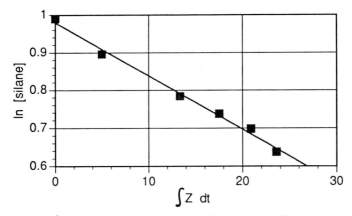

Figure 6. ln[silane] vs. ∫Zdt for *p*-trifluoromethylphenyldimethylethoxysilane.

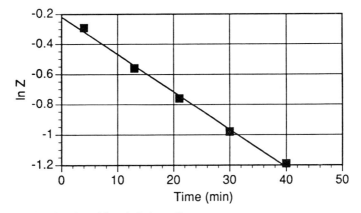

Figure 7. ln Z vs. time for phenyldimethylethoxysilane.

A plot of $\log_{10} k/k_0$ vs. σ gives a Hammett ρ of +0.43 with a correlation coefficient of 0.96 (Fig. 8) for the alcoholysis reaction, where k is the rate constant for the phenyl-substituted material and k_0 is for the reference phenyl compound.

The progress of hydrolysis of the various silanes in an ethanol solution was determined by following the drop in the ethoxysilane and the rise of siloxane dimer with GC. Sulfuric acid was used to catalyze this series. During the reaction, the only silane species present in observable quantities were the ethoxysilane and the dimer. No measurable amounts of silanol were observed (the silanols were prepared and their retention times are known). This fits with the idea of the hydrolysis being the rate determining step.

Early in the reaction, plots of the natural log of the para substituted phenyl-ethoxysilane concentration vs. the time yields relatively straight lines (Fig. 9). As

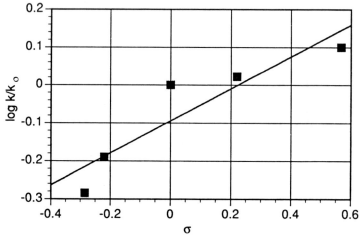

Figure 8. Hammett plot. The determination of ρ for the acid-catalyzed methanolysis of phenyl-dimethylethoxysilanes.

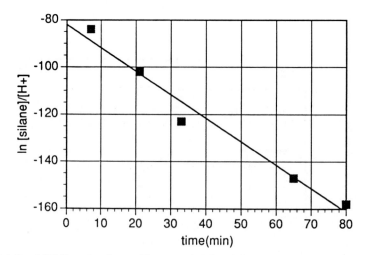

Figure 9. ln[silane]/[H$^+$] vs. time for p-trifluoromethylphenyldimethylethoxysilane.

the concentration of the products increases, the reverse reactions grow in importance and the data begin to tail away fro the line.

If we assume that the reaction mechanism for the hydrolysis goes through a siliconium ion and that its formation is the rate-determining step, the following should be the rate expression of the forward reaction:

$$R'Me_2SiOEt + H^+ \rightarrow R'Me_2Si^+ + EtOH \tag{11}$$

$$-d[silane]/dt = -k[silane][H^+].$$

Since the siliconium ion will then quickly react with water, the H^+ concentration remains constant and we can evaluate the expression:

$$\ln[silane] = -k[H^+]t + \ln[silane]_0.$$

We can determine k by plotting $\ln[silane]/[H^+]$ vs. time. Figure 9 shows the plot for hydrolysis of *p*-trifluoromethylphenyldimethylethoxysilane. Table 7 lists the values of k obtained for each of the silanes. Tables 8 and 9 show the rate constants for the hydrolysis of the silanes with different acid and water concentrations.

Table 7.
Observed rate constants for the hydrolysis of p-$XC_6H_4(CH_3)_2SiOC_2H_5$ in acidic medium at 25°C

X	$[H_2SO_4]$	k	$\log(k/k_0)$
—OCH_3	0.0123 M	3.42	0.175
—CH_3	0.0124	3.27	0.130
—Cl	0.0123	1.60	−0.584
—CF_3	0.0107	1.19	−0.88

Table 8.
Observed rate constants for the hydrolysis of p-$CF_3C_6H_4(CH_3)_2SiOC_2H_5$ at different acid concentrations

$[H_2SO_4]$	k
0.0107 M	1.19
0.0367	1.05
0.0743	1.12
0.110	1.28

Table 9.
Observed rate constants for the hydrolysis of p-$XC_6H_4(CH_3)_2SiOC_2H_5$ with ten-fold increase in water

X	k (1X)	k (10X)
—OCH_3	3.42	3.14
—CH_3	3.27	3.26
—H	2.87	2.23
—Cl	1.60	1.89
—CF_3	1.19	1.15

A plot of the $\log_{10} k/k_0$ vs. σ gives a Hammett ρ value of -1.42 (Fig. 10) for the hydrolysis reaction.

In our experiments, we have determined the activation energy for the hydrolysis of *p*-trifluoromethylphenyldimethylethoxysilane. The reaction was to run at 0°C, 25°C and 50°C. A plot of the ln k vs. 1/T yielded a straight line with a slope of -3061 and a correlation coefficient of 0.996. This translates into an activation energy of 6.1 kcal/mole (see Fig. 11).

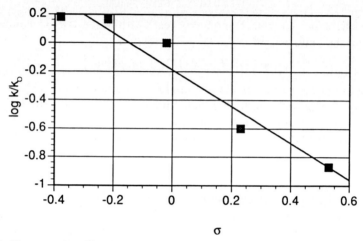

Figure 10. Hammett plot. The determination of ρ for the acid-catalyzed hydrolysis of phenyl-dimethylethoxysilanes.

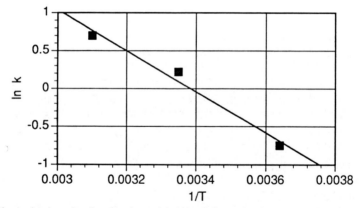

Figure 11. Arrhenius plot for the determination of the activation energy for the acid-catalyzed hydrolysis of *p*-CF$_3$C$_6$H$_4$(CH$_3$)$_2$SiOC$_2$H$_5$.

5. DISCUSSION

In our earlier work [13] we studied the ethanolysis of TMOS. It was noticed that in slow reactions that were catalyzed by carboxylic acids, a change in rate occurred. After an initial induction period, the rate would increase. It was postulated that an initial equilibrium between the alcohol and the catalyst to form some sort of complex was required to occur before the rate of the alcoholysis reaction would become rate determining:

$$ROH + Catalyst \rightleftharpoons Complex\ (OR)$$

$$Si(OR')_4 + Complex\ (OR) \rightleftharpoons \left[Complex-\overset{|}{\underset{/\,\backslash}{Si}}-OR' \right]$$

$$\rightleftharpoons Si(OR')_3OR + Complex\ (OR'). \tag{12}$$

From this mechanism, a rate expression for the first step in the ethanolysis, replacement of the first methoxy group for an ethoxy group and the subsequent reverse reaction, was developed. This was the TMOS based version of Equation (9). Evaluation of this equation for a reaction yields the rate constant k. The equation is based on the assumption that a catalyst complex is formed and also takes into account the slowing of the reaction by the increasing importance of the reverse reaction as the concentration of products becomes significant. As can be seen in Figs 1–6, this treatment yields a very linear relationship.

While our proposed mechanism was interesting, it left some unanswered questions. What was the nature of the catalyst complex and more importantly, why was this not behaving like a classic acid catalysis? Boe [11, 12] had postulated protonation of the alkoxy species followed by SN2 attack by the alcohol nucleophile. This is consistent with the negative value that he found for ρ. This does not agree though with the idea of a catalyst complex, nor does it agree with the findings reported here of a positive value of ρ.

We feel that the apparent contradictions can be explained by looking at the catalysts employed in the reactions and the work of Schowen [9, 10, 16]. Boe used a mineral acid as a catalyst—sulfuric acid—while we used a carboxylic acid—dichloroacetic acid. It is our contention that the two types of acids behave differently as catalysts. When the strong mineral acid is used, the classic acid catalyzed SN2 mechanism occurs. When a carboxylic acid is used, a mechanism more like that of the base catalyzed system occurs.

The finding of a small positive value for ρ was something of a surprise, but together with the idea that a catalyst complex must somehow be formed, we can draw some conclusions. The complex could be some form of protonation of the alcohol, but it is difficult to envision that species carrying out a substitution. A more likely possibility is that the mechanism more closely resembles Schowen's work with buffered carboxylic acid solutions. In this scheme, we propose that the carboxylic acid first protonates either the alcohol or the alkoxysilane. The resulting carboxylate then forms a complex with the alcohol as postulated by both Schowen and Boe for the base catalyzed reaction. This is then followed by a concerted displacement reaction as shown below:

$$RCOOH \underset{}{\overset{H^+}{\rightleftharpoons}} RCOO^-$$

$$RCOO^- + CH_3OH + p\text{-}XC_6H_5(CH_3)_2SiOC_2H_5 \rightleftharpoons \left[\begin{array}{c} CH_3 \\ \backslash \\ \quad O\text{---}\overset{|}{Si}\text{---}O \\ / \quad /\,\backslash \quad \backslash \\ H \qquad\qquad C_2H_5 \\ / \\ RCOO \end{array} \right]$$

$$\rightleftharpoons p\text{-}XC_6H_5(CH_3)_2SiOCH_3 + C_2H_5OH + RCOOH. \tag{13}$$

This mechanism would explain the unusual positive value for ρ. With a small positive value for the Hammett ρ, we would expect a small negative charge to develop on the silicon during the transition state. Electron withdrawing substituents would help to stabilize the transition state and thus speed the reaction. Electron donating substituents would have the opposite effect.

A mechanism similar to Equation (13) could be written when the acid reacts with the silane to give a species, $p\text{-}XC_6H_4(CH_3)_2Si(O_2CR)$, which would be very reactive with alcohols to give the final products. We can now analyze the role played by groups attached to the silicon atom for the hydrolysis reaction. Included in Table 7 are the values for log (k/k_0), where k_0 is the rate constant when the para-substituent is hydrogen. If these are used in the Hammett equation, we find a correlation using σ values with a ρ value of -1.42. The negative values for Hammett ρ indicates that electron donating groups increase the reaction rate constant. The fact that its value is lower than -1 suggests that an ionic intermediate is possible. With this evidence in mind, it seems that the postulate of Schmidt et al. [15] that acid hydrolysis occurs with the formation of a siliconium ion was correct. The mechanism would thus entail the protonation of the oxygen on the alkoxy group. This would be followed by the loss of an alcohol molecule to produce the siliconium ion, the rate determining step. Hydrolysis followed by condensation would then quickly occur. If the reaction involved formation of a penta-substituted silicon in the rate determining step of the reaction, then the electron withdrawing substituents on the phenyl group should have caused an increase in the rate of the reaction. This is just the opposite of what was observed. Therefore, one can disregard the penta-substituted species for the mechanism of the process.

With the advent of the sol-gel process for making ceramic glasses, the eighties saw a large increase in the study of the hydrolysis of alkoxysilanes. Most of these studies deal with the hydrolysis of TEOS [21, 22]. Blum and Ryan [25] studied the acid catalyzed hydrolysis of TEOS by following the formation of ethanol by gas chromatography. They were chiefly concerned about the effect different quantities of water had upon the reaction. They found that when they increased the water four-fold, the reaction still took the same amount of time to complete.

Smith [26] studied the hydrolysis of methylmethoxysilanes and determined the reactions to be first order in both the substrate and the acid. He is one of the few to suggest a mechanism other than the formation of a siliconium ion. He suggested that a five-coordinate reaction intermediate in which the entering and the leaving groups attach to the same face of the silicon:

$$R \underset{R}{\overset{R}{\underset{\diagdown}{\overset{\diagup}{\underset{Si}{\mid}}}}} \overset{H}{\underset{OH}{\overset{OCH_3}{+}}} $$

Jada [27] followed the acid catalyzed hydrolysis of TEOS by NMR. He generated the water *in situ* by the reaction of acetic acid with the ethanol solvent. He was able to follow the changes in the concentrations of the CH_2 and CH_3 of the TEOS and the OH of the ethanol. These changes led him to the conclusion that the mechanism is what he called a 'mixed SN1 and SN2 mechanism' though it appears to be a standard SN1 mechanism.

$$
\begin{array}{ccc}
\text{OEt} & & \text{OEt} \\
| & & | \\
\text{EtO}-\text{Si}-\text{OEt} & \xrightarrow{\text{H}^{\cdot}} & \text{EtO}-\text{Si}-\text{OEt} \\
| & & \quad | \quad \text{H}^{+} \\
\text{OEt} & & \text{OEt}
\end{array}
$$

$S_N^1 \updownarrow$ Slow   (14)

$$
\begin{array}{ccc}
\text{OEt} & & \text{OEt} \\
| & S_N^2 & | \\
\text{H}^{+} + \text{EtO}-\text{Si}-\text{OH} & \underset{\text{H}_2\text{O}}{\rightleftharpoons} & \text{EtO}-\text{Si}^{+} + \text{HO}-\text{Et} \\
| & & | \\
\text{OEt} & & \text{OEt}
\end{array}
$$

$\downarrow$ H·

etc.

The hydrolysis of *p*-trifluoromethylphenyldimethylethoxysilane was carried out at four different acid concentrations and the rate constants were calculated (Table 8). If the reaction is first order in acid, the rate constant should not change as shown in Table 8. Each of the alkoxysilanes was hydrolyzed with a ten-fold excess of water. If the rate equation is correct, water does not react until after the rate determining step and should not affect the rate constants (Table 9). Again, this is the case. If the mechanism proceeded through a pentavalent state such as Smith suggested, we would expect to see water involved in the rate equation. At this point, Jada's mechanism seems more likely.

To understand this reaction more fully, we performed some additional rate measurements. In an attempt to confirm the order of the reaction with respect to the catalyst and water, rates were measured while changing their concentration. The hydrolysis of *p*-trifluoromethylphenyldimethylethoxysilane was carried out at four different acid concentrations and the rate constants were calculated, see Table 8. If the reaction is first order in acid, the rate constant should not change as shown in Table 8.

As stated earlier, we have determined the activation energy for the hydrolysis of *p*-trifluoromethylphenyldimethylethoxysilane to be 6.1 kcal/mole with a correlation coefficient of 0.996.

By comparing the slopes of the Hammett plots for the alcoholysis and the hydrolysis reactions, one can see that there is a different type of intermediate in each case. For the alcoholysis reaction, one obtains a positive slope with a Hammett ρ value of $+0.43$; this suggests a negative charge in the transition state. For the hydrolysis reaction, one obtains a negative slope with a Hammett ρ value of -1.42; this suggests a positive charge in the transition state. One possible reason for the change of charge in these two reactions could be the change in polar character of the media. The addition of sulfuric acid gives a much more polar medium in the case of the hydrolysis.

6. SUMMARY

The kinetics of the acid catalyzed hydrolysis of ethoxysilanes has been studied. Each of the silanes that were used had a phenyl or para-substituted phenyl group attached to the silicon atom. This permitted a study of the linear free energy relationships of this reaction. The reaction is of interest because of its role in silane coupling agent chemistry, in the preparation of zinc-rich silicate coatings, in the sol-gel process and in the preparation of silicones in general.

The hydrolysis reaction was first order in acid and zero order in water. This reaction was found to have a Hammett ρ value of -1.42. This suggests that the electron donating groups speeded up the reaction, presumably by stabilizing the developing positive charge of the siliconium intermediate.

The kinetics of the acid catalyzed alcoholysis of alkoxysilanes was studied. The influence of attaching a para-substituted phenyl group to the silicon was explored.

The alcoholysis reaction was found to be first order in both the silane and the catalysts, and of intermediate order in the alcohol when catalyzed by carboxylic acids. The reaction was found to have Hammett ρ value of $+0.43$. This is consistent with a concerted displacement reaction between the alkoxysilane and the complex involving the alcohol and the carboxylic anion. The intermediate is a negatively charged intermediate, probably a penta-substituted silicon species.

REFERENCES

1. E. P. Plueddemann, *Silane Coupling Agents.* Plenum Press, New York (1982).
2. H. Ishida (Ed.), *Interfaces in Polymer, Ceramic and Metal-Matrix Composites.* Elsevier, New York (1988).
3. B. Arkles, *Silanes and Silicones.* Petrarch Systems (1987).
4. D. M. Burger, *Metal Finishing* **27** (April 1979).
5. (a) J. F. Montle, H. J. Markowski, P. D. Lodewyck and D. F. Schneider, III, US Patent #4323690 (April 6, 1982).
   (b) J. Stoffer, J. Montle and N. Somasiri, US Patent #4778910 (1988).
6. L. W. Kelts, N. J. Effinger and S. M. Melpolder, *J. Non-Cryst. Solids* **83**, 35 (1986).
7. M. G. Voronkov, V. P. Mileshkevich and A. Yuzhelevskii, *The Siloxane Bond,* pp. 340–355. Consultants Bureau, New York (1978).
8. R. L. Schowen and K. S. Lantham, Jr., *J. Amer. Chem. Soc.* **89**, 4677 (1967).
9. A. Modro and R. L. Schowen, *J. Amer. Chem. Soc.* **95**, 6980 (1973).
10. R. L. Schowen, C. G. Swain, K. R. Porschke and W. Ahmed, *J. Amer. Chem. Soc.* **96**, 4700 (1974).
11. B. J. Boe, *J. Organometal. Chem.* **43**(2), 275 (1972).
12. B. J. Boe, *J. Organometal. Chem.* **57**(2), 255 (1973).
13. D. J. Oostendorp and J. O. Stoffer, *J. Coatings Technol.* **57**, 55 (1985).
14. I. G. Khaskin, *Dokl. Akad. Nauk SSSR* **85**, 129 (1952).
15. H. Schmidt, H. Scholze and A. Kaiser, *J. Non-Cryst. Solids* **63**, 1 (1984).
16. R. L. Schowen and K. S. Latham, Jr., *J. Amer. Chem. Soc.* **88**, 3795 (1965).
17. E. R. Pohl and F. D. Osterholtz, in: *Molecular Characterization of Composite Interfaces,* H. Ishida and G. Kumar (Eds), pp. 157–170. Plenum Press, New York (1985).
18. E. R. Pohl and F. D. Osterholtz, in: *Chemically Modified Surfaces in Science and Industry,* D. E. Leyden and W. T. Collins (Eds), 2, pp. 545–558. Gordon and Breach Science Publishers, New York (1987).
19. E. R. Pohl and F. D. Osterholtz, in: *Silanes, Surfaces, and Interfaces,* D. E. Leyden (Ed.), pp. 481–485. Gordon and Breach Science Publishers, New York (1986).
20. Partial hydrolysis occurred after the preparation of the samples since the analysis was done at a later date.

21. Y. Paoting, L. Hsiaoming and Y. Yuguang, *J. Non-Cryst. Solids* **52**, 511 (1982).
22. S. Sakka, Y. Tanaka and T. Kokubo, *J. Non-Cryst. Solids* **82**, 25 (1986).
23. J. Jonas, *Physica* **139** & **140** (ser. B&C), 673 (1986).
24. C. W. Turner and K. J. Franklin, *J. Non-Cryst. Solids* **91**, 402 (1987).
25. J. B. Blum and J. W. Ryan, *J. Non-Cryst. Solids* **81**, 221 (1986).
26. K. A. Smith, *J. Org. Chem.* **51**, 3827 (1986).
27. S. S. Jada, *J. Amer. Ceram. Soc.* **70**, C298 (1987).

Silanes and Other Coupling Agents, pp. 181–198
Ed. K. L. Mittal
© VSP 1992.

Hydrolysis, adsorption, and dynamics of silane coupling agents on silica surfaces

FRANK D. BLUM,* WIRIYA MEESIRI,† HYE-JUNG KANG and
JOAN E. GAMBOGI‡

*Department of Chemistry and Materials Research Center, University of Missouri-Rolla,
Rolla, MO 65401, USA*

Revised version received 18 March 1991

Abstract—The hydrolysis of alkoxysilane coupling agents has been followed using proton NMR. The disappearance of the silane ester and the appearance of the alkoxy group were observed to follow pseudo-first-order kinetics. The apparent rate constant was found to be dependent on the concentration of the silane coupling agent and the pH. The adsorption of the coupling agents onto Cab-O-Sil silica was quantified using FT-IR spectroscopy. It was found that monolayer coverage was obtained with about 0.65 mmol/100 m². Solid-state deuterium NMR was used to probe the behavior of the coupling agent on the silica. The coupling agents adsorbed on the surface from deuterated aminopropyltriethoxysilane (DAPES) and deuterated aminobutyltriethoxysilane (DABES) showed similar motional characteristics. In contrast, the mobility of condensed polymers from DABES was much less restricted than that from DAPES. The amount of hydrolysis in acetone–water mixtures increased the amount of surface-adsorbed material and lessened surface mobility. Overpolymerization with bismaleimide increased the rigidity of the surface layer of the coupling agent.

Keywords: Silane coupling agent(s), deuterium NMR, molecular motion.

1. INTRODUCTION

Silane coupling agents are often used to improve the mechanical properties of composite materials [1]. They have been used to treat silica and other surfaces in order to enhance the bonding between the solid and the matrix resin. Even small amounts of coupling agents may have a profound effect on the performance of composites. In order to understand the mechanisms by which coupling agents operate, it is necessary to characterize the material at the interface. This interface is microscopic, so microscopic probes are needed. A review of microscopic studies on coupling agents and their importance in multiphase materials has been carried out by Ishida [2]. Both microstructure and dynamics together control the behavior of the interfacial material. To date, most coupling agent studies reported have focused on interfacial structure and not dynamics. A few recent studies have attempted to probe the dynamics (molecular motion) using NMR spectroscopy [3–13]. However, in order to probe the dynamics of coupling agents on surfaces effectively, further studies of the basic reactions and adsorption of the coupling agents are needed as many of the effects observed may be dependent on the processes by which they are adsorbed on the surface.

---

*To whom correspondence should be addressed.
†Present address: Royal Thai Airforce, Bangkok 10210, Thailand.
‡Present address: Chemistry Department, Princeton University, Princeton, NJ 08544, USA.

Silane coupling agents are normally hydrolyzed to silanols before adsorption onto a surface. There are several factors which influence the hydrolysis, stability, and adsorption behavior of the silanes, including pH, type of silane used, concentration, and water content [1]. Knowledge of the stability of the silane solution is also important and may be used as a guideline in many applications. Adsorption of silane coupling agents on solid surfaces is affected by the type of surface and methods of application as well.

Silane coupling agents—from initial hydrolysis in the treating solution to adsorption onto the filler surface—have been studied by several techniques, including radioisotope assay, infrared (IR), ultraviolet, Raman, and NMR spectroscopy [14–21]. The kinetics of the hydrolysis reaction of silane coupling agents in aqueous solutions have been studied by Raman spectroscopy [14]. The results indicated that the rate of hydrolysis depended on the primary structure of the coupling agents. The use of ultraviolet spectroscopy for trimethoxyphenylsilane [15] and an extraction method with IR spectroscopy for glycidoxypropyltrimethoxysilane (GPS) [16] followed the rates of hydrolysis for silane esters. It was reported that the hydrolysis reactions followed pseudo-first-order kinetics where the rate appeared to be both acid- and base-catalyzed. The hydrolysis proceeded through a stepwise process in which the first step, the formation of a mono-ol, was the slowest.

Condensation reactions of the silanols may occur after hydrolysis to form dimers, higher molecular weight oligomers, and polymers. It was reported [22] that dimethylsilanediol monomers condensed to a dimer, followed by slower reaction of the resulting oligomers with either acid or base catalysts. Condensation products often separate from the treating solution as agglomerated particles or gels [1]. If the silane coupling agents are most effective when adsorbed as monomers the optimum time for adsorption from a treating solution is after hydrolysis and before oligomerization.

In hydrolyzed silane coupling agent solutions, monomeric or oligomeric silanols may either react to form siloxane bonds with other coupling agents or surface silanols, or exist as free silanols. A variety of factors determine adsorption processes [1] including (i) the rate at which the active species are formed, (ii) their relative ability to be attracted to the particular surface, (iii) their orientation with respect to the surface, and (iv) the type of layer formed on the surface. Ishida and Koenig [23] reported evidence for covalent bonding between vinylsilane and the silica surface through silanol groups. Sindorf and Maciel [24] found ^{29}Si-NMR to be sensitive to surface bond formation and the type of silicon substitution. Kaas and Kardos [25] suggested intermolecular condensations of aminopropylsilane (APS), forming a siloxane polymer film on a silica wafer. In most applications, silane molecules are chemisorbed, forming covalent Si–O–surface bonds in the first layer and physisorbed as oligomers in the outer layers. Most of these physisorbed layers can be washed away with a suitable solvent while the chemisorbed species remain attached to the surface. On silica surfaces, where silanol groups are moderately far apart, it seems reasonable that one surface–siloxane bond (Si–O–Si) per silane molecule occurs. Additional surface–siloxane bonds per silane molecule may be formed, for example, on a surface of silica gel where there is a high concentration of surface silanols [26].

Johannson *et al.* [21] investigated the nature of adsorbed silane coupling agents on E-glass using ^{14}C-labeled amino- and glycidoxy-propyltrimethoxy-silanes. They obtained an adsorption isotherm by immersing heat-cleaned E-glass fibers in a solution of coupling agent at various concentrations, and measured the amounts of silane deposited on the glass surface. They also reported that the amounts of methacryloxypropyltrimethoxysilane (MPS) adsorbed on silica surfaces were much smaller than on the E-glass surface, probably due to different properties of the two surfaces. Glass surfaces are typically more hygroscopic than silica surfaces, due to the presence of alkali and alkaline earth oxides [27]. Silane coupling agents were reported to be adsorbed at monolayer coverages or less on silica surfaces [17, 28]. De Haan *et al.* [29] reported that amino- and methacryloxy-silanes from toluene and water solutions were adsorbed at less than monolayer coverages. Apparently, not only the hydrolyzed silanes, but also the unhydrolyzed molecules were adsorbed by the glass surface [14]. The unhydrolyzed silanes are, however, less effectively adsorbed by surface hydroxyl groups [1].

Most of the previous adsorption studies of silane coupling agents have been done in aqueous and aqueous–alcoholic solutions which are commonly used in the application of coupling agents. Quantitative study of these solutions by IR spectroscopy is difficult due to strong absorption of the solvents. This is especially problematic at very low coupling agent concentrations. Acetone is a good solvent for coupling agents, and is relatively transparent in the appropriate IR range of interest. Therefore, it was chosen as a co-solvent for our studies, although there is the possibility of a reaction between it and the aminofunctional group of this silane [30].

In this paper, we report results on the hydrolysis of silane coupling agents in acetone–water mixtures, their adsorption on silica surfaces, and the dynamics of one of the coupling agents on the surface. The hydrolysis of silane coupling agents was followed by proton NMR spectroscopy, the adsorption isotherms were determined using IR spectroscopy, and the dynamics probed with solid-state deuterium NMR. There has been considerable use of solid-state NMR techniques on surface-bound coupling agents through the use of [13]C [8, 24, 31–36], [29]Si [10, 37–40], or both [4, 29, 41–44]. These studies have helped to provide an excellent view of the structure of the surface bound coupling agent. As previously mentioned, however, only a few studies have focused on the molecular motion of these species [3–13].

2. EXPERIMENTAL

2.1. Materials

The silane coupling agents used in this study were aminopropyl- (APMS), aminoethylaminopropyl- (AAPS), methacryloxypropyl- (MPS), vinyl- (VS), glycidoxypropyl- (GPS), and mercaptopropyl-trimethoxysilane (MrPS), deuterated aminopropyltriethoxysilane (DAPES), and deuterated aminobutyltri-ethoxysilane (DABES). Except for DAPES and DABES, the coupling agents were purchased from either Aldrich Chemical Co. (Milwaukee, WI) or Petrarch Systems (Bristol, PA) and used without further purification. The structures of these silanes are shown in Table 1. When DABES or DAPES is adsorbed onto the silica surface or polymerized, the resulting coupling agent is then referred to

Table 1.
Structures of the silanes used

DAPES	$(CH_3CH_2O)_3Si-CH_2CH_2CD_2NH_2$
DABES	$(CH_3CH_2O)_3Si-CH_2CH_2CH_2CD_2NH_2$
APMS	$(CH_3O)_3Si-CH_2CH_2CH_2NH_2$
AAPS	$(CH_3O)_3Si-CH_2CH_2CH_2NHCH_2CH_2NH_2$
MPS	$(CH_3O)_3Si-CH_2CH_2CH_2OC\overset{\overset{\displaystyle O}{\|\|}}{-}C=CH_2$
VS	$(CH_3O)_3Si-CH=CH_2$
GPS	$(CH_3O)_3Si-CH_2CH_2CH_2OCH_2-HC\overset{O}{\overset{\diagup\diagdown}{-}}CH_2$
MrPS	$(CH_3O)_3Si-CH_2CH_2CH_2SH$

as DABS or DAPS, respectively. The loss of the E is consistent with the loss of the ethoxy group.

High surface area fumed silica (about 99.8% SiO_2), Cab-O-Sil grade M5, was donated by Cabot Corp. (Tuscola, IL). It was reported, and verified by N_2 adsorption, to have a surface area of 200 m^2/g. The silica was heated to about 120°C for at least 24 h under vacuum before use, which did not change the surface area. Distilled acetone and water were used as solvents for the coupling agents. Acetic acid and hydrochloric acid were used as catalysts for the non-aminofunctional silane coupling agents.

2.2. Hydrolysis studies

The hydrolysis rates were determined using a JEOL FX-100 NMR spectrometer at ambient temperature (21°C) using protons at 100 MHz. A weighed amount (*ca.* one drop) of coupling agent was placed in a 5 mm NMR tube. Weighed amounts of solvent (acetone–water solution) were added and mixed with the coupling agent just before insertion into the NMR probe. Spectra were collected using the kinetics program on the instrument which collected spectra at pre-determined time intervals. Intervals of 5, 10, or 20 min were typically used for the hydrolysis reactions. Typically, a 16 μs pulse width (90° flip), an 8 s pulse delay, 1000 Hz spectral width, 8192 data points, and 4 scans were used for each spectrum. The intensities of the methanol resonances were determined from the peak heights and plotted vs. time in order to observe the amount of methanol produced.

2.3. Adsorption isotherms

A stock solution of the coupling agent was prepared by mixing the trimethoxy-silane with 10:1 acetone–water solutions. The stock solution concentrations were 2–3 g/100 ml for APMS and AAPS. The solution pH was not adjusted and the experiments were done at ambient temperature (21°C). From our kinetics studies we were convinced that the hydrolysis was completed under these conditions and that no appreciable polymerization occurred. Various concentrations of hydrolyzed coupling agents were prepared by diluting the stock solutions. These dilutions were used for surface treatments (treating solutions) and for Beer's law plots (standard solutions). Weighed amounts of Cab-O-Sil silica were placed in 15 ml screw-cap tubes to which 8 ml aliquots of the treating solution were added using a syringe. The tube was then shaken on a vortex mixer for about half a minute, tumbled at a rate of about 20 rpm for 15 min, and centrifuged at 3500 rpm for 45 min. The supernatant treating solution was pipetted from the tube and saved for concentration measurement by FT-IR. The treated silica was washed with distilled water, centrifuged, decanted, and heated to about 110°C for at least 3 h under vacuum.

In order to determine the concentration in the supernatant solutions, absorbance measurements were carried out using a Perkin-Elmer 1750 FT-IR spectrometer. The solution was injected into a demountable liquid cell fitted with calcium fluoride windows and a 0.1 mm Teflon spacer. Spectra of the supernatant solutions were collected in 'single-ratio' mode with 4 cm^{-1} resolution and 3 scans. A spectrum of the solvents used was also collected and subtracted from these spectra. The absorbance values were determined by measuring the absorbance at 2877 cm^{-1} (methylene C—H stretch) in the difference spectra. This frequency has a minimum contribution from the methyl groups of the acetone. The baseline was chosen from the range at about 2650–2600 cm^{-1}, which was consistently flat due to complete subtraction of the spectra. A second absorbance reading was taken at about 1030 cm^{-1}, where intensity was linear with methanol concentration. Since the methanol concentration was the same in both solutions, the absorbance readings were used as a reference for the standard and corresponding supernatant solutions. In the event that the cell spacings changed slightly, the absorbance due to methanol at this frequency was used as an internal standard. The ratio of the methanol absorbance from the spectra of the standard and supernatant solutions was used as a multiplying factor for the resonance at 2877 cm^{-1}.

Beer's law plots were constructed from data taken from the standard solutions at 2877 cm^{-1} and used to calculate the concentrations of coupling agents in the supernatant solutions. These were linear and fitted using the least-squares method for calibration. An absorbance correction was made by subtracting absorbance due to the methyl groups from methanol at 2877 cm^{-1} from a separate Beer's law plot. The amounts of silane coupling agent adsorbed by the silica were determined by taking the difference between the concentrations of the initial (standard) and supernatant (equilibrium) solutions. An adsorption isotherm was constructed by plotting the amount of silane coupling agent adsorbed per surface area of silica vs. the equilibrium concentration.

We have also compared the determination of the adsorption isotherms with those from a standard gravimetric technique. The precision of the data obtained by the gravimetric technique was found to be lower than that obtained by the FT-IR technique, particularly at very low silane concentrations (about 5%, and 3% at 0.4 g/100 ml concentration for the gravimetric and the FT-IR techniques, respectively). This is because very small weights of the silanes cannot be measured accurately and the silane is volatile. The effects of suspended silica particles in the solution were also examined by centrifuging methanol solutions with suspended silica for different periods of time. FT-IR spectra were taken of the supernatant solutions and compared. The results showed that suspended silica was responsible for less than 1% error in the FT-IR absorbance values.

2.4. Solid-state NMR studies

The solid-state deuterium NMR experiments were performed on surface-adsorbed material from triethoxyaminopropylsilane-d_2 (DAPES) and triethoxy-aminobutylsilane-d_2 (DABES) prepared as described previously [9]. The samples were heated to 90°C at 10 mmHg for 12 h after adsorption. These deuterated coupling agents have been characterized by IR and NMR spectroscopy. The solid-state deuterium NMR spectroscopy was done on a Varian VXR-200 at 30.7 MHz. A quadrupole echo pulse sequence was used with a 2 s interval, 2 μs 90° pulse, 30 μs echo time, 2 MHz sweep width, and, typically, overnight accumulation.

2.5. Overpolymerization of treated surfaces

Cab-O-Sil treated with DABES was overpolymerized with bismaleimide, 1,1'-(methylenedi-4,1-phenylene)bismaleimide (BMI) from Aldrich. A mole ratio of 5:1 (BMI–DABES) was used and the BMI was added to the surface in 2-butanone, heated at 80°C for 90 min, dried, and then heated at 185°C for about 2 h [13].

3. RESULTS AND DISCUSSION

3.1. Hydrolysis

The hydrolysis of different silane coupling agents in various acetone–water mixtures was studied by monitoring the amount of methanol produced using proton NMR. The NMR spectra showed an increase in the methanol and a decrease in the methoxy-resonance intensity as a function of time. A series of proton-NMR spectra of aminopropyltrimethoxysilane (APMS) in acetone–water solution is shown in Fig. 1 as a function of time. The spectra show the decay of the methoxysilane resonance (A, 3.56 ppm) and the build-up of the methanol resonance (B, 3.35 ppm) with increasing time. A variety of coupling agents have been studied by this method. It was found that APMS hydrolyzed to completion in about 200 min (5.2% APMS, 10:1), 100 min (8.5% APMS, 10:1), and less than 5 min (8.6% APMS, 1:1). (The ratios are the volume ratios of acetone–water.) The completion of the hydrolysis was suggested from the appearance of the maximum in the intensity of the methanol. It was often very difficult to quantify this, since other reactions like alcoholysis and condensation reactions could occur during the hydrolysis as previously reported [45]. No evidence of reaction of the amine with acetone [30] was observed in the proton spectra.

In some cases, such as the hydrolysis of glycidoxypropyltrimethoxysilane (GPS) in a 10:1 acetone–water mixture at pH 3.9, the silanol intensity did not reach the maximum even after several hours. The spectra also suggested that there was some methanol present initially, probably due to a small amount of hydrolysis. When the hydrolysis reactions were very slow, some of the hydrolysis plots showed an initial slow rate followed by a faster rate. This is consistent with

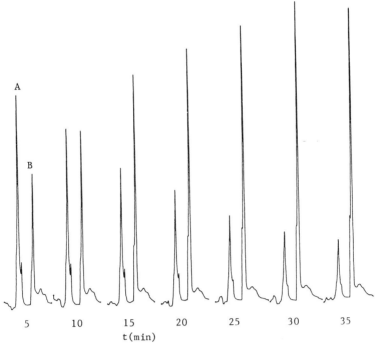

A

B

5   10   15   20   25   30   35

t(min)

Figure 1. A series of partial proton-NMR spectra for the hydrolysis of 4.8% aminopropyltri-methoxysilane (APMS) in 5:1 acetone–water as a function of time. (A) methoxysilane resonance (3.56 ppm); (B) methanol resonance (3.35 ppm).

earlier kinetics studies [15, 16]. In the faster reactions, this was not observed, possibly due to the time intervals used in data collection, which were long compared with the hydrolysis time.

The analysis of the hydrolysis data, in terms of the fundamental reactions, is made quite complicated because of the large number of simultaneous condensation and hydrolysis reactions which can occur [45]. It is possible to express the rates of these reactions in terms of the disappearance of the methanol species (MeOH), which should be equal to the amount of silanol species or

$$\text{rate} = \text{d[MeOH]}/\text{d}t = k'[\text{W}][\text{C}][\text{A}] = k[\text{A}], \tag{1}$$

where the latter three terms represent the concentration of water (W), catalyst (C), and coupling agent (A), respectively. The rate constant k' is for the fundamental reaction, but if the water and catalyst concentrations are kept constant the rate law may be rewritten in terms of an apparent rate constant, k. The catalyst could be added acid or even the coupling agent itself. Integration of this equation suggests that a plot of $\ln (I_{max} - I)$ vs. time should be linear. I_{max} was taken as the asymptotic value in the intensity vs. time plots. Within experimental error, these fit the pseudo-first-order model well.

It is an oversimplification to fit the hydrolysis reactions with a pseudo-first-order rate analysis, but this does allow comparison among different systems. Even with these complications, some general observations can be made. Shown in Table 2 are the results from different systems at ambient temperature under

Table 2.
Hydrolysis of silane coupling agents in acetone–water

	Acetone–water volume ratio	pH	Silane conc. (M)	k (s⁻¹) (×10⁻⁴)
APMS	10:1		0.23	2.2
			0.38	3.8
	5:1		0.22	6.5
	1:1		0.14	Fast
			0.29	Fast
			0.43	Fast
AAPS	10:1		0.084	3.2
	5:1		0.085	4.8
	1:1		0.085	18
MPS	10:1		0.098	Very slow
			0.42	Very slow
		3.9	0.19	0.38
		3.7	0.052	0.57
		3.3	0.046	1.8
		1.0 (HCl)	0.052	Fast
	5:1	4.5	0.11	0.95
		3.6	0.13	2.8
	1:1	3.6	0.044	Fast
VS	10:1	3.9	0.082	1.0
		3.7	0.11	1.4
		1.0 (HCl)	0.082	Fast
	5:1	4.5	0.095	3.3
		3.6	0.12	3.5
	1:1	3.6	0.097	Fast
GPS	5:1	4.5	0.22	0.63
		3.6	0.27	2.0
	1:1	3.6	0.098	Fast
	Water[a]	5.00	0.030	23
		6.13	0.030	2.2
		7.16	0.0210	0.52
		8.39	0.030	0.73
		8.78	0.030	10
MrPS	10:1	3.9	0.16	0.60
		3.7	0.11	0.67
	5:1	3.6	0.097	1.4
	1:1	3.6	0.096	Fast

[a] Data from Pohl and Osterholtz [16].

different conditions. In all cases, the hydrolysis rate increased with the amount of water present. For aminofunctional silanes (APMS and AAPS), the apparent hydrolysis rate constant increases with the methanol (and consequently silanol) concentration and the reaction proceeds at a reasonable rate without pH control. The increase in the hydrolysis rate of APMS and AAPS is probably due to self-catalysis. Ammonia has been reported to act as a catalyst for the adsorption of trimethylchlorosilane onto E-glass and Aerosil (a fumed silica) [46]. Quali-tatively, AAPS appears to hydrolyze faster than APMS under similar conditions. This faster reaction is possibly due to the second catalytic amine group in the AAPS molecule. The apparent rate constant also usually increased with the coupling agent concentration, suggesting a higher order rate law, but we were unable to quantify the exponent reliably.

Non-aminofunctional silanes did not show an observable methanol resonance after about 5–6 h when no catalyst was added. This suggests that hydrolysis was very slow under these conditions. An acid was added to aid in the hydrolysis of MPS, VS, GPS, and MrPS. The results show that the rate constant increases with a decrease in the pH of the solution, similar to previously reported results [16, 47]. The addition of HCl also significantly increased the reaction rates, as reported by Pohl and Osterholtz for the hydrolysis of GPS in water [16].

It must be re-emphasized that the use of these apparent rate constants is somewhat tentative. For example, different methoxy groups attached to silicon atoms hydrolyze at different rates: the first methoxy groups hydrolyze most slowly. These competing reactions may occur simultaneously, and consequently mask the true rate of silanol production. This is consistent with hydrolysis rate constants of the silane coupling agents obtained from previous authors [16].

3.2. Adsorption isotherms

The adsorption isotherms of completely hydrolyzed APMS and AAPS samples from an acetone–water (10:1) mixture are shown in Fig. 2. The isotherms show low adsorbed amounts at very low concentrations and a leveling-off at higher values. The leveling-off in the higher concentration range is believed to be due to the formation of a 'monolayer'. Approximately 20 samples for APMS were run and most isotherms fell between the two shown for APMS. The behavior of the

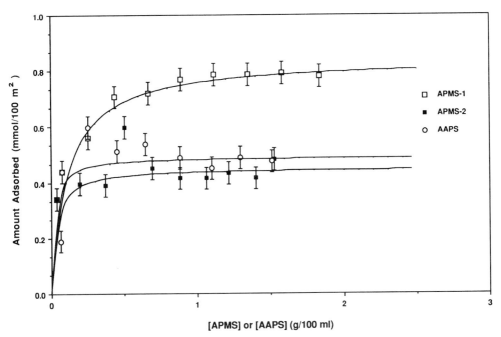

Figure 2. Adsorption isotherms of APMS (1 and 2) and AAPS from 10:1 acetone–water mixtures onto silica. The curves are drawn from a Langmuir-type isotherm. The two extremes for the APMS adsorption roughly span a large number of measurements done.

APMS and AAPS isotherms was similar. It was observed that the precision of a single adsorption experiment is very good, but the reproducibility of the entire isotherm is poor. The lack of reproducibility is again probably due to the wide variety of reactions undergone by the silanes. The partial formation of oligomers also makes the hydrolysis and adsorption process rather complicated.

The calculated adsorption isotherm based on the Langmuir equation for adsorption of gases is

$$\theta = bP/(1 + bP),\tag{2}$$

where θ is the fraction of surface covered, b is the ratio of the equilibrium constants, and P is the gas pressure. Since the value of θ is determined by the availability of adsorption sites, which may vary under different conditions, an arbitrary constant, K, is introduced into the Langmuir equation. The Langmuir equation is then expressed as

$$L = KbC/(1 + bC),\tag{3}$$

where L and C are defined as the amount of adsorption (in mmol/100 m²) and the concentration of the supernatant solution (in g/100 ml), respectively. The values of b, K are 9, 0.84 (APMS-1), 30, 0.45 (APMS-2), and 50, 0.49 (APMS-3) for the three isotherms shown in Fig. 2, respectively. It should be noted that the Langmuir isotherms, normally used to describe a monolayer-type adsorption, are shown only for comparison.

Naviroj et al. [28] reported that VS and MPS were adsorbed on silica (surface area of 130 m²/g) at 0.35 and 0.24 monolayer equivalents, respectively. Space projections of MPS molecules in the perpendicular and parallel orientations of the extended form were reported to be about 24 and 55 Å², respectively [48]. Therefore, it is expected that a single molecular layer coverage of MPS would differ by a factor of about 2 for the two different orientations. The difference by a factor of about 2 from our experiments may also be due to similar effects.

For APMS, the adsorption isotherms roughly average about 0.65 mmol/100 m² at 'monolayer' coverage. This would be lower, about 0.45, for a 2% solution of coupling agent. Cab-O-Sil silica has about four SiOH groups present per 100 Å² of surface [49]. Full coverage would correspond to just under 1 coupling agent molecule per surface hydroxyl group. At 2%, our data suggest there are about 0.7 coupling agents per surface hydroxyl. At approximately 25 Å² per coupling agent molecule, silanes positioned perpendicular to the surface are expected.

It is not only the difference in activities of the silanes that determines how they can be adsorbed, but also the surface and solvent effects. Our silica was dehydrated before use. Silica is known to rehydrate more slowly than E-glass; thus, it might have fewer surface silanols to react with the coupling agents [17]. E-glass surfaces also adsorb about twice as much water as silica surfaces [50], suggesting that the silane molecules may be more strongly attracted by the E-glass surface than by the silica surface. Aminopropylsilane (APS) on iron coupons exists as an incompletely hydrolyzed and strongly bound film of about 60 Å thick, i.e. about 4–6 monolayer equivalents [51]. Bascom [52] reported that MPS films on silica were about 5–10 Å and about 200 Å thick when adsorbed from methylethylketone and water solutions, respectively (a monolayer film of MPS was estimated to be about 10–15 Å thick). These results suggest that some polar solvents may keep the silane molecules in solution. The amount of water present may also yield different coverages. With acetone as a co-solvent, the adsorption of the silanes on silica is expected to be at monolayer and sub-monolayer coverages.

Although the FT-IR technique described here appears to be simple, it is very sensitive to many correction and calculation factors. These factors may produce considerable error in the adsorption isotherms. At the present time, we have not been able to obtain results on the adsorption isotherms which have the desired accuracy. However, the method still provides reasonable guidelines for adsorption experiments. Bascom also reported that some adsorption isotherms under particular conditions were moderately reproducible in some cases, but differed in other cases [52].

3.3. NMR dynamics studies

Solid-state NMR studies have proven to be very valuable in the study of dynamics of surface-bound coupling agents [3–13]. These studies have shown differences in mobility depending on the type and amount of coupling agent used. In our laboratory, we have used deuterated coupling agents to probe this mobility. In Fig. 3 the spectra of deuterated aminopropyltriethoxysilane (DAPES) and aminobutyltriethoxysilane (DABES) both in bulk polymerized form and adsorbed in monolayer coverage on silica surfaces are shown. The coupling agent was labelled with deuterons on the methine position next to the amine. There are several interesting things about the spectra. First, it was noted, on the basis of the line shape, that the DABES polymerized to the bulk polymer has faster and/or isotropic motion than the DAPES polymer [53]. This was easily observed by the line width of the narrow portion of the spectra. In addition, both materials consist of a superposition of a broad and a narrow component. The broad component was that given by a static powder pattern spectrum. These spectra have been fitted in terms of molecular motional parameters [12]. For the condensed polymers, DABES was less restricted than DAPES. The spectra may be quantitatively fitted based on a rigid plus a mobile (undergoing anisotropic rotational diffusion) component [53]. The spectra of the surface-adsorbed materials for both polymers are broader than those of the condensed bulk polymer. This was indicative of slower and/or less isotropic motion of the coupling agent adsorbed on the surface. Quantitative estimates of the motion responsible for these spectra based on line-shape simulations have been made [12] and the line shapes were simulated with a rigid component plus a mobile component. The mobile component undergoes both anisotroic rotation and two site conformational jumps separated by an angle of 109°. The fact that both surface species yield similar spectra, even though they have different chain lengths, may be due to the packing of the chains at the interface. For perpendicularly packed chains, there did not appear to be a significant difference in mobility. This could be because the chain behaved as if it were confined to a cylinder.

The matching of the experimental and simulated spectra for coupling agents as well as other materials provides evidence for certain modes and rates of motion.

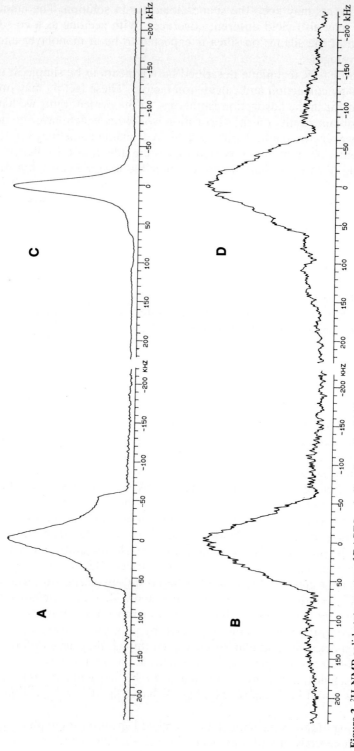

Figure 3. ^{2}H-NMR solid-state spectra of DAPES as bulk polymerized (A) and on silica (B), and DABES as bulk polymerized (C) and on silica (D).

In order to probe the limits of these, the experimental spectrum and some simulated spectra of deuterated surface-bound DABES at monolayer coverage are shown in Fig. 4. In this figure, the 'best-fit' spectrum is shown to be approximately that with a mobile component of 80% and a rigid component of 20%.

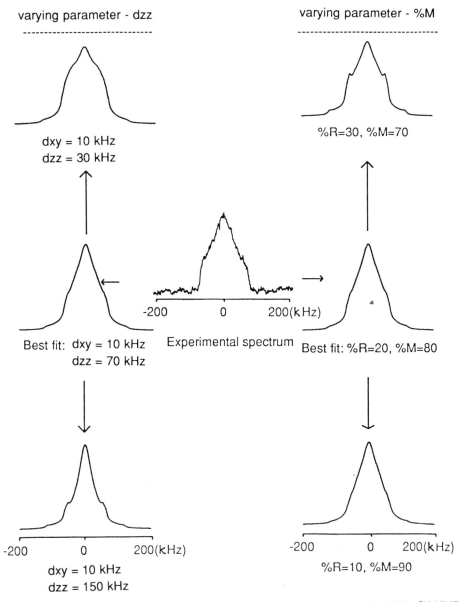

Figure 4. Experimental (center) and 'best-fit' simulated (right middle and left middle) ²H-NMR spectra of DABES on silica from a 2% treating solution (approximately monolayer coverage). On the right-hand side, the fractions of rigid and mobile components were varied from 30% rigid (upper) to 10% rigid (lower). On the left-hand side, the anisotropic rotational rates for the faster rotation (d_{zz}) were varied from 30 kHz (upper) to 150 kHz (lower). These magnitudes of changes roughly bracket the uncertainty in the simulations.

This is shown (twice) on the right and left middle of the figure. The mobile component is simulated with an anisotropically diffusing C–D bond vector ($d_{zz} = d_{\parallel} = 10$ kHz and $d_{xy} = d_{\perp} = 10$ kHz), plus jump-type motions between two sites with the C–D bond vectors separated by the angle of 109° (rate = 1 MHz) [12]. On the right-hand side of the figure, we have altered the amounts of rigid and mobile components of the spectrum by 10% (absolute) in each direction and the resulting simulated spectra are shown in the right-hand corners. At 30% rigid component, the simulated spectrum clearly shows features that are not evident in the experimental spectrum (e.g. peaks at ±60 kHz). At 10% rigid fraction, the simulated spectrum does not fit the experimental spectrum as well, but the simulated spectrum still borders on an acceptable fit. We therefore estimate the uncertainty in the amounts of rigid and mobile components to be approximately 10% (absolute). On the left-hand side, we have altered the rate of the slow diffusion by about a factor of 2 and the simulated spectra are shown in the left-hand corners. The simulated spectra also do not match the experimental spectra. Consequently, we believe that we can estimate the rotational rates within about a factor of 2. A second iteration of these simulated spectra, where the other variable has been altered (for example, the percent rigid and mobile components of the upper left-hand spectrum) to achieve the best-fit spectrum, did not match the experimental spectrum as well as the original best-fit (right and left middle). Consequently, while it is not possible to rule out all other possible fits of this experimental spectrum, it appears that we have good physically reasonable fits of the data which seem to be relatively sensitive to the parameters used in the fitting procedure.

We have previously shown the sensitivity of the dynamics of the coupling agent to the concentration of the solution used to deposit the coupling agent [12]. In addition, the spectra are also sensitive to the amount of hydrolysis which occurs. The spectra of DABS adsorbed on silica from DABES at various hydrolysis times are shown in Fig. 5. As previously noted [9], the hydrolysis of ethoxylated species occurs significantly more slowly than that of the methoxylated analog. The 10:1 ratio of acetone–water allows significant control of the hydrolysis, and consequently the adsorption process. As noted in the figure, the width of the spectrum increases at longer hydrolysis times and it is believed to be indicative of more material adsorbed on the surface. At submonolayer coverage, there is more room for the coupling agent to move and consequently its motion is more isotropic and faster. At concentrations which would correspond to the deposited layer being approximately a monolayer, the mobility of the surface-bound coupling agent is minimized. After approximately 1 day of hydrolysis in a 10:1 acetone–water ratio, the spectra are similar to those produced by hydrolysis with only water.

Finally, the mobility of the coupling agent can be significantly reduced by reaction and overpolymerization with a reactive species. In the case of DABES adsorbed on silica and overcoated with bismaleimide, the spectrum in Fig. 6 shows a broadening upon overpolymerization. Comparison of the native surface species and overpolymerized species suggests a significant reduction in mobility. Detailed studies which have simulated the line shapes for these materials suggest that significant rotational diffusion is not present in the overpolymerized systems. In the overpolymerized systems, the line shapes were fitted with two-site

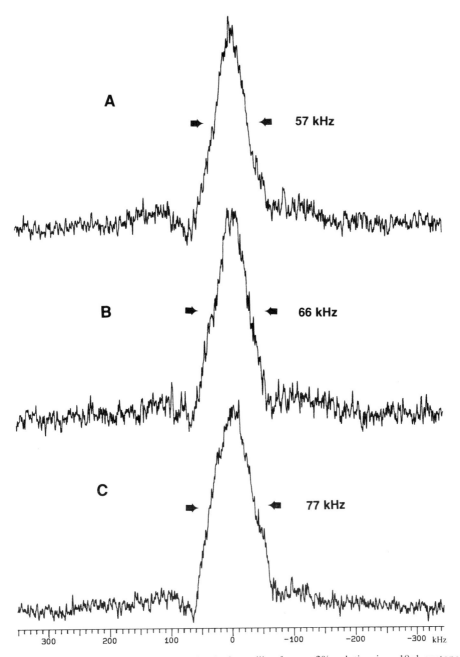

Figure 5. ^{2}H-NMR spectra of DABES adsorbed on silica from a 2% solution in a 10:1 acetone–water mixture for hydrolysis times of (A) 1.75 h, (B) 3 h, and (C) 1 day. The full-width at half-height of each is given in the figure.

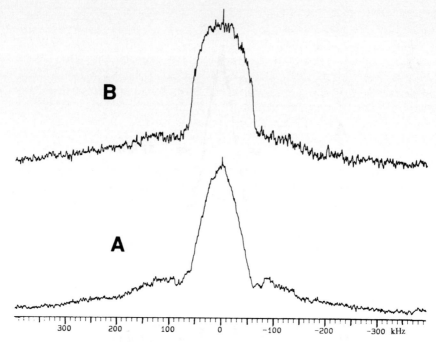

Figure 6. ^{2}H-NMR spectra of DABES on silica: (A) treated surface and (B) treated surface over-polymerized with bismaleimide.

jump-type motions (conformational jumps) with the jump angle approximately 109°. The anisotropic motion observed in the bare coupling agents on the surface was absent. Simulations of these spectra are in good agreement with the echo time dependence of the quadrupole echo spectra [13]. The fact that these spectra no longer require anisotropic diffusion in order to fit the spectrum suggests that in the overpolymerized system, this rotational diffusion is probably no longer occurring at rates that are comparable to the NMR time-scale, which is the reciprocal of the static quadrupole splitting. Consequently, we may view the mobility of the species when overpolymerized as undergoing only internal motions without large to medium scale rotations. This qualitative difference with the overpolymerization is significant in terms of the mobility of the coupling agent.

4. CONCLUSIONS

We have shown how NMR and IR are used to probe the hydrolysis and adsorption isotherms of silane coupling agents. It was observed that the pseudo-first-order hydrolysis rate constants of aminofunctional silane coupling agents increase with increasing amounts of the silanes, probably due to the self-catalyzing properties of these silanes. The hydrolysis rates of non-aminofunctional silanes increased with decreasing pH. In all cases, the hydrolysis rate constants increased with increasing amount of water in the solution.

The adsorption isotherms of silane coupling agents from acetone–water solutions onto silica appear to show less coupling agent on the surface compared with those from aqueous solutions onto an E-glass surface reported by other authors. The amount adsorbed on the surface for aminopropylsilane corresponds to about 1 molecule per surface hydroxyl, which is roughly 4 molecules per $100 \, \text{Å}^2$. This corresponds to packing perpendicular to the surface. Inaccuracy in these experiments may be due to either experimental uncertainties or differences in the packing of the silane molecules on the silica surface. Our results appear to have greater precision than accuracy. It is possible to control the adsorbed amounts of silane coupling agents on the silica surface by controlling the hydrolysis and treating conditions of the silanes. The mobility of the coupling agent on the surface is also controlled by similar effects.

We have shown that the deuterium NMR technique is very powerful when applied to these materials adsorbed on surfaces and that it can probe the interfacial layer in a composite. This technique does not require optical clarity and in general should be amenable to a large number of systems. The similarity of the aminopropylsilane and aminobutylsilane dynamics on the surface may be rationalized if these molecules are fairly well extended when adsorbed at monolayer coverages on the surface. The packing for these relatively small molecules may be such that conformational changes as well as some wobbling on the surface dominate the motional dynamics, as observed by deuterium NMR. It was also observed that the mobility of coupling agents in interfacial layers is reduced when overpolymerized with a polymer.

Acknowledgements

We wish to thank the Office of Naval Research and the Royal Thai Air Force (WM) for financial support of this project.

REFERENCES

1. E. P. Plueddemann, *Silane Coupling Agents*, Ch. 1. Plenum Press, New York (1982).
2. H. Ishida, *Polym. Composites* **5**, 101 (1984).
3. E. C. Kelusky and C. A. Fyfe, *J. Am. Chem. Soc.* **108**, 1746 (1986).
4. G. S. Caravajal, D. E. Leyden, G. R. Quinting and G. E. Maciel, *Anal. Chem.* **60**, 1776 (1988).
5. F. D. Blum, R. B. Funchess and W. Meesiri, in: *Interfaces in Polymer, Ceramic, and Metal Matrix Composites*, H. Ishida (Ed.), p. 205. Elsevier, New York (1988).
6. F. D. Blum and W. Meesiri, *Polym. Preprints* **29**, 44 (1988).
7. M. Gangoda, R. K. Gilpin and J. Figueirinhas, *J. Phys. Chem.* **93**, 4815 (1989).
8. T. P. Huijgen, H. A. Gaur, T. L. Weeding, L. W. Jenneskens, H. E. C. Schuurs, W. G. B. Huysmans and W. S. Veeman, *Macromolecules* **23**, 3063 (1990).
9. H.-J. Kang, W. Meesiri and F. D. Blum, *Mater. Sci. Eng.* **A126**, 265 (1990).
10. K.-P. Hoh, H. Ishida and J. L. Koenig, *Polym. Composites* **11**, 121 (1990).
11. F. D. Blum, R. B. Funchess and W. Meesiri, in: *Solid State NMR of Polymers*, L. Mathias (Ed.). Plenum Press, New York (1991).
12. H.-J. Kang and F. D. Blum, submitted.
13. J. E. Gambogi and F. D. Blum, submitted.
14. P. T. K. Shih and J. L. Koenig, *Mater. Sci. Eng.* **20**, 137 (1975).
15. K. J. McNeil, J. A. DiCaprio, D. A. Walsh and R. F. Pratt, *J. Am. Chem. Soc.* **102**, 1859 (1980).
16. E. R. Pohl and F. D. Osterholtz, in: *Molecular Characterization of Composite Interfaces*, H. Ishida and G. Kumar (Eds). Plenum Press, New York (1985).
17. O. K. Johannson, F. O. Stark, G. E. Vogel and R. M. Fleischmann, *J. Composite Mater.* **1**, 278 (1967).
18. N. Nishiyama and K. Horie, *J. Appl. Polym. Sci.* **34**, 1619 (1987).
19. M. E. Schrader, *J. Adhesion* **2**, 202 (1970).

20. H. Ishida, S. Naviroj, S. K. Tripathy, J. J. Fitzgerald and J. L. Koenig, *J. Polym. Sci., Polym. Phys. Ed.* **20**, 701 (1982).
21. O. K. Johannson, F. O. Stark and R. Baney, AFML-TRI-65-303, Pt. I (1965).
22. Z. Lasocki and S. Chrzczonowicz, *J. Polym. Sci.* **59**, 259 (1962).
23. H. Ishida and J. L. Koenig, *J. Colloid Interface Sci.* **64**, 555 (1978).
24. D. W. Sindorf and G. E. Maciel, *J. Am. Chem. Soc.* **105**, 1848 (1983).
25. R. L. Kaas and J. L. Kardos, *Polym. Eng. Sci.* **11**, 11 (1971).
26. K. Tanaka, S. Shinoda and Y. Saito, *Chem. Lett.* 179 (1979).
27. D. E. Leyden and W. T. Collins (Eds), *Silylated Surfaces*. Gordon & Breach, New York (1980).
28. S. Naviroj, R. Culler, J. L. Koenig and H. Ishida, *J. Colloid Interface Sci.* **97**, 308 (1984).
29. J. W. de Haan, H. M. van den Boguert, J. J. Ponjee and J. M. van de Ven, *J. Colloid Interface Sci.* **110**, 591 (1986).
30. J. D. March, *Advanced Organic Chemistry*, p. 796. Wiley Interscience, New York (1985).
31. D. E. Leyden, D. S. Kendall and T. G. Waddell, *Anal. Chim. Acta* **126**, 207 (1981).
32. D. E. Leyden, D. S. Kendall, L. W. Buggraf, F. J. Pern and M. DeBello, *Anal. Chem.* **54**, 101 (1982).
33. C.-H. Chiang, N.-I. Liu and J. L. Koenig, *J. Colloid Interface Sci.* **86**, 26 (1982).
34. G. R. Hays, A. D. H. Clague, R. Huis and G. van der Velden, *Appl. Surf. Sci.* **10**, 247 (1982).
35. A. M. Zaper and J. L. Koenig, *Polym. Composites* **6**, 156 (1985).
36. A. M. Zaper and J. L. Koenig, *Adv. Colloid Interface Sci.* **22**, 113 (1985).
37. G. E. Maciel and D. W. Sindorf, *J. Am. Chem. Soc.* **102**, 7606 (1980).
38. G. E. Maciel, D. W. Sindorf and V. Bartuska, *J. Chromatogr.* **205**, 438 (1981).
39. D. W. Sindorf and G. E. Maciel, *J. Phys. Chem.* **86**, 5208 (1982).
40. C. A. Fyfe, G. C. Gobbi and G. J. Kennedy, *J. Phys. Chem.* **89**, 277 (1985).
41. D. W. Sindorf and G. E. Maciel, *J. Am. Chem. Soc.* **105**, 3767 (1983).
42. E. Bayer, K. Albert, J. Reiners, M. Neider and D. Muller, *J. Chromatogr.* **264**, 197 (1983).
43. E. J. Sudholter, R. Huis, G. R. Hays and N. C. Alma, *J. Colloid Interface Sci.* **103**, 554 (1985).
44. J. M. Vankan, J. J. Ponjee, J. W. DeHaan and L. J. van de Ven, *J. Colloid Interface Sci.* **126**, 604 (1988).
45. J. C. Pouxviel and J. P. Boilot, *J. Non-Cryst. Solids* **89**, 345 (1987).
46. O. K. Johannson, F. O. Stark, G. E. Vogel, R. M. Lacefield, R. H. Baney and O. L. Flaninham, in: *Interfaces in Composites*, p. 168. ASTM STP 452, ASTM (1969).
47. C. G. Swain, R. M. Esteve, Jr. and R. H. Jones, in: *Molecular Characterization of Composite Interfaces*, H. Ishida and G. Kumar (Eds). Plenum Press, New York (1985).
48. J. D. Miller and H. Ishida, *Surf. Sci.* **148**, 601 (1984).
49. *CAB-O-Sil Properties and Functions*, Cabot Corporation, Tuscola, IL (1983).
50. W. D. Bascom, *J. Adhesion* **2**, 161 (1970).
51. F. J. Boerio, L. H. Schoenlein and J. E. Greivenkamp, *J. Appl. Polym. Sci.* **22**, 203 (1978).
52. W. D. Bascom, *J. Colloid Interface Sci.* **27**, 789 (1968).
53. H.-J. Kang, Ph.D. Thesis, University of Missouri-Rolla (1990).

Silanes and Other Coupling Agents, pp. 199–213
Ed. K. L. Mittal
© VSP 1992.

New evidence related to reactions of aminated silane coupling agents with carbon dioxide

K. P. BATTJES, A. M. BAROLO and P. DREYFUSS*

Michigan Molecular Institute, 1910 W. St. Andrews Road, Midland, MI 48640, USA

Revised version received 10 May 1991

Abstract—The nature of the product of the reaction between an aminated silane and carbon dioxide was re-examined with the aid of simple model compounds, several amines, and several aminosilanes. Since the reaction products previously proposed include the amine bicarbonate and a carbamate derived from the amine, ammonium bicarbonate and ammonium carbamate were studied as models for the anions. Carbon dioxide adducts of neat model amines were prepared and studied. Results from a variety of techniques are summarized. Among the most useful was Fourier transform infrared (FTIR) spectroscopy of fluorolube mulls. FTIR spectra were distinctive and assignments characteristic of the two species were extracted from the spectral data. Comparisons of these assignments with the products of the reaction between carbon dioxide and various amines were made. The results indicate that alkylammonium carbamates are the principal product. Nuclear magnetic resonance (NMR) spectra in D_2O indicated much dissociation and were not helpful in defining the products.

Keywords: Silane coupling agents; alkylammonium carbamates; aminosilane–carbon dioxide salts; amine–carbon dioxide salts.

1. INTRODUCTION

It is fitting and a real pleasure to dedicate this paper in memory of Dr. E. P. Plueddemann on the occasion of an international symposium in his honor. The paper is related to silane coupling agents, an area to which Dr. Plueddemann has made monumental contributions, and presents new insights into a controversial issue.

Amine-containing coupling agents have played an important role in commerce for more than 30 years [1], but even now in 1990 scientists still do not agree on the mechanism that makes amine groups helpful. Note that two distinct sets of experimental conditions must be considered:

(1) On standing in air, surfaces covered with aminopolysiloxanes become coated with a strongly adhering white powder that has been shown to be the product of the interaction of the amine with CO_2 in the air. Similar products are isolated from reactions of neat, dry model amines with dry ice [2].

(2) Aminosilanes are most often reacted with surfaces either by dipping into a dilute solution of aminosilane in 95% ethanol or water, or by spraying a dilute solution of aminosilane onto the surface [3]. These conditions differ from (1) in two ways; namely, the reactants are in solution, and water is present in large excess of the concentration of amine groups. Theoretically,

---

*To whom correspondence should be addressed.

these amine groups can react with CO_2 dissolved either in the solvent or in the silane itself. Experiments indicate that the reaction with CO_2 does not occur in solution. If the silane-treated surfaces are dried in a nitrogen atmosphere or under vacuum with heat to form the polysiloxane layer, the reaction with CO_2 is not observed [2, 4, 5]. As stated in (1), the white powder forms on standing in air. (Silane compounds absorb air very readily and surprisingly quickly.) Also, the powder forms very quickly (within 5 min) if the samples are dried in an atmosphere of gaseous CO_2 [4, 5].

Therefore, this paper is primarily concerned with what happens under the first set of conditions. Some data and discussion related to the second set of conditions will also be presented, but we will always be careful to point out when solutions are being discussed. If no solvents are mentioned, the reader should assume that the reactants are neat and that care to minimize moisture has been taken.

Probably the most studied coupling agent of this type is the widely used 3-aminopropyltriethoxysilane (**1**) (also known as A1100, 3-APS, and γ-APS). Zisman first reported in the early 1960s [6], and it has been observed regularly since then [2, 4, 5, 7–10], that the propylamine group interacts with carbon dioxide from the air. Most investigators have used infrared (IR) spectroscopy to study the product. A typical spectrum of the product from **1** and CO_2 is shown in Fig. 1 along with spectra of the CO_2 adducts from two other amines. Note that the spectra shown are of KBr pellets. This may have affected the results in ways that will be discussed in the Results and Discussion. The formation of bicarbonate salt, $-CH_2CH_2CH_2NH_3^+ HCO_3^-$), was initially postulated [6] and has been invoked repeatedly by spectroscopists in assignments of observed IR bands near 1640 and 1310 cm^{-1} and in explanations of other observed phenomena such as the pH dependence of the appearance of the above IR bands [4, 5, 7–10]. The presence of bicarbonate groups has not been proven.

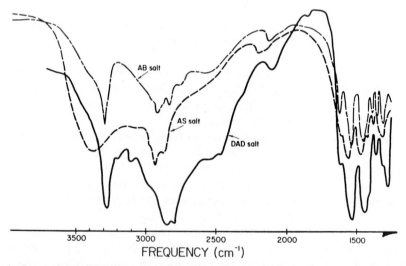

Figure 1. Comparison of the IR spectra of amine salts taken in KBr pellets on a Perkin-Elmer 337 Grating Infrared Spectrometer. AB, AS, and DAD are 1,4-diaminobutane, 3-aminopropyltriethoxysilane, and 1,12-diaminododecane, respectively [2].

The formation of bicarbonate salts proved inadequate to explain some previously reported results from model amines and CO_2 obtained using non-spectroscopic techniques [2]. In particular, elemental analyses and fragments observed by mass spectrometry of the purified CO_2 adduct of 1,4-butanediamine were consistent with the formation of a 1:1 CO_2:1,4-butanediamine carbamate salt, $^+NH_3CH_2CH_2CH_2CH_2NHCO_2^-$. (Calculated for $C_5H_{12}N_2O_2$, %C 45.44, %H 9.15, %N 21.20, %O 24.21; observed, %C 45.84, %H 9.61, %N 21.40, %O 23.01. The monobicarbonate salt, $NH_2CH_2CH_2CH_2CH_2NH_3^+HCO_3^-$, $C_5H_{14}N_2O_3$, would have %C 39.99, %H 9.40, %N 18.65, %O 31.96, and the bis-bicarbonate salt, $C_6H_{12}N_2O_6$, %C 33.96, %H 7.60, %N 13.20, %O 45.24.) Accordingly, the observed IR band near $1620 \, cm^{-1}$ was reassigned to the carbonyl of the carbamate salt [2]. Both the bicarbonate and the carbamate salt would show a similar pH dependence and release CO_2 at pHs below 7. Moreover, the same report [2] emphasized a change in the $3400-3200 \, cm^{-1}$ region of the IR spectrum on formation of the CO_2 adduct of an amine. The change in question was from the doublet of the free amine to a singlet. Apparently, this change was not observed or was considered unimportant by other investigators [4, 5, 7–10]. Yet the change is consistent with the formation of an alkyl-ammonium carbamate and cannot be explained by the formation of an alkyl-ammonium bicarbonate.

Naviroj *et al.* [5] reported preliminary results which indicate that only about half of the amine groups on an E-glass surface treated with **1** react with CO_2. The conclusion is based on the amount of CO_2 liberated on heating the treated E-glass after treatment with **1**. They postulate that half the amine groups form the bicarbonate salt and that the other amine groups interact with silanol groups on the surface of the glass. A simpler explanation occurred to us. One amine group is consumed in the formation of the comparatively unstable carbamic acid, which quickly reacts with another amine to form the more stable alkylammonium salt of the carbamic acid. Thus, 2 mol of a monoamine, such as *n*-propylamine, and 1 mol of a diamine, such as 1,4-butanediamine, would be required for each mol of CO_2:

$$2 -CH_2CH_2CH_2NH_2 + CO_2 \rightarrow -CH_2CH_2CH_2NHCO_2^- \, ^+NH_3CH_2CH_2CH_2-$$

The symposium honoring Dr. Plueddemann seemed like an appropriate occasion at which to re-examine this longstanding controversy. Accordingly, new experimental data consisting of an examination of spectra of known ammonium salts and comparison with the spectra of CO_2 reaction products of aliphatic amines and of aminosilanes were obtained. These data are presented below.

2. EXPERIMENTAL

Model compounds with the two anions being considered in this paper, HCO_3^- and $RNHCO_2^-$, and the same cation, NH_4^+, were selected for study. Ammonium bicarbonate, $NH_4^+HCO_3^-$, and ammonium carbamate, $NH_4^+NH_2CO_2^-$, were freshly purchased from Aldrich Chemical Co. in 99+% purity and used as received. Sodium bicarbonate, from Arm and Hammer, was also examined, because its spectrum is described in several compilations of spectra [e.g. 11–13]. *n*-Butylamine ($CH_3CH_2CH_2CH_2NH_2$), isopropylamine [$(CH_3)_2CHNH_2$], and

1,4-butanediamine [$NH_2(CH_2)_4NH_2$], all from Aldrich, were examined neat and after reaction with dry ice. The aminosilanes studied were **1** from Aldrich, aminobutyldimethylmethoxysilane, $NH_2(CH_2)_4Si(CH_3)_2(OCH_3)$, from Huls America, Inc. (formerly Petrarch Systems), and Z-6020, ethylenepropylene-diaminotrimethoxysilane, $NH_2(CH_2)_2NH(CH_2)_3Si(OCH_3)_3$, from Dow Corning. All CO_2 adducts were stored in a desiccator over silica gel until used.

NMR spectra in D_2O were recorded on a Bruker WM-360 NMR spectrometer. IR spectra were recorded neat with liquids or as fluorolube mulls with solids using a Nicolet 20DXB Fourier Transform Infrared Spectrophotometer.

3. RESULTS AND DISCUSSION

3.1. Studies of ammonium bicarbonate and ammonium carbamate

3.1.1. Synthesis and properties. A comparison of the methods of synthesis and some of the properties of ammonium bicarbonate and ammonium carbamate gives clues about the probability of formation of each anion under different experimental conditions. Ammonium bicarbonate is usually prepared by passing an excess of carbon dioxide through concentrated ammonia water [14]. This means that the pH is quite high and very high concentrations of all reactants are used. These conditions are very different from typical silylation experiments, which are carried out in very dilute solution. Ammonium bicarbonate is comparatively stable at room temperature. It decomposes above 60°C to give NH_3, CO_2, and H_2O. This suggests that the typical temperature (110°C) used to cure the silane layer [3] could be high enough to decompose any bicarbonate ion formed.

Ammonium carbamate is prepared from dry ice and liquid ammonia [14]. These conditions are very similar to the conditions under which we have observed the formation of amine salts. To some readers, ammonium carbamate may seem to be an exotic compound. In fact, it is manufactured industrially on a multiton scale, because on heating (usually at 100–185°C) ammonium carbamate is converted to urea and water [14–16]. Urea is important for both the agricultural and the plastics industries. The ammonium carbamate is not always isolated during urea preparation. Instead, the reactions are carried out under conditions where the carbamate is just an intermediate. Ammonium carbamate is only moderately stable and it gradually loses ammonia in air. Although the data are sparse, the rate of decomposition of carbamates in solution seems to decrease as the volatility of the parent amine decreases [17]. Free carbamic acids in solution do not decompose spontaneously to free amine and CO_2. Instead, the acid ionizes by reaction with water; the proton is transferred from the hydronium ion to the amine; and then decomposition occurs [17]. Acids catalyze the decomposition.

3.1.2. NMR spectra. NMR spectra of ammonium bicarbonate and ammonium carbamate are shown in Fig. 2. The spectra of the two compounds are identical, for all practical purposes. The differences seen are possibly due to small differences in pH or temperature. Spectra recorded after fewer scans, and shorter times, were also identical. A spectrum of a 50:50 mixture of the two compounds gave the same number of peaks. Apparently, in solution in the time

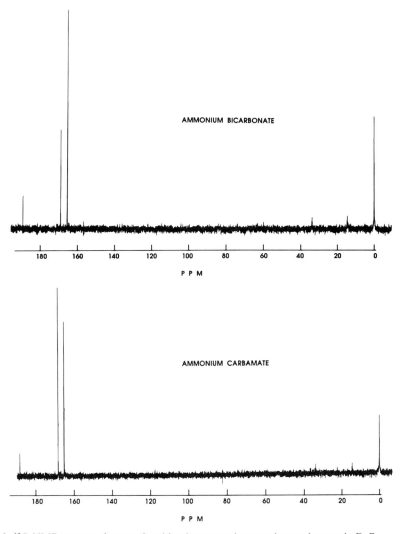

Figure 2. ¹³C NMR spectra of ammonium bicarbonate and ammonium carbamate in D₂O.

needed to record the spectra, the compounds equilibrate to a similar mixture. The ^{13}C-NMR spectra are difficult to explain. The ammonium bicarbonate and the ammonium carbamate each contain only one carbon atom. Thus, one would expect to see only one peak in each ^{13}C-NMR spectrum of the pure compounds if no dissociation occurs in D$_2$O. Since we observe two principal peaks and several smaller ones, we conclude that considerable dissociation occurs. We made some preliminary attempts to assign the peaks with the aid of ^{13}C-NMR literature [18], but the results were inconclusive. Since we are not studying the dissociation of the compounds in D$_2$O at this time, we did not continue the NMR studies.

3.1.3. Infrared spectra. As stated above, IR spectra were recorded, when needed, as fluorolube mulls rather than in the KBr pellets used for most earlier studies. This choice was made because KBr can interact with amine salts, causing the spectrum of a given salt to vary from sample to sample over the entire 2–16 μm (5000–625 cm^{-1}) range [13]. Figure 3 shows the IR spectrum of sodium bicarbonate in fluorolube and the IR spectrum of fluorolube alone. Since the fluorolube spectrum is essentially featureless from 4000 to 1280 cm^{-1}, comparisons with literature spectra were readily made in this region and the peaks were assigned as indicated on the spectrum and in Table 1. Even the peak around 1300 cm^{-1} is only partially obscured. Significant peaks below 1280 cm^{-1} are readily visible except for the one around 970 cm^{-1}, which overlaps with a fluorolube peak in the same region. The observed spectrum of sodium bicarbonate is in excellent agreement with the one published in ref. 11 and our assignments agree with the ones published there. A report by Little [12] lists absorptions for sodium bicarbonate at 1660, 1630, 1295, 1000, 838, and 698 cm^{-1}, but makes no mention of the strong OH absorption at about 2600–2400 cm^{-1}.

Figure 4 and Table 1 compare the IR spectra of ammonium bicarbonate and ammonium carbamate in fluorolube mulls. The spectrum for ammonium bicarbonate has absorptions due to the ammonium ion plus absorptions due to bicarbonate anion. The bicarbonate absorptions are similar to those observed for sodium bicarbonate but the absorptions are shifted a little. Although the ammonium and OH absorptions overlap somewhat, the OH peak at 2569–2540 cm^{-1} is clearly separated from the ammonium absorption region and

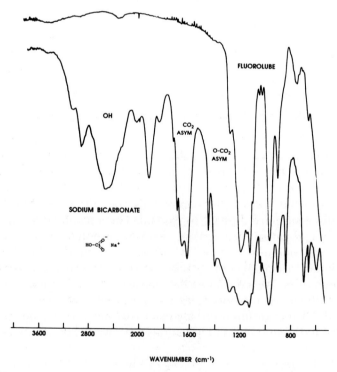

Figure 3. FTIR spectra of fluorolube and of sodium bicarbonate as a fluorolube mull.

agrees well with the 2607–2480 cm^{-1} OH absorption region of sodium bi-carbonate.

The ammonium carbamate also shows ammonium absorptions but the OH absorption at about 2550 cm^{-1} is missing and instead there are two amide-like NH$_2$ peaks at about 3470 and 3300 cm^{-1}. The NHCO$_2^-$ spectral characteristics are similar to those of an amide, NHCO [19]. Thus, the carbonyl group gives rise to two absorptions: one at about 1625 cm^{-1} and the other at about 1530 cm^{-1}. The higher absorption is the carbonyl absorption. Since ammonium carbamate is analogous to an unsubstituted amide, the lower one at about 1530 cm^{-1} is fused to the NH$_2$ deformation around 1615 cm^{-1} and contributes to its broadness. From comparison of the spectra and of the data for these two compounds in Table 1, we conclude that the most distinctive IR peaks for distinguishing between compounds with bicarbonate and carbamate anions lie in the NH–OH region, i.e. about 2500–3500 cm^{-1}, and in the carbonyl region, i.e. about 1500–1700 cm^{-1}. Thus, these are the 'fingerprint' regions that have been examined most carefully in the spectra of the other amines and aminosilanes examined in this study.

Note that the carbonyl assignments above are for carbamate salts with the general structure RNHCO$_2^-$RNH$_3^+$. In the case of ammonium carbamate, R=H. These carbamate *salts* should be distinguished from urethanes, which are carbamate *esters*, RNHCO$_2$R'. For urethanes, the carbamate bands are a combination of the amide and ester absorptions. The carbonyl stretch band shifts to about 1695 cm^{-1}, a position between that of the normal ester and amide. The NH and NH$_2$ stretch bands remain essentially unaltered as in corresponding amides. Absorptions between 1250 and 1000 cm^{-1} are a result of the $\overset{\overset{\textstyle O}{\|}}{C}-O-C$ stretch vibrations of the ester portion of the carbamate [19]. Data for three aliphatic urethanes, taken from *The Aldrich Library of Infrared Spectra*, are included in Table 1 [20].

3.2. Infrared spectra of amines and carbon dioxide adducts of amines

Figure 5 and Table 1 show a comparison of the IR spectra of neat iso-propylamine, a primary amine, and of the fluorolube mull of its CO$_2$ adduct. The spectrum for the amine agrees well with the one shown in spectrum number 300 on p. 419 of ref. 11. Assignments for most of the observed absorptions are shown on the spectrum. In our fingerprint regions, the spectrum for the amine has a doublet from the primary NH$_2$ group with peaks at about 3409 and 3280 cm^{-1}, a rather broad NH$_2$ deformation at about 1609 cm^{-1}, and, of course, no carbonyl peaks. The spectrum for the CO$_2$ adduct has an NH singlet at about 3370 cm^{-1} due to the NH of the monoalkyl-substituted carbamate, a broad ammonium peak superimposed on the carbon–hydrogen region at about 3000–2100 cm^{-1}, and two absorptions in the carbonyl region at about 1649 and 1561 cm^{-1}, respectively. Following ref. 19, we assign these absorptions in the carbonyl region to the carbonyl group of NHCO$_2^-$ and to the hydrogen of the NHCO in the *trans* position to the carbonyl, respectively. Similar definitive evidence for the presence of bicarbonate salt is absent. Unfortunately, the strong OH peak of an ammonium bicarbonate around 2500 cm^{-1} and indeed all the other strong

Table 1.
Comparison of selected IR absorptions (cm^{-1}) ^{a,b}

Compound	Formula	RNH
Sodium bicarbonate	$HO-C\underset{O}{\overset{O^-}{\lessgtr}}$ Na$^+$	—
Ammonium bicarbonate	$HO-C\underset{O}{\overset{O^-}{\lessgtr}}$ NH$_4^+$	—
Ammonium carbamate	$H_2N-C\underset{O}{\overset{O^-}{\lessgtr}}$ NH$_4^+$	3469, 3302 doublet
n-Propylammonium chloride	$CH_3(CH_2)_2NH_3^+Cl^-$	—
i-Propylamine	$(CH_3)_2CHNH_2$	3409, 3280 doublet
i-Propylamine · CO$_2$	$(CH_3)_2CHNHCO_2^- {}^+NH_3CH(CH_3)_2$ ^{d}	3370 singlet
n-Butylamine	$CH_3(CH_2)_3NH_2$	3462, 3370 doublet
n-Butylamine · CO$_2$	$CH_3(CH_2)_3NHCO_2^- {}^+NH_3(CH_2)_3CH_3$	3332 singlet
1,4-Diaminobutane	$NH_2(CH_2)_4NH_2$	3361, 3284 doublet
1,4-Diaminobutane · CO$_2$	$^+NH_3(CH_2)_4NHCO_2^-$	3332 singlet
n-Butylaminodimethylmethoxysilane	$NH_2(CH_2)_4Si(CH_3)_2(OCH_3)$	3371, 3292 weak doublet
n-Butylaminodimethylmethoxy-silane·CO$_2$	$^-O_2CNH(CH_2)_4Si(CH_3)_2(OCH_3)$ $^+NH_3(CH_2)_4Si(CH_3)_2(OCH_3)$	3298 singlet
n-Propylaminotriethoxysilane	$NH_2(CH_2)_3Si(OCH_2CH_3)_3$	3368, 3295 doublet
n-Propylaminotriethoxysilane · CO$_2$	$^-O_2CNH(CH_2)_3Si(OCH_2CH_3)_3$ $^+NH_3(CH_2)_3Si(OCH_2CH_3)_3$	3302 singlet
Ethylenepropylenediamino-trimethoxysilane	$NH_2(CH_2)_2NH(CH_2)_3Si(OCH_3)_3$	3339, 3280, 3240, triplet
Ethylenepropylenediamino-trimethoxysilane · CO$_2$	$^-O_2CNH(CH_2)_2NH(CH_2)_3Si(OCH_3)_3$ $^+NH_3(CH_2)_2NH(CH_2)_3Si(OCH_3)_3$	3295 singlet
Ethyl carbamate (urethane)	$NH_2CO_2CH_2CH_3$	3413, 3345, 3267, 3209 split doublet
Ethyl *N*-ethylcarbamate (*N*-ethylurethane)	$CH_3CH_2NHCO_2CH_2CH_3$	3300 singlet
N,N'-Methylene-bis-(ethylcarbamate)	$CH_2(NHCO_2CH_2CH_3)_2$	3344 singlet

^{a} Absorptions that aid in the identification of the CO$_2$ adduct of amines are listed. For most compounds other characteristic absorptions are described and assigned in the text and figures. For very weak or overlapping absorptions, a best guess for the location of a particular absorption has been made on the basis of peak picking by the FTIR instrument's computer and by analogy to similar comounds.

^{b} Below about 1200 cm^{-1}, overlap among bicarbonate, carbamate ion, and fluorolube absorptions makes assignments somewhat ambiguous and so these are omitted here.

^{c} A distinct peak is visible at about 2500 cm^{-1} as part of the CH region. Thus, the 2500 cm^{-1} peak is not definitive for OH in the presence of an alkylammonium ion.

NH$_3^+$	CH	OH	NCO$_2^-$	CO$_2$ asymmetric	NCO$_2^-$	OCO$_2$ asymmetric	Ref.
	—	2600–2475	—	1700–1600	—	1400–1300	This work
200–2850	—	2570–2540	—	1600	—	1500–1370	This work
200–2800	—	—	1625	—	1530	—	This work
200–2250^c	3100–2700	—	—	—	—	—	20
	2960–2740	—	—	—	—	—	This work
000–2190^c	2980–2120	—	1649	—	1561	—	This work
	3050–2860	—	—	—	—	—	This work
174–2140^c	3000–2170	—	1635	—	1561	—	This work
	3055–3029	—	—	—	—	—	This work
000–2150^c	2985–2790	—	1660, *1636^f*	—	1556	—	This work
	2960–2750	—	—	—	—	—	This work
000–2170	2960–2800	—	1650	—	1508	—	This work
	3000–2850	—	—	—	—	—	This work
100–2180	3000–2850	—	1660, 1620	—	1580	—	This work
	2950–2800^g	—	—	—	—	—	This work
950–2180	2950–2800	—	1620 very weak shoulder	—	—	—	This work
	2950–2856	—	1662, 1621	—	—	—	20
	2950–2860	—	1695	—	1517	—	21
	3075–2780	—	1680	—	1526	—	21

^{d}We have written the carbamate alkylammonium structure as an aid in comparing assignments of observed absorptions.
^{c}The CH region is clearly visible, as can be seen from the figures, but the CH region is within the ammonium range.
^{f}The italicized absorption is the stronger one.
^{g}Includes SiOCH$_3$.

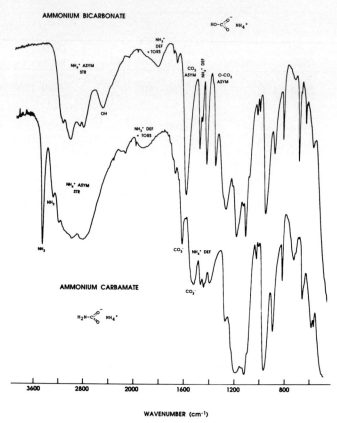

Figure 4. FTIR specra of ammonium bicarbonate and ammonium carbamate as fluorolube mulls.

ammonium bicarbonate peaks overlap with peaks that can equally well be assigned to a monoalkylammonium salt. (See published spectra for $RNH_3^+A^-$, where A^- is an anion, in refs 11–13 and Table 1.) Thus, further study was indicated.

Two different experiments were carried out to determine where a bicarbonate salt would absorb, if one were present in our samples. In the first, about 20% by weight of sodium bicarbonate was added to a fluorolube mull of the CO_2 adduct of isopropylamine, and the IR spectrum of the mixture was taken in the usual way. The only definitive evidence for the presence of the bicarbonate was at about 1920 cm^{-1}. All the other absorptions from the sodium bicarbonate were buried under the carbamate absorptions. The 1920 cm^{-1} absorption is not present in the ammonium bicarbonate spectrum.

In the second, really a series of experiments, ammonium bicarbonate was added to an excess of neat isopropylamine. Evidence for reaction included the formation of bubbles in the amine and the smell of ammonia. Several such mixtures were allowed to stand with variable amounts of stirring until the excess amine had evaporated. Solid powders remained. IR spectra of fluorolube mulls of the powders were neither identical nor were they simple additions of the spectra

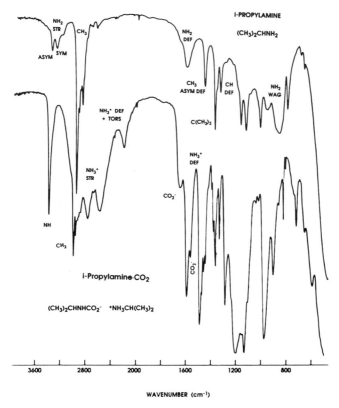

Figure 5. FTIR spectra of neat isopropylamine and its carbon dioxide adduct as a fluorolube mull.

of isopropylamine and ammonium bicarbonate. One was consistent with an isopropylamine carbamate containing something that did not appear to be ammonium bicarbonate. Another seemed to be primarily of ammonium bicarbonate. Yet another showed a very broad ammonium band containing methyl groups; no free amine; little, if any, amine carbamate (no absorptions in the region 3400–3200 cm⁻¹); and perhaps some ammonium bicarbonate. The latter spectrum was about what one would expect to see for slightly contaminated isopropylammonium bicarbonate. It was clear from these experiments that unless a very pure isopropylammonium bicarbonate was prepared and characterized not only by IR spectroscopy, but also by other methods, convincing evidence for the formation of bicarbonate would not readily be obtained. Even then, most peaks would coincide with those of other possible products. At present, only the rather broad and weak peak around 1800 cm⁻¹ seems promising as a peak that is characteristic of ammonium or alkylammonium bicarbonate alone.

Figure 6 and Table 1 show comparisons of the IR spectra of CO₂ adducts of the other amines examined in this study. Data for the corresponding free amines are given only in Table 1. As can be seen from Fig. 6 and Table 1, the evidence from the fingerprint regions is also consistent with the presence of carbamate salts.

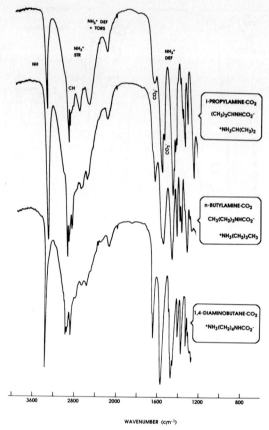

Figure 6. Comparison of the FTIR spectra of the carbon dioxide adducts of isopropylamine, *n*-butylamine, and 1,4-diaminobutane as fluorolube mulls.

3.3. Aminosilanes and carbon dioxide adducts of aminosilanes

IR spectral results for a model, nonpolymerizable silane (*n*-butylamino-dimethylmethoxysilane, $NH_2(CH_2)_4Si(CH_3)_2(OCH_3)$) and for its CO_2 adduct are given in Fig. 7 and in Table 1. The relative concentration of the amine is now made more dilute by the presence of the other substituents on the silicon atom and by the absorptions due to the substituted silicon atom itself, so the absorptions in the fingerprint regions are relatively weaker. Nevertheless, again the evidence is consistent with the formation of a carbamate adduct. Finally, Fig. 8 shows the spectra for the CO_2 adducts obtained from the coupling agents **1** and Z-6020. As summarized in Table 1, the spectra of the coupling agents show the same features of alkylammonium carbamate formation that were obtained with all the other amines examined in this study. The doublet for the free primary amine stretching frequency at about 3300 cm^{-1} becomes a singlet on reaction with CO_2 and an ammonium band appears at a slightly lower frequency. The two peaks characteristic of the carbonyl peak appear above and below 1600 cm^{-1}. Evidence for alkylammonium bicarbonate formation is missing.

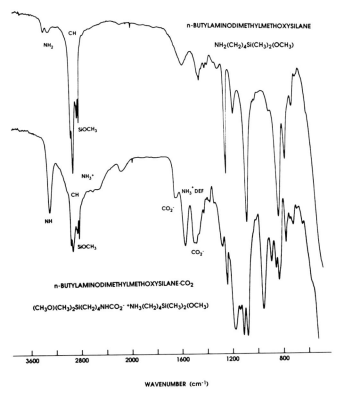

Figure 7. FTIR spectra of neat *n*-butylaminodimethylmethoxysilane and its carbon dioxide adduct as a fluorolube mull.

4. CONCLUSIONS

The evidence presented here overwhelmingly supports the formation of alkyl-ammonium carbamates as the primary products of the reaction between gaseous carbon dioxide and neat amines, including aminosilane coupling agents. The amines can be present as liquid model compounds or as functional groups within a surface. The IR absorption near 1630 cm^{-1} and the one near 1540 cm^{-1} are related to the carbonyl of carbamate salts of the amines. Absorptions at 3400–3200 cm^{-1} correspond to N—H stretching frequencies. Finally, there is a broad ammonium band overlapping the carbon–hydrogen region around 3000 cm^{-1}. In aqueous solutions alkylammonium bicarbonate may also form, but evidence for it is difficult to obtain in the presence of the corresponding carbamate salt.

5. CONCLUDING REMARKS

A few comments about the possible consequences of the formation of alkyl-ammonium carbamates on the behavior of aminosilanes as coupling agents are appropriate at this point.

The following advantages come to mind. The increased interfacial bond strength often observed when an aminosilane coupling agent is used may be explained by the *in situ* formation of an interpenetrating network. This would

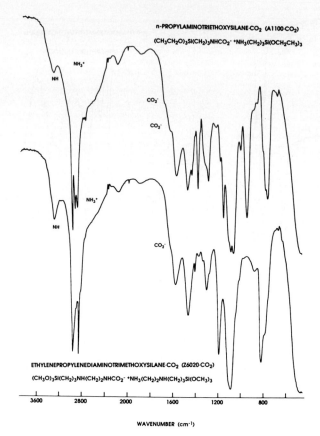

n-PROPYLAMINOTRIETHOXYSILANE-CO₂ (A1100-CO₂)

(CH₃CH₂O)₃Si(CH₂)₃NHCO₂⁻ ⁺NH₃(CH₂)₃Si(OCH₂CH₃)₃

ETHYLENEPROPYLENEDIAMINOTRIMETHOXYSILANE-CO₂ (Z6020-CO₂)

(CH₃O)₃Si(CH₂)₃NH(CH₂)₂NHCO₂⁻ ⁺NH₃(CH₂)₂NH(CH₂)₃Si(OCH₃)₃

WAVENUMBER (cm⁻¹)

Figure 8. FTIR spectra of carbon dioxide adducts of **1** and Z-6020.

happen if a free alkylamino group from the polysiloxane surface coating diffused into the organic layer and reacted with CO_2 to form the carbamic acid, which then reacted with another amine from the coating to form an alkylammonium carbamate. Based on our experience, some kinds of coupling reactions would be facilitated. For example, reactions of carbamate salts with acids seem to proceed faster and more readily than those with the free amines. This was observed during the preparation of picric acid derivatives from both kinds of compounds [2].

Some disadvantages include the following. When carbamates react with acidic substances, carbon dioxide is released. This could result in the formation of bubbles (flaws) at an interface and concomitant weakening of the interfacial bond. The insolubility of a carbamate salt in organic media could retard coupling reactions. The stability of the carbamate salt under nonacidic conditions could retard or inhibit coupling reactions.

REFERENCES

1. E. P. Plueddemann, *Silane Coupling Agents*. Plenum Press, New York (1982).
2. P. Dreyfuss and Y. Eckstein, *J. Adhesion* **15**, 163–178 (1983).

3. B. Arkles, in: *Silicon Compounds*, S-4, pp. 33–36. Petrarch Systems, Levittown, PA (1979).
4. S. R. Culler, M. S. Thesis, Case Western Reserve University, Cleveland, OH (1982).
5. S. Naviroj, S. R. Culler, J. L. Koenig and H. Ishida, *J. Colloid Interface Sci.* **97**, 308–317 (1984).
6. W. A. Zisman, *Proc. 20th Annu. Conf., SPI Reinforced Plast./Composites Inst.*, Section 9-B (1964).
7. F. J. Boerio and C. A. Gosselin, *Proc. 36th Annu. Conf., SPI Reinforced Plast./Composites Inst.*, Section 2-G (1981).
8. F. J. Boerio and J. W. Williams, *Proc. 36th Annu. Conf., SPI Reinforced Plast./Composites Inst.*, Section 2-F (1981).
9. S. Naviroj, J. L. Koenig and H. Ishida, *Proc. 37th Annu. Conf., SPI Reinforced Plast./Composites Inst.*, Section 2-C (1982).
10. F. J. Boerio and D. J. Ondrus, in: *Surface and Colloid Science in Computer Technology*, K. L. Mittal (Ed.), pp. 155–170. Plenum Press, New York (1987).
11. N. B. Colthup, L. H. Daly and S. E. Wiberley, *Introduction to Infrared and Raman Spectroscopy*, p. 308. Academic Press, New York (1975).
12. L. H. Little, *Infrared Spectra of Adsorbed Species*, p. 76. Academic Press, New York (1966).
13. C. J. Pouchert, *The Aldrich Library of Infrared Spectra*, Edition III, p. xix. Aldrich Chemical Co., Milwaukee, WI (1981).
14. M. Windholz, S. Budavari, L. Y. Stroumtsos and M. N. Fertig (Eds), *The Merck Index*, 9th edn, Merck, Rahway, NJ (1979).
15. (a) R. C. Fuson and H. R. Snyder, *Organic Chemistry*, p. 125. John Wiley, New York (1942); (b) P. Karrer, *Organic Chemistry*, p. 217. 2nd English edn, Elsevier, New York (1946); (c) R. T. Morrison and R. N. Boyd, *Organic Chemistry*, p. 686. Allyn and Bacon, Boston, MA (1973).
16. M. Grayson, D. Eckroth, E. Graber, A. Klingsberg and P. M. Siegel (Eds), *Concise Encyclopedia of Chemical Technology*, p. 1209. John Wiley, New York (1985).
17. M. Caplow, *J. Am. Chem. Soc.* **90**, 6795–6803 (1968).
18. E. Breitmaier, G. Haas and W. Voelter, *Atlas of Carbon-13 NMR Data*. Heyden, Philadelphia, PA (1979).
19. C. J. Pouchert, *The Aldrich Library of Infrared Spectra*, Edition III, p. 435. Aldrich Chemical Co., Milwaukee, WI (1981).
20. *Nicolet/Aldrich Condensed Phase Library*. Nicolet Analytical Instruments, Madison, WI (1982).
21. C. J. Pouchert, *The Aldrich Library of Infrared Spectra*, Edition III. Aldrich Chemical Co., Milwaukee, WI (1981).

Silanes and Other Coupling Agents, pp. 215–228
Ed. K. L. Mittal
© VSP 1992.

Silane coupling agents: the role of the organofunctional group

GIULIANA TESORO* and YULONG WU

Department of Chemistry, Polytechnic University, 333 Jay Street, Brooklyn, NY 11201, USA

Revised version received 14 May 1991

Abstract—Some highlights of recent research on silane coupling agents are reviewed. Studies of silanes in solution, of chemical bonding in the substrate/matrix interphase, and of the interfacial bond in composites have provided new insights into the mechanism of effectiveness of silane adhesion promoters. Examples are presented of new chemical structures and of new concepts of bonding designed for optimal performance of silane coupling agents in advanced composites.

Keywords: Coupling agents; silanes; interphase in composites; bonding; adhesion; adhesion promoters.

1. INTRODUCTION

Interest in the chemical structure and mode of action of organofunctional silanes dates back to an observation made by Ralph K. Witt *et al.* at Johns Hopkins University in 1947 [1]. These investigators found that polyester/fiberglass composites where the fibers were finished with allyltriethoxysilane exhibited a much higher strength than those in which the glass was treated with ethyl-trichlorosilane. Extensive evaluations of silane coupling agents in glass-reinforced polyester and epoxy composites in subsequent years were motivated primarily by the objective of increasing the wet strength in glass-reinforced polymer composites. The general definition of coupling agents [1] as 'materials that improve the chemical resistance (especially to water) of the bond across the interface in composites consisting of an organic polymer and a mineral substrate' evolved during that period, and hundreds of compounds were screened—over 100 by Plueddemann *et al.* by 1962 [2]. The experimental evidence resulting from these studies was essentially circumstantial, but sufficiently consistent to allow a generalization stating that 'in general, the effectiveness of the silane as a coupling agent paralleled the reactivity of the organofunctional group with the resin' [1]. In the seventies and eighties, new techniques for analysis and charac-terization made it possible to characterize the chemistry of silane solutions, and surface-bound species, as well as the microstructure of the interphase in re-inforced composites with greater precision, and to advance scientific insight [4–8]. However, the working hypotheses of the sixties for the design of organo-functional silane molecules that would be effective coupling agents for specific polymer composite systems remain valid to this day. This paper will discuss

---

*To whom correspondence should be addressed.

some highlights of recent progress in the characterization of molecular structure and reactions of silane coupling agents in solution and in composite interfaces, and the relationship of chemical interactions at the interface to the effectiveness of the silanes as adhesion promoters. Some specific examples of the role of the organic moiety in organofunctional silanes designed for advanced polymer composites will also be discussed in the context of current knowledge.

2. SILANE COUPLING AGENTS IN SOLUTION

Silane coupling agents have the structure $R-Si-X_3$, where R represents an organofunctional group and X is generally alkoxy. They are generally applied from water solutions or from organic solvent–water mixtures. In the presence of water, hydrolysis occurs step-wise to form alkoxysilanols, and eventually silane triols. Condensation of the silanols to siloxanes also occurs. Reactions shown schematically below:

Hydrolysis

$$R'Si(OR)_3 \underset{}{\overset{H_2O}{\rightleftharpoons}} R'Si(OR)_2(OH)$$

$$\big\updownarrow H_2O$$

$$R'Si(OH)_3 \underset{H_2O}{\overset{}{\rightleftharpoons}} R'Si(OR)(OH)_2$$

Condensation

$$-\overset{|}{\underset{|}{Si}}OH + HO\overset{|}{\underset{|}{Si}}- \rightleftharpoons -\overset{|}{\underset{|}{Si}}-O-\overset{|}{\underset{|}{Si}}- + H_2O$$

$$-\overset{|}{\underset{|}{Si}}OH + RO\overset{|}{\underset{|}{Si}}- \rightleftharpoons -\overset{|}{\underset{|}{Si}}-O-\overset{|}{\underset{|}{Si}}- + ROH$$

dimer $\longrightarrow$ oligomer

proceed at rates which depend on the organofunctionality, water content, temperature, and pH of the solution [9].

Under optimum conditions, hydrolysis is rapid, and precedes slower condensation reactions. However, the silanol groups formed in hydrolysis are very reactive, and condensation can be a significant factor even in dilute aqueous solutions. Understanding the kinetics of the hydrolysis and condensation reactions for a specific silane in solution is important because its effectiveness as an adhesion promoter is influenced by the extent of condensation, which, in turn, affects the structure of the interphase in composites.

The parameters governing the reaction kinetics of silane coupling agents in solution include the organofunctionality, the concentration of silane, the concentration of water, the pH of the solution, and the aging time in solution.

2.1. Effect of the pH and organofunctional group

The work of Pohl in 1983 [10] showed that the hydrolysis and condensation of trialkoxysilanes were strongly dependent on the solution pH. Pohl reported that for γ-substituted coupling agents—γ-aminopropyltriethoxysilane (γ-APS), γ-methacryloxypropyltrimethoxysilane (γ-MPS), and γ-glycidoxypropyltrimethoxy-silane (γ-GPS)—the rate of hydrolysis shows a minimum at pH 7 and the rate of condensation shows a minimum at pH 4.3, where the mechanism of condensation changes from an acid-catalyzed to a base-catalyzed reaction.

In a study of the steric and inductive effects of the alkyl group on the rate of hydrolysis of alkyltrialkoxysilanes, it was also shown [11] that the alkyl substituent on silicon had a significant effect on the rates of hydrolysis under basic and acidic conditions in aqueous medium.

The hydrolysis of aminopropyltrialkoxysilanes was recently studied by Kang *et al.* [12] using NMR spectroscopy. The rates of hydrolysis were shown to be dependent on the water and silane concentrations in solution as well as on the chemical structure. Under comparable conditions, the trimethoxysilane hydrolyzed much more rapidly than the triethoxy compound.

Condensation reactions of aqueous solutions of methylsilanetriol with silica have been studied by NMR [13] at pH values 3.2–3.9, at various concentrations of triol and in the presence of a large excess of water. It was shown that the rates at which the methylsilanetriol polymerized with itself, and with the silica surface, were dependent on the pH of the solution and the concentration of methyl-silanetriol. Concurrent with the reactions of the methylsilanols, the surface hydroxyl groups of the silica disappeared, suggesting that the methylsilanetriol adsorbs onto the silica surface.

2.2. Effect of the concentration of water

The effect of the concentration of water on the hydrolysis and condensation reactions of silanes has been investigated by Kang *et al.* [12] in conjunction with recent studies of the adsorption of silane coupling agents on silica surfaces from acetone–water mixtures as a function of the solvent composition [14]. Adsorption isotherms were determined by infrared (IR) spectroscopy and hydrolysis was followed by NMR.

2.3. Effects of the silane concentration and aging

The concentration of silane and the age of the solution have a significant effect on the condensation reaction for a given compound, and the structure of the organic group is an additional factor. For example, neutral silanes at concentrations below 1% in water yield primarily silanetriols when first hydrolyzed. On the other hand, aqueous solutions of γ-APS at concentrations exceeding 0.2% contain primarily oligomeric polysiloxanols formed in self-catalyzed condensation reactions [4, 5]. Furthermore, even in the concentration range where silanetriols dominate, there exists a concentration of silane at which aggregated monomers are formed by hydrogen bonding, and this concentration, termed the 'onset of association' [4], depends on the organofunctionality and must be determined for each silane compound.

2.4. The special case of γ-APS

An extensive effort has been made in the past decade to elucidate the structure of

aminosilanes, and particularly that of γ-APS, in order to understand their somewhat anomalous behavior in solution and on substrates. It has been reported, for
example, that a 50% solution of an aminosilane in alcohol was found to gel
immediately upon addition of 2 mol of water (based on the silane), but to
become liquid again after standing overnight [15].

Modern spectroscopic methods have been applied to the study of the structure
of aminosilanes, with a focus on the amine group of the molecule. The classic
paper by Plueddemann [16], which hypothesized either an intramolecular five-
membered or six-membered ring for the aminosilane molecule, strongly
influenced later studies, which were clearly and concisely summarized by Ishida
[5]. In conclusion, the complexity of partially reacted γ-APS has been well
documented, and it is postulated that some amines may form intramolecular
hydrogen-bonded structures while others may be intermolecularly hydrogen-
bonded. The mechanism of the effectiveness of γ-APS as a coupling agent and
adhesion promoter is the subject of continuing investigations, particularly in
conjunction with the role of silanes in the adhesion of polyimide thin films to
mineral surfaces.

3. THE SUBSTRATE/COUPLING AGENT INTERPHASE

In recent years, new insights have been gained by employing advanced analytical
techniques (solid-state NMR, FTIR, XPS) for the study of silane adsorption on
solid surfaces and for the characterization of the substrate/coupling agent
interphase, particularly for metal substrates.

3.1. Adsorption

The IR spectra of polymers and coupling agents onto oxidized aluminum have
been studied by Boerio et al. [17, 18]. It has been shown that an external
reflection technique utilizing large, almost grazing angles of incidence and
radiation polarized parallel to the plane of incidence leads to rather large electric
field amplitudes at the surface of a metal. Significant interaction between the
surface films and the IR radiation is obtained under these conditions. The
structures of thin films formed by the adsorption of epoxy polymers and
organosilane coupling agents (such as γ-APS) from solution onto the oxidized
surface of aluminum have been determined. In other work, Boerio et al. used IR
and XPS techniques to determine the chemical reactions of γ-APS adsorbed
onto titanium, iron, and aluminum mirrors from dilute aqueous solutions [19].
The results indicated that γ-APS was adsorbed onto titanium and iron as highly
hydrolyzed oligomers that polymerized further to form siloxane polymers during
drying. Deuterium and ^{13}C-NMR studies of deuterated coupling agents adsorbed
on silica (Cab-O-Sil) surfaces have also been reported [20]. The synthesis of
deuterated aminopropylsilane and deuterated aminobutylsilane was described as
part of this work. The NMR studies showed that the molecular motion of the
coupling agent on the solid surface increases with the addition of water.

3.2. Silane bonding

The chemistry of the silane interphase and of the substrate coupling agent interface has been investigated for silanes containing different organofunctionalities in attempts to provide a basis for generalizing observations made with a specific compound to other compounds of related structure. The compounds studied as prototypes have been primarily γ-MPS, γ-GPS, γ-APS, VS (vinyltrimethoxysilane), and CS (cyclohexyltrimethoxysilane) [5].

The reaction of hydrolyzed coupling agents with surface hydroxyls, and the formation of chemical bonds (oxane bonds) or hydrogen bonds have been studied for many years and are generally accepted as an essential element of the theory of bonding by silane coupling agents in composites [1]. In addition to reactions of silanols with surface hydroxyls on the substrate and condensation reactions in solution, the silanols may condense on the substrate to form polysiloxane structures. Thus, the substrate surfaces may be modified by reaction with hydrolyzed silane monomers, with polysiloxane oligomers, and by formation of a polysiloxane coating. The study of chemical reactions at the substrate/ coupling agent interphase, to determine their relative extent and dependence on the silane structure and on the conditions used for application of specific silane compounds to the substrate, has been the subject of continuing research.

The substrate/silane interphase and the silane/matrix interphase are equally important in considering the mechanism of reinforcement by silane coupling agents in composites. The mineral oxide/silane interphase is more well defined than a metal/silane or a silane/matrix interphase. For example, in the case of a metal substrate, surface oxides may dissolve into the silane layer or form a complex. In the case of the silane/matrix interphase, a diffuse boundary layer may exist due to dispersion of physisorbed silanes in the matrix phase or penetration of the matrix resin into chemisorbed silane layers. Many features of the interaction of a silane coupling agent with a polymer matrix are specific to the system, and thus the chemistry of the silane/matrix interphase must be characterized and defined for each system.

The problem of a definitive theory for the mechanism of bonding by organofunctional silanes in composites is complex. Results of research for numerous specific systems have been explained by the chemical bonding theory. However, the chemical bonding theory alone may not be sufficient, and consideration has been given to other concepts [1], including the formation of interpenetrating networks [21], the effect of surface modification of the substrate on the surface energy and wetting phenomena [22], acid–base reactions at the interface, and the morphology of the interphase [23]. The chemical interactions at the coupling agent/matrix interphase play an essential role in the interfacial bond attained by employing silane coupling agents in composites and are the subject of continuing study.

4. THE COUPLING AGENT/MATRIX INTERPHASE

Organofunctional silanes have been developed for bonding virtually any polymer to mineral in reinforced composites. A great deal of indirect evidence obtained from performance data has suggested that silanes form stable chemical bonds with the polymer as well as with the substrate. Advanced analytical techniques have been used to characterize the interphase and to provide direct evidence regarding the mechanism of bonding in a particular system. For bonding a given polymer to a mineral reinforcement, the silane is selected such that the organo-functional group is capable of reaction or interaction with the polymer. This interaction may result in covalent bonding of the organofunctional group to the polymer, or it may consist of a co-polymerization reaction at the interphase, or of interdiffusion and formation of an interpenetrating polymer network (IPN).

Current knowledge of the coupling agent/substrate interphase and of the coupling agent/matrix interphase for different compounds enables us to state that the molecular structures of coupling agents that exhibit maximum effective-ness as adhesion promoters in polymer/mineral composites should be designed to provide (1) maximum mineral—O—Si— (oxane) bond formation, (2) an organic group capable of interaction with the matrix polymer, and (3) bonding of the organofunctional group to the polymer. Practical experience has shown that composite performance can be correlated with the chemical structure at the interface or interphase region, and that this, in turn, is governed by the organo-functional groups and by the reactivity of the silane. Representative organo-functional groups in commercial silane coupling agents are shown in Table 1. It is evident that they include those capable of reaction with the polymer and formation of covalent bonds, of co-polymerization, and/or of formation of IPNs during the curing reaction, and those in which the organofunctional moiety would have an effect on the critical surface tension of the silane-treated substrate and enhance wetting by the resin. Reactions at the silane/matrix interphase have

Table 1.
Illustrative commercially available silane coupling agents

Organofunctional group	Chemical structure	Abbreviation
Vinyl	$CH_2=CHSi(OCH_3)_3$	VS
Chloropropyl	$ClCH_2CH_2CH_2Si(OCH_3)_3$	—
Epoxy	$\overset{O}{\overset{\diagup\diagdown}{CH_2CHCH_2OCH_2CH_2CH_2Si(OCH_3)_3}}$	γ-GPS
Methacrylate	$\underset{\underset{O}{\parallel}}{CH_2=C-\overset{\overset{CH_3}{\mid}}{C}-OCH_2CH_2CH_2Si(OCH_3)_3}$	γ-MPS
Primary amine	$H_2NCH_2CH_2CH_2Si(OC_2H_5)_3$	γ-APS
Diamine	$H_2NCH_2CH_2NHCH_2CH_2CH_2Si(OCH_3)_3$	—
Methyl	$CH_3Si(OCH_3)_3$	MS
Cationic styryl	$CH_2=CHC_6H_4CH_2\overset{+}{N}H_2(CH_2)_3Si(OCH_3)_3\overset{-}{Cl}$	—
Phenyl	$C_6H_5Si(OCH_3)_3$	—
Carboxy	$HOC\underset{\underset{O}{\parallel}}{C}C_6H_4\overset{-}{C}OH_3\overset{+}{N}(CH_2)_2NH(CH_2)_3Si(OCH_3)_3$	—
Mercapto	$HSCH_2CH_2CH_2Si(OC_2H_5)_3$	—

been investigated for a γ-MPS/polyester system [24] and for a γ-APS/epoxy system [25, 26] as representative of compounds designed for co-polymerization (γ-MPS) and for covalent bond formation (γ-APS) at the interface, respectively. The results of these investigations have provided evidence in support of the theory of chemical bonding as the primary mechanism of effectiveness for silane coupling agents in polymer composites. The elements of this theory have been defined with greater precision and augmented by results of recent investigations [27, 28].

4.1. Chemical bonding and adhesion

For example, in an investigation of silanes as adhesion promoters for ethylene/vinyl acetate co-polymer encapsulants reported by Koenig *et al.*, the organo-silanes (referred to as 'primers') were shown to generate primary chemical bonds at the polymer/substrate interface [27].

The molecular structure of epoxy/metal interphases in the presence of an amino coupling agent was studied by Boerio and co-workers [28] by IR and by XPS. The formation of amide and imide groups in the interphase provided evidence of chemical reaction between the silane primer and the curing agent for epoxy resin.

4.2. Hydrothermal stability of the interfacial bond

The characteristics of silane primer films applied to metal and the use of silane primers to enhance the hydrothermal stability of interfacial bonds to metals have been extensively studied by Boerio and co-workers and are described in detail [29].

5. SOME EXAMPLES OF RECENT RESEARCH

5.1. Methacrylate silanes

The study of methacrylate functional silanes containing different spacer groups [30] was undertaken in order to establish whether the performance obtained with the methacrylate ester γ-MPS, which has been a 'standard' coupling agent for glass-reinforced unsaturated polyester resins, could be improved by spacing the methacrylate group at a greater distance from the silicon, increasing the length of the hydrocarbon chain R in the ester shown schematically below:

$$CH_2=\underset{\underset{O}{\overset{\|}{}}}{\overset{\overset{\displaystyle CH_3}{|}}{C}}-C-O-R-\underset{\underset{O}{|}}{\overset{\overset{\displaystyle O}{|}}{Si}}-O-$$

$$R = (CH_2)_n \qquad n = 3\text{--}11$$

The compounds become more hydrophobic with increasing chain length and changes in orientation at the substrate surface would be expected with a change in the chain length of the esters.

The chemical nature of the experimental methacrylate functional silanes on silica was studied by FTIR, and the performance of the silanes in improving the chemical resistance of glass fiber/polyester composites was evaluated.

The conclusion was reached that the mechanical properties of glass-reinforced unsaturated polyester are influenced by the chemical structure of the spacer groups in the methacrylate functional silane. Effective factors include hydrophobicity, reactivity of the double bond, chain flexibility of the backbone, and adsorption behavior.

5.2. Azidosilanes

In the case of methacrylate coupling agents in glass-reinforced unsaturated polyesters, one invokes co-polymerization at the interphase as the prevailing mechanism for interphase structure and chemical bonding. In the absence of reactive functional groups in the matrix, as is the case for many thermoplastics, it has been suggested [31, 32] that the thermal decomposition behavior of sulfonyl azides would allow covalent bonding of azidosilane coupling agents of the formula $N_3SO_2{-}R{-}Si\,(OR')_3$ to the C$-$H bonds of polyolefins by a nitrene insertion reaction which would apply more generally to the C$-$H bond of other thermoplastic polymers:

$$-R{-}SO_2N_3 \xrightarrow{\ \Delta\ } -R{-}SO_2N: + N_2$$

$$-R{-}SO_2N: + H{-}\overset{/}{\underset{\backslash}{C}}{-} \longrightarrow -R{-}SO_2{-}NH{-}\overset{/}{\underset{\backslash}{C}}{-}$$

Recent studies of azidosilane coupling agents by DRIFT (diffuse reflectance infrared Fourier transform spectroscopy) and by XPS and comparative evaluations of composite performance [32] have provided evidence of covalent coupling of azidosilane-treated glass to polypropylene and have shown that these coupling agents improve the composite properties of glass and mica-reinforced polyolefins. Selected performance results are shown in Table 2.

5.3. Imide silanes

The need for adhesion promoters which can maintain coupling effectiveness for silicon with polyimide films upon exposure to elevated temperatures in nitrogen and in air motivated a study of thermo-oxidative stability for a class of organofunctional silanes (imide silanes) specifically designed for polyimide systems [33]. Several aspects of the results obtained in this research are of interest. The effectiveness of the experimental 'hexafunctional' alkoxysilanes (CA-11 and CA-25, see Fig. 1) was demonstrated. The study also included the effect of application conditions for the coupling agents, and correlations of variables for the coupling agents in solution (amount of water, pH, time of aging) with the relative extent of hydrolysis and condensation prior to application and curing.

The thermo-oxidative stability of a silane where silicon is linked to imide nitrogen by an aromatic radical (CA-25) was greater than that of a compound where silicon is linked to nitrogen by an aliphatic group (CA-11). Upon

Table 2.
Improvement in the mechanical properties of glass/
polypropylene composites by using an azidosilane
coupling agent [32]

	A	B
Tensile strength (MPa)		
As molded	63.46	79.48
72 h water boil	55.87	77.50
Flexural strength (MPa)		
As molded	88.59	115.62
72 h water boil	78.69	108.95
HDT^a at 1.82 MPa (°C)		
As molded	140	137
72 h water boil	141	136

A—20% glass/80% polypropylene resin; B—20%
glass/80% polypropylene resin/0.6% azidosilane
$[N_3SO_2-R-Si-(OR')_3]$.
^{a}Heat distortion temperature.

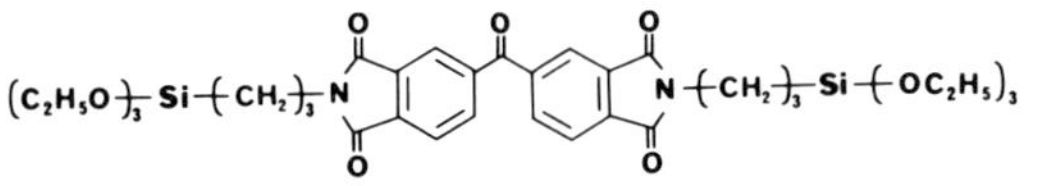

CA 11

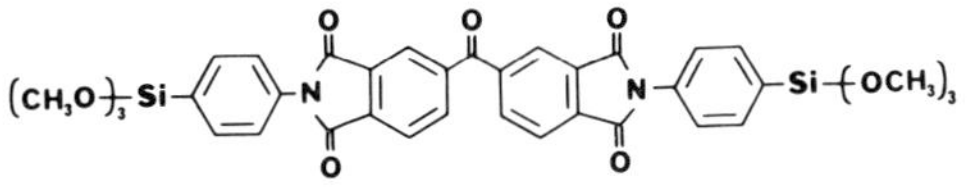

CA 25

Figure 1. Structures of thermally stable coupling agents [33].

prolonged exposure of the composites to elevated temperature (400°C), retention of the interfacial bond obtained with CA-25 was higher than that obtained with CA-11, its aliphatic analogue, and far superior to that shown by γ-APS (Table 3).

5.4. Polymeric silanes: trimethoxysilyl-terminated poly(N-acetylethyleneimine (PAEI)

The problems in bonding polymers to hydrophilic mineral surfaces were discussed and a theory of adhesion in which silane coupling agents provide a bond at the interphase was proposed by Plueddemann in 1974 [3]. However, silane coupling agents specifically designed to impart hydrophilicity to mineral surfaces have been proposed only recently [34]. A trimethoxysilyl-terminated poly(N-

Table 3.
Thermal stability of the interfacial bond: polyimide on silicon wafers with CA-11, CA-25, and γ-APS [33]

| Heat treatment (h at 400°C) | % Peel strength retained[a] | | | | | |
| | In air | | | In nitrogen | | |
	CA-11	CA-25	γ-APS	CA-11	CA-25	γAPS
0	100	100	100	100	100	100
1	40	62	38	87	99	88
2	28	52	28	—	95	—
2.5	—	—	—	81	—	81
3	18	50	—	—	97	—
4	—	41	—	81	99	75

[a] Peel strength by 90° peel test on 5 mm wide strips at a rate of 5 mm/min, as per ASTM D-903-49-1978.

acetylethyleneimine) (PAEI) was prepared from 2-methyl-2-oxazoline and reacted with hydroxyl groups on the surface of silica gel to impart hydrophilicity (Fig. 2). The approach was proposed as the first step in the development of 'polyoxazoline coupling agents', where a broad spectrum of properties (from hydrophilic to hydrophobic) could be obtained depending on the structure of the N-acyl group in the trimethoxysilyl-terminated polymer.

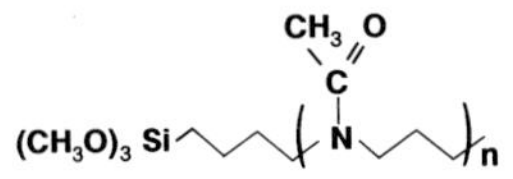

Figure 2. Trimethoxysilyl-terminated poly(N-acetylethylimine) [34].

5.5. Multifunctional polymeric silanes

In the area of polymeric multifunctional alkoxysilanes, the design and synthesis of a polymer film surface which would form a strong interfacial bond to glass have been based on the application of silane coupling agent chemistry to the synthesis of the solid–solid interfaces shown schematically in Fig. 3 [35]. Poly-chlorotrifluoroethylene film (PCTFE) was treated to form surface hydroxyl groups which were reacted with γ-isocyanatopropyltriethoxysilane to produce a polymer surface containing the alkyltriethoxysilane functionality—[PCTFE—OCONH(CH$_2$)$_3$Si(OC$_2$H$_5$)$_3$]. The modified film, characterized by ATR-IR and XPS, adhered tenaciously to glass slides when bonding was carried out at 80°C for 12 h. The film could not be peeled from the glass without tearing it, indicating cohesive failure: the adhesion was attributed to covalent bond formation between the glass and the polymer film.

5.6. The thiol group in organofunctional silanes

Some interesting features of organofunctional silanes containing thiol groups have been documented at various times. The first report of silanes in this class

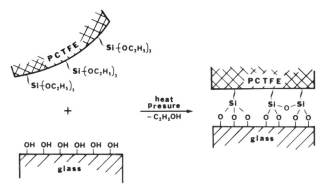

Figure 3. Poly(chlorotrifluoroethylene) surface containing alkyltriethoxysilane functionality: schematic of bonding to glass [35].

was in a patent [36] issued in 1969 covering, for example, silanes of the formula $(RO)_3—Si—(CH_2)_3NHCH_2CH_2SH$, prepared by mercaptoethylation of γ-APS for bonding glass to rubber and other 'sulfur-vulcanizable' polymers. The concept underlying the invention was that covalent bonds with rubber would be formed by the thiol group. The strength of the interfacial bond obtained with the coupling agents in the glass/rubber composites provided evidence in support of this premise. γ-Mercaptopropyltrimethoxysilane became a commercial product in the sixties (as Dow Corning A-189) and continues to be used. Its effect on the strength of silica-filled epoxidized natural rubber (ENR) has been the subject of a recent investigation [37] comparing the effect of γ-MPS to that of bis(tri-ethoxysilylpropyl)tetrasulfide in enhancing the dynamic and mechanical properties of the silica-filled rubber.

The examples summarized above are but indicative of the increasing body of evidence in support of the chemical bonding theory, and of the role of the silane organofunctional group in the formation of covalent bonds at the coupling agent/ matrix interphase by reaction or by co-polymerization with the formation of an interpenetrating polymer network.

5.7. New concepts of bonding

New concepts of bonding across the interphase in composites have also been explored in recent years. Plueddemann has proposed a new mechanism for bonding to thermoplastics under conditions of high shear. The property desired in the interphase region of thermoplastic composites under these conditions 'fits the description of commercial ionomers' [38].

Commercial ionomers are believed to consist of ion clusters dispersed in a neutral thermoplastic matrix. They are thermoplastic polymers modified with up to 10% acrylic or methacrylic acid and neutralized to the extent of about 30% with sodium or zinc cations. Ionomers are fluid under conditions of high temperature and high shear, but reform at room temperature to yield tough, water-resistant structures. An interphase region which is highly fluid during the molding operation (high temperature and high shear rate) but sets to a tough polymer when cooled is desirable in thermoplastic composites. The new approach of providing ionomer bonds at the interface is thus based on a general concept of ionomer bonding in which an acid-functional silane bonded to glass is modified with a carboxylated polymeric film former in the presence of metal cations to obtain a tightly adhering film which is resistant to debonding in the presence of moisture. This film bonds well to matrix thermoplastic polymers if the film former and matrix polymer are compatible. Optimum water resistance was observed when the acid-functional film formers were modified with sodium or zinc ions sufficient to neutralize 20–40% of the acid groups.

Interest continues to stimulate new research on the role of silanes in adhesion. For example, the dynamic mechanical properties of the interfacial region have been explored in recent work [39]. An extensive investigation was carried out in which the dynamic mechanical properties in the interfacial region of glass fiber/unsaturated polyester composites in the presence of silane coupling agents were determined in order to elucidate the role of γ-MPS and of MS (methyltrimethoxysilane) in the mechanism of adhesion (γ-MPS can react with the polyester while MS does not). The dynamic mechanical properties of the 'silane coatings only' on a glass surface as a function of the silane concentration in solution had been measured in previous work [40]. The effects of the organofunctional group of the silane and the silane concentration in solution on the mechanical properties of the polyester region ranging from 200 to 1600 Å removed from the glass surface were determined [39]. Some examples of the results of this work are shown in Table 4.

Table 4.

Some dynamic mechanical properties of glass fiber/unsaturated polyester composites [39]

Glass treatment		Interfacial shear strength[a] (MPa)	E″ (GPa)
γ-MPS	0.2 wt%	38.6	0.17
	1.0 wt%	26.2	7.6
	2.5 wt%	22.1	12.9
H$_2$O		21.4	2.3
MS	0.2 wt%	18.6	15.0

[a] Measured by the single filament test [41].

The conclusions based on the results of this research were that:

(a) a 'good' coupling agent for glass/polyester composites (such as γ-MPS) interacts with the resin through chemical bonding and interpenetration. Depending on the concentration of the silane in solution, the silane coating on glass may induce the formation of either a 'soft' or a 'rigid' boundary layer. A 'soft' layer is formed on treating the glass surface with dilute γ-MPS (0.05–0.5% wt). The T_g and modulus of this layer are lower than those of the bulk resin. In contrast, a concentrated solution of γ-MPS yields a 'rigid' boundary layer with T_g and modulus above those of the matrix;

(b) 'poor' coupling agents such as MS, or γ-MPS applied from concentrated solution, always produce a 'rigid' boundary layer;

(c) the loss modulus and the T_g of the boundary layer are indicators of efficient stress transfer in glass fiber/polymer composites. A good correlation has been found between these properties and the magnitude of the glass/resin interfacial shear strength; and

(d) chemical bonding between the silane-treated glass and the polymer matrix is not sufficient to ensure a strong bond. A strong joint requires a synergistic effect of chemical bonding and a soft boundary layer.

6. SUMMARY

The results of the research investigations summarized here, and of numerous other recent studies, have contributed greatly to current views of the chemical interactions of silane coupling agents in mineral-reinforced composites. However, concepts and hypotheses based on results of earlier work [1, pp. 16–27] have been confirmed in most instances.

Salient considerations for the design of the molecular structure and conditions of application for organofunctional silane coupling agents include the following elements:

(a) For silanes in solution, maximum hydrolysis and minimum condensation are desirable. The kinetics of hydrolysis and condensation depend on the amount of water in the solvent, the aging time, and the pH of the solution. Critical parameters are the structure of the organic moiety and the concentration of silane for the 'onset of association' [4, 5] (e.g. 0.15% for γ-APS, 0.40% for γ-MPS).

(b) The theory of chemical bonding through coupling agents in composites is based on oxane bond formation between hydroxyl groups on the mineral surface and reactive (e.g. alkoxy) groups in the silane *and* on the interaction of the organofunctional group in the silane with the matrix by covalent bond formation, co-polymerization, or interpenetrating polymer network formation.

(c) The microstructure, morphology, and mechanical properties of the interphase region adjacent to the silane-modified surface of the substrate and to the matrix are important considerations.

REFERENCES

1. E. P. Plueddemann, *Silane Coupling Agents.* Plenum Press, New York (1982).
2. E. P. Plueddemann, H. A. Clark, H. A. Nelson and K. R. Hoffmann, *Mod. Plast.* **39**, 135 (1962).
3. E. P. Plueddemann (Ed.), in: *Interfaces in Polymer Matrix Composites*, pp. 173–216. Academic Press, New York (1974).
4. H. Ishida, in: *Adhesion Aspects of Polymeric Coatings*, K. L. Mittal (Ed.), pp. 45–106. Plenum Press, New York (1983).
5. H. Ishida, *Polym. Composites* **5**, 101–123 (1984).
6. E. P. Plueddemann, in: *Silanes, Surfaces and Interfaces*, D. E. Leyden (Ed.), pp. 1–24. Gordon & Breach, New York (1986).
7. J. L. Koenig, in ref. 6, pp. 43–57.
8. C. D. Batich, in ref. 6, pp. 215–234.
9. P. T. K. Shih and J. L. Koenig, *Mater. Sci. Eng.* **20**, 137 (1975).

10. E. R. Pohl, in: *Proc. 38th Annu. Tech. Conf. Reinf. Plast./Composites Inst.*, *SPI*, Section 4-B (1983).

11. E. R. Pohl and F. D. Osterholtz, *Am. Chem. Soc. Polym. Div.*, *Prepr.* **24**, 200–201 (1983).

12. H. J. Kang, W. Meesiri and F. D. Blum, *Proc. Office Naval Res. Conf. Sci. Composite Interfaces*, Leesburg, VA, 18–21 April, pp. 265–270 (1989).

13. E. R. Pohl and C. S. Blackwell, in: *Controlled Interphases in Composite Materials*, H. Ishida (Ed.), pp. 37–49. Elsevier, New York (1990).

14. W. Meesiri and F. D. Blum, *Am. Chem. Soc., Polym. Div.*, *Prepr.* **29**, 196–197 (1988).

15. E. P. Plueddemann, in: *Silylated Surfaces*, D. E. Leyden and W. Collins (Eds), Midland Macromolecular Monographs No. 7. 31–53. Gordon and Breach, New York (1980).

16. E. P. Plueddemann, *Proc. 24th Annu. Tech. Conf. Reinf. Plast./Composites Inst.*, *SPI*, Section 19-A (1969).

17. F. J. Boerio, *Am. Chem. Soc. Org. Coating Plast. Div.*, *Prepr.* **44**, 625–629 (1981).

18. F. J. Boerio and C. A. Gosselin, *Adv. Chem. Ser.* **203**, 531–558 (1983).

19. F. J. Boerio, R. G. Dillingham, J. W. Williams and C. A. Gosselin, *Proc. 38th Annu. Tech. Conf. Reinf. Plast./Composites Inst.*, *SPI*, Section 4-C (1983).

20. F. D. Blum and W. Meesiri, *Am. Chem. Soc. Polym. Div.*, *Prepr.* **29**, 44–45 (1988).

21. E. P. Plueddemann and G. L. Stark, *Proc. 35th Annu. Tech. Conf. Reinf. Plast./Composites Inst.*, *SPI*, Section 20-B (1980).

22. W. A. Zisman, *Ind. Eng. Chem. Res. Dev.* **8**, 108 (1969); see also ref. 1, pp. 20–22.

23. Ref. 1, pp. 22–27.

24. H. Ishida and J. L. Koenig, *J. Polym. Sci. Polym. Phys. Ed.* **17**, 615 (1979).

25. Ch. Chang and J. L. Koenig, *J. Colloid Interface Sci.* **83**, 361 (1981).

26. Ch. Chang and J. L. Koenig, *Polym. Composites* **1**, 88 (1980).

27. J. L. Koenig, F. J. Boerio, E. P. Plueddemann, J. Miller and P. B. Willis, Report Nos. DOEJPL-1012-120, JPL-PUB-86-6, NTIS (Jan. 1986).

28. D. J. Ondrus, F. J. Boerio and K. J. Grannen, in: *Sci. Technol. Adhesive Bonding, Proc. 35th Sagamore Army Mater. Res. Conf.*, L. H. Sharpe and S. E. Wentworth (Eds), Manchester, NH, 26–30 June, pp. 27–42 (1988).

29. F. J. Boerio, in: *Treatise on Adhesion and Adhesives*, Vol. 6, R. L. Patrick (Ed), pp. 255–283. Marcel Dekker, New York (1989).

30. J. Jang, H. Ishida and E. P. Plueddemann, *Proc. 41st Annu. Tech. Conf. Reinf. Plast./Composites Inst.*, *SPI*, Section 2-C (1986).

31. G. A. McFarren, T. F. Sanderson and F. C. Schappell, *Polym. Eng. Sci.* **17**, 46–49 (1977).

32. F. J. Kolpak, *Proc. 41st Annu. Tech. Conf. Reinf. Plast./Composites Inst.*, *SPI*, Section 26-C (1986).

33. G. C. Tesoro, G. P. Rajendran, C. Park and D. R. Uhlmann, *J. Adhesion Sci. Technol.* **1**, 39–51 (1987); see also US Patent 4,778,727 (1988).

34. Y. Chujo, E. Ihara, H. Ihara and T. Seagusa, *Macromolecules* **22**, 2040–2043 (1989).

35. K. W. Lee and T. J. McCarthy, *Macromolecules* **21**, 3353–3356 (1988).

36. J. P. Stevens, US Patent 3,468,751 (1969).

37. M. Nasir, B. T. Poh and P. S. Ng, *Eur. Polym. J.* **25**, 267–273 (1989).

38. E. P. Plueddemann, *J. Adhesion Sci. Technol.* **3**, 131–139 (1989).

39. Y. Eckstein, *J. Adhesion Sci. Technol.* **3**, 337–355 (1989).

40. Y. Eckstein, *J. Adhesion Sci. Technol.* **2**, 339 (1988).

41. L. T. Drzal, M. J. Rich and P. F. Lloyd, *J. Adhesion* **16**, 1 (1982).

Silanes and Other Coupling Agents, pp. 229–240
Ed. K. L. Mittal
© VSP 1992.

Glass fiber surface effects in silane coupling

C. G. PANTANO,* L. A. CARMAN and S. WARNER
Department of Materials Science and Engineering, The Pennsylvania State University, University Park, PA 16802, USA

Revised version received 16 September 1991

Abstract—Glass fibers with the nominal E-glass composition were synthesized with 0, 2, 4, or 6% B_2O_3. The influence of the B_2O_3 content on aminosilane adsorption was evaluated using X-ray photoelectron spectroscopy (XPS) and zeta potential measurements. The water adsorption capacity of the silane-treated fibers was measured using a gravimetric technique. It was found that the presence of B_2O_3 enhanced the adsorption of aminosilane, but, more significantly, it greatly increased the water adsorption capacity of the silane-treated fibers. This effect of the B_2O_3 on water adsorption was considerably less evident in the untreated fibers. The solubility of boron and aluminum appears to play a role in determining the chemical and physical structure of the silane overlayer.

Keywords: Glass surface; E-glass; glass fiber; XPS; zeta-potential; water adsorption.

1. INTRODUCTION

Many theories have been proposed over the years to explain the molecular mechanisms of silane coupling to glass surfaces [1–4]. It is a complex phenomenon that depends on the specific silane, the glass, and the treating solution. Most of the fundamental studies have been focused on aminosilane and the influence of the solution pH, silane concentration, and curing conditions [1, 2]. The purpose of this study was to examine more specifically the effect of the glass composition. Silica and commercial E-glass have been used in the past. But silica may represent an oversimplification of E-glass, while commercial E-glass may be too complex to establish the role of specific oxides in the surface coupling mechanism.

E-glass is a calcium–boroaluminosilicate glass of nominal composition 56% SiO_2, 14% Al_2O_3, 22% CaO, 6% B_2O_3, and fractional percentages of Na_2O, MgO, TiO_2, Fe_2O_3, and F (by weight). Each of these constituents may influence the surface chemistry of the glass. Some will introduce basic sites, and others acidic sites. Their relative activity at the surface may or may not be in direct proportion to their relative concentration in the bulk glass. Thus, simpler glasses were prepared for this study using a reference composition of 60% SiO_2, 25% CaO and 15% Al_2O_3. The concentration ratios in this reference glass are consistent with E-glass. CaO and Al_2O_3 are expected to render the surface more basic relative to pure silica. Systematic additions of B_2O_3 and Na_2O were then made to this calcium–aluminosilicate reference glass. B_2O_3 is expected to introduce acidic sites, while Na_2O is expected to provide additional basicity. But

---

*To whom correspondence should be addressed.

of more concern in the case of B_2O_3 and Na_2O is their surface activity. These constituents can lower the surface energy of the (molten) glass, and, thereby, may be driven to concentate in the surface layers of the glass fiber. In competition with this driving force, however, is the volatility of B_2O_3 and Na_2O during fiber drawing; this can tend to decrease their surface concentration. It is especially noteworthy that these species are exceedingly reactive with water. This hydrolytic susceptibility will influence the glass surface chemistry in the fiber drawing, fiber sizing, and curing stages of commercial fiberglass processing. So, too, this susceptibility may be superimposed upon the behavior of composites that are used in aqueous environments. It is for all of these reasons that variations in the B_2O_3 and Na_2O contents are of interest. Here, the behavior of B_2O_3 in the glasses is emphasized.

It has been well established that the primary effect of silane coupling agents in composites is in wet strength retention [1, 2]. In practice, one can imagine a local equilibrium between the glass fiber and the aqueous sizing, and subsequently between the glass fiber, the interphase, and the polymer matrix. Obviously, this local equilibrium is complicated by the presence of polymer species, by the processing history of the fibers and the composite, and finally by the environmental history of the composite. But water is a principal reactant in all cases. Thus, it can be expected that water will drive the glass surface/silane interphase to a state of local equilibrium.

The evaluation and behavior of the silane-treated glass surfaces in *aqueous environments* were the focus of this study. The glass fibers were aged for long times in γ-aminopropyltriethoxysilane solutions (γ-APS) so that an equilibrium between the glass surface and the aqueous solution could be achieved. Some fibers were treated at pH 4 to simulate the condition during the commercial treatment of fibers. Others were treated at pH 10, which is the natural pH of γ-APS in water. These treated fibers were dried in air and then their water adsorptivity was evaluated. In a parallel set of experiments, interactions between the glass fibers and treating solution were examined using zeta-potential measurements. XPS (X-ray photoelectron spectroscopy) and static, neutral-beam SIMS (secondary ion mass spectroscopy) were used to verify the glass surface compositions and adsorption of silane wherever possible.

2. EXPERIMENTAL

2.1. Materials

The glass compositions were similar to commercial E-glass: 60% SiO_2, 25% CaO, and 15% Al_2O_3 with 0, 2, 4, or 6% B_2O_3 additions (by weight). Two other glasses were synthesized using the 0% and 6% B_2O_3 compositions but with 1% Na_2O additions. The glasses were melted in platinum crucibles at $>1400°C$ and then cast on a stainless steel plate. The cast glass was fragmented and remelted two or three times to ensure homogeneity. Finally, the glass was drawn into a continuous fiber ($\sim 5\ \mu m$ diameter) using a single-tip platinum bushing at $\sim 1300°C$. The fibers were immediately stored in a vacuum desiccator. All of the glass and fiber fabrication was carried out at PPG Fiberglass Research Center, Pittsburgh, PA, and the glass compositions were independently verified using spectrochemical techniques.

2.2. Fiber aging

Some of the fibers in each batch were aged in humid air, in deionized (DI) water, or in silane solution. All of the aging was carried out at 35°C for ~30 days. A commercial (Blue-M) controlled atmosphere chamber held the relative humidity (RH) at 70% and the temperature at 34°C. The silane solutions were prepared by adding the requisite amount (1% by volume) of unhydrolyzed γ-APS (A-1100; Union Carbide) to triply distilled water or to acetic acid solution. A hydrolysis time of approximately 8 h was allowed prior to adding fibers to the solution. This yielded solutions at pH 10 and pH 4, respectively. The fibers were suspended in Nalgene containers, where the glass surface area-to-solution volume ratio was fixed at 50 cm^{-1}. In all cases, the fibers were dried in air at 75°C and the solutions were analyzed for their Si, Al, Ca, and B contents. Table 1 presents the results of these solution analyses.

Table 1.
Solution compositions after 30 days exposure to the glass fibers

Treatment solution	% B$_2$O$_3$ in the glass fiber	pH	Solution concentration (ppm)[a]			
			Si	Al	Ca	B
Water	0	7.36	10.5	0.21	8.2	0.11
	2	7.43	7.9	0.23	7.1	0.31
	4	7.36	7.7	0.26	7.3	0.54
	6	7.36	7.5	0.24	7.1	0.83
	Control	4.08	<0.2	0.03	<0.02	<0.02
γ-APS (ph ~ 4)	0	4.14	1040	0.87	10.1	0.17
	2	4.16	1090	0.81	9.0	0.26
	4	4.18	1140	0.84	9.8	0.43
	6	4.19	1160	0.88	9.2	0.70
	Control	4.19	1180	0.08	0.10	0.04
γ-APS (pH ~ 10)	0	10.47	1070	2.06	8.4	0.15
	2	10.52	1140	2.01	9.4	0.43
	4	10.57	1120	2.23	9.5	0.69
	6	10.55	1030	2.00	7.0	0.97
	Control	10.34	1190	<0.02	<0.02	0.03

[a] By DC plasma emission spectroscopy.

2.3. Water-vapor adsorption

The water-vapor adsorption system consisted of a Cahn 2000 Recording Electrobalance, an MKS (Model 390HA) Absolute Pressure Sensor, a Spectrum Scientific (Series SM1000) Residual Gas Analyzer, a deionized water adsorbate reservoir, and a vacuum system capable of obtaining pressures of <10^{-6} Torr. Water-vapor adsorption isotherms were measured gravimetrically for the glass fibers in the as-drawn state, and after treating the fibers in the silane solutions. The samples consisted of a bundle of fibers which were dried (if necessary) and then loaded into the adsorption system where they were degassed at 150°C until the base pressure of the system was <10^{-5} Torr. The adsorption isotherms were constructed from a series of equilibrium weight changes obtained by monitoring the sample mass change accompanying water-vapor adsorption at each partial pressure p/p_0 (where p_0 is the saturation pressure at 25°C). Equilibrium was defined as the point at which the absolute value of the slope of the mass change was <0.05 μg/h; it typically took about 12–24 h to reach equilibrium. Approximately 15–30 data points were recorded for each adsorption isotherm. Most of the measurements were repeated on two or more samples, and in the case of the 0% and 6% B_2O_3 compositions, the reproducibility was established with an additional batch of glass fibers. The overall precision of the gravimetric measurements was ~4%.

2.4. Inert gas adsorption

The specific surface area of the fibers was determined using inert gas adsorption in a commercial volumetric adsorption system (Micromeritics Instrument Corp.). Krypton gas was used because of its sensitivity to the small specific surface areas of the glass fibers (~0.2 m^2/g). The fibers were degassed at 100°C to a pressure of 80 mTorr before introducing the adsorbate gas into the sample chamber. Several samples were also outgassed at 80 and 200°C (to 80 mTorr) to confirm that outgassing was sufficiently complete under the standard test conditions. A standard five-point surface area determination was made for each inert gas adsorption experiment.

2.5. Surface analyses

The untreated and treated fibers were analyzed for their elemental surface compositions. Fiber bundles approximately 2 cm long by 1 cm wide were mounted on the sample holder for each analysis. All fibers (~5 μm in diameter) were well aligned in each bundle, resulting in a high packing efficiency, and this provided a relatively uniform surface for analysis. Surface roughness contributes primarily to a loss of signal reaching the detector, as well as to slight variations in the sample depth probed; but, in either case, an average surface composition representative of the fiber surfaces was assumed.

The XPS data were obtained with an extensively modified AEI ES-100 photo-electron spectrometer. The samples were analyzed at a pressure typically <10^{-9} Torr. A magnesium anode (1253.6 eV) was used as the excitation source. The analyzed sample area was of the order of 5 mm^2. Survey scans from 0 to 1000 eV were first obtained for each sample to confirm that only the expected elements were present on the fiber surfaces. Subsequently, high resolution spectra were obtained by slowly scanning ~20 eV binding energy windows that included the Si $2p$, Al $2p$, Ca $2p$, B $1s$, O $1s$, and C $1s$ photoelectrons, respectively. Integrated peak areas of the photoelectron spectra were determined. The sensitivity factors, which were independently obtained on this spectrometer with oxide standards, were then utilized in the determination of the surface atomic percent compositions of the fibers.

The SIMS analyses were performed using a neutral-atom beam of Ar (Kratos Macrofab) and a quadrupole mass spectrometer (Extranuclear). The beam was ~2 mm in diameter and the effective beam intensity was ~1 nA/cm². The base pressure in the vacuum system was 1×10^{-9} Torr, but this increased to 1×10^{-6} Torr during the analysis due to the Ar gas. The spectra were normalized to the $^{29}Si^+$ signal to provide semiquantitative analysis of the Na and B contents.

2.6. Electrophoretic mobility

The glass fibers and fused-silica glass (Thermal American Fused Quartz Co.) were crushed and then dispersed in water. The pH of this near-neutral suspension was varied using KOH or HNO_3. In some experiments, a hydrolyzed solution of γ-APS was added to this suspension. Here, the initial pH was 10. The electrophoretic mobilities of glass fragments suspended in these solutions were measured without any further treatment except for the addition of electrolyte (10^{-3} M KNO_3). These analyses were performed using a Rank Brothers Particle Micro-Electrophoresis Apparatus Mark II or a Pen Kem System 3000 Automated Electrokinetics Analyzer.

3. RESULTS

The quantitative compositional surface analysis of the untreated glass fibers using XPS was complicated by the presence of carbon [5]. Nevertheless, the plot in Fig. 1 reveals that the boron concentration on the untreated fiber surfaces did, in fact, increase in proportion to its addition to the glass formulation, while the

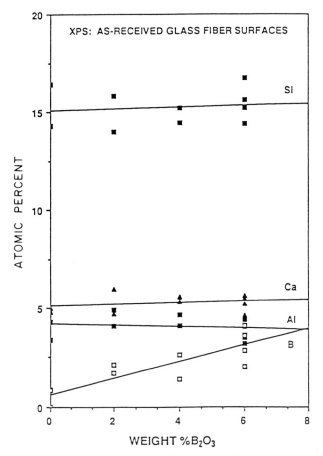

Figure 1. XPS analyses of the metal ion concentrations on glass fiber surfaces plotted vs. the bulk concentration of B_2O_3 in the glass fiber.

levels of the other metal ion constituents were reasonably constant. Additional surface analyses of the fibers were performed using SIMS because of the limited sensitivity of XPS to the low-levels of boron and sodium. The $^{11}B^+/^{29}Si^+$ ratios measured on the fibers were compared with those measured on fracture surfaces of the bulk glasses [6]. These data showed that the boron concentration at the fiber surfaces was equal to the bulk concentration. These Na^+/Si^+ ratios observed on the glasses prepared with 1% Na_2O in the batch formulation were about ten times the values observed on these glasses; thus, the impurity level of Na_2O is of the order 0.1%.

Figure 2 is a plot of the nitrogen concentration measured on the silane-treated glass fibers using XPS. The untreated fibers, and the fibers soaked in water, have been used as control specimens. It can be seen that the levels of nitrogen on these specimens is low ($<1\%$ N) relative to the silane-treated fibers. The C and N concentrations on the treated fiber surfaces were greatly increased, while the signals due to Ca, Al, and B were attenuated. The C 1s binding energies indicated nearly complete hydrolysis of the ethoxy groups. Qualitatively, these observations are consistent with the adsorption of hydrolyzed γ-APS onto the fiber surfaces. Thus, Fig. 2 reveals that there is a dramatic effect on the γ-APS adsorption due to the pH of the silane solution, and a less obvious influence of the % B_2O_3 in the glass composition. The N 1s binding energies were ~ 1 eV higher after the pH 4 treatment than after the pH 10 treatment. This indicates the presence of protonated amines in the silane adsorbed at pH 4. There was no apparent effect of B_2O_3 on any binding energy.

An estimate of the silane coverage could be made on the basis of the glass substrate signal attenuation and the carbon or nitrogen signal. This method of

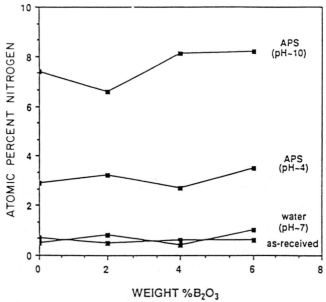

Figure 2. XPS analyses of the nitrogen concentration on glass fiber surfaces after various treatments (APS = 1% solution of γ-aminopropylsilane) plotted vs. the bulk concentration of B_2O_3 in the glass fiber.

quantitative analysis has already been reported [7]. In the case of fibers treated at pH 10, a uniform overlayer of silane could be assumed; that is, there was a uniform attenuation of the apparent Ca and Al surface concentrations and corresponding increases in the C and N concentrations. The calculated overlayer thicknesses—assuming uniform silane coverage—were 2.7, 2.7, 3.0, and 3.5 nm for the 0%, 2%, 4%, and 6% B_2O_3 fibers, respectively, after the 1-month soak in the silane solution.

Some fibers were immersed and immediately withdrawn from the pH 10 solutions. In this case, the overlayer was only 1.5 nm. This value of 1.5 nm corresponds to monolayer coverage with this aminosilane. Thus, lengthy aging of fibers in the pH 10 silane solution leads to multilayer adsorption of γ-APS, and this effect is enhanced in the case of the fibers that contain 4% and 6% B_2O_3.

The fibers aged at pH 4 were incompletely covered by silane, i.e. the N and C concentrations are considerably less than expected for uniform coverage. Interestingly, the Al surface concentration was attenuated by only 20%, whereas the Ca surface concentration was decreased by 75%. Thus, it seems that at pH 4 there was a preferential adsorption of silane on the Ca sites (relative to the Al sites).

Figure 3 shows the water-vapor adsorption isotherms measured on the silane-treated fibers along with those obtained on the untreated fibers. A complete discussion of the water adsorption isotherms of the untreated fibers has already been reported [8]. Here, two new features are immediately evident. First, the presence of the silane overlayer has greatly enhanced the water adsorption capacity of the fibers, and, second, the silane-treated fibers that contain 4% and 6% B_2O_3 adsorb significantly more water than the 0% B_2O_3 fibers. It is important to note that these data have been normalized to the specific surface areas of the

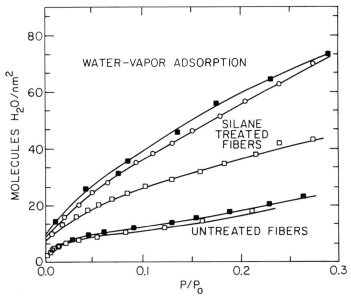

Figure 3. Water-vapor adsorption isotherms for glass fibers after the silane treatment; the isotherms for the untreated fibers with 0% and 6% B_2O_3 are included. APS = 1% solution of γ-aminopropylsilane at pH 10; 0% = □, 4% = ○ and 6% = ■.

various fibers as measured independently using Kr gas adsorption. This normalization eliminates any effects of fiber diameter or surface roughness in the interpretation of water adsorption capacity. In fact, the difference in specific surface area between the silane-treated fibers and the untreated fibers was small and showed no dependence on the B_2O_3 content. The untreated fibers were found to have specific surface areas of ~0.20–0.25 m^2/g, while the silane-treated fibers exhibited specific surface areas of ~0.30–0.40 m^2/g. This means that the differences in water-vapor adsorption capacity revealed in Fig. 3 are due to real differences in the chemical and structural nature of the silane overlayer.

The data in Fig. 3 were analyzed according to the BET equation. A summary is provided in Table 2. The data for fibers that had been aged in water vapor and liquid water are also included to provide reference levels for the effect of hydration (in the absence of silane adsorption). The *active* surface area refers to the surface area determined using the water adsorption isotherms; so, it is a quantitative representation of the water adsorption capacity at low pressure. In the case of the untreated fibers, there is reasonable agreement between the specific surface areas and the geometric surface areas; but the active surface areas are always greater. Thus, it seems likely that there is a tendency for multi-layer adsorption of water. Most evident, however, is the fact that this tendency is enormous in the case of the silane-treated fibers. There, the water adsorption capacity at low pressure exceeds the absolute surface area by 5–10 times. The effect is greatly enhanced by the % B_2O_3 in the glass. Correspondingly, the BET interaction parameter (C) and the adsorption energy (ΔE) for the silane-treated

Table 2.
BET analyses of the water-vapor adsorption isotherms

% B$_2$O$_3$ in the glass	Geometric area (m^2/g)	Kr specific area (m^2/g)	Water-vapor adsorption		
			Active area (m^2/g)	BET parameter (C)	ΔE (cal/mol)
Untreated fibers					
0	0.239	0.226	0.291	83.98	2622
2	0.256	0.243	—	—	—
4	0.213	0.219	—	—	—
6	0.259	0.244	0.391	51.53	2333
Aged in water vapor					
0	0.239	0.284	0.541	2565.9	4646
2	0.256	0.318	—	—	—
4	0.213	0.260	0.599	250.42	3269
6	0.259	0.322	0.662	245.88	3258
Aged in liquid water					
0	0.239	0.317	1.01	39.39	2174
2	0.256	0.306	—	—	—
4	0.213	0.292	—	—	—
6	0.259	0.298	1.14	60.95	2433
Aged in γ-APS at pH $\sim$ 10					
0	0.239	0.320	1.37	20.63	1791
2	0.256	0.389	—	—	—
4	0.213	0.327	2.54	12.77	1508
6	0.259	0.412	3.34	9.204	1314

fibers are considerably lower than the values for the untreated or hydrated fibers, and these numbers, too, scale with the % B_2O_3. It is worth noting, finally, that the desorption isotherms exhibited a strong hysteresis. This would be consistent with the adsorption of a highly structured form of water, or with the chemical incorporation of adsorbed water. The possibility of water condensation in micropores of the silane overlayer cannot be excluded, but does not seem likely at these low pressures.

Figures 4 and 5 summarize the zeta potentials that were measured on pure silica, 0% B_2O_3 glass, and 6% B_2O_3 glass—in water and silane solutions. The zeta

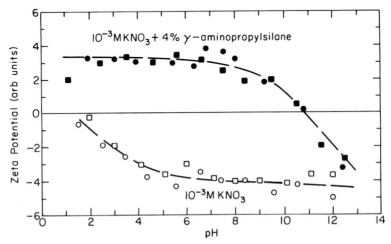

Figure 4. The zeta potential of crushed, untreated 6% B_2O_3 E-glass fibers (○) and silica glass (□) in 10^{-3} M KNO_3 solution and in a 4% solution of γ-aminopropylsilane (●) and (■) measured using a Rank Brothers Electrophoresis analyzer.

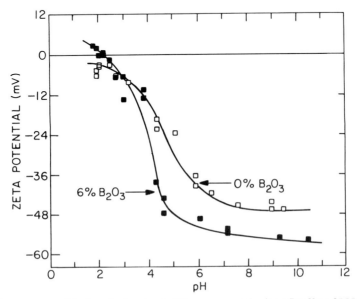

Figure 5. The zeta potential of crushed, untreated fibers measured using a Pen Kem 3000 Analyzer.

potential seems to be independent of the glass composition. It is negative over most of the pH range. In Fig. 5, a more precise measurement of the zeta potential in the 0% and 6% B_2O_3 glasses suggests that there may be a difference at pH 4. In general, though, the B_2O_3 content does not seem to be a determining factor of the surface charges. More evident is the fact that the introduction of aminosilane creates a positive surface potential that is stable up to pH $\approx$ 10 (see Fig. 4). This is consistent with the protonation of amine groups that are associated with silane adsorbed on the glass fiber surface. It is likely that most of these silanes are oriented with the amine group at the solution/adsorbate interface. There is no apparent effect of B_2O_3 on the electrostatic nature of the adsorbed silane.

4. DISCUSSION

The primary effect of B_2O_3 in the silane-treated glass fibers was found to be an enhancement of the water adsorptivity. This dependence on B_2O_3 was also observed in the water adsorption behavior of untreated fibers, water-vapor hydrated fibers, and water-leached fibers [8], but was significantly greater after the silane treatment. It was also found that the presence of B_2O_3 influenced the amount of silane adsorption *per se*. Altogether it can be concluded that there is a direct effect of B_2O_3 on water adsorption. There is also an influence of B_2O_3 on the adsorption and condensation of aminosilane which determines the water adsorptivity of the silane-treated fiber.

The fact that all the fibers adsorb water in excess of the expected monolayer amounts suggests that the water adsorption is multilayer in nature, or that there is pore space which is accessible to water molecules—but not accessible to the Kr used to measure the specific surface area. The XPS analyses showed that the silane overlayers increased in thickness in the 4% and 6% B_2O_3 fibers. But the increase in water adsorptivity with % B_2O_3 is not in direct proportion to the increase in silane overlayer thickness; it is considerably larger. This suggests that B_2O_3 has influenced the chemical and physical structure of the adsorbed silane overlayer. It is likely that there is microporosity, free volume, and/or reactive sites within the silane overlayer, in general.

The fact that these silane overlayers possess a degree of *bulk* adsorptivity due to their intrinsic chemical structure—and further that this structure may depend on the substrate composition—is not surprising. It is already established [2] that γ-APS will hydrolyze and condense on solid substrates through the combined effects of hydrogen bonding and siloxane bond formation. These interactions are quite complex, and, in general, depend on the substrate surface, the solution concentration and pH, and the drying conditions. At pH 4, the silane overlayer was incomplete on the fibers. This limited silane adsorption at low pH has been observed by others [9]. It is attributed to protonation of the amine group, which limits hydrogen bonding of the silane, and to hydroxylation of the glass surface, which limits siloxane condensation. At pH 10, these limitations in bond formation are removed. A 1.5 nm silane overlayer (~one monolayer) forms at short times, and after lengthy exposure, the overlayers are in the range 2.7–3.5 nm. Thus, the presence of microporosity and reactive sites within the silane overlayer is a reasonable expectation. The porosity or free volume could be due to water evaporation and condensation reactions during drying, but must be in the size range 2.7–4.5 Å, i.e. sufficient to permit access to molecular water (~2.7 Å) but smaller than the Kr atom (4–4.5 Å). The reactive sites could be associated with free amine, silanols, hydrogen bonds, or strained siloxane bonds.

The zeta-potential measurements—although very sensitive to the presence of the aminosilane—were insensitive to the glass composition. This has also been observed on many pure oxides [10]. It is not likely, therefore, that the glass composition influences the adsorption through electrostatic effects. More important is the role played by soluble constituents of the glass in the hydrolysis, adsorption, and condensation of the silane overlayer. The analyses of the DI water and silane solutions after exposure to the fibers (Table 1) showed some release of Al, Ca, and B. The release was greater in the silane solution than in the DI water for all species except boron. In the pH 10 silane solution, the release of Al, and to some extent B, was further enhanced. And naturally, the concentration of soluble B increased with the % B_2O_3 in the glass. Altogether, this suggests that condensation of the silane overlayer occurs in the presence of soluble Al, B, and Ca. The XPS analyses did, in fact, show the presence of these species on the treated surfaces, but their relative distribution between the glass fiber substrate and silane overlayer could not be determined. (In principle, variable angle XPS could provide this distribution, but the fiber geometry creates a problem.) Chiang *et al.* [11] have reported that the condensation of silane is faster on E-glass than on pure silica. More recently, Irwin *et al.* [12] showed that boron influences the hydrolysis and condensation of alkoxysilane through the formation of $\equiv$Si—O—B$=$ (boroxane) linkages.

It can be proposed that even trace amounts of soluble B, Ca, and Al—local to the fiber surface—could become incorporated in the silane, and, thereby, enhance hydrogen bonding and/or condensation of the overlayer. The formation of $\equiv$Si—O—B$=$ and $\equiv$Si—O—Al ligands would be expected in the drying stages. The $\equiv$Si—O—B$=$ boroxane ligands would be especially reactive to water, and, thereby, could be responsible for the enhanced water adsorptivity of the B_2O_3-containing fibers after the silane treatment. Even small amounts of B could exert disproportionate effects on the reactivity of the condensed silane overlayer because each B would contribute three (or four) ligands to the chemical structure. In fact, the result is very similar to the situation reported by Hair and Hertl [13] for water adsorption on silica gel that had been exposed to BCl_3 vapor. They found that the Si—O—B sites created by dehydration of the gel were exceedingly reactive to the adsorption of water, i.e. the boron content disproportionately enhanced the adsorption of water.

5. SUMMARY

The water adsorption capacity of the silane-treated fibers—at low pressure where only the most reactive sites are sampled—was 5–10 times the specific surface area. It is proposed that the excess water adsorption capacity of the silane-treated fibers is associated with sites within the silane overlayer. These may be hydroxyls, strained siloxanes, boroxanes, free volume, or microporosity. These sites provide a mechanism for physical and chemical adsorption, swelling, and rearrangement of the adsorbed silane in the presence of water. These observations are consistent with the idea that the adsorbed silane is a chemically dynamic entity whose equilibrium with water is the determining factor in silane coupling [14]. It remains to be determined whether this behavior and structure of silane are exhibited in the presence of polymer species. The study of water adsorption on sized fibers (polymer + silane) seems warranted. Also, the wet-strength behavior of composites containing glass fibers without B_2O_3 additions should be evaluated.

Acknowledgements

We gratefully acknowledge the support of the PPG Fiberglass Research Center.

REFERENCES

1. E. P. Plueddemann, in: *Inerfaces in Polymer, Ceramic and Metal Matrix Composites*, H. Ishida (Ed.), p. 17. Elsevier, New York (1988).
2. H. Ishida, in: *Molecular Characterization of Composite Interfaces*, H. Ishida and G. Kumar (Eds), p. 34. Plenum Press, New York (1985).
3. F. M. Fowkes, D. W. Dwight, D. A. Cole and T. C. Huang, *J. Non-Cryst. Solids* **120**, 47 (1990).
4. E. R. Pohl and F. O. Osterholtz, in: *Silanes, Surfaces and Interfaces*, D. E. Leyden (Ed.), p. 481. Gordon and Breach, New York (1986).
5. C. G. Pantano, *Riv. Staz. Sper. Vetro* **6**, 123 (1990).
6. C. G. Pantano, *Rev. Solid State Sci.* **3**, 379 (1989).
7. C. G. Pantano and T. N. Wittberg, *Surf. Interface Anal.* **15**, 498 (1990).
8. L. A. Carman and C. G. Pantano, *J. Non-Cryst. Solids* **120**, 40 (1990).
9. S. Naviroj, S. R. Culler, J. L. Koenig and H. Ishida, *J. Colloid Interface Sci.* **97**, 308 (1984).
10. K. Esumi, *Bull. Chem. Soc. Jpn.* **56**, 331 (1983).
11. C. H. Chiang, H. Ishida and J. L. Koenig, *J. Colloid Interface Sci.* **74**, 396 (1980).
12. A. D. Irwin, J. S. Holmgren, T. W. Zerda and J. Jonas, *J. Non-Cryst. Solids* **89**, 191 (1987).
13. M. L. Hair and W. Hertl, *J. Phys. Chem.* **74**, 91 (1970).
14. E. P. Plueddemann, in: *Fundamentals of Adhesion*, L. H. Lee (Ed.), pp. 279–289. Plenum Press, New York (1992).

Silanes and Other Coupling Agents, pp. 241–262
Ed. K. L. Mittal
© VSP 1992.

An XPS investigation of the adsorption of aminosilanes onto metal substrates

M. R. HORNER,[1] F. J. BOERIO,[1,*] and HOWARD M. CLEARFIELD[2]

[1] *Department of Materials Science and Engineering, University of Cincinnati, Cincinnati, OH 45221, USA*
[2] *IBM T. J. Watson Research Center, P.O. Box 218, Yorktown Heights, NY 10598, USA*

Revised version received 15 July 1991

Abstract—Spectroscopic evidence has been obtained showing that γ-aminopropyltriethoxysilane (γ-APS) is partially adsorbed onto the oxidized surfaces of certain metals through the amino group. Results obtained from X-ray photoelectron spectroscopy (XPS) indicate that the amino groups on γ-APS are protonated by interaction with hydroxyl groups on the surfaces of silicon, iron, titanium, aluminum, nickel, and magnesium. The extent of protonation is greatest on silicon, aluminum, and titanium; intermediate on iron and nickel; and least on magnesium; and thus correlates well with the predicted isoelectric points of the oxides. The extent of protonation also depends on the pretreatment of the substrate. Thus, the extent of protonation of the amino groups is greater on aluminum surfaces that have been treated by the Forest Products Laboratory (FPL) etching process than on aluminum surfaces that have been mechanically polished. γ-APS interacts with the oxidized surface of copper by formation of complexes in which the lone pair of electrons on the amino nitrogen atoms is coordinated to copper ions.

Keywords: Aminosilanes; acid–base interactions; X-ray photoelectron spectroscopy.

1. INTRODUCTION

Organofunctional silanes such as γ-aminopropyltriethoxysilane (γ-APS) are widely used as coupling agents to enhance the durability of glass fiber-reinforced composites and adhesive joints during exposure to moisture at elevated temperatures. Numerous models have been proposed to explain the mechanisms by which silanes function as coupling agents [1]. However, the most widely discussed model considers that silanes form covalent bonds across the interface, thus binding the adhesive or matrix resin to the substrate [1].

Many investigations of the molecular structure of thin films formed by γ-APS deposited onto inorganic substrates from aqueous solutions have been carried out. Ondrus and Boerio [2] used reflection–absorption infrared spectroscopy (RAIR) to determine the structure of γ-APS films deposited on iron, 1100 aluminum, 2024 aluminum, and copper substrates from aqueous solutions at pH 10.4. They found that the as-formed films absorbed carbon dioxide and water vapor to form amine bicarbonate salts which were characterized by absorption bands near 1330, 1470, 1570, and 1640 cm^{-1}. γ-APS films had to be heated to temperatures above about 90°C in order to dissociate the bicarbonates, presumably to free amine, carbon dioxide, and water. Since the amine bicarbonates failed to react with epoxies, the strength of adhesive joints prepared

---

*To whom correspondence should be addressed.

using low temperature (under 90°C) cured epoxy-based adhesives was predicted to be reduced.

Ondrus and Boerio also found that metallic substrates had a significant effect on the molecular structure of γ-APS films cured against them [2]. When γ-APS films were dried against iron substrates at 110°C for 30 min, the extent of polymerization increased and the bicarbonates dissociated. Similar behavior was observed for films formed on commercially pure 1100 aluminum substrates. However, when γ-APS films were dried against 2024 aluminum substrates which contain about 4.6% copper, a new band attributed to an imine formed by copper-catalyzed oxidation of the propylamine group appeared near 1660 cm^{-1}.

When γ-APS films were formed by deposition from aqueous solutions at pH 10.4 and dried at 150°C for 30 min, considerably different results were obtained. The band near 1660 cm^{-1} was strong in spectra obtained from films dried against iron and 2024 aluminum substrates, indicating almost complete conversion of the amino groups to imine. However, there was little evidence for the formation of imine in films dried against 1100 aluminum, indicating that the oxide inhibited the oxidation of the amino groups.

Boerio and Dillingham [3] investigated the effect of γ-APS on the durability of adhesive bonds between epoxies and the oxidized surfaces of iron and titanium. They suggested that the orientation of the silane on the surface could be controlled by the pH of the solution from which the silane was deposited and that the silane/oxide interface would be stable during exposure to water as long as the pH of the water was between the isoelectric point of the oxide and the pK_a of an ionizable group on the silane. Boerio and Dillingham assumed that the oxidized surface of iron was basic and that γ-APS would adsorb from aqueous solutions at pH 8.0 onto such a surface through acidic silanol groups. They predicted that the interface would be stable during exposure to moisture at near neutral pH values and that is what was observed. However, γ-APS should adsorb from basic solutions onto iron substrates through the amino groups. Such interfaces should be less stable when exposed to water at near neutral pH values, since in this case the pH of the water would not be between the isoelectric point of the oxide and the pK_a of the ionizable amino groups, and that was also observed. Boerio and Dillingham also predicted that γ-APS would be more effective as a primer for improving the moisture resistance of epoxy/titanium adhesive bonds when deposited from aqueous solutions at high pH values than at low pH values. That was also observed.

Theidman *et al.* investigated the effect of silanes on the durability of aluminum/epoxy wedge test specimens exposed to an atmosphere of 95% relative humidity at 56°C [4]. They found that the use of an epoxy-functional silane as a primer led to crack growth rates that were lower than those obtained when the adherends were pretreated by the Forest Products Laboratory (FPL) etching technique. When γ-APS primers were applied to the adherends prior to bonding, the crack growth rates were much higher unless the silane was deposited from aqueous solutions that were acidified with hydrochloric acid. In that case, the crack growth rates were even lower than those obtained using the epoxy silane.

Films formed by γ-APS deposited onto inorganic substrates from aqueous acidified solutions have a different molecular structure from those deposited at the natural pH of 10.4 [5]. The extent of polymerization is higher in films deposited from solutions acidified by addition of HCl and the amino groups are protonated to form amine hydrochlorides. However, little spectroscopic evidence has been presented to indicate that the orientation of silanes such as γ-APS can be controlled by the pH of the solutions from which they are deposited.

Several groups have used XPS to determine the molecular structure of films formed by γ-APS adsorbed onto inorganic substrates. Moses *et al.* [6] investigated the adsorption of γ-APS onto metal oxide electrodes. They observed that the N(1s) peak of γ-APS adsorbed onto SnO$_2$ electrodes consisted of two components near 400.3 and 401.9 eV which were attributed to free and protonated amino groups, respectively. Some of the free and protonated amino groups could be interconverted by rinsing the films with acidic or basic solutions. However, some amino groups were irreversibly protonated while others were irreversibly free. It was suggested that irreversibly protonated amino groups were associated with hydrogen bonding between amino groups and silanol groups or hydroxyl groups on the oxidized metal surfaces.

Very recently, Hook *et al.* [7] used XPS and ATR infrared spectroscopy to investigate the silanization of polytetrafluoroethylene (PTFE) which was surface-modified by treatment in a H$_2$/H$_2$O radio-frequency glow-discharge and then refluxed with hexane solutions of γ-APS. Analysis by XPS showed that the amino groups near the γ-APS/PTFE interface were protonated while those in the bulk of the films were not. Protonation of amino groups near the interface was attributed to hydrogen bonding with silanol groups. The NH$_2$ deformation mode of the amino groups was observed near 1575 cm^{-1} in ATR spectra of γ-APS films deposited on the plasma-modified PTFE, about 25 cm^{-1} less than expected for free primary amino groups. The shift in the NH$_2$ deformation mode to lower frequencies was attributed to coordination of amino groups to silicon atoms.

Fowkes *et al.* [8] used XPS to show that γ-APS could adsorb onto acidic glass surfaces through the amino groups. They reported that the N(1s) band in XPS spectra of γ-APS adsorbed onto a glass substrate consisted of two components separated by about 2 eV. The component at lower binding energy was assigned to free amino groups but the component at higher binding energy was assigned to protonated amino groups. Using angle-resolved XPS, Fowkes *et al.* showed that the protonated amino groups were located adjacent to the surface and that the free amino groups were near the free surface of the films.

Protonation of amino groups adjacent to the oxidized surfaces of metals has been observed for compounds other than silanes. Dillingham and Boerio used XPS to characterize the fracture surfaces of aluminum/epoxy adhesive joints that were cured with aliphatic amines [9]. They observed that the N(1s) spectrum obtained from the substrate fracture surface consisted of components near 398.0 and 399.5 eV having similar intensities. Both components were observed in the XPS spectrum of the adhesive fracture surface but the component at the higher binding energy was very weak in a spectrum of the bulk adhesive. The component near 398.0 eV was attributed to free amino groups but that near 399.5 eV was attributed to amino groups protonated by surface hydroxyl groups.

There is some spectroscopic evidence that γ-APS interacts with mineral surfaces by forming metallosiloxane bonds. Naviroj *et al.* [10] investigated the adsorption of aminopropyldimethylethoxysilane (APDMES) onto several metal oxides using diffuse reflectance infrared spectroscopy. They observed bands near 950 and 963 cm^{-1} for APDMES adsorbed onto TiO_2 and Al_2O_3, respectively, which were assigned to metallosiloxane stretching modes. They also observed that the NH_2 deformation mode near 1600 cm^{-1} shifted somewhat, depending on the substrate, indicating a likely interaction between the amino groups of the silane and the oxides.

There is considerable indirect evidence that silanes interact with mineral surfaces by formation of siloxane bonds. Waddell *et al.* [11] investigated the stability of silane to silica bonds formed when silica was treated with solutions of amino silanes in toluene and subsequently dried at 80°C. Before drying, all of the silanes were easily desorbed by ethanol. After drying for at least 3 h, γ-APS and γ-aminopropyl(methyl)silane were strongly adsorbed and resisted desorption by ethanol, but γ-aminopropyl(dimethyl)silane was easily desorbed. The differences were related to the bonding of the silanes to silica. It was suggested that γ-APS formed three bonds with the substrate but γ-aminopropyl(dimethyl)silane and γ-aminopropyl(methyl)silane formed only one or two bonds with the substrate, respectively. Results obtained from infrared spectroscopy showed that the NH_2 deformation was near 1600 cm^{-1} in all cases, indicating that there was little hydrogen bonding of the amino groups. The amino groups did, however, form complexes when the silanated silica was treated with aqueous solutions containing Cu(II) ions.

Considerable evidence exists indicating that the acidity of an oxide surface can vary according to the pretreatment. For example, Finlayson and Shah [12] used flow microcalorimetry to characterize the oxidized surfaces of three aluminum specimens that had received different pretreatments. They found that the surface chemistry of the three samples was considerably different but was dominated by Lewis base sites in all cases. The peel strength of ethylene/acrylic acid co-polymers laminated against the substrates increased as the basicity of the substrate and the acrylic acid content of the co-polymer increased.

Mason *et al.* [13] used indicator dyes to determine the effect of pretreatment on the surface acidity of Ti-6,4. They found that surfaces treated with a basic etchant left a basic surface with an isoelectric point between 7.3 and 9.2. A phosphate–fluoride etch left the surface acidic with an isoelectric point between 5.4 and 7.3.

The purpose of this paper is to report results that we have obtained using XPS to determine the molecular structure of thin films formed by the adsorption of γ-APS onto the oxidized surfaces of metals having a range of isoelectric points. In particular, we show that some of the amino groups are protonated when γ-APS is deposited on the oxidized surfaces of metals, and that the extent of protonation is greatest on acidic surfaces such as silicon, titanium, and aluminum; least on basic surfaces such as magnesium; and intermediate on surfaces such as iron and nickel which are intermediate in acidity. We show that the extent of protonation is greater on aluminum that has been pretreated by FPL etching than on aluminum that has been mechanically polished. Finally, we also show that the interaction of γ-APS with copper involves the formation of co-ordination complexes between amino groups and Cu(III) ions rather than the protonation of amino groups.

2. EXPERIMENTAL

γ-Aminopropyltriethoxysilane (γ-APS) was obtained from Huls America, Inc. (Bristol, PA). Magnesium, 1018 cold-rolled steel, 1100 aluminum, and Titanium-6 Al,4 V (Ti-6,4) were obtained from Dow Chemical Company, Armco, Inc.; Alcoa; and Timet, Inc., respectively. High-purity (99.99+%) chromium and nickel were obtained from the Aldrich Chemical Co. Oxygen-free high conductivity (OFHC) copper was obtained from AMAX Copper, Inc. A band saw was used to cut the metal substrates into 1.0 cm^2 coupons, which were then polished to a mirror finish by successively grinding with silicon carbide papers followed by wet polishing with alumina compounds. A final wet polishing step was performed using magnesium oxide. Between each grinding step, the coupons were cleaned with detergent, rinsed with distilled water, and blotted dry with a paper towel. After each polishing step, the coupons were rinsed free of excess polishing compound using distilled water and then blown dry in a stream of nitrogen. In some cases, 1100 aluminum was etched according to Forest Products Laboratory (FPL) specifications [14].

γ-APS was applied to the polished substrates under a clean nitrogen environment by immersing them in a 1% (v/v) aqueous solution. After 1 min, the substrates were removed and excess solution was gently blown off the surface using a stream of nitrogen gas. Coated coupons were allowed to cure under a humid nitrogen environment for 24 h before being loaded into the test chamber of the X-ray photoelectron spectrometer.

XPS analysis was mostly carried out using a Physical Electronics Model 5300 ESCA system and Mg K_a X-rays. The X-ray source had an energy of 1253.6 eV and was operated at 300 W and 15 kV DC. Pass energies for survey and high resolution spectra were 44.75 and 17.90 eV, respectively. High resolution spectra were corrected for charging by assigning a value of 284.6 eV to the C(1s) peak for saturated hydrocarbons. Atomic concentrations were obtained from the high resolution spectra using sensitivity factors provided with the software. Information about the change in structure of γ-APS films as a function of depth was obtained by collecting spectra at take-off angles of 15° and 75°, since the depth of analysis increases with the take-off angle. (The take-off angle is defined as the angle between the sample surface and the direction of emission of the photoelectrons.)

XPS analysis of γ-APS applied to nickel and silicon substrates was also carried out using a Surface Science Instruments SSX-100-03 instrument equipped with a monochromatic Al K_a source. The X-ray source had an energy of 1487 eV and the instrument operated at a spot size of 600 μm. Pass energies for survey and high resolution spectra were 150 and 52 eV, respectively. Atomic concentrations were once again obtained from the high resolution spectra using sensitivity factors provided with the software.

3. RESULTS

Although high resolution spectra were obtained for all of the relevant elements, including carbon, oxygen, silicon, nitrogen, and the substrate metal, only the nitrogen spectra differed significantly for the various substrates. Therefore, the discussion will mostly concern the N(1s) high resolution spectra.

3.1. 1100 aluminum substrates

The N(1s) high resolution spectra obtained from polished 1100 aluminum which had been coated with γ-APS from a 1% aqueous solution at pH 10.4 are shown in Fig. 1. When the take-off angle was 75°, two components were observed near 399.4 and 401.3 eV which were attributed to free and protonated amino groups, respectively. Based on the relative intensities of the two components, it was determined that about 42% of the amino groups were protonated. When the take-off angle was decreased to 15°, the band positions and the relative intensities of the components changed only slightly. The relative intensities of the two components in the N(1s) spectra of γ-APS deposited on all of the substrates examined are listed in Table 1.

After the samples used to obtain the spectra in Fig. 1 were rinsed with distilled water, the spectra shown in Fig. 2 were obtained. Two components were still observed in the N(1s) spectra but the component at the higher binding energy increased considerably in intensity, especially when the take-off angle was 75°. When the take-off angle was 15°, only about 45% of the amino groups were protonated. However, when the take-off angle was 75°, about 57% of the amino groups were protonated. Since the extent of amine protonation increased as the

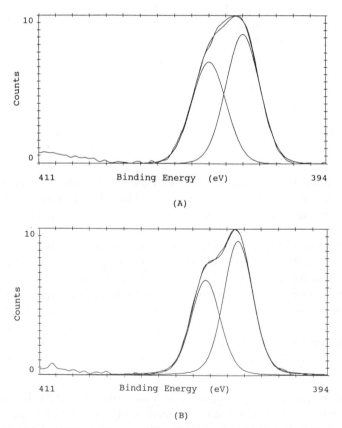

Figure 1. N(1s) high resolution spectra of polished 1100 aluminum coated with γ-APS from a 1% aqueous solution at pH 10.4. The take-off angles were (A)—15° and (B)—75°.

Table 1.

Degree of amine protonation on substrates as a function of the take-off angle

	Take-off angle	
	75°	15°
Substrate	N/N$^+$ (%)	N/N$^+$ (%)
Aluminum		
As-prepared	58/42	57/43
Rinsed	43/57	51/49
Aluminum–FPL etch		
As-prepared	52/48	53/47
Rinsed	36/65	35/66
Silicon	57/43	73/28
Titanium	68/32	74/26
Copper	Coordination	
Chromium	74/26	87/13
Nickel	78/22	81/19
Iron	79/21	87/13
Magnesium	91/9	95/5

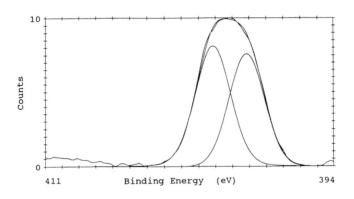

(A)

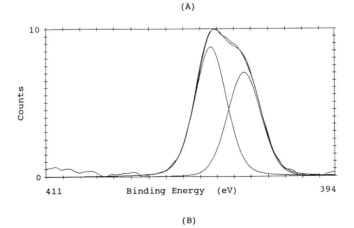

(B)

Figure 2. N(1s) high resolution spectra of polished 1100 aluminum coated with γ-APS from a 1% aqueous solution at pH 10.4 and rinsed with distilled water. The take-off angles were (A)—15° and (B)—75°.

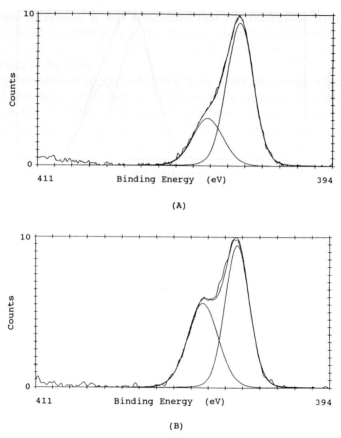

Figure 5. N(1s) high resolution spectra of polished silicon coated with γ-APS from a 1% aqueous solution at pH 10.4. The take-off angles were (A)—15° and (B)—75°.

considerably in intensity when the take-off angle was decreased to 15°. Otherwise, there was little change in the spectra. Considering the relative intensities of the two components in the N(1s) spectra, it was determined that about 43% and about 28% of the amino groups were protonated when the take-off angles were 75° and 15°, respectively.

3.3. Titanium-6 Al, 4 V substrates

The N(1s) high-resolution spectra obtained from polished Ti-6,4 which had been coated with γ-APS from a 1% aqueous solution at pH 10.4 are shown in Fig. 6. When the take-off angle was 75°, two components were observed near 399.7 and 401.6 eV. The component at the higher binding energy decreased somewhat in intensity when the take-off angle was decreased to 15°. Otherwise, there was little change in the spectra. Considering the relative intensities of the two components in the N(1s) spectra, it was determined that about 32% and about 26% of the amino groups were protonated when the take-off angles were 75° and 15°, respectively.

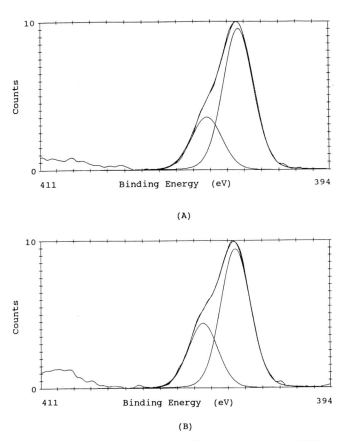

Figure 6. N(1s) high resolution spectra of polished Ti-6,4 coated with γ-APS from a 1% aqueous solution at pH 10.4. The take-off angles were (A)—15° and (B)—75°.

3.4. Chromium

N(1s) high-resolution spectra obtained from polished chromium which had been coated with γ-APS from a 1% aqueous solution at pH 10.4 are shown in Fig. 7. When the take-off angle was 75°, components were observed near 399.2 and 400.9 eV. The relative intensities of the components indicated that about 26% of the amino groups were protonated. When the take-off angle was decreased to 15°, the relative intensity of the component near 400.9 eV decreased and it was determined that only about 13% of the amino groups were protonated.

3.5. Nickel substrates

The position of a nickel Auger peak superimposed on the N(1s) photoelectron peak was detected when spectra of coated nickel samples were collected on the Physical Electronics Model 5300 ESCA system using Mg K_α X-rays. Therefore, XPS spectra of nickel samples were obtained using a Surface Science Instruments SSX-100-03 instrument equipped with a monochromatic Al K_α source. The N(1s) high-resolution spectra obtained from polished nickel which had been coated with γ-APS from a 1% aqueous solution at pH 10.4 are shown in Fig. 8.

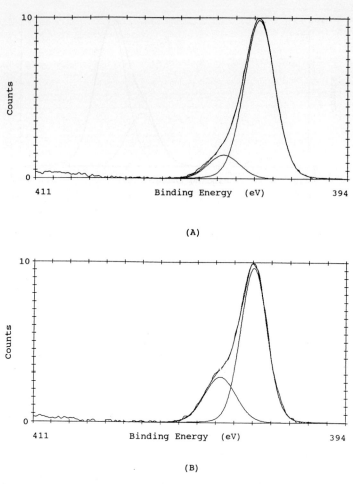

(A)

(B)

Figure 7. N(1s) high resolution spectra of polished chromium coated with γ-APS from a 1% aqueous solution at pH 10.4. The take-off angles were (A)—15° and (B)—75°.

When the take-off angle was 75°, components were observed near 399.2 and 400.9 eV. The relative intensities of the components indicated that about 22% of the amino groups were protonated. When the take-off angle was decreased to 15°, the relative intensity of the component near 400.9 eV decreased and it was determined that about 19% of the amino groups were protonated.

3.6. Iron substrates

N(1s) high-resolution spectra obtained from polished iron which had been coated with γ-APS from a 1% aqueous solution at pH 10.4 are shown in Fig. 9. When the take-off angle was 75°, components were observed near 399.2 and 400.9 eV. The relative intensities of the components indicated that about 21% of the amino groups were protonated. When the take-off angle was decreased to 15°, the relative intensity of the component near 400.9 eV decreased and it was determined that only about 13% of the amino groups were protonated.

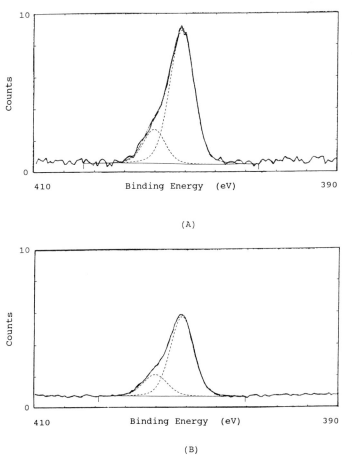

(A)

(B)

Figure 8. N(1s) high resolution spectra of polished nickel coated with γ-APS from a 1% aqueous solution at pH 10.4. The take-off angles were (A)—15° and (B)—75°.

3.7. Magnesium substrates

The N(1s) high-resolution spectra obtained from polished magnesium which had been coated with γ-APS from a 1% aqueous solution at pH 10.4 are shown in Fig. 10. Once again, two components were observed, near 399.3 and 401.2 eV. When the take-off angle was 75°, about 9% of the amino groups were protonated. However, when the take-off angle was decreased to 15°, only about 4% of the amino groups were protonated.

3.8. Copper substrates

The N(1s) high-resolution spectra obtained from polished OFHC copper which had been coated with γ-APS from a 1% aqueous solution at pH 10.4 are shown in Fig. 11. Only one component near 399.8 eV was observed in the N(1s) spectra regardless of the take-off angle. The strong peak near 409.0 eV in Fig. 11 was related to a copper LMV Auger line.

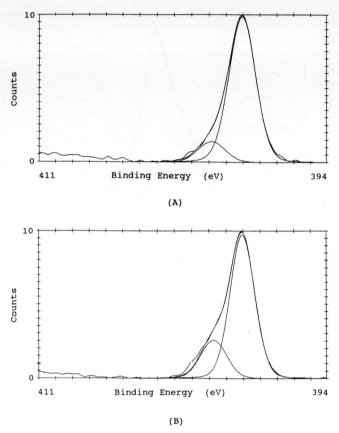

Figure 9. N($1s$) high resolution spectra of polished iron coated with γ-APS from a 1% aqueous solution at pH 10.4. The take-off angles were (A)—15° and (B)—75°.

The Cu($2p_{3/2}$) photoelectron and LVV Auger spectra obtained from polished OFHC copper before coating with γ-APS are shown in Fig. 12. Only one line was observed near 932.4 eV in the photoelectron spectrum. However, two strong components were observed near 335.0 and 337.0 eV in the Auger spectrum. Combining the component near 932.4 eV in the photoelectron spectrum with the component near 335.0 eV in the Auger spectrum resulted in a modified Auger parameter equal to 1851.0 eV, which was attributed to metallic copper. Combining the photoelectron line near 932.4 eV with the Auger line near 337.0 eV resulted in a modified Auger parameter of 1849.0 eV, which was characteristic of Cu_2O.

The Cu($2p_{3/2}$) photoelectron and LVV Auger spectra obtained from a copper mirror treated with an aqueous solution of γ-APS at pH 10.4 are shown in Fig. 13. Two components were observed near 932.4 and 934.9 eV in the Cu($2p_{3/2}$) photoelectron spectrum, clearly indicating the presence of Cu(I) and Cu(II), respectively. The presence of Cu(II) was confirmed by a broad, weak shake-up satellite near 944.0 eV.

The Auger spectrum consisted of a broad, weak band near 340.0 eV and a stronger, sharper band near 336.8 eV. Combining the photoelectron line near

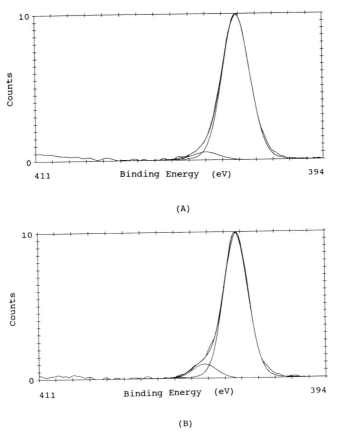

Figure 10. N(1s) high resolution spectra of polished magnesium coated with γ-APS from a 1% aqueous solution at pH 10.4. The take-off angles were (A)—15° and (B)—75°.

932.4 eV with the Auger line near 336.8 eV resulted in a modified Auger parameter of 1849.2 eV, which was attributed to the oxide Cu_2O [16]. Combining the photoelectron line near 934.9 eV with the Auger line near 340.0 eV resulted in a modified Auger parameter of 1848.5 eV, which was characteristic of the Cu(II) complex of γ-APS [16].

These conclusions were supported by results obtained from angle-resolved XPS. The band near 932.4 eV in the $Cu(2p_{3/2})$ photoelectron spectrum of the mirror coated with γ-APS increased in intensity relative to that near 934.9 eV when the take-off angle was increased from 15° to 75°. Similarly, the band near 336.8 eV in the Auger spectrum also increased in intensity relative to that near 340.0 eV. Such behavior would be expected if the bands near 932.4 eV in the photoelectron spectrum and near 336.8 eV in the Auger spectrum were related to an oxide that was covered by a thin film of silane.

4. DISCUSSION

The results described above indicate that some of the amino groups were protonated when γ-APS films were adsorbed onto most metals from aqueous solutions at pH 10.4. Moreover, the extent of protonation depended on the

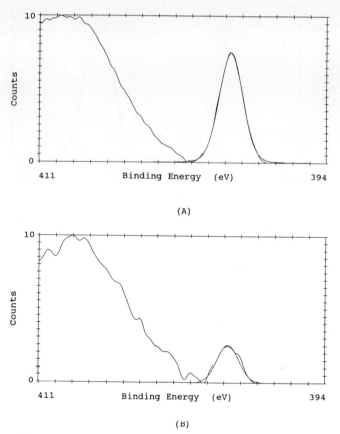

(A)

(B)

Figure 11. N(1s) high resolution spectra of polished OFHC copper coated with γ-APS from a 1% aqueous solution at pH 10.4. The take-off angles were (A)—15° and (B)—75°.

specific metal and on the surface pretreatment. Generally speaking, the extent of protonation was greatest on silicon, aluminum, and titanium; least on magnesium; and intermediate on iron, chromium, and nickel. The protonated amino groups were mostly adjacent to the oxidized surfaces of the substrates. It should be noted that the substrate material (in elemental form) was detected by high resolution spectra obtained from the above coated substrates at take-off angles of 75°. The single exception was for coated iron substrates, where only oxidized metal was detected.

Perhaps the simplest interpretation is to attribute protonation of the amino groups to the formation of bicarbonates or to intramolecular hydrogen bonding between amino and silanol groups. Some evidence for the formation of bi-carbonate and carbonate species on the metal substrates during polishing was observed in the C(1s) high-resolution spectra. However, the formation of signifi-cant amounts of amine bicarbonates was unlikely since the films investigated here were deposited in a nitrogen atmosphere. It is unclear as to why the extent of bicarbonate formation would vary according to the substrate.

Results that we have obtained using infrared spectroscopy showed that amine bicarbonates in γ-APS films deposited on metal substrates exposed to the

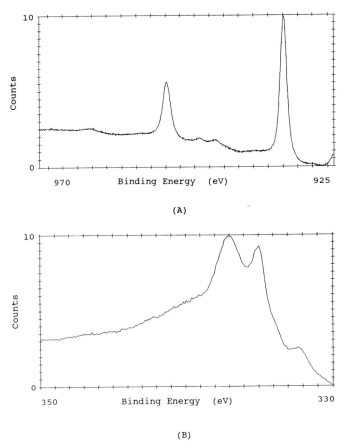

Figure 12. (A)—$Cu(2p_{3/2})$ photoelectron and (B)—LVV Auger spectra obtained with polished OFHC copper before coating with γ-APS.

atmosphere dissociated when exposed to UHV and X-ray irradiation during XPS experiments (see Fig. 14). This effect has probably been observed by others, although they interpreted it differently. As noted earlier, Hook *et al.* [7] used XPS and ATR to determine the molecular structure of films formed by γ-APS adsorbed onto polytetrafluoroethylene (PTFE) which was surface-modified by treatment in a H_2/H_2O radio-frequency glow-discharge. They clearly observed that the NH_2 deformation mode of the amino groups was near 1575 cm^{-1}, about 25 cm^{-1} less than the value expected for free primary amino groups. However, their XPS results clearly showed that the amino groups in the bulk of the γ-APS films were not protonated. As a result, they attributed the shift in the NH_2 deformation mode to lower frequencies to coordination of amino groups to silicon atoms, which was not expected to have a significant effect on the shape or position of the $N(1s)$ peak in XPS spectra. However, the band near 1575 cm^{-1} in the ATR spectra was undoubtedly characteristic of bicarbonate salts which dissociated during the XPS experiment.

Protonation of amino groups at the interface could also be related to intra-molecular hydrogen bonding of amino and silanol groups. The interaction of the

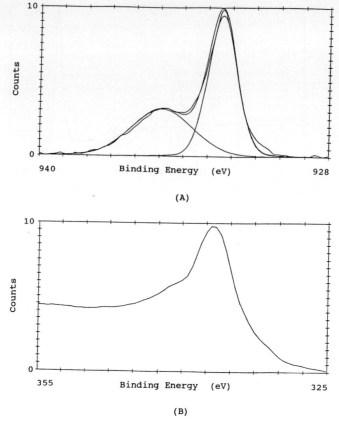

Figure 13. (A)—Cu($2p_{3/2}$) photoelectron and (B)—LVV Auger spectra of polished OFHC copper coated with γ-APS from a 1% aqueous solution at pH 10.4.

silane with the substrate may result in more opportunities for intramolecular hydrogen bonding at the interface than in the bulk and this may certainly vary from one substrate to another.

However, Dillingham and Boerio [9] showed that the amino groups near the interface were protonated when an epoxy was cured against polished aluminum using amine curing agents whereas amino groups in the bulk of the epoxy were not. They attributed the protonation to interaction of the amino groups with hydroxyl groups on the oxidized aluminum surface. Several other groups have reported similar results. Horner and Boerio (unpublished results) have shown that the amino groups are protonated when amines such as 2,4,6-tridimethyl-aminomethylphenol are applied to the oxidized surface of a polished metal. Kelber and Brow (personal communication) investigated the adsorption of diethanolamine, a simple model compound for amine-cured epoxies, onto air-oxidized aluminum and copper. They observed protonation of the model compound on aluminum, but not on copper. Bolouri *et al.* [17] prepared a model compound for amine-cured epoxies by reacting the diglycidyl ether of bisphenol-A with diethylamine. They observed that the N($1s$) peak in the XPS spectrum of the bulk compound consisted of a single component near 399.2 eV.

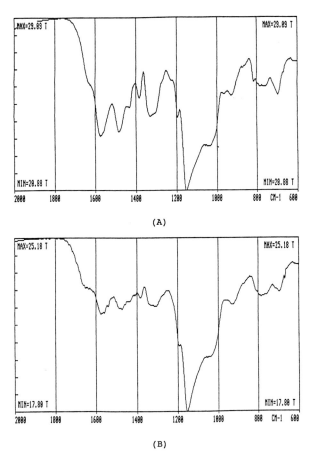

(A)

(B)

Figure 14. Reflection-absorption infrared spectra obtained from a γ-APS film deposited on an iron substrate from a 1% aqueous solution at pH 10.4 (A)—before and (B)—after exposure to X-ray irradiation and a UHV environment for 45 minutes in an XPS experiment.

However, a second component shifted to higher binding energy was observed for thin films of the compound adsorbed onto aluminum that had been cleaned by ion bombardment and then exposed to oxygen. Finally, Linde [18] has suggested that polyamic acids of pyromellitic dianhydride and oxydianiline do not react with films formed by γ-APS adsorbed onto metal substrates such as aluminum and chromium because the silane is adsorbed through the amino groups.

The most likely explanation for the results obtained here is that the amino groups are protonated by surface hydroxyl groups when γ-APS is adsorbed onto oxidized metal substrates. This would explain why the protonation of amino groups is mostly confined to the interface and varies with the substrate. Since the extent of protonation was least on magnesium and greatest on silicon, aluminum, and titanium substrates, this interpretation would also imply that magnesium oxides are very basic and that silicon, aluminum, and titanium oxides are relatively acidic. These implications are consistent with the theory presented by Parks [19], who discussed the isoelectric points of metal oxides. According to the theory, magnesium oxide should be very basic, having an isoelectric point near

12.0. Silicon, titanium, chromium, and aluminum oxides should be relatively acidic, with isoelectric points near 4.0, 6.0, 7.0, and 7.0, respectively. Iron oxide should be intermediate, with an isoelectric point near 9.0. Parks predicted the isoelectric point of nickel oxide to be of the order of 10–12. Other investigators [20] have reported the isoelectric point of nickel oxide to be near 9.4.

The results obtained here also indicate that the acidity of the oxidized surface of a metal depends on the pretreatment of the surface. For a take-off angle of 75°, about 42% of the amino groups were protonated when γ-APS films were applied to mechanically polished 1100 aluminum. That increased to about 57% when the silane films were rinsed, probably because of the dissolution of a small amount of silane from the free surface of the film on which the amino groups were mostly not protonated. However, for a take-off angle of 75°, about 48% of the amino groups were protonated when γ-APS was applied to FPL-etched aluminum substrates. That increased to about 65% when the silane films were rinsed. These results imply that the FPL-etched surface is more acidic than the polished surface.

As indicated above, Mason *et al.* [13] found that the pretreatment affected the surface acidity of Ti-6,4. They found that surfaces treated with a basic etchant had an isoelectric point between 7.3 and 9.2. A phosphate–fluoride etch left the surface acidic with an isoelectric point between 5.4 and 7.3. As noted above, Finlayson and Shah [12] also found that the pretreatment affected the acidity of oxidized aluminum surfaces.

It is interesting to note that the concentration of protonated amino groups decreased as the take-off angle was decreased from 75° to 15° for γ-APS films formed on all the polished substrates except aluminum (see Table 1). These results imply that the protonated amino groups were relatively evenly distributed throughout films formed on aluminum, but were concentrated nearer the silane/oxide interface for the other substrates. The reason for the difference in behavior of films formed on aluminum and those formed on the other substrates may be related to the ease with which aluminum substrates are etched by aqueous solutions of γ-APS at pH 10.4.

Etching of the substrate occurs when γ-APS films are deposited onto aluminum (Horner and Boerio, unpublished results) and aluminum may substitute for silicon during polymerization of the silane on the surface. The resulting alumino-silicate structure would have an overall negative charge which would be balanced by protonated amino groups.

While the extent of protonation was in the range expected for the nickel substrate, the variation with take-off angle was negligible (3%). This was in contrast with results obtained for γ-APS films on other metals when the Physical Electronics Model 5300 ESCA system and Mg K_α X-rays were used for data collection. The most plausible explanation for this is related to the input lens of the Surface Science Instruments spectrometer, which collects electrons within a cone subtending an angle of 30° ($\pm15°$ from the lens axis). As a result, small lineshape variations would not be observed if the (metal) oxide film thickness was of the order of one mean free path length or less. To test this theory, a silicon wafer coated with γ-APS was first analyzed using the Physical Electronics Model 5300 ESCA system followed by analysis with the Surface Science Instruments (SSI) SSX-100-03. N(1s) spectra obtained using the Physical Electronics Model

5300 ESCA system are shown in Fig. 5 for which we previously reported a decrease in the percentage of protonation from 43% to 28% when the take-off angle was decreased from 75° to 15°, respectively. N(1s) spectra obtained using the SSI at take-off angles of 75° and 15° indicated that the extent of protonation was 25%. Therefore, as was seen in the case of γ-APS on nickel, no variation in the extent of protonation with a change in take-off angle was detected using the SSI instrument.

The results described here demonstrate convincingly that metal substrates strongly affect the molecular structure of silanes deposited on them. These effects will have to be considered in developing advanced models for the mechanisms by which silanes function as coupling agents and in interpreting results obtained in investigations on the effect of silanes on the durability of adhesive bonds. The present results also substantiate the earlier results of Dillingham and Boerio [9] which indicated that metal substrates may be expected to have a strong effect on the molecular structure and properties of bonds prepared using amine-cured epoxy adhesives.

5. CONCLUSIONS

When γ-aminopropyltriethoxysilane (γ-APS) is adsorbed from aqueous solutions at pH 10.4 onto oxidized surfaces, a fraction of the amino groups near the silane/oxide interface are irreversibly protonated by interaction with hydroxyl groups on the surfaces of silicon, nickel, chromium, iron, titanium, aluminum, and magnesium. The extent of protonation is greatest on silicon, aluminum, and titanium; intermediate on iron, chromium and nickel; and least on magnesium, which correlates well with the predicted isoelectric points of the oxides. The extent of protonation also depends on the pretreatment of the substrate and is greater on aluminum surfaces that have been FPL-etched than on aluminum surfaces that have been mechanically polished. Most of the amino groups in the bulk of the films react with water and carbon dioxide in the atmosphere to form bicarbonate salts. During exposure to ultra-high vacuum and X-ray irradiation in an X-ray photoelectron spectrometer, carbon dioxide and water are desorbed, leaving mostly free amino groups in the bulk of the films. γ-APS interacts with the oxidized surface of copper by formation of complexes in which the lone pair of electrons on the amino nitrogen atoms is donated to empty orbitals on copper ions.

REFERENCES

1. P. W. Erickson and E. P. Plueddemann, in: *Composite Materials, Vol. 6, Interfaces in Polymer Matrix Composites*, E. P. Plueddemann (Ed.), Ch. 1. Academic Press, New York (1974).
2. D. J. Ondrus and F. J. Boerio, *J. Colloid Interface Sci.* **124**, 349 (1988).
3. F. J. Boerio and R. G. Dillingham, in: *Adhesive Joints: Formation, Characteristics, and Testing*, K. L. Mittal (Ed.), p. 541. Plenum Press, New York (1984).
4. W. Theidman, F. C. Tolan, P. J. Pearce and C. E. M. Morris, *J. Adhesion* **22**, 197 (1987).
5. F. J. Boerio and D. J. Ondrus, in: *Surface and Colloid Science in Computer Technology*, K. L. Mittal (Ed.), p. 155. Plenum Press, New York (1987).
6. P. R. Moses, L. M. Weir, J. C. Lennox, H. O. Finklea, J. R. Lenhard and R. W. Murray, *Anal. Chem.* **50**, 576 (1978).
7. D. J. Hook, T. G. Vargo, J. A. Gardella, K. S. Litwiler and F. V. Bright, *Langmuir* **7**, 142 (1991).
8. F. M. Fowkes, D. W. Dwight, D. A. Cole and T. C. Huang, *J. Non-Cryst. Solids* **120**, 47 (1990).

 9. R. G. Dillingham and F. J. Boerio, *J. Adhesion* **24**, 315 (1987).
10. S. Naviroj, J. L. Koenig and H. Ishida, *J. Adhesion* **18**, 93 (1985).
11. T. G. Waddell, D. E. Leyden and M. T. DeBello, *J. Am. Chem. Soc.* **103**, 5303 (1981).
12. M. F. Finlayson and B. A. Shah, *J. Adhesion Sci. Technol.* **4**, 431 (1990).
13. J. G. Mason, R. Siriwardane and J. P. Wightman, *J. Adhesion* **11**, 315 (1981).
14. H. W. Eichner and W. E. Schowalter, *Forest Products Lab. Rep.* No. 1813 (1950).
15. R. Chen and F. J. Boerio, *J. Adhesion Sci. Technol.* **4**, 453 (1990).
16. F. J. Boerio, J. W. Williams and J. M. Burkstrand, *J. Colloid Interface Sci.* **91**, 485 (1983).
17. H. Bolouri, R. A. Pethrick and S. Affrossman, *Appl. Surf. Sci.* **17**, 231 (1983).
18. H. G. Linde, *J. Appl. Polym. Sci.* **40**, 613 (1990).
19. G. A. Parks, *Chem. Rev.* **65**, 177 (1965).
20. K. B. Yatsimirksii and V. P. Vasil'ev, in: *Instability Constants of Complex Compounds.* Pergamon Press, New York (1960).

Silanes and Other Coupling Agents, pp. 263–276
Ed. K. L. Mittal
© VSP 1992.

Covalent binding of amino, carboxy, and nitro-substituted aminopropyltriethoxysilanes to oxidized silicon surfaces and their interaction with octadecanamine and octadecanoic acid studied by X-ray photoelectron spectroscopy and ellipsometry

KRISHNA M. R. KALLURY, RENO F. DEBONO, ULRICH J. KRULL and MICHAEL THOMPSON*

Department of Chemistry, University of Toronto, Toronto, Ontario, M5S 1A1, Canada

Revised version received 19 April 1991

Abstract—Three silanized silicon substrates were prepared by treating cleaned oxidized silicon wafers with *N*-(3-triethoxysilylpropyl)octadecanamide (**1**), 9-[*N*-(triethoxysilylpropyl)amino]-9-oxononanoic acid methyl ester (**2**), and *N*-(3-triethoxysilylpropyl)-12-nitrododecanamide (**4**). The carbomethoxyl of **2** immobilized on silicon was hydrolysed to yield the corresponding ω-carboxylic surface (**3**) and the nitro group of the surface **4** reduced to afford the ω-amino surface (**5**). All five silanized surfaces were treated with octadecanoic acid and octadecanamine used as models for acidic and basic polymeric adhesives and the interactions were followed by ellipsometry and X-ray photoelectron spectroscopy. While the surfaces **1** and **2** reacted only by physisorption, the carboxylic surface **3** and the amino surface **5** showed a strong reaction with octadecanamine and octadecanoic acid, respectively. The nitro surface **4** exhibited a strong interaction with both probes, but by different pathways.

Keywords: Silanization; silanes (amino, carboxy, and nitro-alkyl functionalized); adhesion; octadecanoic acid; octadecanamine; X-ray photoelectron spectroscopy; ellipsometry.

1. INTRODUCTION

Organofunctional alkoxysilanes are frequently used as adhesion promoters for structural adhesives such as epoxy and acrylic resins which polymerize to give high modulus, high strength materials functioning as load-bearing joints [1]. These silanes impart enhanced durability to the adhesives in wet environments [2, 3]. In addition, silane derivatives find extensive applicability in the manufacture of glass fibre-reinforced plastics as primers to improve the bonding between glass fibre and resin matrix [4] and also for the direct bonding of polymer resins to metals [1, 5]. For example, treatment of oxidized silicon surfaces with silane coupling agents not only improves the adhesion of polyimides to these substrates, but also enhances the overall mechanical stability, and these facors are utilized to advantage in the microelectronics industry for insulation layers in device structures [6]. Another application of silanes in the adhesion field comprises the crosslinking of polyolefins for the production of cable insulation and warm water pipe ducts [7].

---

*To whom correspondence should be addressed.

In all adhesive joints, the interfacial region between the adhesive and the substrate plays an important role in the transfer of stress from one adherend to another [8]. The initial strength and stability of the joint depend on the molecular structure of the interphase after processing and environmental exposure, respectively. Characterization of the molecular structure near the interface is essential to model and, subsequently, to maximize the performance of an adhesive system in a given environment. When deposited on a substrate, the silane primers have a finite thickness and constitute separate phases. If there is interaction between the primer and the adherend surface or adhesive, a new *interphase* region is formed. This interphase has a molecular structure different from the molecular structure of either of the two primary phases from which it is formed. Thus, it is essential to characterize these interphases thoroughly.

A variety of spectroscopic techniques have been utilized to investigate the substrate–silane interfacial region. These include infrared [8–12], Raman [13], solid-state C-13 and Si-29 NMR [14–17], X-ray photoelectron spectroscopy (XPS) [8, 18, 19], inelastic electron tunnelling spectroscopy [20], electron spin resonance [21], and secondary ion mass spectrometry [22–24]. Substrates examined have included silica, aluminium, sapphire, steel, tin oxide, quartz, control pore glass, mica, copper, iron, platinum, and titanium, and the adhesion of polymers such as polyacrylates, polytetrafluoroethylene, and polyimides has been studied. The most commonly employed silane coupling agent is γ-aminopropyltriethoxysilane (APTES; 3-triethoxysilylpropanamine according to IUPAC nomenclature). When reacted in aqueous solution, the ethoxyls on APTES undergo hydrolysis, generating silanols which condense either amongst themselves (intermolecular siloxane formation) or with the hydroxyls on the substrate surface (covalent deposition) [2]. The influences of concentration and pH on this silanization reaction have been thoroughly investigated [25]. It has also been demonstrated that primary and secondary amine-substituted silanes exhibit different adhesion strengths when deposited from aqueous solutions, but non-aqueous conditions resulted in approximately equal adhesion with both functionalities [6]. Evidence from all spectroscopic studies indicates that APTES forms multilayer structures with all of the substrates. The initial layer is covalently bound to the surface through the hydroxyls on the latter and the upper layers consist of oligomeric structures linked more rigidly nearer the surface but more mobile towards the top end. In an earlier study [19], we have shown that a base catalyst like triethylamine has a profound effect not only on the structure of the silane film, but also on the layer build-up. We have also demonstrated [26] that APTES-treated quartz is selective towards nitroaromatics by making use of a surface acoustic wave device as the transducer. Furthermore, our previous experiments [27] on aluminium piezoelectric crystals silanized with APTES under anhydrous conditions revealed that the mobile phase undergoes a change in interfacial viscosity upon hydration.

The importance of experimental parameters such as the nature of the solvent, the age of the silanizing agent solution, the pH, and the drying conditions of the deposited silane film in determining the durability of adhesive-bonded joints has been stressed by Gledhill *et al.* [28]. Comyn [29] has pointed out that improper or inadequate surface treatment is one of the most frequent causes of failure in adhesive binding, while the choice of a good surface treatment can bring about marked improvements in the wet durability of adhesive bonds to metals and glasses and will permit the bonding of otherwise difficult-to-bond materials such as polytetrafluoroethylene or polyolefins. Untreated surfaces may be unsatisfactory for adhesive bonding since they may be contaminated, may lack polar functionalities, or the interface that they make with an adhesive is susceptible to hydrolysis. The requirement of polar functionalities is underlined by Ponjee *et al.* [30], who studied the interaction of a quinone-novolak photoresist with trimethylsilylated silicon surfaces by ToF-SIMS and contact angle measurements and found that the trimethylsilyl moiety causes a deterioration in adhesion due to its low surface energy.

Comyn [1] has pointed out that maximum bond strength and consequently greater adhesion between the substrate and polymer could be achieved with a monolayer of silane bound to both the adherend and adhesive. The current investigation was undertaken to evaluate the possibility of monolayer level depositions on silicon substrates by employing a few ω-functionalized alkanoyl-substituted derivatives of APTES which will provide polar moieties as well. The interactions of these functionalized silanes covalently immobilized on silicon with octadecylamine and octadecanoic acid, used as models for basic and acidic polymeric adhesives, were also examined in this study. Characterization of the silanized surfaces as well as studies on their interactions with the above two chemical probes were carried out through ellipsometric and XPS measurements.

2. EXPERIMENTAL

2.1. Materials

The solvents utilized were all analytical grade reagents further purified by distillation. *N*-(3-Triethoxysilyl)-propanamine (APTES), nonanedioic acid monomethyl ester, 12-nitro-dodecanoic acid, octadecanoic acid, octadecylamine, and lithium iodide were Aldrich samples and were used as received. Silicon wafers were supplied by International Wafer Service, Palo Alto, CA.

N-(3-Triethoxysilylpropyl)octadecanamide, *N*-(3-triethoxysilylpropyl)-12-nitro-dodecanamide, and 9-[*N*-(3-triethoxysilylpropyl)amino]-9-oxononanoic acid methyl ester were synthesized according to procedures reported earlier by us [31].

2.2. Prepration of the ω-carboxy silicon surface (3)

Chemically cleaned silicon wafers (1×1 cm^2 pieces) were treated with 9-[*N*-(3-triethoxysilylpropyl)amino]-9-oxononanoic acid methyl ester in toluene (20 ml of a 2% solution) at room temperature under nitrogen overnight. The wafers were then thoroughly washed with chloroform, methanol, and acetone in that order, and suspended in dimethylformamide (DMF; 10 ml) containing lithium iodide (1 g). The mixture was refluxed for several hours. The wafers were recovered and washed several times with distilled water, once with sodium carbonate (1% aq., 10 ml) and again with distilled water. Finally, the wafers were rinsed with acetone and stored in a vacuum desiccator under nitrogen.

2.3. Preparation of the ω-amino functionalized silicon substrate (5)

Cleaned silicon wafers were silylated with *N*-(3-triethoxysilylpropyl)-12-nitro-dodecanamide in toluene (20 ml of a 2% solution) at room temperature under nitrogen overnight. The wafers were thoroughly washed with chloroform, methanol, and acetone in that order, and suspended in glacial acetic acid (5 ml). Zinc dust (200 mg) was added in small portions and the mixture set aside for 5 h. The wafers were recovered and washed several times with distilled water, sodium carbonate solution (1% aq.), and again with distilled water. After rinsing with acetone, the wafers were stored under nitrogen.

2.4. Treatment of the silanized surfaces 1–5 with octadecanoic acid and octadecanamine guest molecules

The silanized wafers 1–5 were suspended in 10 ml of a 10^{-3} M guest solution in methanol for 15 min, washed extensively with chloroform, and finally rinsed with acetone. The wafers were then dried in a stream of argon for 30 min prior to ellipsometric measurements.

2.5. X-Ray photoelectron spectra

The XPS spectra reported were recorded on a Leybold MAX-200 ESCA spectrometer using an unmonochromated Mg K_a source run at 12 kV and 25 mA. The shape of the spectra indicated that no compensation for differential surface charging was needed. The binding energy scale was calibrated to 285 eV for the main C(1s) (C—C) feature. Spectra were run in both low resolution (pass energy = 192 eV) and high resolution (pass energy = 48 eV) modes for the C(1s), N(1s), and Si(2p) regions. Each sample was analysed at a 90° angle relative to the electron detector. An analysis area of 4×7 mm was used for rapid data collection (typically 5 min for the C(1s) high resolution spectrum).

Elemental compositions were calculated from the satellite-subtracted low resolution spectra normalized for constant transmission using the software supplied by the manufacturer. The sensitivity factors employed in these computations were C(1s) = 0.34, N(1s) = 0.54, Si(2p) = 0.4, and O(1s) = 0.78, empirically derived for the MAX-200 spectrometer by Leybold.

2.6. Ellipsometric measurements

These measurements were made on an Auto EL-II Ellipsometer (Rudolph Research, Flanders, NJ). The laser source was a 1 mW continuous wave helium/neon laser, with a wavelength of 6328 Å. The angle of incidence was 70° and the spot size 2–3 mm. A refractive index of 1.5 was utilized for all the silane layers. The data were analysed on a Hewlett-Packard 85 computer using film 85 software package, version 30, program 13, and the film thickness was calculated using the McCrackin program.

3. RESULTS AND DISCUSSION

3.1. Synthesis and characterization of the silanized surfaces 3 and 5

The protocol for the generation of the ω-carboxy and ω-amino functionalized surfaces **3** and **5**, respectively, is summarized in Fig. 1. The structure of the surface **1** formed by treatment of oxidized silicon with *N*-(3-triethoxysilyl-propyl)octadecanamide is also included in Fig. 1. Initially, alkaline hydrolyses with sodium hydroxide, potassium carbonate, sodium t-butoxide, and methanolic ammonia were tried unsuccessfully to effect the hydrolysis of the methyl ester functionality of the surface **2** (Fig. 1) into a carboxylic acid. In all these trials, the silane film was totally removed from the substrate. Ultimately, neutral conditions were adopted which involved refluxing with lithium iodide in DMF solution. The resulting carboxylic surface **3** was characterized by advancing contact angle measurements and XPS. The contact angle of the methyl ester-terminated surface **2** with water as the probe liquid was found to be $75 \pm 2°$, while the hydrolysed surface **3** exhibited a value of $45 \pm 3°$. In the XP spectra, the ratios between the high resolution C(1s) peaks in the 286–287 eV region (representing C—N and C—O carbons) and 288–289 eV region (representing C=O carbons) were utilized to distinguish between surface structures **2** and **3**. For **2**, this ratio should be 1:1 since there are two carbonyl carbons (amide and ester) and two C-heteroatom-bound carbons (C—N and C—O). In the case of **3**, this ratio works out to 2:1 since the methoxy carbon no longer exists. The experimental values from XPS (see Table 2 for details) were 1.6:1 for **3** and 1:1 for **2**, which are in reasonable agreement with the theoretical values.

Two different sets of conditions were utilized for carrying out the reduction of the nitro-substituted surface **4** for forming the aminated surface **5**. One involved treatment with zinc and acetic acid at room temperature and the other consisted in reacting with aqueous sodium dithionite at room temperature. The former yielded an almost totally reduced amino surface **5**, while the reduction was incomplete with the dithionite reagent. The amino surface **5** was characterized

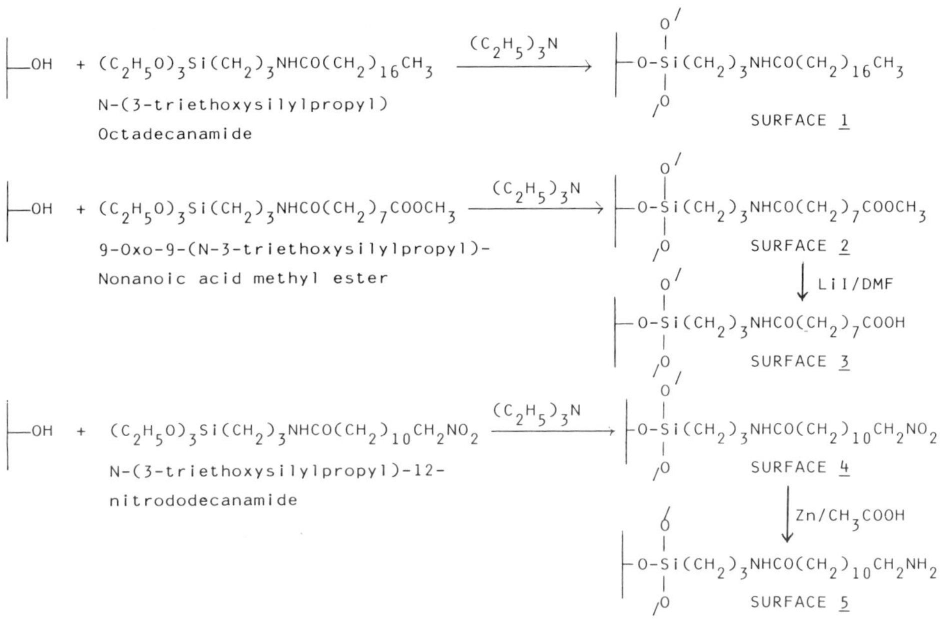

Figure 1. Generation of oxidized silicon surfaces carrying covalently linked functionalized silanes.

by contact angle and XPS measurements. The advancing contact angle (water) of the nitro surface **4** was $50 \pm 1°$ while the reduced product **5** showed a value of $68 \pm 2°$. In the XP spectrum, the N(1s) region clearly showed the total disappearance of the binding energy peak at 406.5 eV corresponding to the nitro group in the reduced surface **5**, while it was present in a 1:1 ratio with the amide nitrogen N(1s) binding energy peak at 400.5 eV in the XP spectrum of **4** (see Fig. 2).

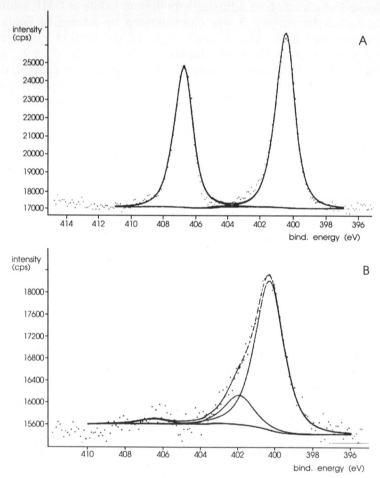

Figure 2. High resolution N(1s) region scan on the nitro surface **4** (A) and the amino surface **5** (B).

3.2. Interaction of the silanized surfaces 1–5 with octadecylamine and octadecanoic acid

3.2.1. Ellipsometric studies. The thicknesses of the unreated silane surface films **1–5** were initially determined and were found to be close to the values expected for monolayer level coverages (see Table 1 for details). Upon treatment of the surfaces with the guest amine or acid solutions, all the surfaces exhibited thickness values corresponding to multilayer coverages by both of these guest molecules. However, when each of the treated surfaces was washed thoroughly

Table 1.

Ellipsometric data on the functional group-substituted silanized surfaces and on surfaces resulting from their interaction with octadecanamine and octadecanoic acid

Surface	Thickness of silane film (Å)
$\vdash$O$-$Si(CH$_2$)$_3$NH CO (CH$_2$)$_{16}$CH$_3$ (**1**)	30 ± 3
$\vdash$O$-$Si(CH$_2$)$_3$NH CO (CH$_2$)$_7$COOCH$_3$ (**2**)	27 ± 2
$\vdash$O$-$Si(CH$_2$)$_3$NH CO (CH$_2$)$_7$COOH (**3**)	31 ± 4
$\vdash$O$-$Si(CH$_2$)$_3$NH CO (CH$_2$)$_{10}$CH$_2$NO$_2$ (**4**)	25 ± 1
$\vdash$O$-$Si(CH$_2$)$_3$NH CO (CH$_2$)$_{10}$CH$_2$NH$_2$ (**5**)	15 ± 3
1 + octadecanoic acid	27 ± 8
1 + octadecanamine	25 ± 5
2 + octadecanoic acid	30 ± 2
2 + octadecanamine	30 ± 4
3 + octadecanoic acid	30 ± 2
3 + octadecanamine	40 ± 4
4 + octadecanoic acid	35 ± 2
4 + octadecanamine	41 ± 5
5 + octadecanoic acid	28 ± 5
5 + octadecanamine	18 ± 3

with chloroform, most of the surfaces returned to their original untreated-level values or even slightly lower figures, especially for the octadecanamide surface **1**. This behaviour clearly shows that physisorption occurs to a considerable extent due to hydrophobic interactions, but it is not strong enough to survive solvent treatment. On the other hand, the carboxylic surface **3** registered a significant enhancement in its thickness upon interaction with octadecanamine and this value was reproducible even after chloroform washing. Similar increases were recorded for the amino surface **5** when treated with octadecanoic acid and for the nitro surface **4** upon interaction with both octadecanamine and octadecanoic acid. From these results, it can be inferred that strong polar interactions occur between the surface amino or carboxyl groups with guest carboxylic or amino moieties, respectively. On the contrary, interactions between like surface groups and guest functionalities are not strong enough to register any significant increases in film thickness. The strong interactions between acidic and basic moieties either on the surface or on the guest molecules are not unexpected [32], but do indicate that these surface functionalities are not involved in polar interactions with surface silanols or the amide function at the other end of the alkyl chain. Such reactions can therefore be utilized for promoting adhesion between a surface silane and a polymer.

The data presented in Table 1 indicate that the thickness of the surface resulting from the interaction of the nitro-functionalized silane surface **4** with octadecanamine is greater than that generated by the treatment of **4** with octadecanoic acid. This could be attributed to the fact that the hydrogen bonding interaction during the addition of octadecanamine is primarily between the nitro group and the amine function of the probe molecule and the amide function is involved only to a small extent, if at all. On the other hand, octadecanoic acid seems to react to a large extent with the amide compared with the nitro group and, consequently, there is less coverage on top of the surface silane film, resulting in the lower thickness value observed. It appears that charge repulsion between the nitro and the carboxyl moieties is responsible for the reduced inter-action at the surface. The affinity of the nitro group towards an amine is well documented in the literature [26, 33]. The above interpretation is supported by the XPS results discussed below.

The hydrogen bonding interactions of the methyl ester moiety on the surface **2** are obviously too weak to be of any consequence.

3.2.2. X-Ray photoelectron spectroscopic studies. The XP spectra [see Table 2 for elemental compositions and binding energy (BE) values] further substantiate the conclusions reached from the ellipsometric studies. Thus, a comparison between the elemental compositions of the aminated surface **5** and the cor-responding surface generated by its interaction with octadecanoic acid indicates that the ratio between the $C(1s)$ and $N(1s)$ binding energy peak is 15:1 for **5** and 30:1 for **5** + SA (SA = stearic acid). Such an increase is in agreement with the addition of a 17-carbon chain of the acid onto a 15-carbon silane surface film [see Fig. 3 for a comparison of the $C(1s)$ region of the two surfaces]. Further-more, the high resolution scan on the $N(1s)$ region of the surface **5** shows a non-hydrogen-bonded peak at 400.4 eV coupled with a hydrogen-bonded peak at 402.0 eV in a ratio of 6:1 (see Fig. 2). It is clear that the amine function is only weakly H-bonded to a similar adjacent group on the surface **5**. In the cor-responding **5** + SA surface, the same ratio was determined to be 2:1, indicating a three-fold increase in the intensity of the H-bonded $N(1s)$ peak, evidently due to the amine–carboxylic acid interaction (see Fig. 4).

A similar detailed comparative analysis of the carboxylic surface **3** and the **3** + octadecanamine surface shows a $C(1s):N(1s)$ ratio of 11:1 for the former and 20:1 for the latter surface. At the same time, the high resolution $N(1s)$ data indicate a 50% enhancement in the H-bonded peak at 402.5 eV (see Fig. 5).

In the case of the nitro-functionalized surface **4**, the $C(1s):N(1s)$ binding energy peak ratio is 10:1 and the high resolution $N(1s)$ narrow region scan indicates two peaks of almost equal intensity at 406.8 and 400.6 eV for the nitro and amide nitrogens, respectively (see Fig. 2). In addition, the $C(1s)(C-N):C(1s)(C=O)$ ratio in the high resolution scan of the $C(1s)$ binding energy region is 2:1, in keeping with the surface silane structure **4** with two $C-N$ carbons and one amide carbonyl carbon. On the other hand, the surface resulting from the treatment of **4** with octadecanamine exhibits a $C(1s):N(1s)$ binding energy ratio of 18:1. In the $N(1s)$ region, three peaks appear at 399.5, 401.3, and 405.8 eV in a ratio of 4:5:1. While the first and last peaks are attributable to the amide and the nitro nitrogens, respectively, the middle peak represents the

Table 2.
X-Ray photoelectron spectroscopic data on the functionalized silane-bound oxidized silicon surfaces and the guest molecule-treated surfaces formed from them

Surface	Elemental composition from low resolution spectra				High resolution data						
					C(1s)		N(1s)		Si(2p)		
	C (%)	N (%)	Si (%)	O (%)	BE (eV)	Area (%)	BE (eV)	Area (%)	BE (eV)	Area (%)	
⊢O—Si(CH₂)₃NHCO(CH₂)₁₆CH₃ (**1**)	68.5	3.7	4.1	23.7	285.0 286.5 288.5	88.0 6.5 5.5	400.1 401.9	87 13	99.1 103.1	10 90	
⊢O—Si(CH₂)₃NHCO(CH₂)₇COOCH₃ (**2**)	67.7	2.9	4.0	25.5	285.0 286.5 288.4	75.1 12.0 13.0	399.8 401.8	90 10	99.2 102.9	14 86	
⊢O—Si(CH₂)₃NHCO(CH₂)₇COOH (**3**)	61.6	5.7	5.8	26.9	285.5 286.9 288.7	76.0 8.6 15.4	400.4 402.3	88 12	99.3 103.4	15 85	
⊢O—Si(CH₂)₃NHCO(CH₂)₁₀CH₂NO₂ (**4**)	65.4	6.1	8.1	20.4	285.0 286.5 288.1	75.0 16.1 8.8	406.4 400.1	48 52	99.1 102.9	35 65	
⊢O—Si(CH₂)₃NHCO(CH₂)₁₀CH₂NH₂ (**5**)	64.7	4.3	3.2	27.8	285.0 286.4 288.5	76.5 16.4 7.6	400.4 402.0	85 15	99.2 103.1	20 80	
3 + octadecylamine	85.5	4.4	3.4	6.6	285.0 286.7 288.4	81.0 12.5 6.5	400.0 402.5	63 37	99.3 103.3	78 22	
4 + octadecylamine	85.4	4.8	2.3	7.5	285.0 286.7 287.9	88.0 8.9 3.1	399.5 401.3 405.8	42 49 9	99.1 102.8	25 75	
4 + octadecanoic acid	83.0	2.4	2.7	11.9	285.0 287.0 289.2	88.0 6.9 5.1	399.8 406.0	78 22	99.0 102.0	10 90	
5 + octadecanoic acid	86.8	3.2	1.2	8.9	285.0 286.4 288.8	81.2 10.0 8.8	400.0 401.8	61 39	99.2 103.1	17 83	

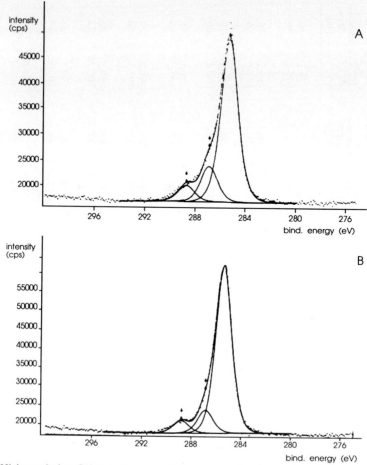

Figure 3. High resolution C(1s) region scan of the untreated amino surface **5** (A) and after treatment with octadecanoic acid (B).

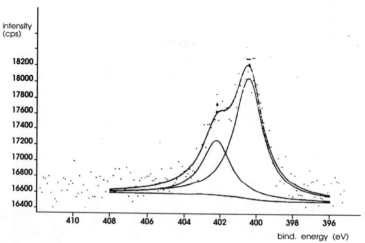

Figure 4. High resolution N(1s) region scan of the amino surface **5** after treatment with octadecanoic acid.

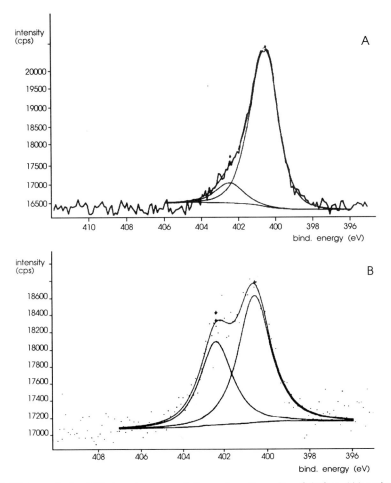

Figure 5. High resolution N(1*s*) region scan of the carboxylic surface **3** before (A) and after (B) treatment with octadecanamine.

H-bonded amino function belonging to octadecanamine. At the same time, the C(1*s*)(C—C):C(1*s*)(C=O) ratio increases from 10:1 for **4** to 30:1 for **4** + octadecanamine, providing further evidence for the addition of the C-18 amine chain to the surface (Fig. 6). The peak widths at half-height of the three nitrogen binding energy peaks in the high resolution N(1*s*) region scan of the **4** + octadecanamine surface are equivalent at 1.67 eV as opposed to the value of 1.38 eV for the amine and nitro nitrogen N(1*s*) peaks in the spectrum of **4** (see Figs 7 and 2, respectively). This observation again supports the H-bonding between the added amine and the amide and nitro moieties of the surface silane **4**.

The interaction of octadecanoic acid with the surface **4**, on the other hand, produces significant differences in the XP spectra compared with octadecanamine. Thus, the amide and carboxylic carbons both merge into a single peak at 289.0 eV in the C(1*s*) region, indicating the equivalence of both functions consequent to H-bonding interaction. In contrast, the **4** + octadecanamine surface shows the amide carbonyl peak at 287.9 eV, indicating much weaker H-bonding

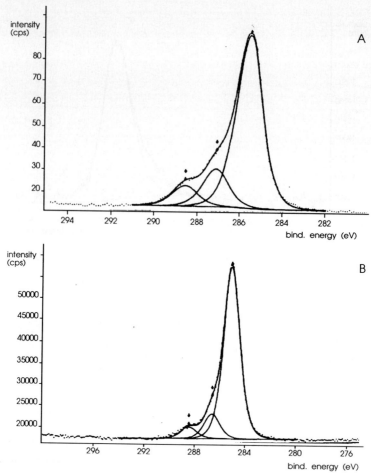

Figure 6. High resolution C(1*s*) region scan of the nitro surface **4** before (A) and after (B) treatment with octadecanamine.

compared with that exhibited by octadecanoic acid. Furthermore, only two N(1*s*) binding energy peaks appear at 399.8 and 406.0 eV with equivalent peak widths at half-height, 2.48 eV, corresponding to the amide and nitro groups, respectively. The ratio between C(1*s*)(C—N) and C(1*s*)(C=O) works out to be 1.1:1, close to the expected value of 1:1 (for two C—N and two carbonyl carbons). It, therefore, appears that the interaction of octadecanoic acid with the nitrosilane surface **4** is stronger than that of octadecanamine with the same surface.

The silane surfaces **1** and **2** display the same XP spectra before and after treatment with either octadecanoic acid or octadecanamine, within experimental error. The carboxylic surface **3** and the amino surface **5** exhibit XP spectra very similar to those of surfaces obtained by treating them with octadecanoic acid and octadecanamine, respectively.

4. CONCLUSIONS

The data collected in the present investigation indicate that a good degree of

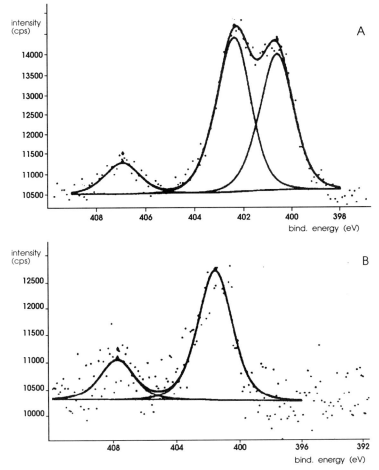

Figure 7. High resolution N(1*s*) region scan of the nitro surface **4** after treatment with octa-decanamine (A) and octadecanoic acid (B).

adhesion could be achieved by utilizing ω-carboxyalkyl and ω-aminoalkyl-substituted silane-treated silicon surfaces with amine and carboxy functionalized polymers, respectively. Silicon surfaces silanized with an ω-nitroalkylsilane show promise as bondable adherends for both amino and carboxy-substituted polymers.

REFERENCES

1. J. Comyn, in: *Structural Adhesives—Developments in Resins and Primers*, A. J. Kinloch (Ed.), Elsevier, London (1986).
2. E. P. Plueddemann, *Silane Coupling Agents*. Plenum Press, New York (1982).
3. P. Walker, *J. Oil Colour Chem. Assoc.* **65**, 415, 436 (1982).
4. M. J. Rosen, *J. Coatings Technol.* **50**, 70 (1978).
5. R. G. Schmidt and J. P. Bell, in: *Advances in Polymer Science*, Vol. 75, pp. 33–71. Springer-Verlag, Berlin (1986).
6. A. V. Patsis and S. Cheng, *J. Adhesion* **25**, 145 (1988).
7. M. Lazar, R. Rado and J. Rychly, in: *Advances in Polymer Science*, Vol. 95, pp. 149–197. Springer-Verlag, Berlin (1990).

8. D. J. Ondrus, F. J. Boerio and K. J. Grannen, *J. Adhesion* **29**, 27 (1989).
9. D. E. Leyden, R. S. S. Murthy, J. B. Atwater and J. P. Blitz, *Anal. Chim. Acta* **200**, 459 (1987).
10. V. M. Rangnekar and P. B. Oldham, *Spectrosc. Lett.* **22**, 993 (1989).
11. C. S. Paik Sung and S. H. Lee, *Polym. Prepr.* **19**, 788 (1978).
12. N. Balachander and C. N. Sukenik, *Langmuir* **6**, 1621 (1990).
13. D. M. Krol, C. A. M. Mulder and J. G. van Lierop, *J. Non-Cryst. Solids* **86**, 241 (1986).
14. T. P. Huijgen, H. Angad Gaur, T. L. Weeding, L. W. Jenneskens, H. E. C. Schuurs, W. G. B. Huysmans and W. S. Veeman, *Macromolecules* **23**, 3063 (1990).
15. G. S. Caravajal, D. E. Leyden, G. R. Quinting and G. E. Maciel, *Anal. Chem.* **60**, 1776 (1988).
16. L. M. Johnson and T. J. Pinnavaia, *Langmuir* **6**, 307 (1990).
17. M. J. J. Hetem, J. W. de Haan, H. A. Claessens, L. J. M. van de Ven, C. A. Cramers and J. N. Kinkel, *Anal. Chem.* **62**, 2288 (1990).
18. J. E. Sandoval and J. J. Pesek, *Anal. Chem.* **61**, 2067 (1989).
19. K. M. R. Kallury, U. J. Krull and M. Thompson, *Anal. Chem.* **60**, 169 (1988).
20. T. Furukawa, N. K. Eib, K. L. Mittal and H. R. Anderson, Jr., *Surf. Interface Anal.* **4**, 240 (1982); *J. Colloid Interface Sci.* **96**, 322 (1983).
21. N. Nishiyama, T. Asakura and K. Horie, *J. Appl. Polym. Sci.* **34**, 1619 (1989).
22. R. A. Cayless and D. L. Perry, *J. Adhesion* **26**, 113 (1988).
23. A. T. DiBenedetto and D. A. Scola, *J. Colloid Interface Sci.* **64**, 480 (1978).
24. K. M. R. Kallury, V. Ghaemmaghami, U. J. Krull, M. Thompson and M. C. Davies, *Anal. Chim. Acta* **225**, 369 (1989).
25. D. Suryanarayana and K. L. Mittal, *J. Appl. Polym. Sci.* **29**, 2039 (1984).
26. W. M. Heckl, F. M. Marassi, K. M. R. Kallury, D. C. Stone and M. Thompson, *Anal. Chem.* **62**, 32 (1990).
27. L. Rajakovic, B. Cavic, V. Ghaemmaghami, K. M. R. Kallury, A. Kipling and M. Thompson, *Anal. Chem.* **63**, 615 (1991).
28. R. A. Gledhill, S. J. Shaw and D. A. Tod, *Int. J. Adhesion Adhesives* **10**, 192 (1990).
29. J. Comyn, *Int. J. Adhesion Adhesives* **10**, 161 (1990).
30. J. J. Ponjee, V. B. Marriott, M. C. B. A. Michielsen, F. J. Touwslager and P. N. Van Velzen, *J. Vac. Sci. Technol. B* **8**, 463 (1990).
31. K. M. R. Kallury, M. Cheung, V. Ghaemmaghami, U. J. Krull and M. Thompson, *Colloids Surf.* (in press).
32. K. L. Mittal and H. R. Anderson, Jr. (Eds), *Acid–Base Interactions: Relevance to Adhesion Science and Technology.* VSP, Zeist, The Netherlands (1991).
33. M. C. Etter, *Acc. Chem. Res.* **23**, 120 (1990); M. C. Etter, Z. Urbanczyk-Lipkowska, M. Zia-Ebrahimi and T. W. Panunto, *J. Am. Chem. Soc.* **112**, 8415 (1990).

Silanes and Other Coupling Agents, pp. 277–287
Ed. K. L. Mittal
© VSP 1992.

Inelastic electron tunneling spectroscopic studies of alkoxysilanes adsorbed on alumina

P. N. HENRIKSEN[1,*], R. R. MALLIK[1] and R. D. RAMSIER[2,†]

[1]*Department of Physics, The University of Akron, Akron, OH 44325-4001, USA*
[2]*Surface Science Center, Department of Chemistry, University of Pittsburgh, Pittsburgh, PA 15260, USA*

Revised version received 5 November 1990

Abstract—Inelastic electron tunneling spectroscopy is used to investigate the adsorption of dimethyl-dimethoxysilane, dimethyldiethoxysilane, and dimethylvinylethoxy silane on alumina at the mono-layer level. Data obtained indicate that different adsorbed layers are produced when the silanes are introduced onto the oxide surface from solution or as a vapor. Silanes introduced in the same way onto different types of oxides suggest that alumina morphology also affects the adsorbed configuration.

Keywords: IETS; vibrational spectra; silane; adsorption; alumina.

1. INTRODUCTION

Inelastic electron tunneling spectroscopy (IETS) is a solid-state electron-energy-loss spectroscopy for obtaining vibrational spectra of adsorbates at the interface of metal/insulator/metal tunnel junctions [1]. The technique is sensitive to sub-monolayer coverages of compounds adsorbed onto a metal oxide surface. Both IR and Raman modes are accessible with an intrinsically high surface sensitivity; in fact, fractional monolayer adsorbate coverage may be sufficient to obtain spectra of certain compounds. It is primarily these features which make IETS particularly attractive for studying reactions of silane coupling agents on alumina, and indeed several articles have appeared over the last decade or so on this subject [2–14].

IETS has enjoyed considerable success in seemingly diverse areas of research including adhesion, biological molecules and polymers, catalysis, corrosion, and lubrication. The reader is referred to several comprehensive review articles [15–17], a compendium of spectra [18], and an excellent book [19] on the subject.

This paper provides a brief overview of the technique and its application to the adsorption of silane coupling agents on alumina.

2. OVERVIEW OF IETS

IETS relies on the quantum mechanical phenomenon of electron tunneling through a potential barrier. When electrons tunnel from one metallic electrode ($M1$) to another ($M2$) in a metal/insulator/metal structure, a tunnel junction,

---

*To whom correspondence should be addressed.
†Present address: Department of Physics, University of Pittsburgh, Pittsburgh, PA 15260, USA.

they can transfer some fraction of their energy to molecules in the insulating barrier. The barrier must be thin (of the order of 2–3 nm) if significant electron tunnel current is to flow. In effect, the electrons are used to excite vibrational modes of the barrier material in a way analogous to photons in IR spectroscopy. However, the intrinsic nature of quantum tunneling of electrons gives IETS certain unique features, not least of which are true surface sensitivity, and an orientational preference for excitation of certain vibrational modes. These aspects will now be discussed in more detail.

2.1. Basic principles of IETS

When a d.c. bias voltage, V, is applied to a metal/insulator/metal tunnel junction the Fermi levels become offset by an amount eV as shown schematically in Fig. 1. Electrons in filled energy states in $M1$ may then tunnel elastically to empty energy states in $M2$. When $eV \geq hv$, where v is the vibrational frequency of a molecular oscillator within the barrier, some electrons may tunnel inelastically by losing a quantum of energy to the oscillator. Only a small fraction of the electrons tunnel inelastically (approximately 1%) with the corresponding I–V characteristic showing a small change in slope commencing at $eV = hv$ as

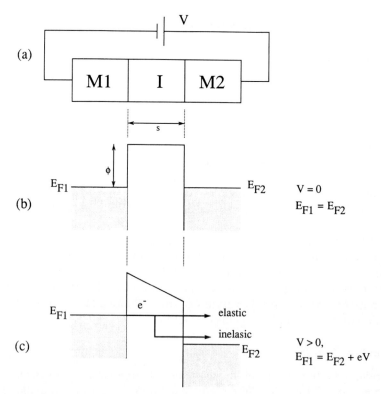

Figure 1. (a) Schematic diagram of a metal/insulator/metal tunnel junction with a variable applied d.c. bias voltage. (b) Partial schematic energy band diagram under zero applied bias conditions, where ϕ and s are the mean barrier height and thickness respectively. (c) Corresponding energy-band diagram where applied d.c. bias V is sufficient to excite a vibrational mode in the barrier thus producing an inelastic tunneling current.

illustrated in Fig. 2a. This change in slope is revealed more easily in the first and second derivative plots, dI/dV and d^2I/dV^2 vs. V (Fig. 2b, c respectively). An IET spectrum is a second derivative plot and shows peaks at voltages corresponding to the energies of various vibrational modes in the barrier material. Since tunneling electrons can excite permanent and induced dipoles, both IR and Raman modes are observed as well as optically forbidden transitions. At a particular bias voltage, the magnitude of the tunneling current decays exponentially with $s\phi^{1/2}$ where s is the barrier thickness, and ϕ is the mean barrier height; consequently, the tunnel current is extremely sensitive to small changes in these parameters. In a typical experimental situation where ϕ is approximately constant, one finds that for barrier thicknesses greater than 2–3 nm the tunneling current is too small to detect. This illustrates clearly why the observed inflections in the tunneling I–V characteristics and hence IET spectral lines are indeed due to interactions with oscillators at or near the adsorbent/adsorbate interface; i.e. *IETS is an intrinsically surface specific technique.*

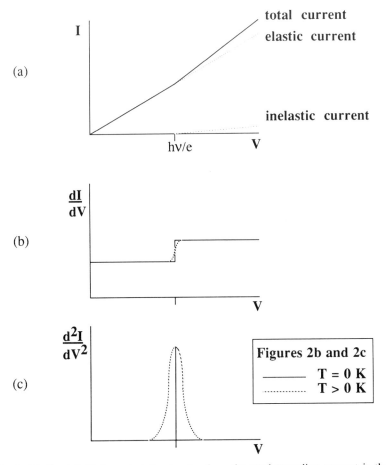

Figure 2. (a) Idealized I–V characteristic showing how the total tunneling current is the sum of elastic, and inelastic currents. A change in slope is observed at a threshold bias, $eV = h\nu$, corresponding to the energy of a vibrational mode. (b) Corresponding first derivative plot. (c) Corresponding second derivative plot.

Another factor to be considered when interpreting IET spectra is an orientational selectivity. It is generally believed that chemical bonds whose dipole moment axes lie perpendicular to the plane of the interface give rise to a greater change in conductance, and hence stronger IET lines, than those lying in plane [15]. This is due to the fact that the components of a dipole moment of the bond and its image normal to the interface are additive while components in plane are subtractive [20]. Thus, the resultant normal component will be a maximum when the bond is oriented perpendicular to the plane of the interface, and a minimum when parallel to this plane. It has been shown that the magnitude of the resultant normal component varies in a complex way as a function of bond angle [15]. From this variation it is inferred that tunneling electrons couple more strongly with bonds perpendicular as opposed to parallel to the surface. Thus, one would expect some correlation to exist between IET peak intensity and bond orientation.

Finally, it should be mentioned that IETS has the additional feature that its applicable energy range is greater than that of conventional IR spectroscopy, extending from very low energies appropriate for investigating metallic phonon modes up to energies high enough to permit electronic excitation of adsorbates.

3. EXPERIMENTAL

3.1. Sample preparation

The first step in sample preparation is the deposition of a thin metal film on an insulating substrate (e.g. a glass microscope slide). This base electrode is deposited by conventional vacuum deposition techniques with the electrode geometry defined by a shadow mask. Next, this electrode is oxidized either by exposing the film to room air or oxygen, or by establishing an oxygen plasma within the vacuum chamber. In the case of Al-electrodes, a remarkably uniform oxide layer is formed, typically 1–2 nm thick. The oxide film may then be dosed with the compound of interest; this is achieved in one of three ways.

3.1.1. Liquid-phase dosing. A dilute solution of the compound in a suitable solvent is poured onto the substrate and any excess is removed using a mechanical spinner.

3.1.2. Vapor-phase dosing. The substrate is exposed to a vapor of the compound under controlled conditions.

3.1.3. Plasma polymerization. The substrate is exposed to a monomer vapor excited by a gaseous plasma to form a polymeric adlayer on the oxide.

In certain cases the latter two methods may be performed within the vacuum chamber where the electrodes are deposited. *In all cases a uniform adlayer approximately one monolayer thick is desirable for optimum spectral sensitivity.*

Finally, a counter electrode (usually Pb) is deposited in a similar way to the base electrode thus completing the tunnel junction. (A fourth dosing method is sometimes used to introduce molecules onto the oxide surface *after evaporation of the lead electrode.* This method, known as infusion dosing [20–22], is used primarily for compounds that interact weakly with the surface). Figure 3 shows the electrode geometry used in our laboratory [23]. The tunnel junction area is of the order of 1 mm^2.

Studies have shown that as few as 10^{10} molecules adsorbed on an area of about 1 mm^2, corresponding to a surface coverage of approximately 0.1 monolayer, can yield acceptable IET spectra [24].

3.2. Recording IET spectra

IET spectra are obtained using voltage modulation techniques and phase-sensitive detection to measure harmonics of a small a.c. voltage developed across the tunnel junction as a function of a slowly swept d.c. bias voltage. By Taylor series analysis it can be shown that the second harmonic voltage is proportional to d^2I/dV^2. Commonly, one of two standard methods are employed to extract the second harmonic voltage. One is to incorporate the tunnel junction in one arm of an a.c. bridge circuit balanced at the fundamental frequency, thus removing unwanted signal at that frequency and leaving mainly the desired second harmonic signal [25–28]. A second method uses a tuned filter to block out signals at the fundamental frequency [23, 29–31]. Both methods have advantages and disadvantages, but we have found that the filter method is

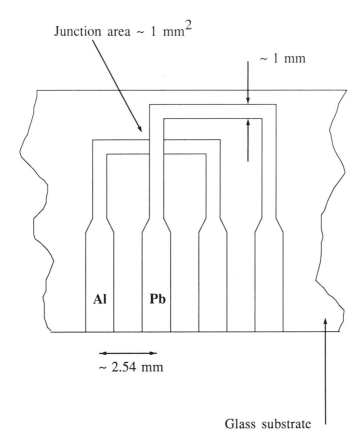

Figure 3. Schematic diagram of electrode geometry facilitating four-point voltage measurements. Three such devices are fabricated simultaneously on pre-cut glass microscope slides.

generally easier to implement and use. A block diagram of our spectrometer which is controlled by a computer and IEEE-488 interface is shown in Fig. 4.

IET line widths are determined by thermal and instrumental broadening, the latter being primarily due to the applied modulation voltage. With a super-conducting Pb counter-electrode, the full-width-at-half-maximum (FWHM) of an IETS peak is given roughly by [15]

$$\text{FWHM} = [(1.73 \, eV_{\text{rms}})^2 + (2.9 \, k_{\text{B}}T)^2]^{1/2} \, (\text{meV})$$

where V_{rms} is the rms modulation voltage in mV, k_{B} is Boltzmann's constant in meV/K, and T is the temperature in K.

Spectra are usually recorded with the samples cryogenically cooled to reduce thermal broadening of the IET lines. In our laboratory this is facilitated by mounting the samples in a PCB edge connector attached to a thin-wall stainless steel tube, which can be inserted into a helium storage Dewar. With a modulation of $\sqrt{2}$ mV (and the junction at 4.2 K) spectra are scanned typically in 1–2 h (signal averaging can be employed to improve the signal-to-noise ratio, but increases the overall scan time). Under these conditions we are able to resolve routinely IET peaks 1–2 mV apart. For IR spectroscopy comparisons note that 1 meV is equivalent to 8.0665 cm^{-1}. IET peak positions are usually quoted after subtraction of 1 mV for the superconducting energy gap of the Pb electrode.

A remaining question to be answered is why should one use IETS to

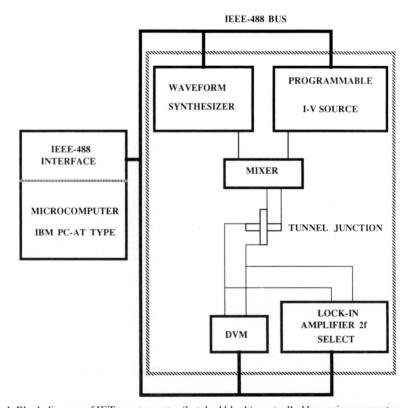

Figure 4. Block diagram of IET spectrometer (hatched block) controlled by a microcomputer.

investigate silane coupling agents? There are several reasons including the following: (i) Many silanes react with aluminum oxide, the most commonly used tunnel barrier in IETS; (ii) Coupling agents by definition provide chemical bonding between two dissimilar materials by acting as an intermediary layer. Aluminum is a widely used component in adhesive bonding, and so IETS is an appropriate choice; and (iii) An understanding of the chemisorption of the first silane monolayer is critical since this will strongly affect subsequent adhesion. Because IETS is inherently surface specific, information regarding surface chemistry (e.g. composition and molecular orientation) of the constituent molecular groups is readily accessible.

4. RESULTS AND DISCUSSION

Several reports have appeared over the last few years on the adsorption of organofunctional alkoxysilanes on alumina studied by IETS [2–14]. Two general observations have emerged, namely:

(1) The chemical nature of the adsorbed silane layer depends in part on the dosing method. For example, many silanes (in particular, triethoxysilanes) are believed to hydrolyze to silanols on alumina and subsequently adsorb via a condensation reaction with surface hydroxyl groups resulting in a siloxane adlayer and an alcohol by-product. However, slightly different spectra are obtained for certain silanes by vapor-phase as opposed to liquid-phase dosing, indicating that the degree of hydrolysis may not be the same in both cases [4, 10, 13].

(2) The adsorption mechanisms depend on the composition and stoichiometry of the alumina surface; slightly different spectra are obtained for silanes adsorbed onto plasma and thermal oxides [4].

We now present IETS data for dimethyldimethoxysilane (DMDMS), and dimethyldiethoxysilane (DMDES) liquid dosed onto thermal and plasma alumina, and for dimethylvinylethoxysilane (DMVES) applied to plasma alumina with a slightly different liquid dosing method. These results illustrate the effect of different dosing methods, and substrate-dependent adsorption effects.

Figure 5 shows a typical IET spectrum obtained from an undosed Al/Al-oxide/Pb junction where the oxide layer was formed by exposing the Al base electrode to an oxygen plasma at a partial pressure of nominally 100 mTorr for approximately 1–2 min. The sample was not removed from the vacuum chamber during fabrication. No surface contamination is evident; the peaks at 945, and 3620 cm^{-1} are due to Al—O bulk phonon modes, and the stretching of surface hydroxyls, respectively.

Figure 6a and b are the IET spectra of DMDMS, and DMDES, respectively, liquid dosed from 2.0 vol.% aqueous solution onto thermally formed alumina (formed by first partially venting the vacuum system with oxygen at a pressure of a few hundred mTorr, and then finally to atmospheric pressure with room air). Comparison of spectra shows that they are essentially identical, indicating that both silanes react with thermal alumina to produce the same adsorbed structure; i.e. nearly complete hydrolysis has occurred. Bands at 1264, 1415, 2898, and 2950 cm^{-1} are assigned to symmetric and asymmetric stretching and deformation respectively of the methyl group [32–34]. Symmetric Si—C stretching

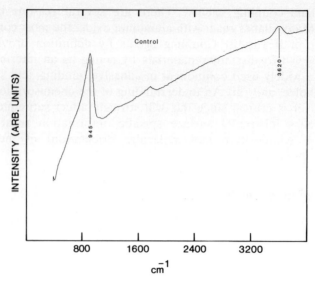

Figure 5. IET spectrum of an undosed Al/Al-oxide/Pb tunnel junction. The oxide was formed by oxygen plasma oxidation.

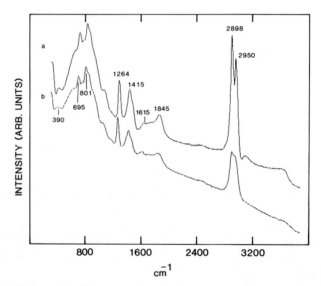

Figure 6. IET spectra of: (a) DMDMS; and (b) DMDES liquid-doped onto thermal alumina from a 2.0 vol.% aqueous solution (after [4]).

produces a peak at 695 cm^{-1} [32, 34–36], while the 801 cm^{-1} band is due to asymmetric Si—C and methyl rocking modes [32–36]. The broad doublet from 1600 to 1900 cm^{-1} is characteristic of a silylated Al-oxide/Pb interface but its origin has yet to be determined.

Figure 7a and b are spectra of DMDES and DMVES, respectively, liquid dosed onto oxides formed by exposure to a d.c. plasma of oxygen at a partial pressure of approximately 100 mTorr for 30 s and then vented to room air. The DMDES was liquid dosed as before, i.e. from a 2.0 vol.% aqueous solution. The

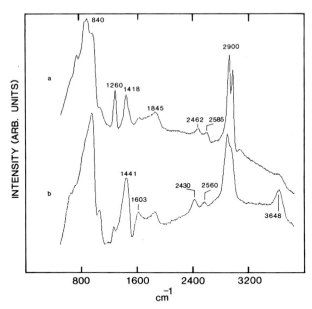

Figure 7. IET spectra of silanes adsorbed onto plasma alumina. Results from: (a) liquid-doping DMDES from 2.0 vol.% aqueous solution; and (b) immersing for 90 s in 0.5 vol.% solution of DMVES in acetone (after [4]).

spectrum of DMDES on plasma alumina (Fig. 7a) has three additional bands. These occur at 840, 2462, and 2585 cm^{-1}. The 840 cm^{-1} band is due to a symmetrical methylene rocking vibration [32, 34–36], and the other two have not yet been definitely assigned, although it has been postulated that they arise as a result of different site selection and/or symmetry of the DMDES on plasma alumina which may increase coupling of methyl vibrations [4]. It is clear from these observations that different adsorbed configurations result on the two oxides.

It should be mentioned that reflection-absorption IR (RAIR) spectra of similarly prepared but thicker DMDMS and DMDES films show incomplete hydrolysis [4]. These results are not contradictory to our IETS findings; IETS probes the first one or two monolayers most effectively, and therefore provides information specific to the interface, whereas the RAIRS data for thicker films (at least several monolayers) will include bulk lines which may obscure those due to near-surface moieties.

Although DMVES reacts on silica surfaces [37], we have found it to adsorb on oxidized Al only under specific conditions [4]. When spin cast on plasma alumina from solutions of either H_2O, acetone, or ethanol at concentrations 2.0 vol.% or greater, prohibitively thick films were obtained which adhered poorly to the alumina surface, evidenced by the fact they could be easily rinsed off with the above solvents. Lower solution concentrations resulted in no detectable adsorption. From these results we concluded that for DMVES to adsorb on alumina, the solutions must be dilute (<2.0 vol.%) and the exposure time increased.

Figure 7b is an IET spectrum of plasma alumina immersed for 90 s in 0.5 vol.% DMVES in acetone. Bands characteristic of methyl and vinyl moieties are present but weak. Bands at 1260, 2430, and 2560 cm^{-1} are again associated with the methyl groups, and the band at 1603 cm^{-1} to the C=C stretching mode of the vinyl group. Further evidence of DMVES adsorption is the prominent 1600–1900 cm^{-1} band mentioned previously and interpreted as being characteristic of the silylated AlO$_x$/Pb interface.

5. CONCLUSIONS

(1) IETS is well suited for investigating the adsorption of organofunctionalalkoxy silanes on alumina. Surface sensitivity at the sub-monolayer level, and orientational selectivity makes this technique particularly attractive, especially when used in conjunction with other techniques such as FT-IR, and SER spectroscopies.

(2) As one might expect, the morphology of the alumina surface also affects the surface chemistry. The silanes investigated here were adsorbed onto plasma alumina in a slightly different way than onto thermal alumina.

(3) IETS data for DMDMS and DMDES indicates that the way in which the silanes are introduced onto an alumina surface (e.g. from solution or from vapor) affects the subsequent surface chemistry, resulting in slightly different adlayers.

(4) Monoalkoxysilanes only react with alumina under a narrow range of conditions, resulting in very low coverage.

(5) A broad band from 1600–1900 cm^{-1} in IET spectra of mono-, di-, and trialkoxysilanes adsorbed on alumina appears to be characteristic of the silylated Al-oxide/Pb interface.

Acknowledgements

This work has been supported in part by grants from NASA Lewis Research Center (NAG 3-813), the Office of Naval Research (ONR 5-33656), the Alcoa Foundation, the Research Faculty Projects Committee of The University of Akron, and an Ohio Board of Regents Fellowship (R.D.R.). The authors gratefully acknowledge helpful suggestions by Professor A. N. Gent.

REFERENCES

1. R. C. Jaklevic and R. C. Lambe, *Phys. Rev. Lett.* **17**, 1139 (1966).
2. A. F. Diaz, U. Hetzler and E. Kay, *J. Am. Chem. Soc.* **99**, 6780 (1977).
3. T. Furukawa, N. K. Eib, K. L. Mittal and H. R. Anderson, Jr., *Surface Interface Anal.* **4**, 240 (1982).
4. R. D. Ramsier, G. R. Zhuang and P. N. Henriksen, *J. Vac. Sci. Technol.* **A7**, 1724 (1989).
5. P. N. Henriksen, A. N. Gent, R. D. Ramsier and J. D. Alexander, *Surface Interface Anal.* **11**, 283 (1988).
6. J. D. Alexander, A. N. Gent and P. N. Henriksen, *J. Chem. Phys.* **83**, 5981 (1985).
7. H. T. Chu, N. K. Eib, A. N. Gent and P. N. Henriksen, in: *Probing Polymer Structures*, Advances in Chemistry series, No. 174, J. L. Koenig (Ed.), pp. 87–98. American Chemical Society (1979).
8. J. Comyn, C. C. Horley, R. G. Pritchard and R. R. Mallik, *Polymer Commun.* **27**, 332 (1986).
9. J. Comyn, A. J. Kinloch, C. C. Horley, R. R. Mallik, D. P. Oxley, R. G. Pritchard, S. Reynolds and C. R. Werret, *Intl. J. Adhesion Adhesives* **5**, 59 (1985).
10. J. Comyn, D. P. Oxley, R. G. Pritchard, C. R. Werret and A. J. Kinloch, *Intl. J. Adhesion and Adhesives* **10**, 13 (1990).
11. J. Comyn, D. P. Oxley, R. G. Pritchard, C. R. Werret and A. J. Kinloch, *J. Adhesion* **12**, 171 (1989).

12. D. M. Brewis, J. Comyn, D. P. Oxley, R. G. Pritchard, S. Reynolds, C. R. Werret and A. J. Kinloch, *Surface Interface Anal.* **6**, 40 (1984).
13. T. Furukawa, N. K. Eib, K. L. Mittal and H. R. Anderson, Jr., *J. Colloid Interface Sci.* **96**, 332 (1983).
14. P. N. T. Van Velzen, *Surface. Sci.* **140**, 437 (1984).
15. P. K. Hansma, *Phys. Lett.* **C30**, 145 (1977).
16. R. G. Keil, T. P. Graham, and K. P. Roenker, *Appl. Spectrosc.* **30**, 1 (1976).
17. C. J. Adkins and W. A. Phillips, *J. Phys. C: Solid State Phys.* **18**, 1313 (1985).
18. D. G. Walmsley and J. L. Tomlin, *Prog. Surface. Sci.* **18**, 247 (1985).
19. P. K. Hansma, Editor, *Tunnelling Spectroscopy.* Plenum Press, New York (1982).
20. R. C. Jaklevic and J. R. Lambe in: *Tunnelling Phenomena in Solids*, E. Burstein and S. Lundqvist (Eds), pp. 233–253. Plenum Press, New York (1969).
21. R. C. Jaklevic and M. R. Gaerttner, *Appl. Surface. Sci.* **1**, 479 (1978).
22. R. R. Mallik, R. G. Pritchard, D. P. Oxley, C. C. Horley and J. Comyn, *Thin Solid Films* **112**, 193 (1984).
23. D. P. Oxley, A. J. Bowles, C. C. Horley, A. J. Langley, R. G. Pritchard and D. L. Tunnicliffe, *Surface Interface Anal.* **2**, 31 (1980).
24. R. M. Kroeker and P. K. Hansma, *Surface Sci.* **67**, 362 (1977).
25. J. G. Adler, T. T. Chen and J. Strauss, *Rev. Sci. Instrum.* **42**, 362 (1971).
26. A. Edgar and A. Zyskowski, *J. Phys. E: Sci. Instrum.* **18**, 863 (1985).
27. S. Colley and P. K. Hansma, *Rev. Sci. Instrum.* **44**, 1192 (1977).
28. P. N. Shott and B. O. Field, *Spectrochim. Acta* **35A**, 301 (1979).
29. S. Reynolds, L. D. Gregson, C. C. Horley, D. P. Oxley and R. G. Pritchard, *Surface Interface Anal.* **2**, 217 (1980).
30. D. G. Walmsley, I. W. N. McMorris, W. E. Timms, W. J. Nelson, J. L. Tomlin and T. J. Griffin, *J. Phys. E: Sci. Instrum.* **16**, 1052 (1983).
31. K. W. Hipps, *Rev. Sci. Instrum.* **58**, 265 (1987).
32. A. Marchand and J. Valade, *J. Organometal. Chem.* **12**, 305 (1968).
33. N. Kozlova, V. Bazov, I. Kovalev and M. Voronkov, *Latv. PSR Ziuat. Akad. Vestis*, Kim. Ser. No. 5, 604 (1971).
34. A. Smith, *J. Chem. Phys.* **21**, 1997 (1953).
35. T. Tenesheva and A. Lazarev, *Zh. Prikl. Spektrosk.* **43**, 91 (1985).
36. A. Smith and D. Anderson, *Appl. Spectrosc.* **38**, 822 (1984).
37. A. N. Gent and E. Hsu, *Macromolecules* **7**, 933 (1974).

Silanes and Other Coupling Agents, pp. 289–294
Ed. K. L. Mittal
© VSP 1992.

Quantitative FT-IR diffuse reflectance analysis of vinyl silanes on an aluminum hydroxide substrate

T. J. PORRO* and S. C. PATTACINI

The Perkin-Elmer Corporation, 761 Main Avenue, Norwalk, CT 06859-0800, USA

Revised version received 7 August 1991

Abstract—This study extends previous work on silanized kaolin clays to other substrates, such as aluminum hydroxide. It will also show that high precision quantitative Fourier Transform Infrared Spectroscopy (FT-IR) diffuse reflectance measurements can be performed on this vinyl silanized substrate and predict that other silanized finely divided powders can be analyzed using these techniques.

Keywords: FT-IR; silanes; aluminum hydroxide.

1. INTRODUCTION

Silanes are commonly used to promote adhesion between inorganic and polymeric materials. Among their applications [1] are to promote adhesion between a polymeric coating and nonpolymeric (ceramic, metal) substrates, or between a filler material and the matrix in reinforced composites. In these applications, it would be very beneficial to know the amount of silane deposited, and how the extent of adsorption changes with their concentration.

The silanization process has also been employed to permit tailoring, for various applications, surfaces such as kaolin clays, Wollastonite, talc, and sodium aluminum silicate with amino, epoxy, mercapto, methacryloxy and vinyl groups.

FT-IR diffuse reflectance measurements have been used to quantitatively determine the amount of silane deposited on a substrate. The advantages of FT-IR are speed and sensitivity, the latter relating to the ultimate analytical precision or the smallest amount of the analyte one can determine. Other classical uses of infrared spectroscopy are as a qualitative identification or structure elucidation tool.

In this work, as in previous investigations [2], we determined the degree of silanization by the coupling agent by quantitatively measuring the carbon-hydrogen stretching absorptions in the region between 3200 cm^{-1} and 2600 cm^{-1}, although Berger and Desmond [3] used the carbonyl band at 1720 cm^{-1} of methacryloxy silane in one instance.

Hanning [4], Miller and Ishida [5], and McKenzie and Koenig [6] favored transmission measurements using FT-IR for monitoring silanized mica quantitatively. Berger and Desmond [3] demonstrated the ability of FT-IR diffuse reflectance measurements to quantitate various silanized substrates including mercapto silane on hydrous clay and epoxy silane on alumina trihydrate. Later Vagberg *et al.*

---

*To whom correspondence should be addressed.

[7] demonstrated similarly the use of FT-IR diffuse reflectance measurements to quantitate aminopropyl silane adsorbed on muscovite mica.

In a previous publication [2], the authors have demonstrated that excellent quantitative FT-IR measurements could be made on various silanized kaolin clays. Figure 1, for example, shows superimposed calibration spectra at various concentrations of vinyl silanized clay from [2], and Fig. 2, the corresponding least squares fit (LSF) plot showing a correlation coefficient of 0.9951 which is comparable to the Kubelka-Munk (KM) corrected LSF data shown in Fig. 3 since the $\Delta \log(1/R)$ values are small.

More recently, the need to analyze vinyl silane on an aluminum hydroxide (alumina trihydrate) substrate has arisen. The silanized aluminum hydroxide substrate system was investigated to determine whether quantitative diffuse reflectance measurements were practical for routine control of the vinyl silanization process. In this paper, we discuss a method to quantify the amount of silane adsorbed on the aluminum hydroxide substrate.

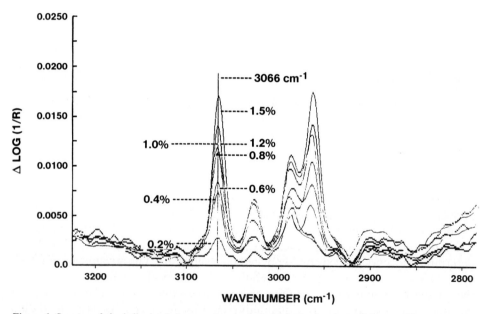

Figure 1. Spectra of vinyl silanized clays used as analytical standards—from Reference 2.

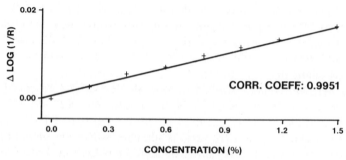

Figure 2. LSF calibration curve of vinyl silanized kaolin clay standards at 3066 cm^{-1} shown in Fig. 1— from Reference 2.

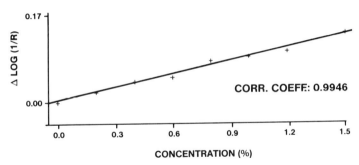

Figure 3. LSF calibration curve of vinyl silanized kaolin clay standards at 3066 cm⁻¹ (Kubelka-Munk corrected)—from Reference 2.

2. EXPERIMENTAL

All measurements were made using a Perkin-Elmer Model 1725X or 1760X FT-IR spectrometer equipped with a Spectra-Tech diffuse reflectance accessory and a deuterated triglycine sulfate (DTGS) detector. Additional instrument conditions were: resolution, 8 cm⁻¹; scans, 50; scan velocity, 0.1 cm sec⁻¹; acquisition time, 3 min. Background scans were obtained using spectral grade potassium bromide (KBr) powder. The aluminum hydroxide samples were received as powders from the J. M. Huber Solem Division in Fairmount, Georgia having an average particle size of 2 μm. No special treatment of the samples were made or required; that is, all silanized aluminum hydroxide samples were run neat (undiluted in KBr) directly as received. Measurements were made on eight concentrations: 0.0, 0.2, 0.4, 0.6, 0.8, 1.0, 1.2 and 1.5% of vinyl silanized aluminum hydroxide.

The vinyl silanized samples from Mr. Robert E. Schultz of J. M. Huber had been prepared as follows: liquid vinylmethoxyethoxy silane (Union Carbide A172) was quantitatively added to a known volume of a mixture of 60% de-ionized water and 40% 2-propanol in a polyethylene beaker via a disposable polyethylene syringe. 100 g of ATH (aluminum trihydrate) was stirred into exactly 300 ml of the vinyl silane mixture for 10 min with a polyethylene rod. The resulting slurry was placed in a 100°C oven and allowed to dry for 24 h.

3. RESULTS AND DISCUSSION

Figure 4 shows diffuse reflectance spectra of a 1.5% vinyl silanized aluminum hydroxide powder (with average particle size of 2 μm), both KM-corrected [8, 9] as well as uncorrected, using KBr powder as a reference. The KM correction compensates for the ordinate reflectance non-linearity induced by the diffuse reflection process and is expressed quantitatively by the expression $R_{km} = (1 - R_{ob})^2/2R_{ob}$ where R_{ob} is the measured uncorrected reflectance and R_{km} is the corresponding KM-corrected reflectance. The ordinates of both spectra were normalized to 100% reflectance at 4000 cm⁻¹.

It is clear from these spectra that the KM correction compensates for the non-linearity of the ordinate but not for the specular component in the spectrum. Close inspection of the C-H stretch region, 3200 cm⁻¹ to 2600 cm⁻¹, shows little evidence of hydrocarbon absorption on this scale and from previous experience

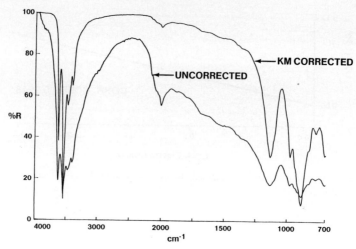

Figure 4. Diffuse reflectance spectral comparison of vinyl silanized (1.5%) aluminum hydroxide powder.

[2] with kaolin clay substrates, the reflectance even at the 1.5% vinyl silane level is quite low relative to that of the untreated aluminum hydroxide powder.

Figure 5 shows the comparison of the KM-corrected diffuse reflectance spectra for the 0.0% and the 1.5% vinyl silanized aluminum hydroxide concentrations. Note again that on this scale and particularly for KM-corrected spectra, the hydrocarbon absorptions between 3200 cm^{-1} and 2600 cm^{-1} are not discernible. However, upon closer inspection of the C-H stretch region and particularly the vicinity of the 3060 cm^{-1}, the absorptions are indeed found.

This is more clearly seen in Fig. 6 with superimposed KM-corrected spectra of the various concentrations of vinyl silanized aluminum hydroxide powder between 0.0% and 1.5% vinyl silane in the range between 3080 cm^{-1} and 3030 cm^{-1}.

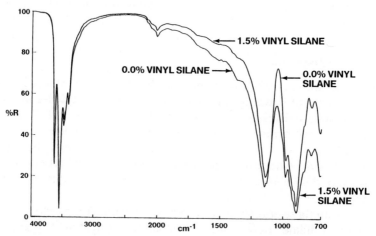

Figure 5. Diffuse reflectance spectral (KM-corrected) comparison of aluminum hydroxide powder silanized at two concentrations.

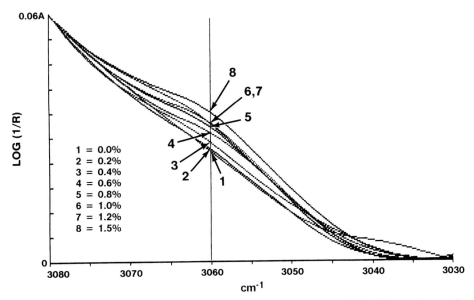

Figure 6. Spectra of various concentrations of vinyl silanized aluminum hydroxide powders used as analytical standards (Kubelka-Munk corrected).

The individual spectra were ordinately computer-expanded such that the highest reflectance was set at 100% R and the lowest at 0% R. This manipulation does not change the original reflectances but normalizes for differences in scattering among the samples. This permits good quantitation consistent with the sampling technique, and of course, the signal-to-noise (S/N) of the corrected spectral data.

In this particular case, we extracted quantitative data directly from the spectra shown in Fig. 6 at the chosen analytical frequency of 3060 cm^{-1}, the peak of the vinyl carbon-hydrogen stretch absorption. We measured the distance of each curve (in mm) from the log (1/R) value of the 0% vinyl silane spectrum from the peak values at the other concentrations and determined the linear LSF among the eight values.

This LSF plot is shown in Fig. 7 with the axes of Δ divisions (mm) vs. concentration of vinyl silane on aluminum hydroxide at 3060 cm^{-1}. Each division, in millimeters, is equal to 0.006 log (1/R) or 'equivalent absorbance' units. The correlation coeficient calculated from these data is 0.9868, which is reasonably good. In fact, the results are excellent considering that the data were obtained from the first set of samples specifically prepared of the vinyl silanized aluminum hydroxide for quantitative FT-IR analysis.

As the authors have shown in [2], another way of measuring the quantitative capability of this system is to apply the QUANT-3 curve fit Q-matrix type software program [10] which measures internal consistency among the standards in addition to determining analytical accuracy.

From a composite report of the QUANT-3 analysis of the 1.0% vinyl silanized aluminum hydroxide sample using the other concentrations as standards, the calculated concentration (1.0%) is within experimental error, while the peak-to-peak (p/p) error of 0.00028 refers to the remaining residuals of the unknown

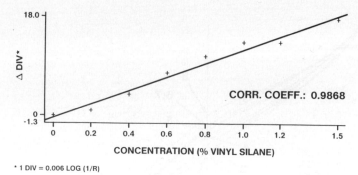

* 1 DIV = 0.006 LOG (1/R)

Figure 7. LSF calibration curve of vinyl silanized aluminum hydroxide powder standards shown in Fig. 6.

using known standards. Since the unknown is accurately known in this case, this error can be viewed as a measure of precision. This measure is comparable to previously obtained values for other reported analyses [2].

4. CONCLUSIONS

We have shown that the determination of vinyl silane on aluminum hydroxide powders is comparable to similar analyses on kaolin clays as reported earlier [2] and suggest with confidence that this technique can be extended to other powder substrate systems with sufficiently small particle sizes, i.e. comparable or less than the wavelength of the analytical frequency radiation as well as to the various other silanes that have been deposited on other substrates.

Acknowledgement

The authors are greatly indebted to Mr. Robert E. Schultz, from the J. M. Huber, Solem Division, Fairmount, GA, who supplied the eight vinyl silanized aluminum hydroxide powder samples used in this study, and details of their preparation.

REFERENCES

1. E. P. Plueddemann, *Silane Coupling Agents*. Plenum Press, New York (1982).
2. T. J. Porro and S. C. Pattacini, *Appl. Spectros.* **44**, 1172 (1990).
3. S. E. Berger and C. T. Desmond, *38th Annual Conference, Reinforced Plastics/Composite Institute.* SPI, Session 8D (1983).
4. A. Hanning, *Appl. Spectros.* **42**, 90 (1988).
5. J. D. Miller and H. Ishida, *40th Annual Conference, Reinforced Plastics/Composite Institute.* SPI, Session 17B (1985).
6. M. T. McKenzie and J. L. Koenig, *Appl. Spectros.* **39**, 408 (1985).
7. L. Vagberg, P. Depotocki and P. Stenius, *Appl. Spectros.* **43**, 1240 (1989).
8. J. Kubelka, *J. Am. Opt. Soc.* **38**, 448 (1948).
9. M. P. Fuller and P. R. Griffiths, *Anal. Chem.* **50**, 1906 (1978).
10. G. L. McClure, P. B. Roush and J. F. Williams, in *Computerized Quantitative Infrared Analysis,* G. L. McClure (Ed.), STP No. 934, p. 147. ASTM, Philadelphia (1987).

Silanes and Other Coupling Agents, pp. 295–304
Ed. K. L. Mittal
© VSP 1992.

Organofunctional silanes as adhesion promoters: direct characterization of the polymer/silane interphase

T. E. GENTLE,[1,*] R. G. SCHMIDT,[1] B. M. NAASZ,[1] A. J. GELLMAN[2] and
T. M. GENTLE[1]

[1] *Dow Corning Corporation, Midland, MI 48686, USA*
[2] *Department of Chemistry, University of Illinois, Urbana, IL 61801, USA*

Revised version received 21 November 1991

Abstract—Organofunctional silanes can be used as coupling agents to promote the adhesion of organic matrices to inorganic substrates. Silane coupling agents are of the general structure $YSi(OR)_3$, where Y is an organofunctional group selected for bonding to organic polymers while (OR) is a hydrolyzable group on silicon which can react with surface hydroxyl groups on the substrate. One mechanism of adhesion promotion by silanes has been postulated to be the interdiffusion of the coupling agent and the polymer to form an interpenetrating network.

This paper reviews some earlier results using X-ray photoelectron spectroscopy to probe the polymer/silane interphase region and gives new results obtained using the technique of sputtered neutral mass spectrometry (SNMS). It was demonstrated that SNMS had the sensitivity necessary to detect the polymer/coupling agent interphase. It was found that interdiffusion of the coupling agent with the polymer was maximized when the solubility parameters of the polymer and coupling agent were matched.

Keywords: Silane; coupling agent; interpenetrating network; adhesion; sputtered neutral mass spectrometry; solubility parameter.

1. INTRODUCTION

While the importance of organofunctional silanes as adhesion promoters has been recognized for many years [1], only recently has an understanding of the nature of the polymer/silane interphase in polymer/metal systems been obtained. Plueddemann's pioneering efforts have led to widespread use of organo-functional silanes as adhesion promoters for organic and inorganic materials. Plueddemann proposed interdiffusion of the silane coupling agent into the polymer matrix to form an interpenetrating network as one possible mechanism of adhesion promotion by silane coupling agents [1]. The goal of the work reported here was to find evidence of interdiffusion of coupling agents into the polymer matrix by direct characterization of the polymer/coupling agent interphase region using surface-sensitive depth profiling techniques.

A review of early studies involving indirect investigation of the polymer/silane coupling agent interphase will be presented before discussing results of experiments involving the direct probing of the polymer/silane coupling agent interphase.

---

*To whom correspondence should be addressed.
This paper was originally presented at the Plueddemann Symposium held in honour of Dr. Edwin Plueddemann on 3–5 April 1991 in Midland, Michigan.

Several analytical techniques have been used to characterize the polymer/silane coupling agent interphase. Culler *et al.* [2] used Fourier transform infrared (FT-IR) spectroscopy to characterize the chemical reactions at the matrix/silane interphase of composite materials. They correlated the extent of reaction of the resin with the coupling agent (as determined by FT-IR) with the extent of interpenetration. Culler *et al.* [2] have also used observations of improved resistance of the interphase region to solvent attack as indirect evidence to support the interpenetrating network theory.

In another study by Koenig *et al.* [3], FT-IR was also used to evaluate the interfacial bonding in ethylene vinyl acetate/glass systems. Koenig *et al.* [3] point out an inherent difficulty in using FT-IR for surface or interfacial studies. The amount of material in the interphase is negligible compared to the bulk, resulting in poor selectivity. Data processing techniques such as spectral subtraction and computer-scale expansion are necessary to enhance selectivity.

Hoh *et al.* [4] used differential scanning calorimetry (DSC), FT-IR, and solid-state ^{13}C-NMR to gain information about the epoxy/silane resin interphase. They used FT-IR to correlate the extent of reaction with the extent of interdiffusion as in the above studies. For both NMR and DSC studies, bulk models were used to study the molecular mobility of interfacial components.

Several studies of polymer/silane coupling agent interphases have involved the use of scanning electron microscopy (SEM) [5–7]. For example, Vaughan and Peek [6] have used SEM to examine fracture surfaces to determine the mode of failure of composite materials and to draw conclusions about interfacial interactions of various coupling agents and epoxide and polymer resin systems.

The characterizations discussed thus far do not involve direct investigation of the actual interphase region although failed surfaces have been analyzed to give indirect evidence for interdiffusion. In other studies, assumptions that the observed properties (extent of reaction, increased solvent resistance, molecular mobility) correlate with the extent of interdiffusion have been made.

There are several methods available to probe the actual interphase to demonstrate the existence of interpenetrating networks directly. Among these techniques are depth profiling by SIMS (secondary ion mass spectrometry) or SNMS (sputtered neutral mass spectrometry), and the use of X-ray photoelectron spectroscopy (XPS) depth profiling or Auger electron spectroscopy (AES) depth profiling.

In a previously reported study, the polymer/silane interphase of a polymer/silane/Ge system was investigated using XPS for depth profiling [8]. The silane in this study was Dow Corning Z6020®, $NH_2(CH_2)_2NH(CH_2)_3Si(OCH_3)_3$. In Fig. 1 [8], XPS depth profiles are displayed for samples prepared using Z6020® to promote adhesion of polyvinyl chloride (PVC) to germanium substrates. The concentration of Z6020® through the laminate was followed by monitoring the concentration of silicon (in atomic percent) as a function of the depth. Sputtering occurred from the germanium side of the laminate. Curve a in this figure shows data for a laminate in which a solution of prehydrolyzed Z6020® was applied to the germanium and allowed to dry at room temperature (25°C) before application of PVC. Curve b shows data for an identical sample except that the Z6020® layer was heated to 175°C before application of the PVC. Heating of the coupling agent promotes condensation of the silanol groups leading to a resin-

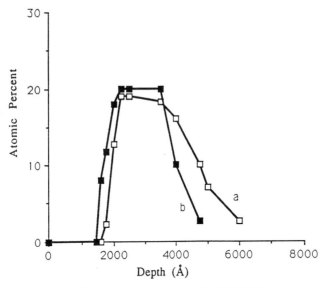

Figure 1. The atomic percent silicon in a laminate of PVC/Z6020®/Ge as a function of the depth when sputtering from the Ge side of the laminate. Curve a corresponds to drying of the coupling agent at 25°C. Curve b corresponds to drying of the coupling agent at 175°C (from ref. [8]).

like network which would limit its mobility into PVC. This is consistent with the decrease in the extent of penetration of the silane into the PVC as shown in curve b of Fig. 1 when compared with curve a for the room temperature sample. Measurements of strength of the bonds for these systems by a 90° peel test revealed that the systems with greater interpenetration of the silane (curve a) were significantly stronger.

In a more recent study, characterization of the polymer/silane interphase was performed using SNMS as the analytical tool [9]. SNMS offers several advantages over XPS depth profiling, such as ease of experimental execution and improved depth profiling resolution. Using SNMS, the polystyrene/Z6020®/Ge system was investigated. The SNMS depth profile for this system is displayed in Fig. 2 [9]. The sample was prepared by room temperature (25°C) drying of Z6020® applied from a solution followed by application of polystyrene. The mass spectral intensities of sputtered fragments arising from Ge, Z6020®, and polystyrene are plotted vs. time. In Fig. 3, the SNMS depth profile of a sample in which the Z6020® layer was heated to 120°C before application of polystyrene is displayed. Comparison of Figs 2 and 3 reveals that the silane is concentrated at the polystyrene/Ge interface when the coupling agent is heat-treated, whereas the silane interdiffuses with the polystyrene to a greater extent when the coupling agent is dried at room temperature.

We have extended these studies of the polymer/coupling agent interphase region using SNMS by examining systems in which the coupling agents and polymers varied in structure and solubility parameter. The coupling agents chosen for this study are listed in Table 1 along with the solubility parameters of their organic side-chains. Aluminum was chosen as the substrate. A series of

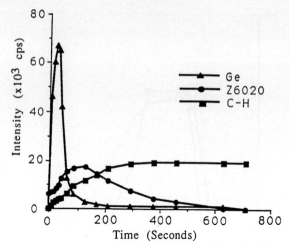

Figure 2. Depth profile of a polystyrene/Z6020®/Ge laminate obtained using sputtered neutral mass spectrometry. The Z6020® coupling agent was dried at room tempreature (25°C) before application of the polystyrene (from ref. [9]).

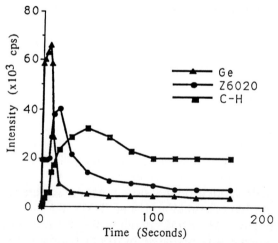

Figure 3. Depth profile of a polystyrene/Z6020®/Ge laminate obtained using sputtered neutral mass spectrometry. The Z6020® coupling agent was dried at 120°C before application of the polystyrene (from ref. [9]).

methacrylate polymers with known solubility parameters was used. The polymers along with their solubility parameters are listed in Table 2.

The effect of the coupling agent and polymer solubility parameters on the extent of interdiffusion in the interphase region, as determined by SNMS, was the focus of this study. Based on solubility parameter principles, if the solubility parameters of the two components are equal, the components are predicted to be miscible in all proportions. Minimizing the difference in the solubility parameters is the basis for matching solubility parameters of the coupling agent and polymer to obtain greater interdiffusion of the two.

Table 1.
Silane coupling agents and solubility parameters
(δ)

Silane	δ
Dow Corning Z6030® 3-Methacryloxypropyltrimethoxysilane	9.0

$$CH_2=C-\overset{\overset{O}{\|}}{C}\cdot O(CH_2)_3Si(OCH_3)_3$$
$$|$$
$$CH_3$$

Silane	δ
Dow Corning Z6040® 3-Glycidoxypropyltrimethoxysilane	9.3

$$CH_2-CHCH_2O(CH_2)_3Si(OCH_3)_3$$
$$\backslash\ /$$
$$O$$

Table 2.
Methacrylate polymers and their solubility
parameters (δ)

Polymer	δ
Poly(isobutyl methacrylate)	8.65
Poly(ethyl methacrylate)	8.95
Poly(methyl methacrylate)	9.3
Poly(benzyl methacrylate)	9.9

2. EXPERIMENTAL

2.1. Sample preparation

Polymer/coupling agent/Al laminates were prepared using slight modifications of a previously described technique [8]. It was preferred that the sputtering be done from the Al side since this layer was thin (approximately 2000–3000 Å) and sputtering from the polymer side would involve sputtering through a relatively thick insulating layer. The general sample preparation method involved the following steps:

(1) A very thin layer of release agent was wiped on a glass substrate (1″ × 3″ slide or 2″ disk).
(2) Al (~3000 Å) was evaporated on the glass substrate.
(3) The coupling agent (Z6030® and Z6040®) solution was spun-coated on the Al/glass using a 10 wt% solution of the coupling agent in isopropanol. Generally a spin speed of 2000 rpm was used to obtain a coupling agent layer of approximately 500 Å in thickness.
(4) The viscous polymer solution (20 wt% in toluene) was cast on the coupling agent layer and the solvent was then evaporated, leaving a polymer coating of approximately 50 μm in thickness.
(5) The polymer/coupling agent/Al laminate was peeled off the glass substrate.

2.2. SNMS experiments

SNMS experiments were performed on a Leybold Heraeus Model INA3 Secondary Mass Spetrometer located in the Materials Research Laboratories of the University of Illinois, Urbana, Illinois. Depth profiling of the polymer/ coupling agent/Al laminate occurred from the Al side with the sample mounted in a standard sample holder. To minimize the generation of excess surface charge on the sample (charging), which results in poor performance of the spectrometer and nonuniform sputtering, a copper grid designed for electron microscopy was placed on the sample before insertion into the sample holder to improve electrical contact. The surface of the sample was sputtered using an argon plasma. The sample was biased negatively with respect to the plasma in order that the argon ions be drawn to and bombard the sample surface. Neutral species sputtered from the sample surface were then ionized in the argon plasma and drawn into the mass spectrometer for analysis. Target potential (biasing of the sample) was varied to determine conditions for obtaining uniform sputtering as evaluated by optical microscopy after sputtering. In the experiments involving methacrylate laminates, the optimum target potential was found to be 500 V. Depth profiles of the laminates were obtained by monitoring the signal of the appropriate mass numbers of the components with sputtering time. Mass spectral intensity vs. sputtering time (proportional to depth of sputtering) plots were obtained. During the depth profiling of the methacrylate laminates, mass number 28 for silicon was followed for the coupling agent, mass number 12 (carbon) for the polymer, and mass number 27 for aluminum. Although there would be some contribution from both the coupling agent and the polymer to mass fragment 12, the contribution from the coupling agent is negligible since it is present in such a low concentration relative to the polymer.

3. RESULTS AND DISCUSSION

3.1. Effect of coupling agent thickness

As a test of the sensitivity of the SNMS technique for the characterization of the polymer/silane interphase, a series of poly(isobutyl methacrylate) laminates in which the thickness of the Z6030® coupling agent (Table 1) was varied were prepared and analyzed using SNMS. Figure 4 shows the SNMS depth profiles for this series of laminates. The thickness of the coupling agent layer was varied by varying the rate at which a substrate was spun during solution application of the coupling agent. Increasing the spin rate decreases the thickness of the coupling agent layer. It was thought that this series of laminates would test whether the SNMS technique was sensitive enough to detect coupling agent thickness differences of the order of 500 Å or less.

By monitoring the signal intensity for mass fragment 28 (silicon in the silane coupling agent) with the sputtering time (proportional to depth into the laminate from the Al side), it was possible to compare the three laminates with different coupling agent layer thicknesses. As the plots in Fig. 4 show, the mass fragment 28 signal intensity and the peak area of the profile increased as the coupling agent layer increased in thickness. This result indicated that it was possible to detect coupling agent thickness differences that occurred with changing spin rate.

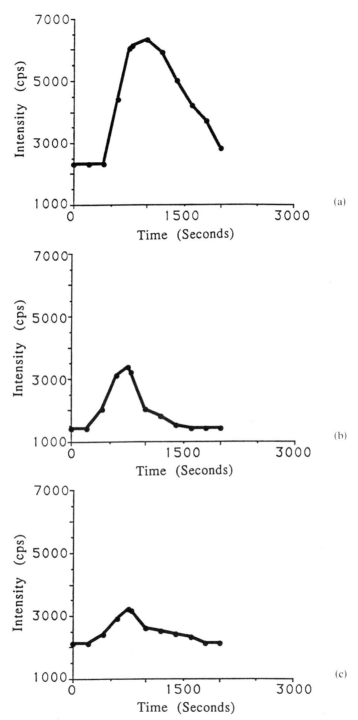

Figure 4. SNMS depth profiles for a series of poly(isobutyl methacrylate)/Z6030®/Al laminates. The thickness of the coupling agent was varied by changing the spin speed at which the coupling agent was applied. The spin speeds utilized were 1500, 2500, and 3500 rpm in profiles a, b, and c, respectively.

3.2. Effect of solubility parameter

As discussed previously, it was thought that compatibility of coupling agent and polymers would vary possibly with their solubility parameters and that this would affect the extent of interdiffusion. When the solubility parameter of the coupling agent is very close to that of the polymer, one might expect a higher degree of interdiffusion than when the solubility parameters are not as closely matched. This hypothesis was tested by carrying out SNMS analysis on a series of laminates in which Z6040® coupling agent (Table 1) with a solubility parameter of 9.3 was used. Plots for this series are shown in Fig. 5. Figure 5a shows the plot for a laminate in which the polymer is poly(isobutyl methacrylate) with solubility parameter 8.65. The solubility parameters of the silane and the polymer are not closely matched. Therefore, it would be predicted that the silane is not highly miscible in the polymer, which would be evident by a peak in the concentration of the silane before the sharp increase in the signal for the polymer. This is consistent with the observed depth profile in Fig. 5a, in which the coupling agent appears to be concentrated in the region before the polymer

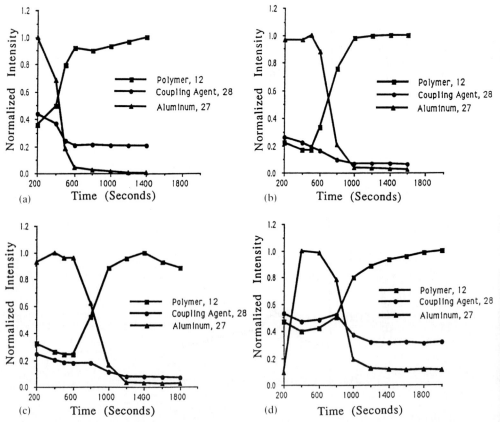

Figure 5. SNMS depth profiles for a series of methacrylate polymer/Z6040® ($\delta = 9.3$)/Al laminates. Profile a is for a laminate in which poly(isobutyl methacrylate) ($\delta = 8.65$) was the polymer. Profiles b, c, and d utilized poly(ethyl methacrylate) ($\delta = 8.95$), poly(methyl methacrylate) ($\delta = 9.3$), and poly(benzyl methacrylate) ($\delta = 9.9$) as the polymer, respectively.

and does not extend into the polymer region to any great extent. When the polymer is poly(ethyl methacrylate) with solubility parameter 8.95 as in Fig. 5b, it appears that the extent of interdiffusion increases as the solubility parameters become more closely matched, as evidenced by a somewhat lower concentration of silane before the polymer region. When the solubility parameters of the coupling agent and polymer are the same as in Fig. 5c (both 9.3), where poly(methyl methacrylate) is the polymer, there is a small peak in the silane concentration that occurs after the rapid increase in the polymer signal, indicating interdiffusion of the silane into the polymer. The solubility parameters are again mismatched in Fig. 5d, where poly(benzyl methacrylate) with solubility parameter 9.9 is the polymer. In this case, the coupling agent is again concentrated in the interface between the polymer and Al rather than diffusing into the polymer.

3.3. Acid–base considerations

It has been suggested that acid–base interactions may be important in predicting physical interactions of polymers [10–14]. Fowkes *et al.* [13] have studied the acid–base properties of glass surfaces and have found that adhesion of polymers to inorganic oxides such as glass depends on the acid–base interactions between acidic or basic sites of the glass and the basic or acidic functional sites of the polymer. They studied the acid–base properties of fiberglass and how these are changed by surface treatment with silane coupling agents. In another study, Fowkes *et al.* [14] examined how the mechanical properties of polymer composites could be enhanced by modification of the surface acidity or basicity of fillers by application of silane coupling agents. They found that by rendering a surface basic by the treatment with a coupling agent, they could promote the adsorption and adhesion of acidic polymers and vice versa. In the present study, both the coupling agents could be considered basic as well as all the methacrylate polymers, which perhaps led to the small amount of interdiffusion seen with that set of laminates. The acidity and basicity of coupling agents, polymers, and the solvents from which they are cast require further investigation to see what effect they have on the interdiffusion of coupling agents and polymers.

4. CONCLUSIONS

The present studies have shown that SNMS can be extended to systems other than those previously studied [9]. Based on the results of this study, it can be concluded that sputtered neutral mass spectrometry is a useful technique for probing the polymer/silane coupling agent interphase in several systems. Uniform sputtering can be obtained through mounting of the samples in such a way that the sample remains grounded while sputtering through the nonconducting polymer. It was found that selection of an appropriate target potential (biasing of the sample during sputtering) is critical for obtaining a sputtering well with walls that are sharply defined allowing for a more accurate depth profile through the interphase region.

With a series of methacrylate polymers whose solubility parameters ranged from 8.65 to 9.9, it was shown that the polymer/silane interphase is influenced by the solubility parameters of the silane coupling agent and polymer. It was determined that more interdiffusion occurred when the solubility parameter of the polymer more closely matched that of the coupling agent.

REFERENCES

1. E. P. Plueddemann, *Silane Coupling Agents*. Plenum Press, New York (1982).
2. S. R. Culler, H. Ishida and J. L. Koenig, *J. Colloid Interface Sci.* **109**, 1–10 (1986).
3. J. L. Koenig, F. J. Boerio, E. P. Plueddemann, J. Miller, P. B. Willis and E. F. Cuddihy, *Chemical Bonding Technology. Direct Investigation of Interfacial Bonds*. Report, DOE/JPL-1012-120; JPL-PUB-86-6; Order No. DE87005424, 52 pp. (1986).
4. K. Hoh, H. Ishida and J. L. Koenig, in: *Composite Interfaces*, H. Ishida and J. L. Koenig (Eds), pp. 251–263. North-Holland, New York (1986).
5. D. J. Vaughan and R. C. Peek, in: *Adhesion Aspects of Polymeric Coatings*, K. L. Mittal (Ed.), pp. 409–419. Plenum Press, New York (1983).
6. D. J. Vaughan and R. C. Peek, 20th Annual Technical Conference, Reinforced Plastics/ Composites Institute, pp. 1–5. The Society of Plastics Industry, New York (1975).
7. J. Denault and T. Vu-Khanh, *Polym. Composites* **9**, 360–367 (1988).
8. M. K. Chaudhury, T. M. Gentle and E. P. Plueddemann, *J. Adhesion Sci. Technol.* **1**, 29 (1987).
9. A. J. Gellman, B. M. Naasz, R. G. Schmidt, M. K. Chaudhury and T. M. Gentle, *J. Adhesion Sci. Technol.* **4**, 597 (1990).
10. F. M. Fowkes, C.-Y. Sun and S. T. Joslin, in: *Corrosion Control by Organic Coatings*, H. Leidheiser, Jr. (Ed.), National Association of Corrosion Engineers, Houston (1981).
11. F. M. Fowkes and D. O. Tischler, *J. Polym. Sci.: Polym. Chem. Ed.* **22**, 547–566 (1984).
12. F. M. Fowkes, in: *Physicochemical aspects of Polymer Surfaces*, K. L. Mittal (Ed.), vol. 2, pp. 583–603. Plenum Press, New York (1983).
13. F. M. Fowkes, D. W. Dwight, D. A. Cole and T. C. Huang, *J. Non-Cryst. Solids* **120**, 47–60 (1990).
14. F. M. Fowkes, D. W. Dwight, J. A. Manson, T. B. Lloyd, D. O. Tischler and B. A. Shah, *Mater. Res. Soc. Symp. Proc.* **119**, 223–234 (1988).

Silanes and Other Coupling Agents, pp. 305–321
Ed. K. L. Mittal
© VSP 1992.

A time-of-flight static secondary ion mass spectrometry and X-ray photoelectron spectroscopy study of 3-aminopropyltrihydroxysilane on water plasma treated chromium and silicon surfaces

B. N. ELDRIDGE,* L. P. BUCHWALTER, C. A. CHESS, M. J. GOLDBERG, R. D. GOLDBLATT and F. P. NOVAK

IBM Research Division, T. J. Watson Research Center, Yorktown Heights, NY 10598, USA

Revised version received 7 September 1991

Abstract—We have used time-of-flight static secondary ion mass spectrometry (ToF-SSIMS), and X-ray photoelectron spectroscopy (XPS), to study films produced by exposure of water plasma treated chromium and silicon surfaces to aqueous solutions of 3-aminopropyltrihydroxysilane (3-APTHS). The chemical structure of positive and negative secondary ions produced by these films was deduced by a combination of exact mass determination and the use of isotopically labelled 3-APTHS. Ions characteristic of the 3-APTHS overlayer were observed for both surfaces. The use of ^{18}O labelled 3-APTHS yields interesting insight into the cross-linking nature of the films studied, suggesting no further cross-linking of the silane as a function of *in situ* thermal exposure. XPS studies of these samples support the ToF-SSIMS data showing similarity of 3-APTHS bonding to the two surfaces studied.

Keywords: ToF-SSIMS; SIMS; adhesion; XPS; aminosilane; water plasma.

1. INTRODUCTION

Silane coupling agents are used extensively to promote adhesion in many systems, including the polymer–metal and polymer–polymer structures important in the fabrication of microelectronic packages and devices. While these coupling agents have been the subject of intense scrutiny, considerable controversy remains regarding the nature of the coupling agent interlayer and the mechanism by which enhanced adhesion is obtained [1–4]. The thickness of the adsorbed layer produced by exposure of a surface to an aqueous solution of 3-APTHS, while dependent on solution concentration, exposure time, and surface pretreatment [5], is thought to be in the order of 10–10 000 Å [1]. For this reason, these layers are most easily accessible by surface analysis techiques such as XPS, or reflection IR spectroscopy. ToF-SSIMS (from hereon referred to as SSIMS) is a surface sensitive technique which potentially offers the ability to study even very thin 3-APTHS overlayers. The sampling depth of the technique is in the order of 10 Å. SSIMS spectra have been shown to contain information relevant to the structure of many organic materials [6], and so may give some insight into the structure of the 3-APTHS overlayer.

---

*To whom correspondence should be addressed.

Figures 1(a) to 1(c) show the steps by which this molecule is thought to attach itself to a surface. The free molecule, shown in Fig. 1(a), is typically introduced in dilute aqueous solution, and initial attachment is via the condensation reaction between the silanol-end of the molecule and the surface hydroxyl group [7], as shown in Fig. 1(b). Heating of a 3-APTHS overlayer on a titanium surface to about 110°C is thought to cause polymerization of the silane to form a high molecular weight siloxane polymer [4, 8] as shown in Fig. 1(c). The studies cited, as well as this work, have been conducted on films prepared under ambient conditions. The behavior of the molecule is thus dictated by its reaction not with a pure metal or silicon surface, but with the native oxide of these materials. Since both the titanium- and silicon-oxide surfaces are acidic in nature [9], it is perhaps logical to expect that the 3-APTHS will show similar behavior the Si and perhaps even on the Cr surfaces examined in this study. It has also been proposed that the molecule may bind to metal surfaces via donation of the nitrogen lone pair electrons to the metal atom [4]. This is shown in Fig. 1(d). Yet another mechanism of bonding of aminosilanes has been suggested [4] and is shown in Fig. 1(e). In this case, the amine is protonated with a substrate surface

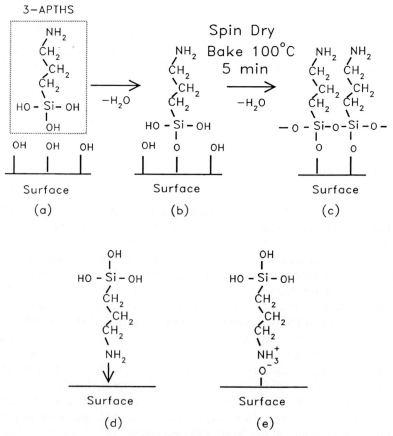

Figure 1. Potential reaction mechanisms for 3-APTHS: (a)–(c) condensation attachment mechanism; (d) attachment of molecule via donation of nitrogen lone pair electrons; and (e) attachment of molecule via protonation of amine with surface hydroxyl group.

hydroxyl group; this process yields a surface bound amine which is not available for further reaction. While some previous work has been done in attempting to use SSIMS to characterize surface layers of 3-APTHS and similar compounds [10], we believe this paper is the first systematic attempt to characterize the layers formed on Si and metal surfaces with SSIMS.

2. EXPERIMENTAL

The time-of-flight mass spectrometer used in these studies has been described in [11]. A sample load lock lets in samples from atmosphere into ultra high vacuum (UHV) in under 15 min. Heating raises the sample temperature while the sample is positioned in front of the mass spectrometer. The primary ion beam species is $^{40}Ar^+$. The primary ion beam energy is 10 keV. The spectrometer resolution is approximately 800 at $m/z = 100$. Experiments with known compounds have demonstrated that for singlet peaks with good signal-to-noise, the mass accuracy is generally within 10 mmu in the range 0–100 amu, and within 20 mmu in the range of 100–300 amu. The presence of doublet or triplet peaks at the same nominal mass which cannot be resolved by the spectrometer may cause a shift in the peak centroid, and an incorrect interpretation of the identity of the peak in question. This could occur more with increasing nominal mass, since the number of possible combinations of even a limited number of elements increases dramatically as the nominal mass increases. The mass assignments given in this paper are based on a combination of exact mass fits, and peak shifts observed when isotopically labelled materials were substituted for unlabelled 3-APTHS. Where isotopic labelling indicated a chemical formula outside the tolerance ranges assigned above, preference was given to the isotopic information, and it was assumed that the mass discrepancy was due to additional peaks at the same nominal mass, or simply poor signal-to-noise for the peak in question. It should be noted that the overall intensity for most of the high mass peaks observed in this study was not particularly high. We have thus limited our assignments to those peaks which we felt most comfortable with, considering all the above criteria. The total ion dose was maintained at less than 10^{13} primary ions cm^{-2}. No evidence of change in the SSIMS data with primary beam dose was observed for the mass spectra acquired.

The precursor used to create the 3-APTHS solutions was 3-aminopropyl-triethoxysilane (3-APTES). Commercial 3-APTES obtained from Huls Petrarch Systems Inc. was twice vacuum-distilled. Individual ampules were filled with the distilled precursor and sealed for later use. Triply-distilled HPLC grade water was used to prepare the 3-APTHS solutions used for sample treatment. For isotopically labelled 3-APTHS studies, we used 99 atom % ^{15}N labelled 3-APTES, and 97.8 atom % ^{18}O labelled water. Both compounds were obtained from MSD Isotopes [12]. The substrates used were polished Si wafers 1.5 cm in diameter. The Cr was sputter-deposited to a thickness of 1000 Å onto one half of the Si wafer so that both the Si and Cr surfaces had identical preparation histories up to and including the analysis for every sample prepared.

Prior to application of the aqueous solution of 3-APTHS, the samples were subjected to a water plasma treatment for 1–3 h. The pressure of water vapor in the plasma chamber was 26 Pa. The plasma was fired by an RF power supply.

During treatment, the samples were held so near to the plasma potential, that physical sputtering of species from the target surface was negligible. Optical emission spectroscopy indicates that the dominant species in the plasma are hydrogen and hydroxyl neutral radicals. The water plasma treatment was done because SSIMS analysis showed this pre-treatment step to be especially effective at removing organic contaminants often found on surfaces exposed to the laboratory ambient. Samples treated in this way produced only ions formed from Si or Cr atoms, alone or in combination with oxygen or hydrogen. The latter were presumably due to the presence of surface oxide species and/or adsorbed water. This treatment may also lead to a higher level of surface hydroxylation relative to that of the untreated surface, but no attempt was made to verify this.

One hour before a sample was removed from the plasma chamber, a fresh working solution was prepared using 0.1 vol % of 3-APTES in water. A fresh ampule of 3-APTES was used for each experiment. When HPLC water was used, the precursor was injected into a fresh bottle of water without removing the seal. When 97.8 atom % ^{18}O water was used, the ampule containing the water was cracked open just before use, and afterwards, immediately sealed with a septum. Working solutions, while not tested, were thought to be at their natural pH (10.4) [13]. The 3-APTHS solution was puddled on the freshly plasma-treated samples for 30 s. The samples were spun at 2500–3000 rpm for 30 s, and then immediately introduced into the vacuum system. The working solution was then discarded. 3-APTHS films were exposed to air for a maximum of 10 min prior to introduction into UHV. Once in the vacuum, positive and negative SSIMS spectra were obtained for each surface. The samples were then raised to 100°C < T < 120°C for 1 h in the vacuum. The heat was turned off, and positive and negative SSIMS spectra were then re-acquired from both sections of the sample. Subsequent to the SSIMS analysis, XPS spectra were obtained from the same samples. Samples were at atmosphere for 1–2 days between the SSIMS analysis and XPS studies.

The XPS spectra were recorded on a Surface Science Laboratories small spot system using a monochromatized AlK_a X-ray radiation source. The take-off angle used for these measurements was 35°. Full details of the methods used in interpreting the XPS data have been described elsewhere [14]. Data reduction was done using Surface Science Laboratories software version 8.0. This software utilizes a least squares curve fitting approach with only chi square statistics for goodness of the calculated fit to the experimental data.

Fourier Transform Infrared Spectroscopy (FT-IR) measurements were made using a Nicolet Instruments 740 FT-IR spectrometer. A horizontal attenuated total reflectance cell equipped with a 45° zinc–selenide crystal trough was used. Spectra of neat solutions were obtained by co-addition of 256 scans at 4 cm^{-1} resolution.

3. RESULTS

3.1. SSIMS

Figure 2 shows positive SSIMS spectra obtained from the Si and Cr surfaces before heating of the substrate in vacuum. Table 1 gives the chemical formulae for the positive ions identified in the figure.

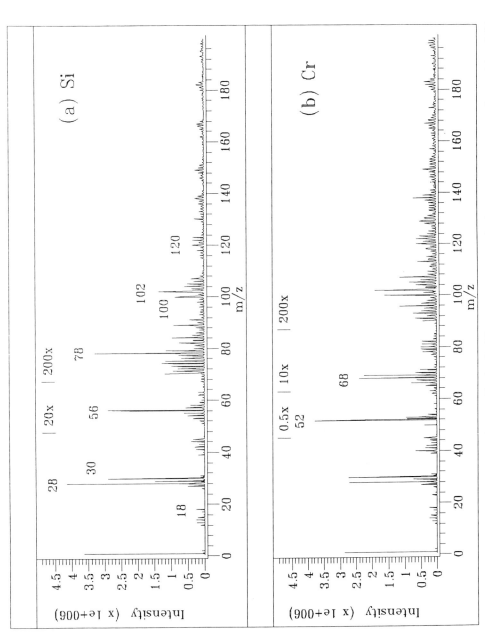

Figure 2. Positive SSIMS spectra for 3-APTHS films studied: (a) Si; (b) Cr. The spectra were acquired at room temperature. Curves (a) and (b) are plotted on the same relative intensity scale. The absolute intensity scale for (a) and (b) is arbitrary.

Table 1.
Chemical formulae and peak fits for the positive ions identified in Figs 2(a) and 2(b). The intense Cr^+ peak shows a shift to lower exact mass characteristic of multiple ions arriving at the detector, and so no exact mass fit was attempted

Surface	m/z	Formula	Fit	Exact	Error
Si and Cr	18	NH_4^+	18.044	18.034	+0.010
	28	Si^+	27.973	27.977	−0.004
	30	$CH_2NH_2^+$	30.031	30.034	−0.003
	45	$SiOH$	44.978	44.980	−0.002
	56	$CH_2(CH)_2NH_2^+$	56.043	56.050	−0.007
	78	$Si(OH)_2NH_2^+$	78.003	78.001	+0.002
	100	$SiOCH_2(CH)_2NH_2^+$	100.05	100.02	+0.03
	102	$SiO(CH_2)_3NH_2^+$	102.06	102.03	+0.03
	120	$Si(OH)_2(CH_2)_3NH_2^+$	120.06	120.05	+0.01
Cr	52	Cr^+	(See table caption)		
	68	CrO^+	67.941	67.934	+0.007
	69	$CrOH^+$	68.933	68.942	−0.009

As stated in Section 2, the ions in Tables 1 and 2 are identified by the exact mass fits and isotopic shifts for labelled species. For example, the peak at $m/z = 56$ has an exact mass fit of 56.0489, using easily identifiable peaks below $m/z = 56$, such as H_2^+, C^+, CH^+, C_2^+, etc. for calibration. Possible chemical formulae for this ion would be $C_3H_6N^+$, $m/z = 56.0491$, $C_2H_4N_2^+$, $m/z = 56.0365$, and $C_4H_8^+$, $m/z = 56.0617$. While the last two formulae lie just outside the range of 10 mmu, the first is the best possibility based on the structure of 3-APTHS. Indeed, when 99 atom % ^{15}N labelled 3-APTHS was used, the peak appeared at $m/z = 57$, confirming the presence of exactly one nitrogen atom in the ion. Similarly, the peak at $m/z = 45$ has an exact mass fit of 44.9847, and is seen to split into two peaks at $m/z = 45$ and 47 when 97.8 atom % ^{18}O labelled 3-APTHS was used. For this peak, there are not many possibilities; however, the exact mass and labelling experiments show unambiguously that the peak is principally composed of $SiOH^+$, $m/z = 44.9787$. The reason for the splitting of this peak, rather than a complete shift to $m/z = 47$, is discussed below. Similar reasoning was employed in identifying the other ions in Tables 1 and 2. The ions visible in Fig. 1 at $m/z = 18$, 30 and 56 are formed by fragmentation of the aminopropyl chain of the molecule. A rather surprising result is that the positive SSIMS spectrum of the Cr surface, shown in Fig. 2(b), is essentially identical to that of the Si surface, shown in Fig. 2(a), with most of the additional ions seen consistent with a Cr native oxide [6]. While the absolute y intensity scale of Fig. 2 is arbitrary, the spectra of Fig. 2 are shown on the same relative scale.

Figure 3 shows the negative SSIMS spectra obtained from Si and Cr surfaces at room temperature. Table 2 shows the chemical formulae for the peaks identified in Fig. 3. The negative ion spectra from these surfaces are again found to be very similar. Comparison of Figs 3(a) and (b) shows that the Cr and Si surfaces yield similar 3-APTHS related ions with roughly comparable intensities. Negative ions consistent with a chromium native oxide were also observed for the Cr surface. Parent ions due to 3-APTHS were absent from both the positive and negative SSIMS spectra. No dimer, trimer, or higher molecular weight

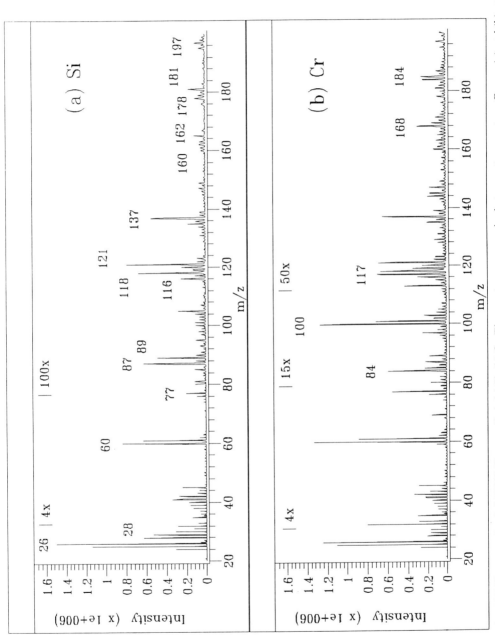

Figure 3. Negative SSIMS spectra for 3-APTHS films studied: (a) Si, (b) Cr. The spectra were acquired at room temperature. Curves (a) and (b) are plotted on the same relative intensity scale. The absolute intensity scale is arbitrary.

Table 2.
Chemical formulae for the negative ions identified in Figs 3(a) and 3(b)

Surface	m/z	Formula	Fit	Exact	Error
Si and Cr	26	CN^-	26.008	26.004	$+0.004$
	28	Si	27.967	27.978	-0.011
	60	SiO_2^-	59.961	59.966	-0.005
	61	SiO_2H^-	60.976	60.975	-0.001
	76	SiO_3^-	75.969	75.962	$+0.007$
	77	SiO_3H^-	76.974	76.970	$+0.004$
	87	$SiO_2C_2H_3^-$	86.995	86.990	$+0.005$
	89	$SiO_2C_2H_5^-$	89.013	89.006	$+0.007$
	116	$SiO_2CH_2(CH)_2NH_2^-$	115.986	116.016	-0.029
	118	$SiO_2(CH_2)_3NH_2^-$	118.047	118.032	$+0.015$
	121	$Si_2O_4H^-$	120.944	120.941	$+0.003$
	137	$Si_2O_5H^-$	136.931	136.936	-0.005
	160	$Si_2O_3CH_2(CH)_2NH_2^-$	159.98	159.99	-0.01
	162	$Si_2O_3(CH_2)_3NH_2^-$	162.01	162.00	$+0.01$
	178	$Si_2O_4(CH_2)_3NH_2^-$	177.97	178.00	-0.03
	197	$Si_3O_7H^-$	196.92	196.90	$+0.02$
Cr	84	CrO_2^-	83.926	83.930	-0.004
	100	CrO_3^-	99.925	99.925	$+0.000$
	117	CrO_4H^-	116.931	116.927	$+0.004$
	168	$Cr_2O_4^-$	167.854	167.859	-0.005
	184	$Cr_2O_5^-$	183.860	183.854	$+0.006$

oligomer species were detected as either positive or negative secondary ions from either surface.

Table 3 and 4 summarize the effects of temperature and substrate composition on the integrated intensity of the ions identified in Tables 1 and 2. Ions at higher m/z of these tables which are not shown in Tables 3 and 4 do not possess sufficient intensity to render the ratios meaningful. Tables 3 and 4 are representative of a single sample in this study. Repeated trials with a number of samples did not demonstrate repeatable variations with sample temperature in either the positive or negative SSIMS spectra from either surface. The trend for lower intensity of the $Si_xO_yR^-$ type ions shown in Table 4 for Si was observed consistently from sample to sample; however, the average value of the ratio of observed intensities for these ions for Si/Cr varied between 0.3 and 0.5. Comparable intensity for both substrates of ions with $m/z = +18$, $+30$, $+56$ and -26, was observed for all samples. It was also observed that the integrated intensity of both Si^+ and Si^- ions was higher for the Si surface, although the extent of enhancement was greater for the Si^- species, as is shown in Tables 3 and 4.

The insensitivity of the spectra to sample heating, and in particular the insensitivity of the $Si_xO_yR^-$ species, suggests that the films as prepared are not subject to further cross-linking. To investigate this possibility, 3-APTES was hydrolyzed in 97.8 atom % ^{18}O labelled water, with the result that the hydroxyl groups of the molecule are fully substituted with ^{18}O [15]. This was verified using FT-IR spectroscopy; the results of the FT-IR experiments are shown in Fig. 4. All the spectra of Fig. 4 were produced by subtraction of reference spectra of $H_2^{16}O$, or where appropriate, $H_2^{18}O$, from the raw data.

Table 3.
Summary of the effect of temperature and substrate composition on the observed intensity of positive secondary ions identified in Table 1

m/z	Formula	100°C/RT		Si/Cr	
		Si	Cr	RT	100°C
18	NH_4^+	0.96 ± 0.06	0.90 ± 0.06	1.00 ± 0.06	1.06 ± 0.06
28	Si^+	0.99 ± 0.01	1.07 ± 0.01	1.56 ± 0.02	1.45 ± 0.02
30	$CH_2NH_2^+$	1.07 ± 0.02	1.02 ± 0.02	1.11 ± 0.02	1.17 ± 0.02
45	$SiOH^+$	0.77 ± 0.05	0.91 ± 0.05	0.80 ± 0.05	0.68 ± 0.06
56	$CH_2(CH)_2NH_2^+$	0.93 ± 0.07	0.90 ± 0.07	1.00 ± 0.07	1.02 ± 0.07
78	$Si(OH)_2NH_2^+$	0.68 ± 0.20	0.78 ± 0.20	0.56 ± 0.20	0.49 ± 0.20
100	$SiOCH_2(CH)_2NH_2^+$	0.88 ± 0.38	0.64 ± 0.27	0.57 ± 0.26	0.79 ± 0.35
102	$SiO(CH_2)_3NH_2^+$	0.70 ± 0.29	0.62 ± 0.24	0.67 ± 0.25	0.76 ± 0.32
120	$Si(OH)_2(CH_2)_3NH_2^+$	0.60 ± 0.50	0.80 ± 0.40	0.40 ± 0.30	0.30 ± 0.20

Table 4.
Summary of the effect of temperature and substrate composition on the observed intensity of negative secondary ions identified in Table 2

m/z	Formula	100°C/RT		Si/Cr	
		Si	Cr	RT	100°C
26	CN^-	1.06 ± 0.02	1.06 ± 0.02	1.18 ± 0.03	1.18 ± 0.03
28	Si^-	1.02 ± 0.03	0.94 ± 0.05	4.26 ± 0.18	4.65 ± 0.19
60	SiO_2^-	0.97 ± 0.04	0.99 ± 0.03	0.55 ± 0.02	0.54 ± 0.02
61	SiO_2H^-	0.90 ± 0.04	0.95 ± 0.03	0.59 ± 0.02	0.55 ± 0.02
76	SiO_3^-	0.96 ± 0.11	1.04 ± 0.08	0.40 ± 0.04	0.40 ± 0.04
77	SiO_3H^-	0.86 ± 0.07	0.93 ± 0.04	0.27 ± 0.02	0.26 ± 0.02
87	$SiO_2C_2H_3^-$	1.17 ± 0.30	0.98 ± 0.13	0.27 ± 0.05	0.31 ± 0.06
89	$SiO_2C_2H_5^-$	1.00 ± 0.28	1.10 ± 0.15	0.29 ± 0.07	0.29 ± 0.06
116	$SiO_2CH_2(CH)_2NH_2^-$	0.75 ± 0.28	0.85 ± 0.15	0.28 ± 0.07	0.29 ± 0.07
118	$SiO_2(CH_2)_3NH_2^-$	0.71 ± 0.14	0.83 ± 0.10	0.42 ± 0.06	0.36 ± 0.06
121	$Si_2O_4H^-$	0.85 ± 0.16	0.93 ± 0.10	0.41 ± 0.06	0.38 ± 0.06
137	$Si_2O_5H^-$	0.85 ± 0.19	0.87 ± 0.11	0.36 ± 0.06	0.35 ± 0.06

Figure 4(a) shows the spectrum of a 1 vol % solution of ^{16}O ethanol in water. The peak at 1088 wavenumber has been attributed to the C—O stretching vibration, and the peak at 880 wavenumber is attributable to out-of-plane bending of the hydrocarbon backbone of the ethanol molecule [16]. The strong peak at 1050 wavenumber is attributed to the C—OH deformation mode of ethanol [16]. Figure 4(b) is the FT-IR spectrum of a 1 vol % solution of 3-APTHS in $H_2^{16}O$. Figure 4(c) is the result of subtracting Fig. 4(a) from Fig. 4(b) using nulling of the ethanol related features as the criterion for subtraction. This process reveals a broad feature at about 1090 cm^{-1}, and a sharper feature at 1000 cm^{-1}. These features are consistent with those attributed to the Si—O portion of dimethylsilanediol [17], and of alkoxysilanes [18]. Both bands are thus attributed to the Si—O portion of 3-APTHS. Figure 4(c) confirms that the peaks attributable to ethanol in Fig. 4(a) may be removed completely by subtraction from a combined solution of 3-APTHS and ethanol. Figure 4(d) is the FT-IR spectrum of a 1 vol % solution of 3-APTHS in $H_2^{18}O$. Figure 4(e) is the result of subtracting Fig. 4(a) from Fig. 4(d), again using nulling of the

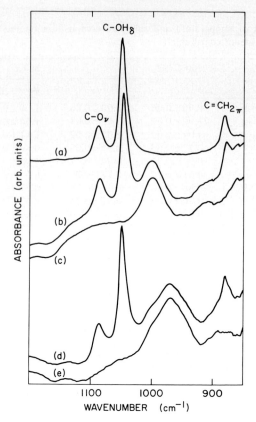

Figure 4. FT–IR absorption spectra for: (a) 1 vol % solution of ^{16}O ethanol in water; (b) 1 vol % solution of 3-APTHS in $H_2{}^{16}$O water; (c) result of subtracting (a) from (b); (d) 1 vol % solution of 3-APTHS in $H_2{}^{18}$O water; and (e) result of subtracting (a) from (d).

ethanol-related features as the criterion for subtraction. Two results are apparent from Fig. 4(e). First, all the features of Fig. 4(a) are completely removed by subtraction, indicating that no isotopic shift in the position of the C—OH deformation mode of ethanol at 1050 wavenumber or the C—O stretching vibration at 1080 wavenumber is observed for the spectrum of Fig. 4(d). Second, there are sizable shifts associated with the silanol features of Fig. 4(c), indicating a strong isotopic shift for the 3-APTHS features due to Si—O structures. The data show unambiguously that only ^{16}O ethanol is formed as a result of the hydrolysis reaction, and that consequently, the 3-APTHS formed during this reaction is fully labelled with ^{18}O. The shoulder at 1000 cm^{-1} visible in Fig. 4(e) is probably due to subsequent reaction of the fully labelled 3-APTHS with the IR cell.

Labelling the 3-APTHS molecule with ^{18}OH groups provides an isotopic tag at the site on the molecule where cross-linking should take place. 3-APTHS hydrolyzed in 97.8 at % ^{18}O labelled water was spun onto sample substrates in the manner described above and SSIMS spectra were acquired both at room temperature and after heating of the sample to 100°C in vacuum for 1 h. The relative intensity of the oxygen containing ions produced by these samples can be used to determine the relative abundance of ^{16}O and ^{18}O in the ions observed. Figure 5 shows this for the Si and Cr surfaces, highlighting the portions of the negative SSIMS spectra produced by these samples containing the SiO_2^- ions. The peaks at $m/z = 60$, 62 and 64 in Fig. 5(b) are predominantly due to SiO_2^- ions containing 0, 1 and 2 ^{18}O atoms, respectively. A similar effect is observed for the higher mass $Si_xO_yR^-$ ions shown in Table 2; however, peak interferences due to poor signal levels make determination of ^{16}O and ^{18}O abundances difficult for these species. Table 5 shows the ratio of ^{16}O to ^{18}O in these ions both before and after heating for both the Cr and Si surfaces. Also shown is the ratio of $^{16}O^-$ to $^{18}O^-$ ions obtained from these surfaces. The ratio of ^{16}O to ^{18}O in the SiO_2^- ions and the overall ratio of ^{16}O to ^{18}O at the surface, as given by the ratio of $^{16}O^-$ to $^{18}O^-$ ions, is an indication that the SiO_2^- ions convey information about the local chemistry of the silanol end of the 3-APTHS molecule at the surface.

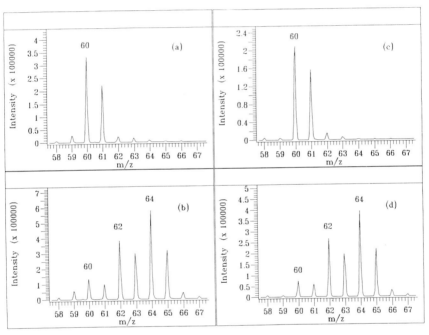

Figure 5. Comparison of SiO_2^- ion intensities for: (a) $Cr + ^{16}O$ 3-APTHS; (b) $Cr + ^{18}O$ 3-APTHS; (c) $Si + ^{16}O$ 3-APTHS; and (d) $Si + ^{18}O$ 3-APTHS. Peaks at $m/z = 60$, 62 and 64 correspond to SiO_2^- ions containing 0, 1 and 2 ^{18}O atoms, respectively. The spectra were acquired at room temperature. Curves (a) and (c) are plotted on the same relative intensity scale. Curves (b) and (d) are plotted on the same relative intensity scale. The absolute intensity scale for the figures is arbitrary. Similar results were obtained at 100°C and are summarized in Table 5.

As shown in Table 5 heating of the sample in vacuum produces almost no change in the ratio of ^{16}O to ^{18}O in the SiO_2^- ions for either the Cr or Si surface. Cross-linking as detailed in Fig. 1(c) should lead to selective elimination of ^{18}O from the silanol network, with a resultant shift in the ratio of ^{16}O to ^{18}O in the SiO_2^- ions. The fact that this effect was not observed is strong evidence that the films as prepared do not undergo additional cross-linking upon sample heating.

Table 5.
Summary of observed $^{16}O:^{18}O$ ratios in SiO_2^- and O^- ions from 3-APTHS films on Si and Cr

		SiO_2^- $^{16}O:^{18}O$	O^- $^{16}O:^{18}O$
Si	Room temp.	0.28:0.72	0.44:0.56
	100°C	0.29:0.71	0.44:0.56
Cr	Room temp.	0.30:0.70	0.56:0.44
	100°C	0.30:0.70	0.56:0.44

3.2. XPS

XPS analysis results for 3-APTHS on H_2O-plasma treated Si and Cr surfaces are summarized in Tables 6 and 7. Figure 6 shows the comparison of typical N 1s XPS data for both surfaces.

Based on the elemental composition of both surfaces analyzed, it appears that the 3-APTHS coverage is comparable. This is particularly clear from the N atom % concentration, which is consistently 6–7 atom % independent of the surface. Table 6, which shows the oxidation state analysis of the surfaces studied, indicates that some minor differences do exist between the two surfaces. There are two N 1s peaks on Si, while Cr exhibits three peaks in this binding energy

Table 6.
Elemental composition of 3-APTHS on H_2O-plasma treated Si and Cr surfaces

Surface	Thermal exposure °C	C %	O %	N %	Si %	Cr %
Si	100	30	38	7	25	—
	100	21	45	6	27	—
	120	34	33	7	26	—
Cr	100	32	43	7	6	11
	100	26	45	7	7	16
	120	33	43	6	6	12

Table 7.
Elemental oxidation state analysis (%) results for 3-APTHS on H_2O-plasma treated Si and Cr surfaces

Surface/°C	C 1s 285.0	286.1	288.6	O 1s 532.5	531.4	401.1	N 1s 399.3		Si 2p 102.3
Si/100	59	33	8	83	17	16	84		14
Si/100	61	33	6	86	14	18	82		11
Si/120	67	27	6	91	9	23	77		13
	285.0	286.2	288.5	532.1	530.4	401.2	399.6	397.5	102.6
Cr/100	64	30	6	51	49	25	72	4	100
Cr/100	61	31	8	47	53	20	76	4	100
Cr/120	80	13	7	46	54	21	73	6	100

BE repeatability between ± 0.1 to 0.5 eV for a given peak.

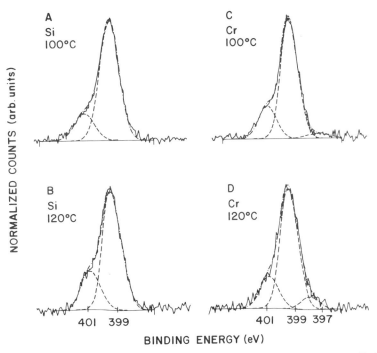

Figure 6. N 1*s* XPS data for 3-APTHS on H$_2$O-plasma treated Si and Cr surfaces. (a) Si, 100°C; (b) Si, 120°C; (c) Cr, 100°C and (d) Cr, 120°C.

(BE) region. The ΔBE between the two N 1*s* peaks on the 3-APTHS treated Si surface, and the ΔBE between the two higher BE N 1*s* peaks on the treated Cr surface, are between 1.6–1.8 eV. These values fall within the reported ΔBE for 3-APTHS N 1*s* peaks on a number of different metal surfaces [4, 19]. The peak at 399.3–399.6 eV may be assigned to a free amine (—NH$_2$), and the peak at 401.1–401.2 eV to a protonated amine (—NH$_3^+$). The lower BE N 1*s* peak observed on the 3-APTHS treated Cr surface at 397.5 eV may be assigned to a Cr-nitride type of N 1*s* electrons [20, 21].

4. DISCUSSION

The SSIMS spectra of the Cr surface show a sizable signal from the Cr substrate. Since the sampling depth of SSIMS is ~ 10 Å, we conclude that the films under study are either near monolayer and uniform, or patchy, so as to allow the Cr substrate signal to be visible by SSIMS. Tables 3 and 4 show that both Si$^+$ and Si$^-$ ions are observed with greater intensity from the Si surface. This probably indicates that the Si substrate contributes to the observed intensity for these species. This conclusion is supported by the somewhat surprising result that the overall yield of other Si bearing positive and negative ions is smaller for the Si surface relative to Cr. A simple increase in total ion yields for Si over Cr thus cannot account for this behavior. XPS shows that the 3-APTHS coverage on both surfaces is comparable. Detection of substrate ions from both surfaces is thus expected, provided the morphology of the surface layer is similar for both substrates. The higher intrinsic secondary ion yield for Cr^+ species relative to Si^+ in oxide matrices [22] may be why the former is easily seen in the spectra of Figs 2 and 3, while the latter is less obvious on first inspection. As noted previously, peaks at $m/z = +18$, $+30$, $+56$ and -26 are observed with roughly comparable intensity from both surfaces. XPS information showing comparable surface coverage allows us to conclude that for films prepared in the manner described, the relative intensity of these species is an indicator of the relative coverage of 3-APTHS for the Si and Cr substrates. The insensitivity of these species to the substrate composition suggests that the aminopropyl end of the molecule is not adjacent to the substrate surface, and thus, no substrate effect on secondary ion formation behavior is observed. Conversely, the apparent sensitivity of the Si_xO_yR type ions to substrate composition may indicate that these ions contain structural elements which are initially at or near the overlayer/ substrate interface.

The most plausible explanation for the similarity in the observed ratio of ^{16}O to ^{18}O in the SiO_2^- ions for both Si and Cr is that these species arise entirely from the 3-APTHS overlayer for both surfaces. This indicates that the 3-APTHS overlayer, and by implication, the interaction between the molecule and the surface, is highly similar for these two materials. The appearance of ions containing both ^{16}O and ^{18}O in the SSIMS spectra of Figs 5(b) and 5(d), is consistent with the reaction scheme shown in Figs 1(a) to 1(c). In the case of Si, it is reasonable to assume that the condensation reaction of Fig. 1(a) will proceed randomly with respect to the hydroxyl groups involved. That is, there is a 50:50 chance of incorporating either the 3-APTHS oxygen atom or the surface oxygen atom in the final network. Following this assumption, Fig. 7 shows a simple model to predict the expected $^{16}O:^{18}O$ ratio for the configurations of Fig. 1. The final stoichiometry for attachment via condensation falls between the two extremes shown in Fig. 7(a). Figures 7(b) and 7(c) show that for the donation and protonation attachment schemes, no mixed ions containing ^{16}O would be expected. The observed ratios of ^{16}O to ^{18}O for both Si and Cr, shown in Table 5 fall nicely between the two extremes of Fig. 7(a). While this does not rule out the existence of some fraction of the molecules being attached via the mechanisms shown in Figs 1(d) and 1(e), it indicates that these mechanisms are not predominant on either the Si or Cr surface. Inherent in the model is the assumption that the SiO_2^- secondary ions observed do not represent the result of recombination processes involving atoms which are initially widely spaced within the solid matrix. While the processes involved in molecular ion formation are complex, previous work has suggested that a fragment model [23–25], in which the SiO_2^- molecular ions are present initially as a localized fragment at the surface of the solid, should be operative in this case.

The XPS data provide information complementary to that provided by SSIMS. The XPS data suggest that pure 3-APTHS applied to H_2O-plasma treated Si and Cr surfaces from freshly prepared 0.1 vol % solution in ultrapure water binds to these surfaces in a similar fashion with the primary nitrogen component in the form of a free amine (72–84 atom %). This agrees with the SSIMS results, which suggest that the majority of molecules are bonded to the surface via the silanol group, leaving the amine end free and unreacted. The protonated nitrogen species is a minor component on both surfaces (16–23 atom %). The nitride

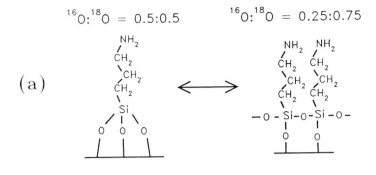

Figure 7. Simple model for predicting the $^{16}O:^{18}O$ ratio in observed secondary ions as a function of bonding configuration. (a) Condensation allows for a range of abundances depending on the number of bonds between the silane molecule and the surface. Electron donation (b), and protonation (c), both predict that no mixing of $^{16}O:^{18}O$ should be observed.

species is found only on the Cr surface and at concentrations of 4–6 atom % of the total nitrogen.

From the XPS data it is not clear if the protonated amine is due to a bicarbonate salt (reaction with moist ambient CO_2), as suggested by some authors [26, 27], or perhaps due to protonation achieved via surface hydroxyl-groups (Fig. 1(e)). A bicarbonate, if present, represents a rather favorable 'preformed ion' and should lead to a sizable CO_3^- or HCO_3^- ion signal in the negative SSIMS spectrum [25]. While this peak would interfere with the SiO_2^- ion at $m/z = 60$, or the SiO_2H^- ion at $m/z = 61$, the lack of any sensitivity to surface temperature for these peaks for any of the samples studied, as well as excellent exact mass fits for both peaks, appears to indicate that a bicarbonate is not present.

The highly similar results obtained by both SSIMS and XPS suggest that 3-APTHS applied to H_2O-plasma treated surfaces in the manner described above binds to the Si and Cr surfaces in a similar fashion. In repeated trials, both the surface coverage of 3-APTHS, as given by XPS, and the configuration of the molecules at the surface, as inferred from XPS and SSIMS, were found to be reproducible. Our interpretation of the information provided by these two techniques is that the strong mixing of the ^{16}O and ^{18}O isotopes in the SSIMS spectra, indicates that the 72–84% of the nitrogen detected as a free amine is due to 3-APTHS bonded in the configuration of Fig. 7(a). The much smaller fraction detected as a protonated species by XPS is probably in the configuration of Fig. 7(c). The picture is thus one of a mixed phase of 'amine up' and 'amine down' molecules at the surface, with the dominant configuration being that with the amine group up and available for further reaction. The excellent adhesion which has been observed between Si surfaces primed with 3-APTHS and subsequent polymer overlayers is further support for this picture [28]. The repeatable size of the fraction of 'amine down' species relative to free amine may be due to a trapping mechanism which freezes a reproducible portion of the surface population in this 'amine down' configuration during the spinning step. The coordination of the amine nitrogen lone pair of electrons to metal atom in the Cr case is unlikely. The ΔBE between coordinated amine N $1s$ peak and free amine N $1s$ peak is about $+2$ eV [29]. This would result in a N $1s$ peak overlapping the protonated amine, thus making it impossible to separate the two. It is known, however, that Cr does not preferably form coordination complexes with amines [30].

5. SUMMARY

The following statements can be made based on the SSIMS and XPS results:
(1) The coverage of 3-APTHS is comparable on Si and Cr surfaces.
(2) The 3-APTHS interacts in a similar fashion with the surfaces studied. The only observed difference is a small inorganic nitrogen component on the Cr surface which is tentatively assigned to Cr—N structures.
(3) No further cross-linking of the 3-APTHS is observed as a function of thermal exposure to temperatures up to 120°C.
(4) The primary bonding mode of 3-APTHS to the Si and Cr surfaces is via the silanol-end of the molecule, leaving the amine end free. A smaller nitrogen component at about 401 eV is associated with protonated amine species. It is proposed that this feature is due to interaction with surface hydroxyl groups. The results indicate that this feature is not due to interaction of the surface layer with moist ambient CO_2.

Acknowledgements

The authors would like to thank Dr. P. Murphy for the use of isotopically labelled materials and Dr. R. Thomas for help in plasma-treating samples for this study.

REFERENCES

1. H. Ishida, in: *Adhesion Aspects of Polymeric Coatings*, K. L. Mital (Ed.), p. 45. Plenum Press, New York (1983).
2. E. P. Plueddemann, *Silane Coupling Agents.* Plenum Press, New York (1982).
3. L. P. Buchwalter, *J. Adhesion Sci. Technol.* **4**, 697 (1990).
4. M. R. Horner, F. J. Boerio and H. Clearfield, *J. Adhesion Sci. Technol.* **6**, 1 (1992).
5. L. P. Buchwalter, T. S. Oh and J. Kim, *J. Adhesion Sci. Technol.* **5**, 333 (1991).
6. D. Briggs, A. Brown and J. C. Vickerman, in: *Handbook of Static Secondary Ion Mass Spectrometry (SIMS).* John Wiley and Sons (1989).
7. H. Linde and R. T. Gleason, *J. Polym. Sci. Polym. Chem. Ed.* **22**, 3043 (1984).
8. F. J. Boerio, *Annual Report, Office of Naval Research, Arlington, VA* **NR-4313-202**, 1 (1986).
9. G. A. Parks, *Chem. Rev.* **52**, 177 (1965).
10. M. Gettings and A. J. Kinloch, *J. Mater. Sci.* **12**, 2511 (1977).

11. B. N. Eldridge, *Rev. Sci. Instrum.* **60**, 3160 (1989).
12. *MSD Isotopes*, Div. of Merck Frosst, Inc., Montreal, Quebec, Canada.
13. F. J. Boerio and D. J. Ondrus, in: *Surface and Colloid Science in Computer Technology*, K. L. Mittal (Ed.), p. 155, Plenum Press, New York (1987).
14. L. P. Buchwalter, B. D. Silverman, L. Witt, and A. R. Rossi, *J. Vac. Sci. Technol.* **A5**, 226 (1987).
15. A. R. Bassindale and P. B. Taylor, *The Chemistry of Organic Silicon Compounds*, Part 1, p. 887. John Wiley, New York (1989).
16. L. J. Bellamy, in: *The Infrared Spectra of Complex Molecules*. John Wiley, New York (1959).
17. S. W. Kantor, *J. Amer. Chem. Soc.* **75**, 2712 (1953).
18. R. E. Richards and H. W. Thompson, *J. Chem. Soc.* 124 (1949).
19. P. R. Moses, L. M. Weir, J. C. Lennox, H. O. Finklea, J. R. Lenhard and R. W. Murray, *Anal. Chem.* **50**, 576 (1978).
20. C. D. Wagner, W. M. Riggs, L. E. Davis, J. F. Moulder and G. C. Muilenberg, in: *Handbook of X-ray Photoelectron Spectroscopy*, G. C. Muilenberg (Ed.), p. 72. Perkin-Elmer Corporation, Eden Prairie, MN (1979).
21. M. J. Goldberg, J. G. Clabes and C. A. Kovac, *J. Vac. Sci. Technol.* **A6**, 991 (1988).
22. J. A. McHugh, in: *Methods of Surface Analysis*, A. W. Czanderna (Ed.), p. 233. Elsevier Scientific Publishing Company, New York (1975).
23. A. Benninghoven, *Z. Naturforsch* **24a**, 859 (1969).
24. A. Benninghoven, *Surface Sci.* **35**, 427 (1973).
25. A. Benninghoven, in: *Secondary Ion Mass Spectrometry SIMS II*, A. Benninghoven, C. A. Evans, R. A. Powell, R. Shimizu and H. A. Storms (Eds), p. 116. Springer-Verlag (1979).
26. F. J. Boerio and J. W. Williams, *Proc. 35th Ann. Tech. Conf., Reinforced Plastics/Composites Div.* SPI, Section 2-F (1981).
27. S. Naviroj, J. L. Koenig and H. Ishida, *Proc. 37th Ann. Tech. Conf., Reinforced Plastics/Composites Div.* SPI, Section 2-C (1982).
28. H. R. Anderson, Jr., M. M. Khojasteh, T. P. McAndrew and K. G. Sachdev, *IEEE Trans. CHMT* **9**, 364 (1986).
29. J. R. Lusty and J. Peeling, *Arabian J. Sci. Eng.* **7**, 3 (1982).
30. J. R. Lusty and J. Peeling, *Inorg. Chem.* **24**, 1179 (1985).

Silanes and Other Coupling Agents, pp. 323–343
Ed. K. L. Mittal
© VSP 1992.

Characterization of films of organofunctional silanes by TOFSIMS and XPS. Part I. Films of N-[2-(vinylbenzylamino)-ethyl]-3-aminopropyltri-methoxysilane on zinc and γ-aminopropyltriethoxysilane on steel substrates

W. J. VAN OOIJ* and A. SABATA

Armco Research and Technology, 705 Curtis Street, Middletown, OH 45043, USA

Revised version received 11 June 1991

Abstract—The structures of thin films formed by the silanes N-[2-(vinylbenzylamino)-ethyl]-3-amino-propyltrimethoxysilane (SAAPS) and γ-aminopropyltriethoxysilane (γ-APS) deposited onto mechanically polished zinc or mild steel from dilute aqueous solutions were determined using time-of-flight (TOF) SIMS and XPS. TOFSIMS gave structural information which was highly complementary to the XPS data. Aspects such as silane condensation and crosslinking, oxidation at elevated temperatures, the formation of metallosiloxane bonds, and incomplete hydrolysis were detected by TOFSIMS by virtue of its high mass resolution and unlimited mass range. The structures of the films were found to be strongly dependent on the nature of the substrate, the deposition conditions, and heat treatment of the films.

Keywords: Silanes; time-of-flight SIMS; XPS; SAAPS; γ-APS; steel; zinc.

1. INTRODUCTION

Many different silane coupling agents are available for improving the adhesion between polymers and metals or fillers and fibers [1]. Although organofunctional silanes have been reported to be very effective in such adhesion applications, the exact mechanism by which they function has not been determined conclusively in certain cases. Several techniques have been used to characterize silane films formed on various substrates, such as glass, aluminum, and steel. Some of these methods are XPS [2, 3], FTIR [4], IETS [5, 6], NMR [7], and Raman spectroscopy [8].

Static SIMS would be particularly appropriate to study films formed by silane coupling agents by virtue of its high surface sensitivity, its capability to detect all elements and isotopes, and because it provides molecular structural information on organic materials [9–12]. So far, SIMS has rarely been used to study silane films. It was used, however, to detect covalent bonds between iron or chromium and silane films deposited on stainless steel [13, 14]. Such bonds were indeed reported; this claim was based on the presence of ions in the spectrum which were assumed to be $FeOSi^+$ and $CrOSi^+$. However, these SIMS analyses were not done under static conditions, which had not yet been established at the time of these studies.

---

*To whom correspondence should be addressed.

With the advent of the newer generation of time-of-flight mass spectrometers, the SIMS technique should allow detailed fingerprinting of silane films because of its much higher mass resolution and mass range. The former aspect enables a more accurate and hence unambiguous peak identification, and the latter is very useful for the detection of crosslinking and condensation reactions. Further time-of-flight spectrometers are more sensitive than quadrupole instruments, which enables spectra to be acquired at a much reduced total ion dose, thus avoiding ion beam damage.

In this series of papers we will report on the use of TOFSIMS for the characterization of films of a number of commonly used silanes deposited on various substrates, mainly metals. The background of this interest in metals is the possible application of silanes as corrosion-inhibiting metal pretreatments.

It is known that the molecular structure of silane films is dependent on the deposition conditions [15]. Therefore, it is one of our objectives to demonstrate that changes in these conditions can be detected qualitatively in the SIMS spectra and that films of the same silane but deposited on different substrates may have different structures, as has been reported [16, 17]. Another objective of this study is to demonstrate that the fingerprinting capability of SIMS can be used to distinguish between different silanes.

The silane films were prepared from aqueous solutions of SAAPS and γ-APS and the variables were the pH of the solution, the type of cleaning process of the substrate, and the effect of heating the films after deposition. Some of the films were also analyzed by XPS for comparison.

In this paper we report on TOFSIMS and XPS analyses of thick and thin films of SAAPS on zinc in order to demonstrate the effect of film thickness. We also present some TOFSIMS results obtained with films of γ-APS on mild steel in which the effects of the cleaning process of the substrate and the pH of the silane solution are demonstrated.

2. EXPERIMENTAL

2.1. Materials

γ-APS was obtained from Aldrich Chemicals and SAAPS (Z6032) from Dow Corning Corporation. The substrates used were ASTM B69 standard grade of cold-rolled zinc (99.9%, type 303, Platt Brothers, Waterbury, CT) and cold-rolled, low carbon steel of automotive grade, manufactured by Armco ASC, LP (Middletown, OH).

2.2. Sample preparation

The metals were all cut to coupons of 2×2 cm^2 size and then polished mechanically to a mirror finish using silicon carbide papers followed by diamond paste polishing. The final finish was 1 μm. Prior to silane deposition, the coupons were cleaned ultrasonically in reagent-grade methanol for 2 min and then blown dry with nitrogen. In order to improve the wettability of the substrates by the silane solutions, some samples were further cleaned in a hot alkaline solution for 30 s at 50°C. An industrial cleaning bath (Parker 338® from Parker + Amchem, Madison Heights, MI) was used for this purpose. Following this step, the samples were rinsed with distilled water and immediately dipped into the freshly hydro-lyzed silane solution. The surfaces were analyzed by TOFSIMS before and after alkaline cleaning and immediately after the silane treatment. All substrates were treated with the silane solutions immediately after cleaning.

2.3. Deposition conditions

The silanes were hydrolyzed in water containing 0.1% acetic acid. The solutions (30 vol% for SAAPS and 75 vol% for γ-APS) were stirred until clear and then diluted with water to 1 vol% concentration. In certain cases, γ-APS was hydrolyzed in distilled water. The pH was then adjusted by adding HCl or NaOH.

The silane films were prepared by dipping the substrates in 1 vol% aqueous solutions for 30–60 s. Some were then blown dry with nitrogen and introduced into the SIMS instrument within 2 h. In general, relatively thick films were prepared by dipping for 60 s, followed by dripping for 60 s. The films were then dried in air and not blown dry. Thin films were prepared by dipping for 30 s after which they were blown dry immediately. Some films were heated for 30 min at 150°C in air prior to analysis.

2.4. Analysis of silane films

XPS analysis was carried out on a Kratos integrated imaging instrument (AXIS). Spectra were recorded at a 90° take-off angle using 300 W monochromatized Al K_α radiation. Deconvolution was done using the software available in the VISION work station. Gaussian peak shapes were assumed. Binding energies (BEs) were calibrated using a binding energy of 285.0 eV for the C 1s line.

TOFSIMS analyses were performed on a Kratos PRISM instrument. It was equipped with a reflectron-type time-of-flight mass analyzer and a pulsed 25 kV liquid metal ion source of monoisotopic $^{69}Ga^+$ ions with a minimum beam size of 500 Å. Positive and negative spectra were obtained at a primary energy of 25 keV, a pulse width of 10–50 ns, and a total integrated ion dose of about 10^{11} ions/cm². This is well below the generally accepted upper limit of 5×10^{12} ions/cm² for static SIMS conditions in the analysis of organic materials [12]. The mass resolution at mass 50 amu* varied from $M/\Delta M = 1000$ at 50 ns pulse width to about 2500 at 10 ns pulse width.

Surface charge neutralization by a pulsed electron flood was not required for any of the materials, either in the positive or in the negative ion mode. The spectra were carefully calibrated using the exact masses of peaks with known composition, such as those from the source, $^{69}Ga^+$ at 68.926 amu; the substrate, e.g. $^{56}Fe^+$ at 55.935 amu; and some siloxane peaks, e.g. $(CH_3)_3Si^+$ at 73.047 or $(CH_3)_5Si_2O^+$ at 147.066 amu.

The relative yields of certain ions from different samples were compared as follows. The intensities of all ions detected in the 0–400 mass range were normalized by dividing their yields by the total integrated positive or negative ion yield from the samples in the same mass range. All yields were expressed as peak areas rather than peak heights, since the widths were not constant for all masses. This method enables a direct comparison to be made of the yields of certain ions

---

*Amu is equivalent to *m/e*, *m/z*, or Daltons used in the literature.

from samples of similar nature, because all ion yields are of the same order of magnitude. It should be noted that this method is not recommended for comparison of totally different types of materials since total ion yields may vary by several orders of magnitude.

3. RESULTS AND DISCUSSION

3.1. XPS and TOFSIMS analyses of SAAPS films on zinc substrates

XPS analysis. In Fig. 1, wide and narrow scan XPS spectra are shown of a thick film of SAAPS deposited on zinc before and after heating at 150°C. The pH of the silane solution was 4.7. This was the pH of the solution after hydrolysis in acetic acid. Before hydrolysis, SAAPS ($M_w = 374.5$) has the following structure:

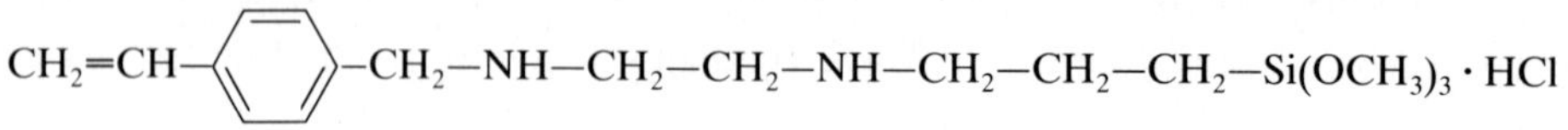

$$CH_2=CH-\langle\!\!\!\bigcirc\!\!\!\rangle-CH_2-NH-CH_2-CH_2-NH-CH_2-CH_2-CH_2-Si(OCH_3)_3 \cdot HCl$$

The elemental compositions calculated from the XPS data are given in Table 1 for the elements C, O, N, and Si. In addition, Cl and Zn were detected. Before heating, an appreciable concentration of C=O groups was present, suggesting partial oxidation of the vinyl groups. The concentration of C=O groups increased further upon heating. Two chemical states of oxygen were distinguished, which cannot easily be assigned because oxygen is bonded to silicon, to the oxidized carbon sites and, after heating, to zinc as well. Upon heating, the concentrations of Cl and Zn increased sharply, those of oxygen and nitrogen increased somewhat, and the total amount of carbon decreased. Two forms of nitrogen were detected, which may result from the partial conversion of amine groups to $-NH_2Cl-$ groups (Table 1). Before heating, no Zn was observed in the spectrum, which, considering the take-off angle of the analysis, indicates that the thickness of the silane film was at least 50Å.

Clearly, a detailed description of this silane film based on XPS data alone is not possible because of the many different structures that may be present and because

Table 1.

XPS data for films of SAAPS on zinc before and after heating[a]

Peak	BE (eV)[b]	Before (at%)	After (at%)
C—C, C—Si	285.0	50.65	45.55
C—O, C—N	286.4	17.68	17.99
C=O	288.1	2.60	4.23
O	531.0	2.87	1.99
O	532.4	12.88	14.42
N	399.7	6.67	7.54
N	401.6	2.55	2.70
Si	102.4	4.09	5.56
		100.00	100.00

[a] 30 min at 150°C in air.
[b] Average of BEs before and after heating.

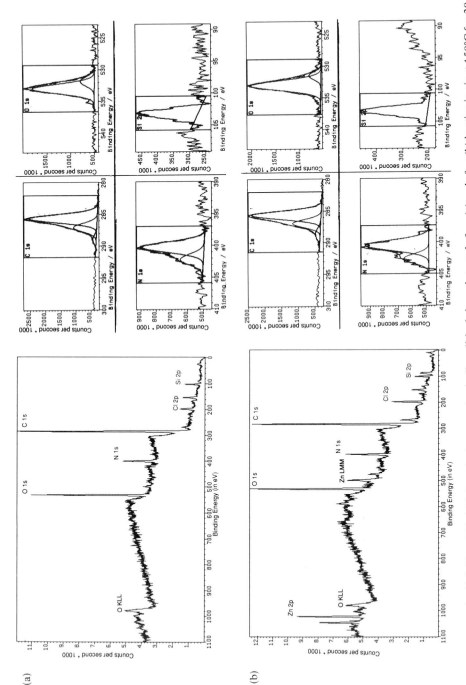

Figure 1. XPS spectra of a thick film of SAAPS deposited on a mechanically polished zinc substrate before (a) and after (b) heating in air at 150°C for 30 min. For conditions see text.

of the oxidation effects. Further, it is not clear whether the increase in zinc concentration upon heating is due to diffusion into the film as a result of an attack of the substrate [3], or whether a reduction of film thickness due to oxidation has occurred. Finally, the purity of the starting material was not more than 98%, the balance being other silanes. As a result, only some general statements can be made about the structure of the film and the reactions that occur during heating.

3.1.2. TOFSIMS analysis. The positive and negative TOFSIMS spectra of the same films are shown in Fig. 2. It should be noted that these were acquired in the low mass resolution mode, hence the peak identification is not completely unequivocal in some cases.

Several highly characteristic peaks are observed in Fig. 2. The compositions of all major peaks are given in Table 2. These are based on their exact mass determination, which is accurate to within 0.01 amu. This is possible in a TOFSIMS instrument because of its higher mass resolution as compared with quadrupole instruments. Therefore, peak identities can be determined with higher certainty than in quadrupole instruments which have only unit mass resolution.

Before heating, the peak with the highest intensity is that of the vinylbenzyl group (+ 117 amu). Other typical aromatic ions are at + 39, + 50/51, + 91, and + 115 amu. In addition, a series of peaks typical of siloxanes were observed, i.e. at + 73, − 75, + 147, + 207, + 221, and + 281 amu. The molecular ion M^+ or $(M \pm 1)^+$ of the completely hydrolyzed species ($M_w = 296$) was not detected, but the ion at +262 amu was identified as the monomer minus two —OH groups. The intense peak at +308 amu (Fig. 2a) and ions at +453 and +569 amu (Fig. 2b) were identified as arising from dimers and trimers. Their likely structures are given below. These structures suggest that polymerization and crosslinking have taken place via the vinyl groups.

Further evidence for these proposed structures is that the masses listed for these three ions are those with the highest intensities in clusters of several peaks (Fig. 2b). This observation indicates that ions with more or less unsaturation are also formed. For instance, ions at + 309, + 310, + 451, + 455, + 570, + 571, and + 572 were also detected in appreciable intensities. Their exact structures can be easily found from those above. The odd-numbered masses contain an even number of nitrogen atoms and vice versa.

The high yields of the ions + 28 and + 59 amu, as compared with siloxanes, indicate that a fraction of the silane is still in the monomeric form. The yields of these ions are lower in highly polymerized materials [9, 10]. The high yield of the ion at + 43 amu may already be indicative of some oxidation (see below). Peaks due to zinc (+ 64 amu and other isotopes) are completely absent in the spectra, indicating complete coverage of the substrate. Chloride ions, detected in high amounts by XPS, are not seen here. Typical fragments from secondary amines, i.e. at + 42, + 44, + 56, and + 58 amu [12], are not detected in high yields. The ion at + 121 amu, identified as $C_8H_9O^+$, may arise from oxidation of the vinylbenzyl group, in agreement with the XPS results.

The TOFSIMS results thus indicate that the surface of the silane film consists mainly of a random mixture of partly oxidized vinylbenzyl groups and the silane end of the molecules. The silanol groups have formed siloxane groups to some extent and some crosslinking of the molecules has occurred via the vinyl groups.

Upon heating, changes were observed in the spectrum (Fig. 3). The intensities

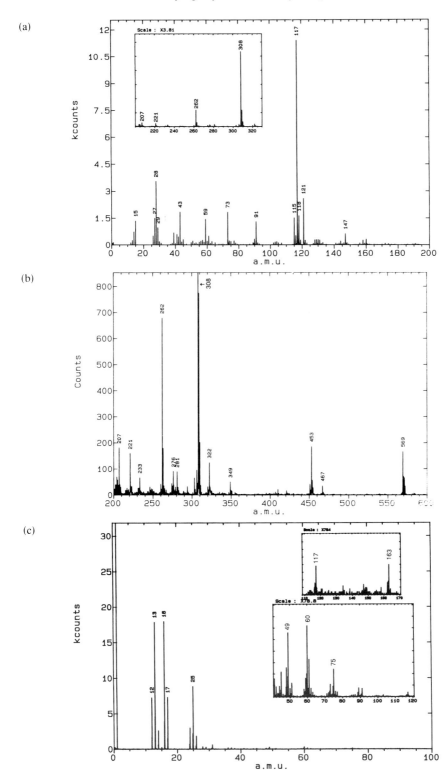

Figure 2. Positive (a, b) and negative (c) TOFSIMS spectra of the SAAPS film of Fig. 1. before heating.

Table 2.
Peak identifications for films of SAAPS on zinc before and after heating[a]

Ion	Composition	Before heating		After heating	
		Δm^b	Counts/yield[c]	Δm^b	Counts/yield[c]
−16	O^-	1	6.70	1	11.00
−17	OH^-	1	5.40	1	6.00
+18	NH_4^+	2	0.14	1	0.04
+27	$C_2H_3^+$	10	3.50	8	7.46
+28	Si^+	1	3.81	0	6.80
+30	CH_4N^+	0	1.02	12	0.81
+39	$C_3H_3^+$	10	3.52	12	5.31
+42	$C_2H_4N^+$	5	1.70	1	1.44
+43	$C_3H_7^+$	0	1.50	0	6.00
+43	$C_2H_3O^+$				
+43	$Si(CH_3)^+$				
+44	$C_2H_6N^+$	9	0.75	8	0.42
−49	C_4H^-	10	0.50	1	0.30
+50	$C_4H_2^+$	3	0.76	8	0.57
+51	$C_4H_3^+$	6	1.61	1	1.06
+55	$C_4H_7^+$	8	1.04	10	3.75
+56	$C_3H_6N^+$	2	0.65	0	0.85
+57	$C_4H_9^+$	4	0.57	6	2.86
+58	$C_3H_8N^+$	8	0.10	7	0.23
+59	$Si(CH_3)_2^+$				
−60	SiO_2^-				
−62	$Si(OH)_2^-$	2	0.064	1	0.19
−62	$C_5H_2^-$				
+67	$C_5H_7^+$	16	0.45	11	1.58
+73	$(CH_3)_3Si^+$	—	0.00	5	0.78
−73	C_6H^-	5	0.09	10	0.06
−75	$CH_3SiO_2^-$	2	0.25	3	0.20
+78	$C_4H_2N_2^+$	8	0.53	0	0.33
−89	$HSi_2O_2^-$	8	0.03	7	0.03
+91	C_7H_7	7	2.53	12	1.28
+115	$C_9H_7^+$	10	7.15	10	1.59
+117	$C_9H_9^+$	9	19.85	10	2.57
−117	$HSi_3O_2^-$	—	0.08	—	0.01
+121	$C_8H_9O^+$				
+147	$C_5H_{15}Si_2O^+$	12	0.06	2	0.31
+149	$C_9H_9O_2^+$	—	0.036	10	0.19
−163	$H_3Si_4O_3^-$				
+207	$C_5H_{15}Si_3O_3^+$	—	0.00	4	0.06
+215	$C_{14}H_{19}N_2^+$	8	0.06	—	0.03
+221	$C_7H_{21}Si_3O_2^+$	—	0.00	2	0.06
+262	$C_{14}H_{22}N_2SiO^+$	10	0.23	—	0.02
+281	$C_7H_{21}SiO_4^+$	—	0.00	5	0.05
+303	$C_{16}H_{23}N_2SiO_2^+$	10	0.06	—	0.00
+305	$C_{16}H_{25}N_2SiO_2^+$	10	0.03	—	0.00
+308	$C_{21}H_{26}N_3^+$				
+453	$C_{26}H_{41}N_4SiO^+$				
+569	$C_{35}H_{49}N_4SiO^+$				

[a] Ions listed without Δm and counts/yield values were observed in spectra from thick film only.

[b] Difference between listed composition and measured mass in milli-amu.

[c] Ratio of counts in peak area and total ion yield.

+308 amu · · · · · + 453 amu · · · · · + 569 amu

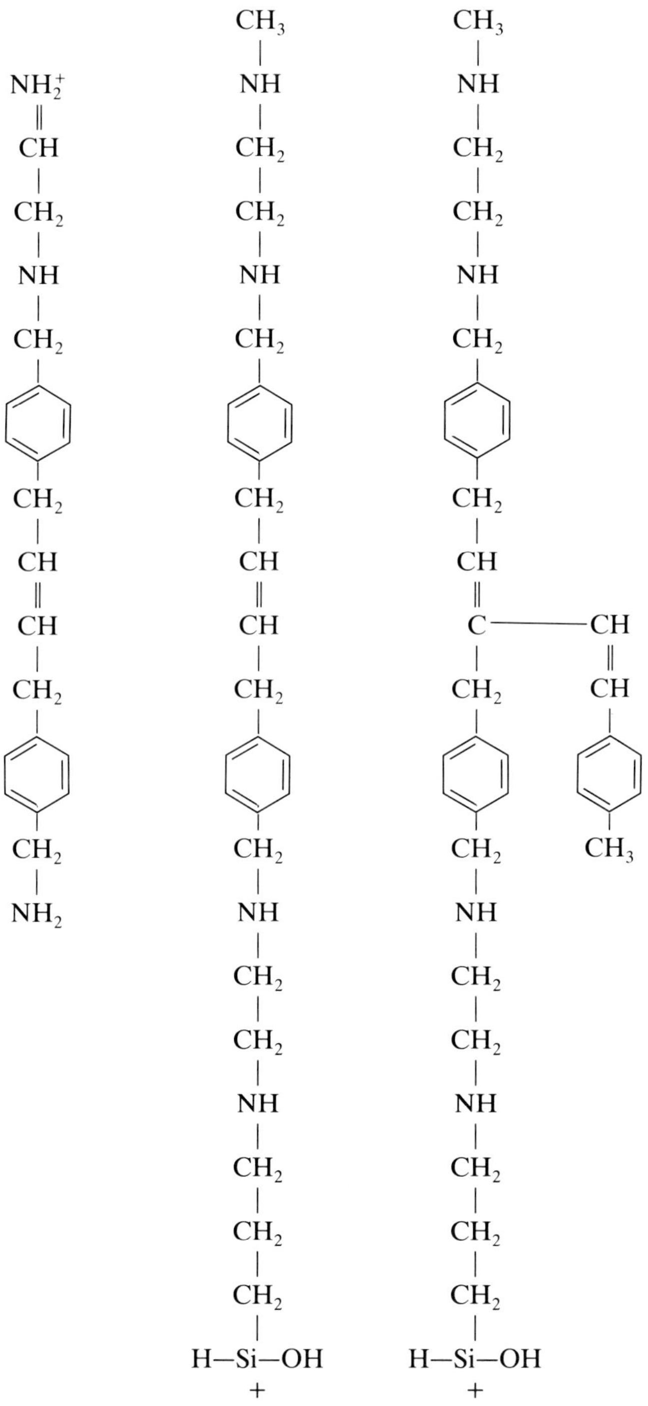

CH₃ · · · · · CH₃

NH₂⁺ · · · · · NH · · · · · NH

CH · · · · · CH₂ · · · · · CH₂

CH₂ · · · · · CH₂ · · · · · CH₂

NH · · · · · NH · · · · · NH

CH₂ · · · · · CH₂ · · · · · CH₂

(benzene ring) · · · · · (benzene ring) · · · · · (benzene ring)

CH₂ · · · · · CH₂ · · · · · CH₂

CH · · · · · CH · · · · · CH

CH · · · · · CH · · · · · C—CH

CH₂ · · · · · CH₂ · · · · · CH₂ · · · · · CH

(benzene ring) · · · · · (benzene ring) · · · · · (benzene ring) · · · · · (benzene ring)

CH₂ · · · · · CH₂ · · · · · CH₂ · · · · · CH₃

NH₂ · · · · · NH · · · · · NH

· · · · · CH₂ · · · · · CH₂

· · · · · CH₂ · · · · · CH₂

· · · · · NH · · · · · NH

· · · · · CH₂ · · · · · CH₂

· · · · · CH₂ · · · · · CH₂

· · · · · CH₂ · · · · · CH₂

· · · · · H—Si—OH · · · · · H—Si—OH
· · · · · + · · · · · +

(a)

(b)

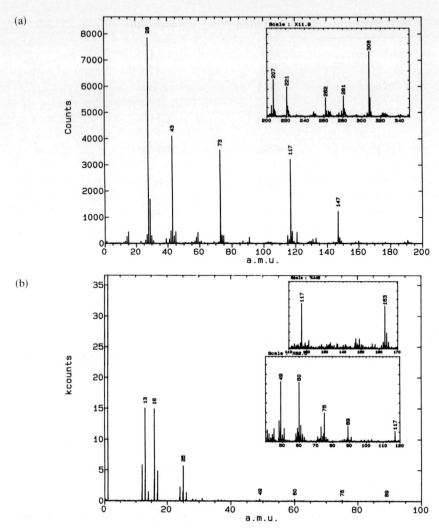

Figure 3. Positive (a) and negative (b) TOFSIMS spectra of the SAAPS film of Fig. 2 after heating in air at 150°C for 30 min.

of all aromatic peaks decreased and the intensities of those typical of polydimethyl siloxane increased sharply. It is not clear what the origin of this siloxane is, nor why its surface concentration increases so strongly upon heating. The most likely explanation is that it is a contaminant in the silane solution. The peaks at $+28$, -89, -117, and -163 amu, which are indicative of inorganic polysilicates, increased also. The sharp increase of the ion at $+43$ amu ($C_2H_3O^+$) indicates oxidation of the vinyl group. The presence of dimers and trimers (i.e. the ions at $+308$ $+453$, and $+569$ amu) is still detected, but their yields have decreased somewhat. It could be argued that polymerization via vinyl groups should increase upon heating. Therefore, the observed decrease may be the result of concurrent oxidation effects. Neither Zn nor Cl, with prominent peaks in the XPS spectrum of the heated film, is observed in the TOFSIMS spectra.

These observations suggest that the organic part of the silanes is oxidized and decomposed, forming a surface which is polymerized to some extent and which contains both siloxanes and inorganic silicates.

The elemental and binding state information obtained from the XPS analysis is in good agreement with, and sometimes complementary to, the information from the TOFSIMS spectra. The information that can be extracted from the SIMS spectra is, however, more detailed in terms of the structure of surface molecules and is also more surface-sensitive. For instance, the high degree of aromaticity in the surface of the film, as seen very clearly in the SIMS spectra (e.g. at $+117$ amu), can be confirmed only with difficulty by XPS, because the only feature indicating aromaticity is the $\pi-\pi^*$ transition, which is not clearly seen in Fig. 1.

3.1.3. High-resolution TOFSIMS. The above experiment was repeated with a very thin film of SAAPS on zinc by decreasing the time of deposition. Since in TOFSIMS the sampling depth is of the order of two molecular layers [12], the spectra obtained from thick films as described above can be expected to be largely independent of the substrate. Orientation effects as a result of different acid–base properties of the substrate can therefore be studied only with very thin films.

Initial experiments with thin films indicated that the surface preparation of the substrate plays an important role. Homogeneous films were obtained on zinc and steel substrates only if the substrate was alkaline-cleaned after mechanical polishing.

The spectra obtained before and after heating of the thin SAAPS film deposited on zinc immediately following alkaline cleaning are shown in Figs 4 and 5, respectively. It should be noted that these spectra were acquired under conditions of higher mass resolution and exact mass determination. Peak identification is therefore more accurate because overlapping peaks are now well resolved in many cases. Further, the peak intensities were quantified in this experiment by normalizing the area of each peak in the range 0–400 amu with respect to the total ion yield in this range. The results are listed in Table 2, which also gives the deviation Δm in milli-amu of the measured mass of each peak from the exact mass calculated from its most likely structure. We feel that if this deviation is 10 milli-amu or less, the peak identification is unique as given in the table. In other cases, a deviation higher than 10 milli-amu may be caused by the presence of one or more additional unresolved structures in the peaks. Very high resolution work, to resolve overlapping peaks further, is currently in progress.

3.1.3.a. Before heating. The following observations can be made:

(1) The peak at $+117$ amu is now of even higher intensity than in the spectrum of the thick film. Typical Si-containing peaks at $+28$ and $+43$ amu are lower. Several N-containing ions are formed in higher intensities. These observations all suggest that the film is now much more oriented, i.e. the silanol groups are predominantly adsorbed by the zinc surface and the aromatic vinylbenzyl groups are oriented away from the substrate, in agreement with the higher yields of the ions -49 and -73 amu.

(2) No siloxane formation can be detected in the spectra (e.g. at $+147$, $+207$, and $+221$ amu).

(3) The ions typical of the monomer molecule at $+215$ and $+262$ amu are still formed to an appreciable extent. However, the intense peak at $+308$ amu and those at $+453$ and $+569$ amu, observed in the spectrum of the thick film and identified as arising from dimers and trimers, are completely absent. It thus seems

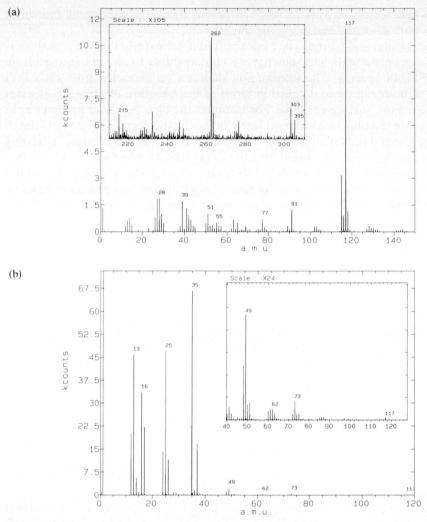

Figure 4. High-resolution positive (a) and negative (b) TOFSIMS spectra of a thin film of SAAPS deposited on an alkaline-cleaned zinc substrate. For conditions see text.

that crosslinking of the silane molecules via the vinyl groups can take place upon drying of a thick film but not in a thin film of highly oriented molecules.

(4) The signal due to Cl-containing groups is much higher than in the thick film. This is in agreement with the substrate orientation effect of the molecules in the thin film, which will contain some $-NH_2Cl-$ groups. In the thick film, Cl was also present (as detected by XPS) but SIMS did not detect it at the surface of the film. Possibly the surface was covered by the aromatic groups as a result of the higher molecular mobility during drying of the thick film and thus the $-NH_2Cl-$ groups escaped detection by TOFSIMS. In other words, in the thick film the Cl^- ions can diffuse away from the surface during drying but not in the thin film which is essentially one molecular layer thick. Recent, unpublished work with SAAPS on Zn, in which the immersion time was systematically varied, confirmed that the Cl^- intensity in TOFSIMS decreased markedly with immersion time.

+ 303 amu   + 305 amu

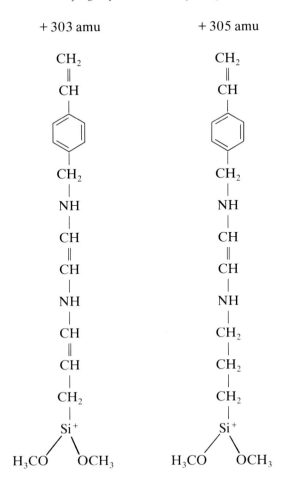

(5) The likely structure of the ions at + 303 and + 305 amu, as shown above, indicates that some molecules which have not completely been hydrolyzed have been adsorbed. It should be noted that the silane was hydrolyzed according to the manufacturer's specifications.

(6) The ion at + 121 amu, interpreted as originating from oxidized vinyl groups, is now absent. This suggests that the highly oriented molecules are more resistant to oxidation at room temperature than the randomly oriented molecules in the thick film.

3.1.3.b. After heating. When comparing the spectra of the heated film with those of the fresh, thin film, the following differences can be noted:

(1) There is a slight increase of some peaks indicative of the presence of siloxane structures, namely at + 28, + 73, and + 147 amu.

(2) The surface concentration of vinylbenzyl groups decreases, as shown by the decrease in intensities of typical ions such as + 77, + 91, + 117, − 49, and − 73 amu. This decrease is the result of at least two effects, namely a conversion to aliphatic unsaturated hydrocarbons and an oxidation. The former effect is evidenced by the increase of the aliphatic ions at + 27, + 55, + 57, and − 62 amu.

(a)

(b)

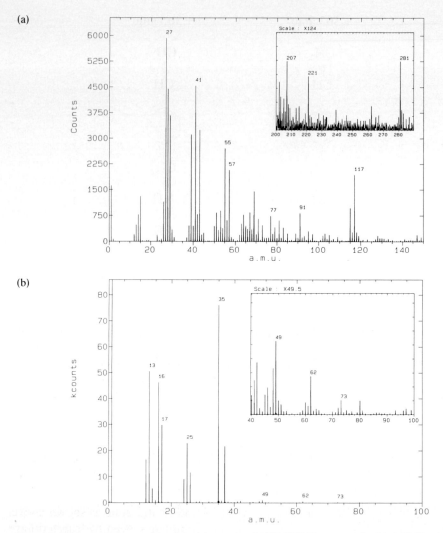

Figure 5. High-resolution positive (a) and negative (b) TOFSIMS spectra of a SAAPS film (of Figure 4) after heating in air at 150°C for 30 min.

The oxidation effect is concluded from the increase of the aromatic oxygen-containing ion at +149, the aliphatic ions at +67 amu, and also from the increase of the ions at −16 and −17 amu.

(3) The concentration of amines in the film surface decreases, as can be concluded from the intensities of the ions at +18, +30, +42, +44, and +78 amu. No specific ions are identified which could be interpreted as arising from oxidized amines, e.g. nitroso, nitro, or nitrate groups.

(4) The concentration of Cl^- increases somewhat; in addition, the zinc isotopes are now also detected, as shown in Fig. 6. It is not clear whether the appearance of zinc at the surface is the result of diffusion or of a decrease in film thickness.

A comparison between the effects of heating the thick and the thin film leads to the following observations:

(1) The thick film forms considerably more inorganic silicates and also more

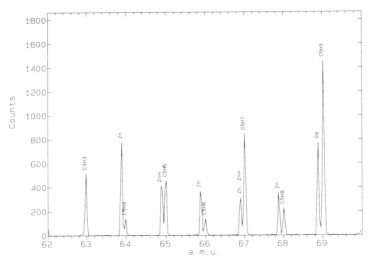

Figure 6. Narrow range of the high-resolution positive TOFSIMS spectrum of Fig. 5 showing zinc isotopes separated from organic ions.

siloxane structures upon heating. The counter-ions of the silicate anions may be Na^+, which is observed to increase (Fig. 3), and Zn^{2+}, detected by XPS but not by SIMS (Fig. 1).

(2) The thin film forms more aliphatic unsaturated hydrocarbons.

(3) Oxidation effects are observed in both films, but it seems that the oxidation products formed in the thin film are different. This may be a result of the differences between the orientation of the molecules in the film surface in the two types of sample.

(4) Zinc is detected only after heating the thin film.

As an interim conclusion from the data presented so far, it can be stated that by TOFSIMS a description can be given, albeit in a qualitative mode only, of the structure of the various molecules adsorbed on a metallic substrate. This information is totally different from that available from XPS analyses and is therefore highly complementary.

3.2. TOFSIMS analysis of films of γ-APS on steel substrates

The purpose of the experiments described below was to study the effect of the pre-treatment of the substrate and of the pH of the silane solution on the structure of the film. Only thin films, as described in the previous section for SAAPS on zinc, were prepared and analyzed. Heating experiments were not performed, but will be reported in a forthcoming paper. All spectra were run under conditions of high mass resolution.

3.2.1. Methanol-cleaned substrate. The film formed here was of irregular coverage, as was evident from the strong variation of the intensity of Fe^+ ions ($+56$ amu). It was also observed that the wettability of the steel by the silane solution was rather poor. The negative spectrum clearly revealed the reason for the poor wettability of this substrate. Almost all peaks in the negative spectrum could be uniquely identified as aliphatic fatty acids, originating most likely from the polishing paste. The predominant peaks were at $-199, -227, -255, -283$, and -325

amu. Ultrasonic cleaning in methanol apparently could not remove these acids, so they must have been chemisorbed. The silane solution did not remove them either and the spectra are thus not typical of a γ-APS film on steel because this substrate seems to have been passivated in the polishing process. Another observation is that ions were detected which seemed to have been formed from the silane monomer ($M_w = 137$) by loss of one and two oxygen atoms, respectively, i.e. at + 121 and + 105 amu.

3.2.2. Alkaline-cleaned substrate. After alkaline cleaning, the steel substrate was completely devoid of aliphatic fatty acids, as was verified by TOFSIMS. The freshly cleaned substrate was immediately immersed in the 1% silane solution and analyzed. The wettability of the cleaned steel by the solution was considerably improved by the alkaline cleaning procedure. The spectra obtained with a film deposited from a pH 10.5 solution are shown in Fig. 7. A rigorous peak identification procedure was applied here. The masses of all major peaks could be

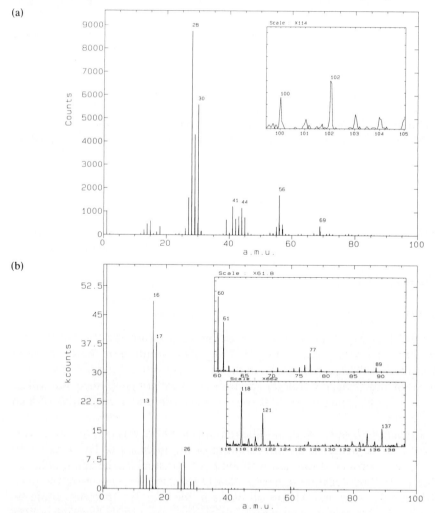

Figure 7. Positive (a) and negative (b) TOFSIMS spectra of a γ-APS film deposited at pH 10.5 on alkaline-cleaned mild steel.

assigned plausible compositions with better than 10 milli-amu accuracy. These compositions are also listed in Table 3.

As compared with the spectrum obtained with the methanol-cleaned sample, the yield of ions at $+28$ amu (Si^+) is considerably higher. However, peaks due to siloxane structures are not observed here. Apparently, the stronger interaction with the activated substrate results in a different film structure. The coverage of

Table 3.
Peak identifications for films of γ-APS formed on steel at pH 10.5 and 8.0[a]

Ion	Composition	pH 10.5		pH 8.0	
		Δm^b	Counts/yield[c]	Δm^b	Counts/yield[c]
-1	H^-	0	50.57	0	28.92
-16	O^-	2	16.70	3	12.60
-17	OH^-	2	10.55	1	5.56
$+17$	NH_3^+	1	0.41	1	1.42
$+18$	NH_4^+	2	0.80	2	2.65
$+28$	Si^+	0	23.30	0	19.29
$+31$	CH_5N^+	1	0.80	3	3.42
-35	Cl^-	—	—	0	28.60
$+39$	[d]$C_2HN^+/C_3H_3^+$	1	1.85	2	4.18
-41	CHN_2^-	7	0.13	5	0.07
$+41$	[d]$C_3H_5/C_2H_3N^+$	3	3.30	3	5.90
$+42$	[d]$C_2H_4N^+/C_3H_6^+$	3	2.06	2	2.48
$+43$	[d]$C_3H_7^+/C_2H_5N^+$	4	2.73	1	2.60
$+44$	SiO^+	10	3.01	10	1.35
$+45$	$Si(OH)^+$	2	2.15	8	2.86
$+55$	$C_4H_7^+$	5	1.14	2	1.03
$+56$	$Fe+$	0	4.76	3	0.44
-60	SiO_2^-	2	0.26	1	0.27
-61	$HSiO_2^-$	11	0.15	10	0.13
$+62$	$Si(OH)_2^+$				
-77	$HSiO_3^-$	6	0.07	5	0.065
$+79$	SiO_3^+	6	0.15	3	0.36
-89	$C_2H_7NSiO^-$	1	0.01	1	0.01
$+100$	$C_2H_2NSiO_2^+$	3	0.60	3	0.10
$+102$	$C_2H_4NSiO_2^+$	1	0.13	2	0.15
$+105$	$C_3H_{11}NSiO^+$				
-118	$C_3H_{10}N_2SiO^-$	4	0.03	4	0.008
$+121$	$C_3H_{11}NSiO_2^+$				
-121	$HSi_2O_3NH_2^-$	0	0.013	2	0.009
-127	$C_2H_3SiOFe^-$	—	—	10	0.028
-137	$HSi_2O_4NH_2^-$	5	0.008	5	0.005
$+147$	$(CH_3)_5Si_2O^+$	—	—	5	0.05
$+149$	$C_3H_{11}NFeO_2^+$	—	—	4	0.17
$+163$	$C_4H_{13}NSi_2O_2^+$	—	—	5	0.06
$+207$	$C_5H_{15}Si_3O_3^+$				
$+221$	$C_7H_{21}Si_3O_2^+$				
$+281$	$C_7H_{21}Si_4O_4^+$				

[a] Ions listed without Δm and counts/yield values are for methanol-cleaned sample only.
[b] Difference between listed composition and measured mass in milli-amu.
[c] Ratio of counts in peak area and total ion yield.
[d] At pH 10.5/pH 8.0

the substrate, as judged by the intensities of the Fe^+ ions at various sample locations, was generally better than after methanol cleaning.

An interpretation of the spectra of Fig. 7 leads to the following conclusions:

(1) All ions can be identified as arising from the monomeric silane molecules. There is no evidence for peaks due to contaminants originally present at the substrate surface or in the silane solution. Further, there are no ions which contain more than three carbon atoms.

(2) Many positively identified ions contain nitrogen and they appear both at odd-numbered and even-numbered masses. This is contrary to most published data on *N*-containing polymers analyzed on quadrupole instruments, in which only even-numbered peaks have been reported [12, 13]. Some of the major *N*-containing peaks here are at $+17$, $+18$, $+39$, $+41$, -41, $+42$, $+43$, -89, -118, -121, and -137 amu.

(3) Higher yields of Si- and Si–O-containing ions are now observed, such as those at $+28$, $+44$, $+45$, -60, -61, $+62$, $+79$, and -77 amu. The yield of hydrogen at -1 amu is very high and so are those of oxygen at -16 and -17 amu. These high yields are indicative of the presence of water molecules at the outermost surface, probably associated with the $-NH_2$ groups.

(4) A peak due to Fe at $+56$ amu is always observed, although its intensity varies somewhat. This suggests that the film formed at this pH is extremely thin.

(5) Several peaks are detected which contain Si and N without C, or peaks with Si and two N atoms. Such ions are interpreted as arising from silane dimers or oligomers in which a $-SiOH$ group of one molecule is associated with an $-NH_2$ group of another molecule. Such internal acid–base interaction can occur in the solution at high pH because the $-SiOH$ group becomes somewhat ionic. However, it cannot be excluded that such ions are actually formed in the ion formation process involving $-SiOH$ and $-NH_2$ groups at the metal surface if they are adsorbed at adjacent surface sites. Some of the ions that are indicative of such a phenomenon are those at -89, $+100$, $+102$, -118, -121, and -137 amu.

(6) The amino group in γ-APS can be assumed to be protonated to a certain extent [3, 16]. The ions at $+17$ and $+18$ amu are indicative of this protonation because it is unlikely that an NH_2 group can form NH_4^+ ions in the SIMS ionization process, i.e. by simultaneous reaction with two H atoms. Further, the ion at $+17$ is never observed in spectra from *N*-containing polymers, probably because it is very unstable [9, 10]. The high yield observed here thus suggests strongly that NH_3^+ ions were already present as such in the surface layers.

The spectra of the film deposited on alkaline-cleaned steel from a solution of pH 8.0 are presented in Fig. 8. The following characteristics and differences are observed when compared with the film deposited at higher pH.

(1) The film seems to be thicker and more uniform, as can be concluded from the lower intensity of the peak at $+56$ amu (Fe^+). There were no areas at the surface of this sample where high Fe signals were observed.

(2) Some siloxanes are detected now. This is concluded from the increase of the peaks at $+147$ and $+163$ amu, both of which contain Si–O–Si structures.

(3) A marked increase of the protonation of amino groups has occurred. This is shown by the higher yields of the ions at $+17$, $+18$, $+31$, and $+41$ amu. The presence of -35 amu (Cl^-) at the surface and the decrease of the peak at -1 amu (H^-) are consistent with this observation. It seems that $-NH_3Cl$ groups are now

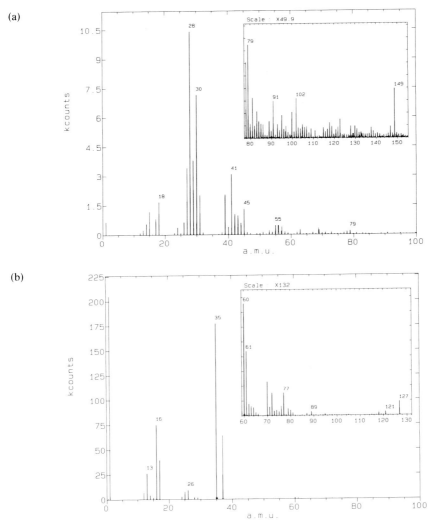

Figure 8. Positive (a) and negative (b) TOFSIMS spectra of a γ-APS film deposited at pH 8.0 on alkaline-cleaned mild steel.

present at the surface as a result of the addition of HCl which was used to lower the pH of the silane solution.

(4) The silane molecules seem to have interacted with Fe to form covalent bonds. This is concluded from the composition of the ions at +149 and −127 amu. Using the accurate mass determination capability, these were uniquely identified as Fe-containing (Table 3). The ion at +149 amu seems to indicate adsorption of amino groups on the steel surface [18]. The ion at −127 amu, which may have a cyclic structure, suggests interaction between silanol groups and the substrate surface. The possibility that these Fe-containing ions are formed by ion mixing processes is very unlikely, as such peaks were completely absent in the spectra obtained from films deposited from solutions of pH 10.5 on methanol- or alkaline-cleaned substrates.

It should further be noted that the Fe—O—Si$^+$ ion at +100 amu, which was reported in the literature as evidence for covalent bond formation [13], is not observed here. The low intensity of the ions at 56 amu (^{56}Fe$^+$) and the detection of iron-containing ions in the spectrum are consistent with an etching of the steel substrate by the silane solution. Such attack of metallic substrates has been reported for Zn and Al [3], but not yet for steel.

(5) Ions originating from the silane dimers, as obtained from the film deposited at pH 10.5, are also observed here, but at lower intensities, e.g. at −89, +100, +102, −118, −121, and −137 amu. The lower occurrence of such dimers can be understood in terms of the higher degree of amino group protonation.

(6) Consistent with the lower degree of association between the silanol and amino groups, an increase in the intensities of the ions at +45, +62 and +79 amu is observed. These are Si(OH)$_x^+$ type ions. However, the Si$^+$ and the SiO$^+$ peak intensities, which are a measure of the total amount of silicon, have decreased. This is to be expected in view of the higher amino group surface concentration.

4. CONCLUSIONS

(1) Static SIMS, especially high-mass-resolution TOFSIMS, is a powerful tool for the study of silane films. The accurate mass determination capability of high-mass-resolution SIMS analysis of organic materials has demonstrated that many peaks from organic materials consist of two or more components. Hence ion identification in SIMS with low mass resolution may be ambiguous.

(2) A thin film of SAAPS on a zinc substrate appears to be highly oriented with the vinylbenzyl groups away from the surface. The orientation is much less pronounced in a thick film.

(3) Upon heating of the SAAPS films on zinc oxidation effects are observed and siloxanes, possibly present as contaminants in the solution, begin to dominate the surface.

(4) Thick films of SAAPS are, to some extent, polymerized and crosslinked via the vinyl groups. These effects are not observed in highly oriented thin films on zinc substrates.

(5) Incomplete hydrolysis of the methoxy groups is detected in a thin SAAPS film adsorbed onto a zinc substrate after hydrolysis according to the manufacturer's specifications.

(6) The nature of the film formed by γ-APS on steel is strongly dependent on the pretreatment of the steel surface. This in itself is not surprising and has been documented in the literature. However, the TOFSIMS data indicate that methanol-cleaned, mechanically polished steel is passivated because it has formed iron soaps by reaction with fatty acids in the polishing process.

(7) Films of γ-APS on steel do not contain siloxanes if very thin, but they become detectable as the thickness increases.

(8) The orientation of the γ-APS films on steel is in part with the silanol groups and in part with the amino groups adsorbed at the film–steel interface.

(9) The amino groups in films of γ-APS are protonated to a certain extent. The protonation becomes more pronounced at lower pH.

(10) Spectroscopic evidence for the formation of covalent metallosiloxane bonds was presented for the case of γ-APS on steel. Two peaks were uniquely

identified as Fe-containing when γ-APS was adsorbed on alkaline-cleaned surfaces from a solution of pH 8.0.

(11) The accurate mass determination capability of TOFSIMS indicated that the ions formed from γ-APS sometimes have an unexpected composition. For instance, many odd-numbered *N*-containing ions were positively detected. It thus seems that ion formation from such highly oriented monomeric molecules is different from that of nonoriented polymeric surfaces.

Acknowledgement

We are indebted to Dr. David Surman of Kratos Analytical Inc. for performing the XPS analyses described in this paper.

REFERENCES

1. E. P. Plueddemann, *J. Adhesion Sci. Technol.* **2**, 179 (1988).
2. C. G. Pantano and T. N. Wittberg, *Surf. Interface Anal.* **15**, 489 (1990).
3. R. Chen and F. J. Boerio, *J. Adhesion Sci. Technol.* **4**, 453 (1990).
4. F. J. Boerio, *Polym. Prepr. Am. Chem. Soc.* **24**, 204 (1983).
5. T. Furukawa, N. K. Eib, K. L. Mittal and H. R. Anderson, Jr., *Surf. Interface Anal.* **4**, 240 (1982).
6. J. Comyn, D. P. Oxley, R. G. Pritchard, C. R. Werrett and A. J. Kinloch, *J. Adhesion* **28**, 171 (1989).
7. C. H. Chiang, N.-I. Liu and J. L. Koenig, *J. Colloid Interface Sci.* **86**, 26 (1983).
8. P. T. K. Shih and J. L. Koenig, *Mater. Sci. Eng.* **20**, 145 (1975).
9. D. Briggs, A. Brown and J. C. Vickerman, *Handbook of Static Secondary Ion Mass Spectrometry.* John Wiley, Chichester (1989).
10. B. A. Carlson, J. G. Newman, W. Katz, R. S. Michael and W J. van Ooij, *Handbook of Static SIMS Spectra of Polymers*, Perkin Elmer (in press).
11. W. J. van Ooij, A. Sabata and A. D. Appelhans, *Surf. Interface Anal.* **17**, 403 (1991).
12. J. C. Vickerman, A. Brown and N. M. Reed (Eds), *Secondary Ion Mass Spectrometry; Principles and Applications.* Clarendon Press, Oxford (1989).
13. M. Gettings and A. J. Kinloch, *J. Mater. Sci.* **12**, 2049 (1977).
14. M. Gettings and A. J. Kinloch, *Surf. Interface Anal.* **1**, 189 (1980).
15. Dow Corning Form No. 23-012B-85, *A Guide to Dow Corning Silane Coupling Agents*, Dow Corning Corporation, Midland, MI (1985).
16. D. J. Ondrus and F. J. Boerio, *J. Colloid Interface Sci.* **124**, 349 (1988).
17. S. Naviroj, J. L. Koenig and J. Ishida, *J. Adhesion* **18**, 93 (1985).
18. F. M. Fowkes, D. W. Dwight, D. A. Cole and T. C. Huang, *J. Noncryst. Solids* **120**, 47 (1990).

Silanes and Other Coupling Agents, pp. 345–364
Ed. K. L. Mittal
© VSP 1992.

TOF SIMS and XPS study of the interaction of hydrolysed γ-aminopropyltriethoxysilane with E-glass surfaces

D. WANG, F. R. JONES* and P. DENISON

School of Materials, University of Sheffield, Sheffield S10 2TZ, UK

Revised version received 21 October 1991

Abstract—The interaction of E-glass surfaces with hydrolysed γ-aminopropyltriethoxysilane (HAPS) has been studied by time-of-flight secondary ion mass spectrometry (TOF SIMS) and X-ray photoelectron spectroscopy (XPS). The SIMS spectrum was found to consist of four series of fragmentation patterns, two of which could be assigned to the poly(aminosiloxane). The remaining two series were consistent with the well-known fragmentation of polydimethylsiloxane. Using aqueous extraction procedures, it was possible to show that the largest observable fragment increased in size after treatment with warm water but subsequently decreased again after hot water extraction. This is considered to demonstrate the removal of low molecular weight oligomeric species overlying a crosslinked network of graded density. The positive aluminium ion intensity remained strong, suggesting its incorporation into the silane film. XPS analysis confirmed the incorporation of aluminium ions from the substrate into the coating.

Keywords: E' glass; surface analysis; silane coupling agent; fibre composites.

1. INTRODUCTION

γ-Aminopropyltriethoxysilane (APS), $NH_2(CH_2)_3Si(OC_2H_5)_3$, is widely used as a promoter to enhance the adhesion between a glass fibre surface and the matrix in a reinforced plastic composite. In most applications, APS is prehydrolysed in aqueous solution to an equilibrium mixture of oligomers and polymers prior to deposition [1]. With a view to understanding the nature of the coupling agent coating, the interaction of hydrolysed γ-aminopropylsilane (HAPS) with the glass surfaces has been studied extensively by various techniques [1–12]. However, the microchemistry is still not fully understood. Because of difficulties (e.g. contamination and geometry) associated with the use of commercial glass fibre substrates, we chose to employ well-characterized laboratory-prepared glass slides for these studies. In a previous study [13], we identified the presence of polydimethylsiloxane (PDMS) contaminants in the hydrolysed APS deposit. In addition, there were other mass fragments which were difficult to assign. In this paper we present the results of a confirmatory study using a freshly opened batch of APS. It was also necessary to ensure that the PDMS contamination was inherent in the APS and not associated with the dosing procedures.

---

*To whom correspondence should be addressed.

2. EXPERIMENTAL

2.1. Glass slide preparation

The glass plates were cast from E-glass cullet, used in the fibre process, and then cut into slides of approximately 8 mm × 8 mm × 2 mm using a rotary diamond-impregnated cutting wheel. These slides were ground to 1 mm thickness, polished to better than 1 μm using a diamond paste polishing wheel, washed many times with deionized water, and then heat-cleaned at 200°C for 0.5 h.

2.2. Silane coating

A freshly opened batch of γ-aminopropyltriethoxysilane (A-1100, supplied by Union Carbide) was used in this treatment. It is reported to be 99% pure.

The glass slides were respectively dipped into a freshly prepared solution of APS at pH values of 1, 3, 5, 7, 9, natural (10.6), and 13 in deionized water for 45 min at room temperature, and then washed several times with deionized water and dried *in vacuo* at 50°C. Adjustment of the pH from 10.6 to 1, 3, 5, 7, and 9 was accomplished using glacial acetic acid, while for pH 13, aqueous sodium hydroxide was used.

Some as-coated slides were conditioned in deionized water at 50°C for 24 h and at 100°C for 4 h. For a control, an as-polished slide was heat-cleaned at 250°C immediately prior to analysis by TOF SIMS and XPS, respectively. Control specimens were also treated similarly with deionized water. In fact, the deionized water-washed slide was checked by SIMS prior to coating with APS.

2.3. TOF SIMS analysis

For the TOF SIMS analysis, only slides treated with a natural pH HAPS solution were used. These were subsequently extracted with warm and hot water. They were mounted into a grid sample holder for transportation into a VG IX23S time-of-flight (TOF) SIMS instrument operating at a vacuum of $<10^{-10}$ Torr with a microfocused liquid Ga metal ion primary beam source (30 keV × 1.0 nA). For charge compensation, an electron flood gun was used. The working resolution of the spectrometer was determined from a lead phthalocyanine spectrum; for Pb$^+$ at $m/z = 208$ and the molecular ion at $m/z = 720$, it was 500 and 1000, respectively.

For calibration of the TOF SIMS spectra, we used hydrogen ($m/z = 1$) and aluminium ($m/z = 27$) peaks to calibrate the range of $m/z = 0$–100, followed by the aluminium and well-known PDMS ($m/z = 147$) peaks to calibrate the range of $m/z = 0$–200, with subsequent rechecking of the masses in the previous range. If all the peak positions did not alter, the procedure moved to the next range. To calibrate the range of $m/z = 0$–300, the aluminium peak and another PDMS peak at $m/z = 281$ were used, with rechecking of the previous ranges. In this manner, the calibration of the range of $m/z = 0$–400 was achieved using aluminium and $m/z = 358$ HAPS peaks. Similarly, this was extended in 100 m/z steps to the range of $m/z = 0$–700 using the HAPS peaks at $m/z = 477$ and 596. Rechecking ensured that all the peaks remained at their known values. In this case, an assignment error of $m/z = 1$–2 in the higher mass region above $m/z = 300$ can arise.

2.4. *XPS analysis*

The XPS spectra were recorded with a VG Microtech CLAM100 XPS spectrometer operating at a vacuum of $<10^{-8}$ Torr. An Al K_a X-ray source (1486.6 eV) was used at a power of 10 kV × 10 mA. The narrow scan spectra for individual elements were obtained at 15°, 30°, and 45° relative to the slide surface. C 1s (285 eV) was employed as a reference.

3. RESULTS AND DISCUSSION

3.1. *XPS study*

3.1.1. Heat-cleaned E-glass slide surface. Table 1 shows that the heat-cleaned glass surface is confirmed to be silica-rich because of higher Si and O surface concentrations compared to the bulk analysis obtained by inductively coupled plasma (ICP). This is accommodated by a lower surface calcium concentration. The SIMS results given below demonstrate that a significant proportion of the oxygen is probably present as silanol.

Table 1.
Relative surface atomic concentrations (in %) of the elements present at the three resolved angles for the uncoated E-glass slide

	Angles (degrees)			
Element	15	30	45	Bulk analysis[a]
Si	29.9	25.1	25.7	22.3
Al	8.0	7.9	7.6	7.4
Ca	7.9	8.1	8.5	16.4
O	54.2	58.9	58.3	49.6
Mg	0.0	0.0	0.0	0.4
B	0.0	0.0	0.0	2.1
Fe	0.0	0.0	0.0	0.6
N	0.0	0.0	0.0	0.0

[a] Analysis using ICP.

3.1.2. Coated E-glass slide surface. From Table 2 it can be seen that the increase in nitrogen concentration indicates the deposition of an HAPS-rich layer onto the glass surface. From Fig. 1 it can be seen that deposition of HAPS onto the glass surface is highly pH-dependent, reaching a maximum at natural pH. This is in good agreement with the observation of Naviroj *et al.* [14]. The concentration of aluminium present on the coated surface is also shown to be a function of the pH, following a similar increasing trend as the N 1s signal reaches a maximum at a pH of 9. Since the N 1s peak can be attributed to the HAPS coating, it is clear that its thickness also increases, reaching a maximum at pH 9. In these circumstances, the surface aluminium concentration should decrease to a minimum, if it remained in the substrate and became progressively buried by the deposit. Therefore, the aluminium ion would appear to have diffused away from the glass into the HAPS coating. Moreover, as shown in Fig. 2, the aluminium

Table 2.
Relative surface atomic concentrations (in %) of the elements present at the three resolved angles for the E-glass slides immersed in 1.5% aqueous solution of APS at natural pH

	Angles (degrees)		
Element	15	30	45
Si	27.5	27.3	27.2
Al	7.5	7.3	7.2
Ca	5.2	5.3	5.8
O	51.6	53.3	54.6
N	8.2	6.8	5.2

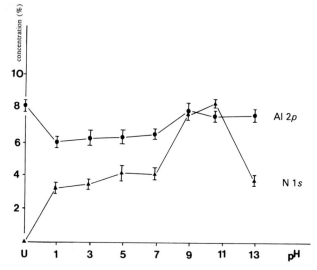

Figure 1. Variation in the surface concentration of aluminium and HAPS (as N 1s) of an aqueous silane-treated E-glass plate at differing media pH. The analysis angle was 15° (see Table 2 for details). U = uncoated surface.

surface concentration did not change significantly after extraction with warm water, yet it increased after hot water extraction. This was accompanied by a remarkable decrease in the N 1s (i.e. HAPS) surface concentration, because a significant proportion of the HAPS deposit had been removed by extractive hydrolysis. This observation could arise from a significantly heterogeneous deposit in which the aluminium component remains uncoated or is incorporated into the deposit. Since, after hot water extraction, the aluminium concentration remains at a higher level than in the original surface (or bulk), it follows that the latter is most probable. Furthermore, imaging TOF SIMS [15] did not provide evidence of heterogeneity at the 0.2 μm scale with apparent coincidence of HAPS specific and Al ions.

Another significant factor shown in Fig. 2 is that of the absence of calcium on the as-coated surface after extraction with warm water. In comparison, the angle-independent calcium concentration for the as-coated slide (Table 2) suggests that

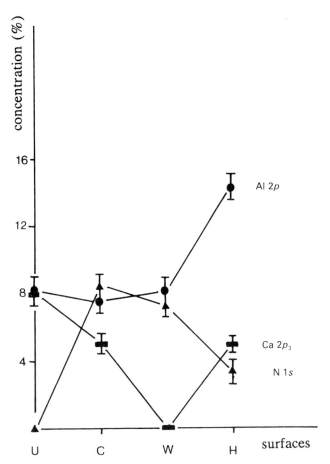

Figure 2. The change in HAPS (as N 1*s*), aluminium, and calcium concentrations after aqueous extraction. The analysis angle was 15°. U = uncoated surface; C = coated surface; W = the silanized surface after extraction with warm water at 50°C; H = the silanized surface after extraction with hot water at 100°C.

some calcium is incorporated into the film but as a water-soluble form, probably precipitated $Ca(OH)_2$. The main implication of these observations is that a uniform aminosilane coating has been deposited but it contains reacted aluminium components.

3.2. TOF SIMS study

3.2.1. Coated E-glass slide surface. (A) Positive ions. As shown in Fig. 3 and Table 3, the major components on the heat-cleaned surface are Si ($m/z = 28$), Al ($m/z = 27$), and Ca ($m/z = 40$). However, from Table 4 it can be seen that the aluminium surface concentration after treatment with HAPS was still very intense and increased after extraction with warm and hot water. These results are consistent with the XPS results shown in Figs 1 and 2, and are further indicative of aluminium becoming an integral part of the surface coating. Moreover, for the HAPS deposited from deionized water onto silver foil (99.99%), the peak at

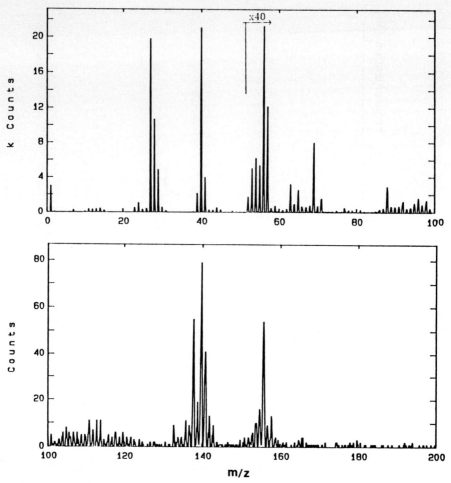

Figure 3. The positive ion TOF SIMS spectra for the deionized water-washed E-glass slides after heat cleaning at 250°C.

Table 3.
The positive ions from the heat-cleaned E-glass slide surface

m/z	Possible ion	m/z	Possible ion
23	Na^+	24	Mg^+, C_2^+
25	Mg^+, CH^+	26	Mg^+, CH_2^+
27	Al^+	28	Si^+
29	Si^+, $CH_3CH_2^+$	39	K^+
40	Ca^+	56	CaO^+
57	$CaOH^+$	63	Cu^+
65	Cu^+	69	Ga^+
71	Ga^+	88	$Al_2(OH)_2^+$
138	Ba^+	140	Ce^+
141	Pr^+	142	Ce^+
157	$(HO)_2-Si^+-O-Si-(OH)_3$		

Table 4.
Relative ratios of the positive ion peak intensities for Al ($m/z = 27$) and Si ($m/z = 28$) species

Surface	Heat-cleaned	Coated	Warm-water extracted	Hot-water extracted
Al/Si	1.67	0.87	1.49	1.89

$m/z = 27$ had a very low intensity (intensity ratio of ions at $m/z = 27/28$ is 0.05) compared with aluminium on the heat-cleaned glass surface (Table 4). Therefore, the peak at $m/z = 27$ for the HAPS-coated glass surface comes predominantly (>90%) from the aluminium. Without knowledge of reference SIMS spectra, interpretation can be achieved only by application of the rules of organic mass spectrometry [16]. Therefore, as demonstrated for published reference spectra, attempts to identify all the secondary ions are not practical [17]. Confidence in the interpretation can be achieved only by identifying a series of mass peaks which can be assigned to an appropriate fragmentation pattern. Therefore, in a previous paper [13], on this basis, only three ions at $m/z = 120$, 239, and 221 could be attributed to the HAPS deposit.

APS is known to condense into a series of linear, branched, and network polymers [18, 19]. The repeat unit of the linear polymer or its branches is given by:

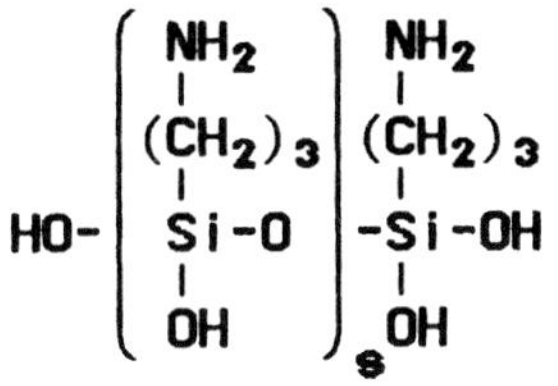

(i)

where s is the structural repeat unit.

Following the accepted assignments for PDMS [17], the ions with mass at $m/z = 120$ and 239 were readily assigned to an analogous linear form for HAPS (ii) with $w = 0$ and 1, which corresponds to $s = 0$ and 1 (i). The ion at $m/z = 221$ was considered to arise from a cyclic structure of x type (iii) with $x = 0$. Ions with more than two structural repeat units were not observed.

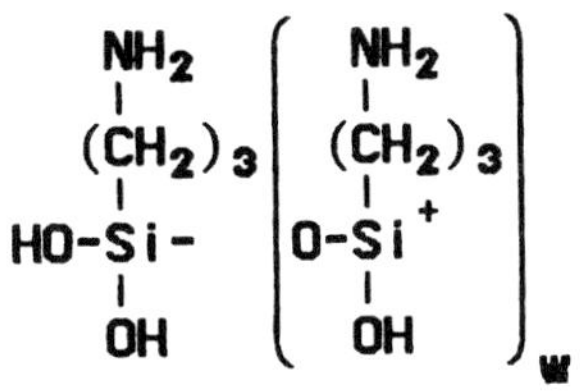

w fragment (ii)

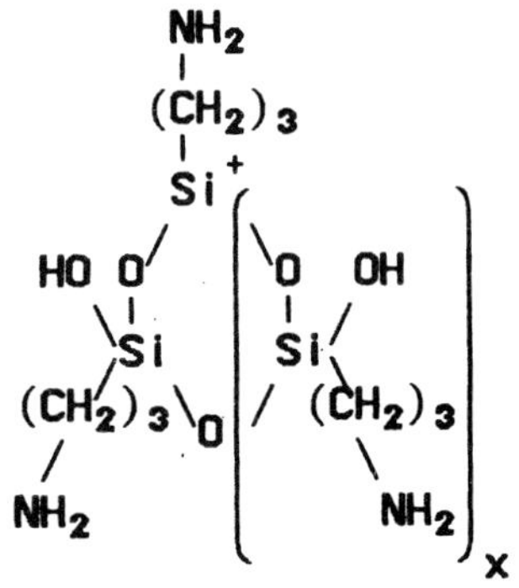

x fragment (iii)

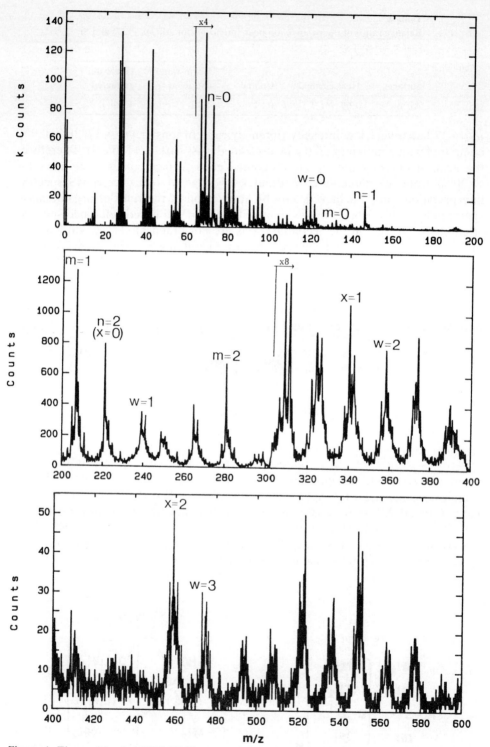

Figure 4. The positive ion TOF SIMS spectra for the HAPS-coated E-glass slides. See text for definitions.

However, as shown in Fig. 4, with a fresh unopened batch of APS (A-1100) additional peaks from these series were observed. Thus, as shown in Table 5 a series of positive ions from $m/z = 120$ to 477 could be assigned to the HAPS deposit.

Table 5.
Positive ions of the linear (w) and cyclic (x) forms of the HAPS component from the HAPS-coated glass slides

	Fragment (m/z)						
	120	239	358	477	221	340	459
Linear, w	0	1	2	3	—	—	—
Cyclic, x	—	—	—	—	0	1	2

The highest mass fragment, obtained from the fresh HAPS-coated glass surface, occurred at $m/z = 477$ and contains four Si—O bonds in its structure. This is twice the size of the largest HAPS fragment ($m/z = 221$) obtained from the as-coated surface prepared with the previously used batch of APS.

Positive ion fragments typical of PDMS are also present in the spectrum presented in Fig. 4. Much experimentation with differing silanes and distilled water- and deionized water-treated substrates confirmed that the source of the PDMS was the APS itself. The control spectra with the heat-cleaned glass plate (Fig. 3) and those washed with deionized water (Fig. 5) using the same drying facilities were devoid of PDMS contaminants (see Fig. 3). In contrast, the distilled water-washed plates consistently exhibited PDMS contamination as shown in Fig. 6. Therefore, in comparison with Fig. 5, it can be concluded that the use of deionized water does not contaminate the surface with PDMS and that the latter is inherent in the original APS. Two series of mass peaks of analogous structure to HAPS and reported by Briggs *et al.* [17] for PDMS were observed as indicated below, with linear (n type) and cyclic (m type) structures (see Table 6).

n fragment (**iv**)   *m* fragment (**v**)

(B) Negative ions. From Fig. 7 and Table 7 it can be seen that the dominant negative ions O^- and OH^- from the untreated glass surface arise from surface oxygen and hydroxyl groups. The typical negative ions from the glass, such as SiO_2, SiO_2H, SiO_3, and SiO_3H, were also observed but ions with mass greater than $m/z = 100$ were not detected. After treatment with the previously used batch of APS, some ions with mass up to $m/z = 128$ were identified but they did not correspond to appropriate assignments for the oligomers or polymers of HAPS molecular structure [13].

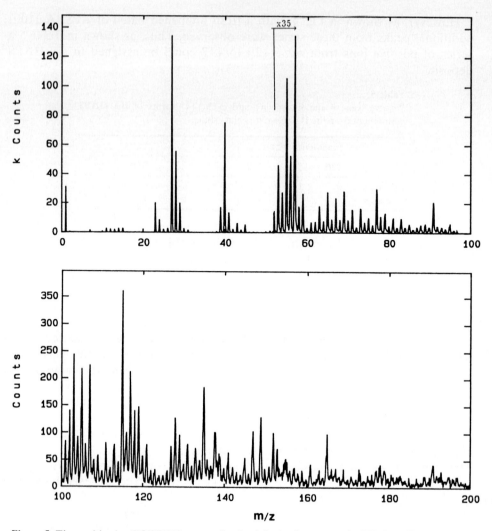

Figure 5. The positive ion TOF SIMS spectra for the deionized water-washed E-glass slides.

Table 6.
Positive ions of the linear (n) and cyclic (m) forms of the PDMS component from the HAPS-coated glass slides

	Fragment (m/z)					
	73	147	221[a]	133	207	281
Linear, n	0	1	2	—	—	—
Cyclic, m	—	—	—	0	1	2

[a] This ion can arise from either APS ($x = 0$) or PDMS ($n = 2$).

However, the spectrum from the fresh HAPS-coated glass surface exhibited a series of negative ions due to the HAPS deposit from $m/z = 136$ to 475 (see Fig. 8) which could be attributed to the fragmentation pattern of the linear (y type) and cyclic (z type) forms (see Table 8).

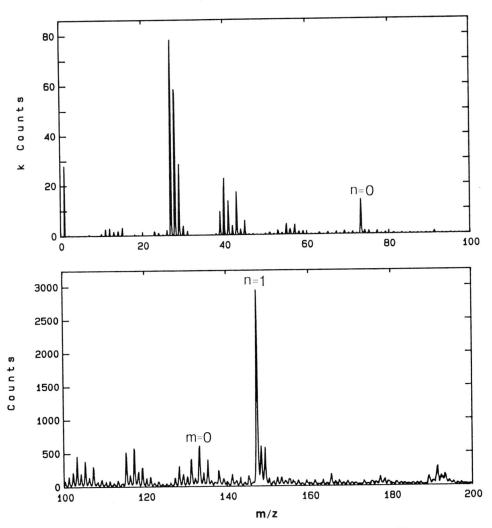

Figure 6. The positive ion TOF SIMS spectra for the distilled water-washed E-glass slides.

y fragment (**vi**)

z fragment (**vii**)

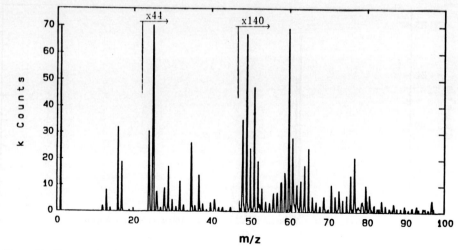

Figure 7. The negative ion TOF SIMS spectra for the deionized water-washed E-glass slides after heat cleaning at 250°C.

Table 7.

The negative ions from the heat-cleaned E-glass slide surface

m/z	Possible ion	m/z	Possible ion
16	O^-	17	OH^-
24	C_2^-	25	C_2H^-
35	Cl^-	37	Cl^-
60	SiO_2^-	61	SiO_2H^-
76	SiO_3^-	77	SiO_3H^-

Table 8.

Negative ions of the linear (y) and cyclic (z) forms of the HAPS component from the HAPS-coated glass slides

	Fragment (m/z)					
	136	255	374	237	356	475
Linear, y	0	1	2	—	—	—
Cyclic, z	—	—	—	0	1	2

Table 9.

Linear negative ions of p type for the PDMS component from the HAPS-coated glass slides

m/z:	89	163	237[a]
p:	0	1	2

[a]This ion can arise from either APS ($z = 0$) or PDMS ($p = 2$).

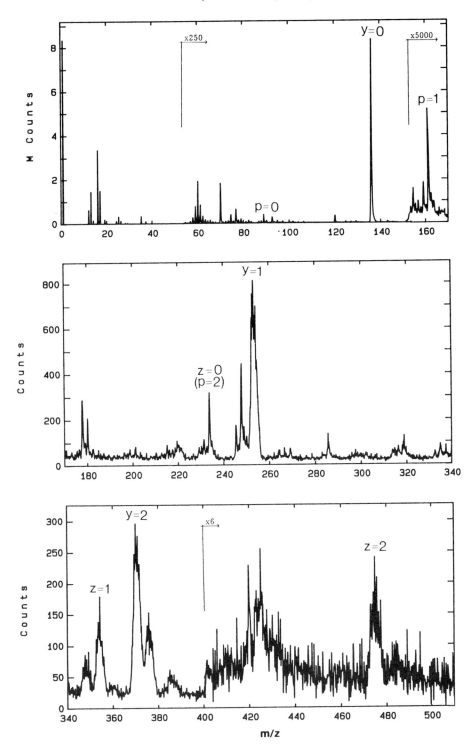

Figure 8. The negative ion TOF SIMS spectra for the HAPS-coated E-glass slides. See text for definitions.

Negative ions which could be assigned to the linear PDMS (*p*-type) structure were also present in the spectrum (Table 9).

$$
\underset{\underset{\text{CH}_3}{|}}{\overset{\overset{\text{CH}_3}{|}}{\text{CH}_3-\text{Si}-\text{O}-}}\left(\underset{\underset{\text{CH}_3}{|}}{\overset{\overset{\text{CH}_3}{|}}{\text{O}-\text{Si}-\text{O}^-}}\right)_p
$$

p fragment (**viii**)

Positive and negative secondary ions with no more than three repeat units were detected. Since the fragment size increased simultaneously with a small reduction in the surface nitrogen concentration (Fig. 2) after extraction with warm water (see Section 3.2.2), it appears that the outer surface layer is dominated by small oligomers. This is in good agreement with the results of Koenig and co-workers [1, 9–11], who also demonstrated the presence of an oligomer-rich surface, using infrared spectroscopy. The presence of a series of ions readily attributable to PDMS was surprising; however, as shown in Figs 3 and 5 this could not be assigned to contaminated deionized water. Not only that, using the same procedures, γ-glycidoxypropyltrimethoxysilane (A-187) yielded a PDMS-free coating [15]. The absence of PDMS fragments in HAPS coatings on steel has also been reported [20]. Similarly PDMS was also absent from HAPS films deposited from γ-APS supplied by a laboratory chemical supplier, using the same techniques [15]. However, it should be remembered that the HAPS deposit is essentially a crosslinked network for which a low ion yield can be expected, whereas PDMS is a linear polymer giving rise to high ion yields. Furthermore, the migration of PDMS to surfaces and interfaces is fully recognized. Since TOF SIMS is highly surface-sensitive to the first monolayer, small degrees of contamination by PDMS will give rise to large spectrum counts.

3.2.2. The as-coated glass slide surface after extraction with warm water. In a previous study [13], it was observed that extraction with warm water led to an increase in the mass of the largest fragment. It was also argued that this was consistent with the exposure of a crosslinked network. Large molecular fragments of PDMS with degrees of polymerization of 5–7 were also present. It was essential, therefore, to examine the effect of aqueous extraction on the as-coated glass plate, derived from the fresh batch of APS. After immersion in warm water at 50°C for 24 h, it was immediately apparent that the series of positive and negative ions indicative of PDMS with more than two repeat units were absent. Furthermore, the low mass fragments were of much lower intensity. Therefore, the discussion will concentrate on the assignments to the HAPS component in the high mass range of $m/z = 300$ to 730 for the positive ions and from $m/z = 200$ to 540 for the negative ions.

(A) Positive ions. In Fig. 9 it is seen that extraction with warm water introduced a series of relatively high mass positive ion fragments from $m/z = 310$ to 650. Most of the fragments extend the *w* and *x* patterns for the

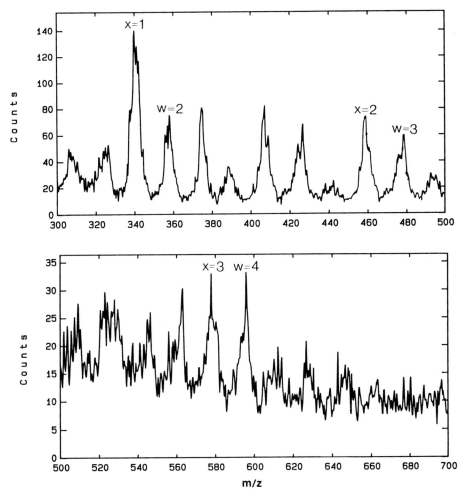

Figure 9. The positive ion TOF SIMS spectra of the warm water-conditioned, as-coated E-glass slides. See text for definitions.

Table 10.
Positive ions of the linear (w) and cyclic (x) forms of the HAPS component from the HAPS-coated glass slides after extraction with warm water

	Fragment (m/z)								
	120	239	358	477	596	221	340	459	578
Linear, w	0	1	2	3	4	—	—	—	—
Cyclic, x	—	—	—	—	—	0	1	2	3

poly(aminosiloxane) to higher molar mass as illustrated in Table 10. In confirmation to our previous work [13], the implication of this result is that the remaining HAPS deposit is more highly polymerized.

(B) Negative ions. From Fig. 10 it can be seen that a series of relatively high

mass ion fragments from $m/z = 237$ to 493 were also observed after extraction. As for the positive ion fragments, these can be attributed to an extension of the y and z patterns for HAPS as illustrated in Table 11.

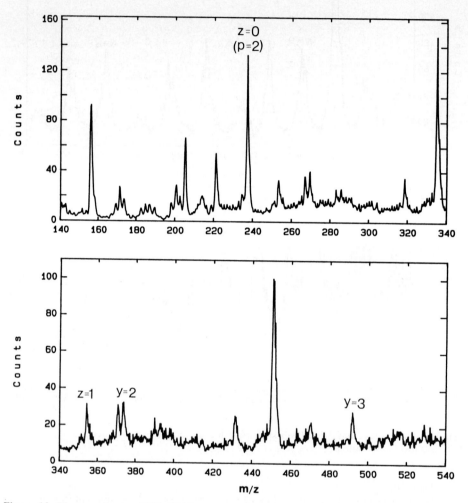

Figure 10. The negative ion TOF SIMS spectra of the warm water-conditioned, as-coated E-glass slides. See text for definitions.

Table 11.
Negative ions of the linear (y) and cyclic (z) forms of the HAPS component from the HAPS-coated glass slides after extraction with warm water

	Fragment (m/z)					
	136	255	374	493	237	356
Linear, y	0	1	2	3	—	—
Cyclic, z	—	—	—	—	0	1

3.2.3. Structure of the three-dimensional network remnant after extraction. An increase in the degree of polymerization of the poly(aminosiloxane) negative and positive ion fragments after warm water extraction can be most readily attributed to the removal of oligomeric HAPS with the simultaneous exposure of a three-dimensional network. Whereas the linear (*w*-type) HAPS fragments probably arise from network ends, the formation of cyclic fragments (*x*-type) probably requires their removal from the middle of a chain. With the absence of PDMS fragments it would appear that the main source of secondary ion is likely to be the crosslinked network. Thus, there is a high probability that these cyclic fragments may reflect the density of the crosslinking. The largest fragment of $m/z = 578$ contains five Si—O bonds and could represent the molecular structure of the network chain (see **ix**).

$$
\left(
\begin{array}{ccccc}
NH_2 & NH_2 & NH_2 & NH_2 & NH_2 \\
| & | & | & | & | \\
(CH_2)_3 & (CH_2)_3 & (CH_2)_3 & (CH_2)_3 & (CH_2)_3 \\
| & | & | & | & | \\
-O-Si- & O-Si- & O-Si- & O-Si- & O-Si- \\
| & | & | & | & | \\
OH & OH & OH & OH &
\end{array}
\right)
$$

(**ix**)

3.2.4. The warm water-conditioned, as-coated glass surface after extraction with hot water. (A) Positive ions. The spectrum in Fig. 11 shows that further extraction with hot water led to the absence of all the relatively high mass fragments from $m/z = 541$ to 707 with the largest one at $m/z = 477$. The assignments were again consistent with HAPS fragmentation patterns of *w* and *x* types (see Table 12).

Following the argument presented above, it is suggested that the linear fragment at $m/z = 477$ consisting of four siloxane bonds probably arises from a network end as a result of hydrolysis of the less densely crosslinked material.

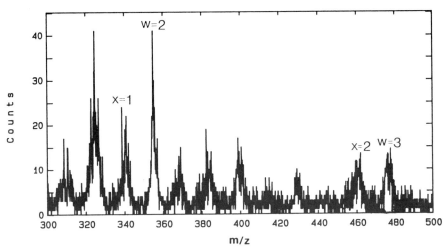

Figure 11. The positive ion TOF SIMS spectra of the hot water-conditioned, as-coated E-glass slides. See text for definitions.

Therefore, the as-coated glass deposit consists of low oligomers, overlying a three-dimensionally polymerized siloxane network which grades into a molecular interfacial layer.

4. CONCLUSIONS

Further confirmation of the presence of components of differing hydrolytic stability in the hydrolysed γ-aminopropyltriethoxysilane (HAPS) deposit on the E-glass surface, from an aqueous solution of γ-aminopropyltriethoxysilane, has been obtained. The incorporation of aluminium from the glass into the HAPS coating also occurred. A gradation in the degree of polymerization from a physisorbed oligomeric component to a three-dimensional network has been demonstrated, but it was complicated by the presence of PDMS impurities. The presence of an interfacial monolayer of high crosslink density on the glass surface after the hot water extraction was indicated.

Acknowledgements

Financial assistance from SERC and GKN plc is gratefully acknowledged. We thank Professor D. Briggs and Dr. M. Hearn (ICI plc) for valuable assistance. We also thank Union Carbide Ltd for supply of γ-aminopropyltriethoxysilane (A-1100).

REFERENCES

1. H. Ishida, S. Naviroj, S. Tripathy, J. J. Fitzgerald and J. L. Koenig, *J. Polym. Sci. Phys.* **20**, 701 (1982).
2. D. Pawson and F. R. Jones, in: *Interfacial Phenomena in Composite Materials '89*, F. R. Jones (Ed.), p. 188. Butterworth, London (1989).
3. H. R. Anderson, Jr., F. M. Fowkes and F.H. Hielscher, *J. Polym. Sci., Phys.* **14**, 879 (1976).
4. R. Bailey and J. E. Castle, *J. Mater. Sci.* **12**, 2049 (1977).
5. P. R. Moses, L. M. Wier, S. C. Lennox, H. O. Finklea, J. R. Lenhard and R. W. Murray, *Anal. Chem.* **50**, 576 (1978).
6. J. F. Cain and E. Sacher, *J. Colloid Interface Sci.* **67**, 539 (1978).
7. O. K. Johannson, F. O. Stark, G. E. Vogal and R. B. Fleischman, *J. Composite Mater.* **1**, 278 (1967).
8. A. T. DiBendetto and E. Scola, *J. Colloid Interface Sci.* **64**, 480 (1978).
9. C. H. Chiang, H. Ishida and J. L. Koenig, *J. Colloid Interface Sci.* **74**, 396 (1980).
10. S. R. Culler, S. Naviroj, H. Ishida and J. L. Koenig, *J. Colloid Interface Sci.* **96**, 69 (1983).
11. C. H. Chiang, N. I. Liu and J. L. Koenig, *J. Colloid Interface Sci.* **86**, 26 (1982).
12. T. L. Weeding, W. S. Veeman, L. W. Jenneskens, H. A. Gaur, H. E. C. Schuurs and W. G. B. Huysmans, *Macromolecules* **22**, 706 (1989).
13. D. Wang, F. R. Jones and P. Denison, *J. Mater. Sci.* (in press).
14. S. Naviroj, S. R. Culler, J. L. Koenig and H. Ishida, *J. Colloid Interface Sci.* **97**, 308 (1984).
15. D. Wang and F. R. Jones, in preparation.
16. D. Briggs, *Surf. Interface Anal.* **5**, 113 (1983).
17. D. Briggs, A. Brown and J. C. Vickerman, *Handbook of Static Secondary Ion Mass Spectrometry (SIMS)*. John Wiley, New York (1989).
18. H. A. Clark and E. P. Plueddemann, *Mod. Plast.* **40**, 133 (1963).
19. E. P. Plueddemann, *Silane Coupling Agents*. Plenum Press, New York (1982).
20. W. J. van Ooij and A. Sabata, *J. Adhesion Sci. Technol.* **5**, 843 (1991).
21. B. N. Eldridge, C. A. Chess, L. P. Buchwalter, M. J. Goldberg, R. D. Goldblatt and F. P. Novak, *J. Adhesion Sci. Technol.* **6**, 109 (1992).

Silanes and Other Coupling Agents, pp. 365–377
Ed. K. L. Mittal
© VSP 1992.

Positron annihilation in polyethylene and azidosilane-modified glass/polyethylene composites

F. H. J. MAURER* and M. WELANDER

DSM Research BV, P.O. Box 18, 6160 MD Geleen, The Netherlands

Revised version received 7 January 1991

Abstract—Positron annihilation lifetime spectra were measured on glass bead-filled high density polyethylene, with a glass content ranging from 0 to 50% by volume. In some cases, the glass beads were surface-modified by different amounts of an azidofunctional alkoxysilane. A composite model for positron annihilation was proposed to account for the observed changes in the relative intensities with increasing amount of filler. A decrease in the intensity of the longest lifetime with increased amount of the silane coupling agent was observed. This suggests a reduction of the free volume in the interphase region between glass and polyethylene, despite a reduction in the crystallinity of the interphase.

Keywords: Positron annihilation; high density polyethylene; interphase; glass-filled; silane coupling agent; heat of fusion.

1. INTRODUCTION

Positron annihilation lifetime spectroscopy (PALS) provides a method for studying changes in free volume and defect concentration in polymers and other materials [1, 2]. A positron can either annihilate as a free positron with an electron in the material or capture an electron from the material and form a bound state, called a positronium atom. *Para*-positroniums (*p*-Ps), in which the spins of the positron and the electron are anti-parallel, have a mean lifetime of 0.125 ns. *Ortho*-positroniums (*o*-Ps), in which the spins of the two particles are parallel, have a mean lifteime of 142 ns in vacuum. In polymers and other condensed matter, the lifetime of *o*-Ps is shortened to 1–5 ns because of pick-off of the positron by electrons of antiparallel spin in the surrounding medium.

An important feature of *o*-Ps in polymers is that these particles are preferentially formed or trapped in holes or regions of low electron density. The annihilation rate of *o*-Ps is proportional to the overlap of the positron and the pick-off electron wavefunctions and therefore the lifetime of *o*-Ps will depend on the size of the hole. The relative number of *o*-Ps pick-off annihilations is related to the number of suitable free volume sites in the polymer [3].

The sensitivity of *o*-Ps to small holes and defects in the material renders PALS a potentially interesting technique for studying composites. If the adhesion between the filler/fibres and the matrix is poor, small voids or defects may be present at the interface which may be detected by the positrons. On the other hand, if the coupling between the constituents in the composite is good, there may be an

---

*Correspondence address:— Chalmers University of Technology, Department of Technology, 41296 Gothenberg, Sweden.

interphase layer with free volume and annihilation properties different from those of the bulk polymer matrix. Only a few studies have been performed on polymer composites using the positron lifetime technique. In the work of West *et al.* [4] on carbon black-filled polybutadiene, it was shown that the intensity of the longest lifetime decreased both with carbon black loading and with increased surface area of the carbon black. The latter effect was explained in terms of an increased amount of bound rubber forming stable free radicals with increased filler surface area. Dale *et al.* [5] have studied carbon fibre-filled epoxy and compared PALS with density measurements to determine the amount of fibres in the composites.

In this work, composites of high density polyethylene (HDPE) filled with glass beads were studied using PALS, differential scanning calorimetry (DSC), and stress–strain measurements. The amount of glass filler was varied from 0 to 50% by volume, and for the composites with 20 and 50 vol% filler the glass filler surface was modified with different amounts of an azidofunctional alkoxysilane coupling agent [6]. Studies of the mechanical performance of polymer composites using organofunctional silanes as well as the chemistries involved have been reviewed by Plueddemann [7] and Ishida [8]. The model composite system used here has already been extensively studied using several techniques, such as dynamic mechanical spectroscopy in the melt state and FTIR [9, 10], scanning electron microscopy of fracture surfaces [11], and non-linear creep and stress relaxation in the solid state [12]. The silane used is known to allow formation of covalent bonds between glass and polyethylene [9, 10] and to promote formation of an interphase region with properties different from those of the bulk polymer matrix [9–12]. The aim of the present investigation was to determine the influence of filler fraction on the positron annihilation properties, and whether the changes effected on the interface by the silane coupling agent and the interphase region formed have an influence on the positron lifetime properties.

2. EXPERIMENTAL

2.1. Materials

The HDPE grade used was Stamylan 9089F (DSM) with a nominal density of 963 kg/m^3, a melt flow index (MFI 190/2) of 8 g/10 min, and an average molecular mass (M_w) of 60 kg/mol. The glass spheres (No. 5000, Potters Industries) had an average diameter of 10–13 μm and a density of 2500 kg/m^3.

In some cases, the glass beads were chemically modified to produce a reactive azide group on the surface. This was accomplished by washing the glass beads in 0.5 N HCl to leach ions and to increase the surface concentration of SiO$_2$. This was followed by reacting the glass with an azidofunctional alkoxysilane (AZ-CUP^{TM} MC, Hercules Inc.) in methylene chloride. The azide moiety was allowed to react to completion with the polyethylene during the compounding and moulding operations [9, 10]. Several amounts of the silane were used, corresponding to 0–100 equivalent molecular layers. One equivalent molecular layer corresponds to a concentration of 0.00024 g azidosilane solution per g glass and the azidosilane solution consisted of 50 vol% active matter. The calculation of equivalent molecular layers assumes that each azidofunctional silane molecule occupies approximately 150 Å^2 of surface. Those glass beads which were not silane-treated were washed with HCl in the same way as the silane-coated ones.

2.2. Sample preparation

Composites containing 0–50 vol% glass were prepared by milling at 170°C for 15 min on an open two-roll mill and compression moulding at 190°C into 3 mm thick plates. The cooling rate was approximately 40°C/min. The samples were stored at room temperature for several years before the measurements.

2.3. Methods

The positron source consisted of *ca.* 0.7 MBq of ^{22}Na deposited between two 7 μm nickel foils. The source was placed between two identical pieces of the material under study. ^{22}Na emits a gamma ray of 1.28 MeV at the same time as the positron is emitted and when the positron annihilates, two gammas of 0.51 MeV each are emitted. The time difference between the start gamma and one of the stop gammas was measured using a fast-fast coincidence system with CsF crystals. The time resolution of the spectrometer, as measured with ^{60}Co, was 280 ps full width at half-maximum (FWHM). The count rate was 2200 cps. Each sample was measured at least three times and each spectrum contained about 2.6×10^6 counts. All measurements were performed at room temperature.

Positron lifetime spectra can normally be resolved into a number of exponentials, each representing an annihilation process with a mean lifetime τ and a relative intensity I. In this case, it was found that four exponentials were necessary to describe the spectra. A fit with only three terms gave an inferior fit to the data. First, all spectra were fitted with four lifetimes using the computer program POSITRONFIT (PATFIT-88, Risø National Laboratory, Denmark). The resolution function was approximated with one Gaussian with FWHM 10% wider than that measured with ^{60}Co, i.e. 310 ps [2]. The amount of positrons annihilating in the nickel foils was estimated to be 21.4% ($\tau = 0.18$ ns) and was subtracted by the fitting program in the numerical analysis. It was then found that the shortest lifetime τ_1 was independent of both the filler content and the amount of surface modification and had a value close to the theoretical lifetime of *p*-Ps. To decrease the scatter in the evaluated lifetimes and intensities, the shortest lifetime was fixed at 0.12 ns and all the spectra were evaluated again.

Pieces of the compression-moulded sheets were pressed into dogbone-shaped specimens of thickness 1.7 mm and a gauge width of 15 mm for the tensile tests. Stress–strain curves were obtained at room temperature with a Zwick tensile tester at a strain rate of 0.0001 s^{-1}. The samples were strained up to 3.6% strain and the strain was measured with an extensometer with a gauge length of 50 mm. The tensile tests were done only on the composites with 50 vol% glass beads, with different amounts of the silane.

The heat of fusion of the composites with silane-coated glass beads was measured with a differential scanning calorimeter (Mettler TA4000) at a heating rate of 5°C/min.

3. RESULTS AND DISCUSSION

3.1. Lifetimes and intensities in unfilled polyethylene

The lifetimes in unfilled HDPE at room temperature were found to be 0.12 (fixed), 0.392 ± 0.009, 1.05 ± 0.17, and 2.53 ± 0.06 ns. The corresponding intensities were 10.0 ± 0.17, 63.8 ± 2.2, 10.6 ± 1.3, and $15.6 \pm 1.3\%$. The standard deviations are from five different measurements on the same sample. Several authors have studied polyethylene by positron annihilation techniques to identify the origin of the lifetime components. The longest lifetime component, with lifetime τ_4 and intensity I_4, is always assigned to pick-off annihilation of o-Ps formed or trapped in the amorphous regions and it has been shown that I_4 decreases with increasing crystallinity in polyethylene [13]. The shortest lifetime (τ_1, I_1) is assigned, at last partly, to the annihilation of p-Ps and the shorter of the two intermediate components (τ_2, I_2) is assumed to be due to positrons which have not formed positroniums [14]. It is still not clear whether the longer of the intermediate components (τ_3, I_3) should be attributed to annihilation of positrons [15–17] or $ortho$-positroniums [13, 18–20]. However, at present it seems that there exists more evidence in favour of the latter interpretation of τ_3. Cornaz $et\ al.$ [20] have compared results from angular correlation measurements with lifetime measurements on polyethylene at different electrical field strengths and have concluded that both τ_3 and τ_4 are due to annihilation of o-Ps.

Recently Abbé $et\ al.$ [13] deconvoluted spectra measured at room temperature on a series of polyethylenes of different crystallinities into both three and four lifetime components. Their values of the longest lifetime τ'_3 from a three-term fit are replotted in Fig. 1 and compared with the two longest lifetimes (τ_3 and τ_4) obtained from our data. It is clear that τ'_3 from a three-term fit is some average of τ_3 and τ_4 from a four-term fit. As mentioned above, the intensity of the longest lifetime when fitting with four lifetimes (I_4) decreases with increasing density, confirming that τ_4 is due to annihilation of o-Ps in the amorphous phase. It is then expected that an extrapolation of τ'_3 to 100% crystallinity would give a value corresponding to τ_3 from a four-term fit. This was also the case here, as shown in Fig. 1. If τ'_3 is due solely to pick-off annihilation of o-Ps, which is commonly

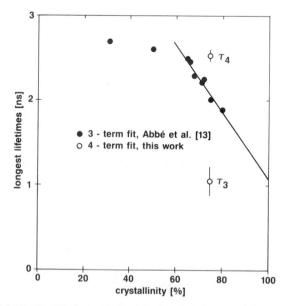

Figure 1. Values of the longest lifetime τ'_3 from a three-component analysis [13] and τ_3 and τ_4 from a four-component analysis of the spectra (obtained from our data) vs. crystallinity.

assumed, this should also be the case with both τ_3 and τ_4 from a four-term fit. A possible assignment of τ_3, on the grounds of Fig. 1, is annihilation of o-Ps in the crystals, but this is probably not the case. Suzuki *et al.* [17] and Reiter and Kindl [19] have recently shown that the third lifetime component τ_3 from a four-term fit also persists in the melt.

3.2. Influence of filler content on the positron lifetime parameters

Figure 2 shows the change of the intensities when the volume fraction of glass beads is increased from 0 to 0.5, corresponding to 0 to 72.2 wt%. All intensities vary linearly with weight per cent filler. The lifetimes changed very little when the amount of glass was increased. A very slight increase in τ_2 was observed with increasing volume fraction glass, from 0.392 ± 0.009 ns for unfilled polyethylene to 0.416 ± 0.006 ns for the composite with 50 vol% glass. Some influence was also noticed on τ_3, which had a constant value of about 1.05 ns for the composites with 0–30 vol% glass but decreased to about 0.94 ns for the composites with 40 and 50 vol% filler.

It can be seen from Fig. 2 that the contribution from the glass to I_1 and I_4 is negligible. When a straight line is fitted through the data points for I_1 and I_4, the intercept at 100% glass is close to zero intensity, these values being 0.05 and 1.05%, respectively. This can be interpreted to mean that no or a negligible amount of positroniums is formed in the glass beads.

A linear relationship of the intensities vs. weight fraction of the second component can be predicted using a simple composite model. The conditions of this

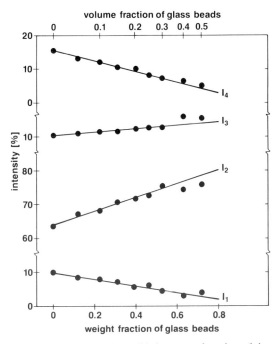

Figure 2. The relative intensities of the positron lifetimes as a function of the weight fraction of glass beads. The straight lines are least-square fitted through the data points.

longest lifetime τ_4, of relative intensity I_4, is related to the size and number of free volume sites in the amorphous parts [3].

The variations in the positron intensities in HDPE with different amounts of glass filler were satisfactorily explained by the proposed composite model for positron annihilation. The composite model takes into account the size of the filler particles and the density difference between the filler and matrix.

The tensile tests on the composites with 50 vol% glass beads indicate that good adhesion was already obtained with a coating of five equivalent molecular layers of the silane, and the fracture behaviour of specimens fractured at $-196°C$ confirmed the earlier observations that an adhering layer of modified polyethylene is formed around the glass when treated with silane.

DSC measurements showed that the crystallization ability of this interphase region was reduced by the silane modification of the glass beads. Despite an increase in the amount of amorphous material with increasing number of silane layers, a decrease in the intensity of the fourth lifetime was observed. This decrease in the free volume is in accordance with the earlier observed reduced mobility in the interphase region measured by dynamic-mechanical spectroscopy in the melt state [9, 10] and creep and stress relaxation measurements in the solid state [12].

Acknowledgements

Thanks are due to R. van Sluijs and P. A. M. Steeman for their valuable contributions to the experiments and to Professor J. D. McGervey for fruitful discussions. M. Welander acknowledges the grant from the Swedish National Board for Technical Development.

REFERENCES

1. W. Brandt and A. Dupasquier (Eds), *Positron Solid-State Physics.* North-Holland, Amsterdam (1983).
2. D. M. Schrader and Y. C. Jean (Eds), *Positron and Positronium Chemistry.* Elsevier, Amsterdam (1988).
3. W. Brandt, S. Berko and W. W. Walker, *Phys. Rev.* **120**, 1289–1295 (1960).
4. D. H. D. West, V. J. McBrierty and C. F. G. Delaney, *Appl. Phys.* **18**, 85–92 (1979).
5. J. M. Dale, L. D. Hulett, T. M. Rosseel and J. F. Fellers, *J. Appl. Polym. Sci.* **33**, 3055–3067 (1987).
6. G. A. McFarren, T. F. Sanderson and F. G. Schappell, *Polym. Eng. Sci.* **17**, 46–49 (1977).
7. E. P. Plueddemann, *Silane Coupling Agents.* Plenum Press, New York (1982).
8. H. Ishida, *Polym. Composites* **5**, 101–123 (1984).
9. J. D. Miller, H. Ishida and F. H. J. Maurer, *Rheol. Acta* **27**, 397–404 (1988).
10. J. D. Miller, H. Ishida and F. H. J. Maurer, *Polym. Composites* **9**, 12–19 (1988).
11. J. D. Miller, H. Ishida and F. H. J. Maurer, *J. Mater. Sci.* **24**, 2555–2570 (1989).
12. C.-G. Ek, J. Kubát, F. H. J. Maurer and M. Rigdahl, *Rheol. Acta* **26**, 474–478 (1987).
13. J. C. Abbé, G. Duplâtre and J. Serna, in: *Positron Annihilation,* L. Dorikens-Vanpraet, M. Dorikens and D. Segers (Eds), pp. 796–798. World Scientific, Singapore (1989).
14. W. Brandt and I. Spirn, *Phys. Rev.* **142**, 231–237 (1966).
15. J. R. Stevens and P. C. Lichtenberger, *Phys. Rev. Lett.* **29**, 166–169 (1972).
16. D. P. Kerr, *Can. J. Phys.* **52**, 935 (1974).
17. T. Suzuki, Y. Oki, M. Numajiri, T. Miura, K. Kondo and Y. Ito, in: *Third International Workshop on Positron and Positronium Chemistry.* Y. C. Jean (Ed.), pp. 39–46. World Scientific, Singapore (1990).
18. P. Kindl and G. Reiter, *Phys. Status Solidi (A)* **104**, 707–713 (1987).
19. G. Reiter and P. Kindl, *Phys. Status Solidi (A)* **118**, 161–168 (1990).

20. A. Cornaz, J. Brunner and E. Cartier, in: *Positron Annihilation*, L. Dorikens-Vanpraet, M. Dorikens and D. Segers (Eds), pp. 793–795. World Scientific, Singapore (1989).
21. W. Brandt and R. Paulin, *Phys. Rev. B* **15**, 2511–2518 (1977).
22. F. H. J. Maurer and M. Welander, in: *Third International Workshop on Positron and Positronium Chemistry*, Y. C. Jean (Ed.), pp. 89–94. World Scientific, Singapore (1990).

APPENDIX

Composite model for positron annihilation

Consider spherical filler particles (phase f) in a matrix (phase m). The probability of a positron hitting a filler particle is proportional to the volume fraction of filler v_f. The probability of the positron thermalizing and annihilating in this filler particle can be written [21] as

$$P_f = 1 - \exp(-a_f x), \tag{A1}$$

where x is the average thickness of the fillers and a_f is the absorption coefficient for positrons. a_f is given by [21]

$$a_f = (16 \pm 1)d_f E_{max}^{-1.43} \, [\text{cm}^{-1}], \tag{A2}$$

where d_f is the density of the filler and E_{max} is the maximum positron energy ($E_{max} = 0.54$ MeV for positrons emitted by ^{22}Na). The probability that a positron, which does not hit a filler particle, annihilates in the matrix can be expressed in the same way. The intensity contribution I_i^* to the spectrum from positrons annihilating with mean lifetime τ_i in the first slice of thickness x of the composite can then be written as

$$I_i^* = v_f P_f I_{f,i} + (1 - v_f) P_m I_{m,i}. \tag{A3}$$

The remaining fraction f of positrons, i.e. the positrons which did not annihilate in the first slice that they encountered, is

$$f = 1 - v_f P_f - (1 - v_f) P_m. \tag{A4}$$

After a large number of composite slices, when all positrons are expected to have annihilated, the relative intensity I_i of a certain lifetime τ_i in the spectrum is

$$I_i = [v_f P_f I_{f,i} + (1 - v_f) P_m I_{m,i}](1 + f + f^2 + \ldots) \tag{A5}$$

The sum in equation (A5) converges to

$$\sum f^n = \frac{1/P_m}{(P_f/P_m - 1)v_f + 1} \tag{A6}$$

and equation (A5) can be written [22] as

$$I_i = \frac{1}{(P_f/P_m - 1)v_f + 1} [P_f/P_m v_f I_{f,i} + (1 - v_f) I_{m,i}]. \tag{A7}$$

The value of x, which is needed to calculate P_f and P_m, was chosen here as the thickness of a disc with the same diameter as the glass spheres, i.e. $4r/3 = 7.7$ μm, where r is the radius of the glass beads. This gave a value of the ratio P_f/P_m of 2.54, which is almost the same as the density ratio $d_f/d_m = 2.60$. It can be seen from equations (A1) and (A2) that P_f/P_m will approach the density ratio d_f/d_m if the filler particles are small enough, i.e. when x approaches zero. Equation (A7) then becomes

$$I_i = w_f I_{f,i} + (1 - w_f) I_{m,i} \tag{A8}$$

where w_f denotes the weight fraction of filler.

The critical filler size for equation (A8) to be valid depends on the densities of the two components. Higher densities or a higher density ratio between the filler and matrix gives a smaller critical filler size.

Silanes and Other Coupling Agents, pp. 379–397
Ed. K. L. Mittal
© VSP 1992.

Acid–base characteristics of silane-treated E glass fiber surfaces

SHELDON P. WESSON,[1,*] JAMES S. JEN[2] and GARY M. NISHIOKA[3]

[1] *TRI/Princeton, P.O.B. 625, 601 Prospect Ave, Princeton, NJ 08542, USA*
[2] *Ashland Chemical Inc., 5200 Blazer Parkway, Dublin, OH 43017, USA*
[3] *H&N Instruments, Inc., P.O.B. 955, Newark, OH 43055, USA*

Revised version received 19 August 1991

Abstract—Acid–base properties of E glass fiber surfaces treated with various commercial organosilane coupling agents were studied with angle dependent X-ray photoelectron spectroscopy (XPS), electrolytic thermodesorption analysis of water (ETA), inverse gas chromatography (IGC) and programmed thermal desorption (PTD). XPS analysis indicates that γ-aminopropyltriethoxysilane and γ-chloropropyltrimethoxysilane show some preference for inverted surface orientation. Monolayer isotherms using Lewis acids and bases as probe adsorbates show silane deposition to attenuate acid–base interaction between probe molecules and weak and medium strength sites on the substrate. Moisture thermodesorption analysis shows that the sorptive capacity for physically bound water was attenuated by all the silane treatments. Desorption polytherms using acidic and basic probes demonstrate that γ-aminopropyltriethoxysilane imparts strongly acidic and basic chemisorptive characteristics to the glass surface. Methyltrimethoxysilane imparts acidic but not basic chemisorptive properties, while γ-chloropropyltrimethoxysilane mitigates the acid–base properties of bare glass without imparting additional chemisorptive character. Chemisorptive properties imparted by silane deposition are not thermally stable, and disappear upon ramping the system temperature to 300°C.

Keywords: glass fiber; inverse gas chromatography; programmed thermal desorption; silanes.

1. INTRODUCTION

While organosilane coupling agents are known to play an important role in interfacial adhesion in many composite systems, the mechanisms by which they improve product properties are not well understood. Sizes and surface treatments that use silanes are often developed entirely empirically as a consequence. The structure of the coupling agent, variables associated with silane deposition, and concentrations of a plethora of additives are adjusted in large matrix experiments, using end product properties as criteria for improved or diminished performance.

Statistical experiments are required to make timely progress in developing complex size systems, where interactions between many components are impossible to understand in detail. Problems arise, however, when product performance changes significantly with seemingly minor adjustments in size composition, or with a variable that was not part of the statistical experiment. It can also be very difficult to achieve incremental improvements in composite performance with a size that has already been optimized for a list of other specifications. These events often provide the incentive for a special study of the silane,

---

*To whom correspondence should be addressed.

which is usually added to improve composite properties, while most other components are present to optimize production and processing.

Plueddemann noted that the diversity of applications for organosilanes precluded any single explanation for their efficacy at improving composite properties, mentioning chemical reactivity, interpolymer network formation, and interphase modificaton as important mechanisms [1]. Previous work has shown instances where chemical reactivity is precluded, and interphase effects are minimized, by application of monoalkoxy silanes to improve composite performance under hygrothermal stress [2]. Other investigations have centered on glass fiber surfaces featuring submonolayer coverage by mono- and tri-alkoxy silanes, all of which rendered the substrates relatively hydrophobic while improving epoxy composite properties in the presence of water [3–5]. The inference is that one effect of silane deposition is to lower the interfacial energy in glass reinforced epoxy composites, thus diminishing the tendency of water to migrate from the matrix to the glass surface.

Another possibility is that silanes modify acid–base characteristics of glass, enhancing secondary glass/matrix attractive forces without necessarily promoting direct reaction of silane with resin. Fowkes improved the toughness of cast polymer films by modifying glass powder fillers with silanes to enhance interfacial acid–base interactions [6]. Fowkes *et al.* measured heats of adsorption using flow microcalorimetry on glass fibers, demonstrated significant enhancement of acid–base character upon silane deposition, and calculated Drago E and C constants for bare E glass [7]. Dwight *et al.* correlated failure modes in glass-filled epoxy and polyester composites with calorimetric data [8].

Gas adsorption complements wetting and calorimetry for investigating solid surface energetics, and IGC is a productive method for obtaining monolayer adsorption isotherms. Saint Flour and Papirer obtained multilayer isotherms on short glass fibers [9], described methods for estimating the retention volumes for skewed chromatograms at low surface coverage [10, 11], and contrasted the sorptive properties of bare, silane-treated, and titanate-treated glass fibers [11]. Papirer developed a protocol for analyzing retention volumes of alkanes and selected Lewis acids and bases on glass fiber substrates [12]. Papirer and Balard presented mechanical properties and fractography data for these bare and treated glass fibers in phenolic resin composites: the silane-treated substrate that rendered the surface more basic than the control improved adhesion with the acidic resin, while acidic character imparted by the titanate had a deleterious effect [13].

Osmont and Schreiber described acid–base properties of E glass fiber treated with various silanes by the ratio of the specific retention volumes of *n*-butanol and *n*-butylamine (Lewis acid and base, respectively) [14]. They calculated activation energies for solid–vapor interactions from the temperature dependence of the specific retention volume, and compared the sorptive capacity of the glass surfaces for adsorption from a solution of PVC (an acidic polymer) and PMMA (a basic polymer). Heats of adsorption for acidic and basic probes on bare and silane-treated glass beads were measured by Tiburcio and Manson, who resolved the enthalpies into dispersive and acid–base components and estimated Drago C and E coefficients for surface acidic sites [15]. They described how water vapor permeability was affected by acid–base interactions between the glass bead fillers and phenoxy resin in solvent cast films [16]. Tsutsumi and Ohsuga investigated changes in dispersive and electron donor–acceptor properties of glass fibers subjected to silane treatment [17].

In this study, we examine the surface properties of water-sized E glass fiber, and fiber treated with dilute aqueous solution of commercially important organosilane coupling agents. Our purpose is to describe glass surface heterogeneity in detail to determine the nature and extent of chemisorptive character imparted to glass fiber by silane deposition. We examine glass surface chemical composition and structure with angular resolved XPS in wide scan and high resolution modes. Acid–base properties of bare and silanized glass are deduced from monolayer isotherms obtained by IGC, using Lewis acids and bases as probe adsorbates. Isothermal adsorption measurements characterize weak solid–vapor interactions, while strong attractions are evinced by thermally desorbing minute quantities of chemisorbed probe. These trends are correlated with results from ETA, which quantifies the effect of silane deposition on the sorptive capacity of glass fiber for water.

2. EXPERIMENTAL

2.1. Materials

2.1.1. Glass. Packages of 200 filament strand pulled from a standard E glass composition were provided by PPG Industries. Density ρ was taken as 2.55 g cm^{-3}; fiber diameter d measured by optical microscopy ranged from 8.9 to 9.1 μm.

2.1.2. Silanes. γ-aminopropyltriethoxysilane and methyltrimethoxysilane were obtained from Union Carbide (A1100 and A163 silanes, respectively). γ-chloropropyltrimethoxysilane was obtained from Petrarch Systems, Inc.

2.2. Size

10^{-2} M silane solutions were made with distilled water acidified to pH 4 with acetic acid. Silanes were dissolved in equal volumes of methanol and slowly stirred into the acidified water. A glass strand was coated with silane solution on line during forming; applicator pads were scrupulously cleaned to minimize glass contamination. Packages were subjected to a standard air drying procedure. The procedure for water-sized glass was the same, except that no silane was added to the acidified water.

2.3. X-ray photoelectron spectroscopy

XPS analysis was performed using a VG ESCALAB spectrometer with an Mg K$_{\alpha}$ source. Instrument pressure was maintained in the low to middle range of 10^{-9} Torr; analyses were done at liquid nitrogen temperature to reduce thermal effects from the X-ray source. Fiberglass filaments were cut to lengths of approximately 1.5 cm and placed parallel to each other on the sample holder. For the grazing angle analysis, the plane of the sample holder was rotated to obtain greater surface sensitivity, which increases as the photoelectron exit angle decreases [18]. The long fiber axis was placed perpendicular to the axis of rotation of the sample holder to minimize interference with the angular analysis from fiber curvature. The adventitious carbon peak was used to correct charging effects for binding energy determinations.

2.4. Electrolytic thermodesorption analysis

Desorption of water from glass fibers was measured with an H&N Instruments Hydromet as follows: approximately 1 g of sample was placed in a silica tube contained in a tube furnace. Humidified nitrogen (2.5% RH) flowed over the sample into an electrolytic cell. The cell continuously electrolyzed the water, generating a current proportional to the water concentration in the nitrogen stream. The sample equilibrated with the humidified nitrogen after several hours, as indicated by a stable reading from the electrolytic cell equivalent to 2.5% RH. The measurement began by flowing dry nitrogen over the sample as the oven was heating at 8°C min^{-1}. The time, temperature, and water evolved were measured continuously up to a temperature of 800°C. This technique has been described previously [19].

Upon completion of a measurement, the raw data were plotted as volts vs. time. The rate of water evolution or the cumulative water evolved was plotted as a function of temperature. The data were normalized by subtracting the corresponding blank measurement, and dividing by the weight of the sample. The quantities of water obtained by integrating under each desorption peak were tabulated as micrograms of water per gram of sample, and as molecules of water per square nanometer of surface.

2.5. Inverse gas chromatography

Nickel columns 55 cm long (0.52 mm i.d.) were loaded with 7 to 10 g of fiber, a quantity that provided about 1.5 m^2 of surface for investigation. A Hewlett-Packard 5880A gas chromatograph fitted with a flame ionization detector and a D/A output board was connected to a 20 MHz 80386/80387 Micronics computer containing a 16 bit Data Translation Series 2801 A/D conversion card. Routines written in ASYST programming language collected detector voltages at frequencies between 1 and 10 Hz.

Injector and detector temperatures were maintained at 150 and 200°C, respectively. Nitrogen carrier flow rates were measured with a Gasmet flow meter and were maintained between 22 and 24 ml min^{-1}. Gas holdup times were measured with 20 μl injections of methane.

Hamilton 7101NCH syringes were used to inject 2 μl volumes of pentane (neutral probe) and t-butylamine (Lewis base); injections of t-butanol (Lewis acid) were restricted to 0.5 μl in order to obtain sharp peak fronts. The method for collecting chromatograms and transforming diffuse profiles into adsorption isotherms has been described previously [20]. Previous work shows that the specific surface of glass fiber is that of a uniform cylinder [21]. The geometric specific surface $\Sigma_{geo} = 4/\rho d$ was therefore computed for each fiber sample using values for density and diameter listed above. Adsorption volume n (μmol g^{-1}) was normalized for Σ_{geo} so that isotherms are displayed as n/Σ_{geo} (μmol m^{-2}) vs. equilibrium pressure in kPa.

2.6. Programmed thermal desorption

Desorption polytherms were obtained by loading the column with 7 to 10 g of fiber and maintaining the carrier gas flow rate at 24 ml min^{-1}. Signals from the flame ionization detector and oven thermometer were collected at 2 Hz as columns were subjected to linear temperature ramping from 30 to 300°C at 5°C. Parameters were entered into the chromatograph control terminal as follows: initial temperature, 30°C; initial time, 0.5 min; program rate, 5°C; final temperature, 300°C; final time, 0.1 min. Detector response was monitored at zero attenuation. Thermal desorption polytherms are presented as detector response normalized for the geometric surface area in the column (volts m^{-2}) vs. column temperature. The method has been described previously [22].

3. RESULTS AND DISCUSSION

3.1. XPS

Surface element concentrations measured at three photoelectron exit angles for water-sized and silane-treated E glass fibers are shown in Table 1. Angular resolved components of high resolution N 1s and Si 2p photopeaks are displayed in Tables 2 and 3, respectively. The photoelectron exit angles correspond to penetration depths of approximately 1 nm at 15° to 5 nm at 90° [7]. Aminosilane-treated fibers were analyzed before and after ramping to 300°C in a programmed thermal desorption experiment.

The carbon surface concentration of 42.5% at 90° exit angle for water-sized fibers in Table 1 represents a high level of adventitious contamination, more typical of plant production than of fibers formed in a relatively clean laboratory environment, which typically show carbon levels of 25% to 35% on water-sized E glass [4, 7, 8]. This high level of adventitious carbon tends to obscure the

Table 1.
Surface element components of E glass fibers, atm %, at three exit angles

	C	O	Si	Ca	Na	Al	N	Cl
Water sized								
15°	55.6	24.4	6.3	3.9	1.4	7.0	0.0	1.5
45°	48.7	29.1	8.9	3.9	1.1	8.2	0.0	0.0
90°	42.5	35.7	9.5	4.0	1.6	6.6	0.0	0.0
γ-Aminopropyltriethoxysilane								
15°	61.0	17.3	8.2	4.7	1.3	0.0	5.1	2.5
45°	51.1	25.6	8.2	3.9	2.5	4.1	3.3	1.4
90°	47.8	29.3	9.6	4.1	2.0	4.0	2.3	0.9
γ-Aminopropyltriethoxysilane, heat treated								
15°	49.5	23.6	6.3	3.3	2.5	7.6	0.0	0.0
45°	38.9	27.3	10.9	4.2	1.9	6.9	0.0	0.0
90°	35.4	32.0	12.8	4.6	1.7	3.6	0.0	0.0
γ-Chloropropyltrimethoxysilane								
15°	55.4	22.0	7.4	3.5	2.9	0.0	6.6	2.2
45°	50.2	27.6	9.2	3.3	5.2	0.0	1.7	2.8
90°	58.8	20.0	11.6	3.6	1.9	0.0	0.0	4.1
Methyltrimethoxysilane								
15°	65.8	24.0	5.5	2.2	2.5	0.0	0.0	0.0
45°	53.5	31.6	10.2	2.6	2.3	0.0	0.0	0.0
90°	52.8	31.7	12.3	2.6	0.7	0.0	0.0	0.0

Table 2.
Components of N 1s photopeaks

		% NH$_3^+$ (401 eV)	% NH$_2$ (399 eV)
γ-Aminopropyltriethoxysilane	15°		100
	45°		100
	90°	25	75

Table 3.
Components of Si 2p photopeaks

		% SiO$_2$ (103 eV)	% SiOH (102 eV)
Water sized	15°		100
	45°	37	63
	90°	50	50
γ-Aminopropyltriethoxysilane	15°		100
	45°	32	68
	90°	88	12
γ-Aminopropyltriethoxysilane, heat treated	15°	100	
	45°	100	
	90°	81	19
γ-Chloropropyltrimethoxysilane	15°	41	59
	45°	100	
	90°	100	
Methyltrimethoxysilane	15°	100	
	45°	100	
	90°	100	

comparison with silane-treated samples; nevertheless, it is apparent that fibers treated with aminopropylsilane and methylsilane show greater surface carbon concentrations at all penetration depths than does water-sized fibers. Methylsilane deposition produced fibers with the highest surface carbon concentration; this treatment also produced a relatively hydrophobic surface in a previous work [4].

Thermal treatment reduced surface carbon on aminopropylsilane-treated fiber, presumably by desorbing adventitious material. Carbon from the organic substituent on the silane desorbed also, since the nitrogen present before thermal treatment was totally absent afterwards. (The wide scan analysis of heat-treated fibers is incomplete because small concentrations of sulfur, fluorine and phosphorus were present on this sample only. They are considered to be artifacts, deposited perhaps from the inner wall of the chromatographic column during ramping, and are not reported).

Aminopropylsilane is known to deposit with some amine groups oriented towards the substrate surface: the conformation can be that of a zwitter ion, with the silicon and nitrogen atoms next to the substrate with the propyl chain outermost [23], or inverted, with the nitrogen next to the substrate and the silicon atom outermost [24]. Either orientation will produce a peak in the N 1s envelope at 401 eV, a binding energy assigned to the protonated form of the amine [7, 25]. Angle resolved components of the N 1s photopeak from aminopropylsilane-treated fibers in Table 2 show a small contribution from the protonated form at the greatest penetration depth. Surface carbon concentration decreases with increasing penetration depth on aminopropylsilane-treated glass (Table 1), indicating that the propyl chains are oriented outermost.

This combination of angle-resolved data from wide scan and high resolution XPS analysis is evidence for the zwitter ion conformation as opposed to the inverted conformation for aminopropylsilane molecules with protonated amine groups. The result of chloropropylsilane treatment is different, in that carbon concentration increases with increasing penetration depth, as does the chlorine concentration. These observations suggest that a significant fraction of the silane is oriented upside down, with the chlorine atom and propyl chain oriented towards the glass. (The appearance of adventitious chlorine on the aminopropylsilane-treated specimen and surface nitrogen on the chloropropylsilane-treated glass may have been caused by contamination of the silane applicator during fiber forming.) Fowkes *et al.* [7] observed a similar mix of orientation resulting from A174 methacryloxypropylsilane deposition; they explained the acidity of the resultant surface by preferential orientation of the carbonyl group towards the glass, and attributed C 1s peak broadening at low exit angles to acrylate carbons oriented away from the glass.

Glass surface constituents such as calcium and sodium are present on all the specimens in Table 1, showing that coverage by silane deposition was incomplete, which is in accord with previous work [4, 7, 8] and is consistent with the fact that even a single completed monolayer of silane would inhibit photoelectrons from glass surface constituents from reaching the detector [8]. A study of mechanical properties of silane-treated glass fiber substrates suggests that adsorption from 10^{-2} M silane solutions (about 0.25 wt %) deposits multilayer quantities [26, 27], which is supported by FTIR analysis [28]. These results taken together infer that silane deposits in co-polymerized clumps, forming a patchy network rather than orderly monolayers.

Surface aluminum concentration is severely mitigated by silane deposition (Table 1), suggesting that silanes adsorb preferentially on aluminols, and expose themselves again upon thermal desorption. This is consistent with the finding that silica doped with sodium aluminate ions became strongly acidic upon washing with ammonium hydroxide, which created aluminol groups by ion exchange. This acidity was mitigated by silane adsorption [29]. This effect has not been seen in earlier work, however, [4, 7, 8].

High resolution analysis of the Si 2p photopeak from water-sized glass indicates that silicon at the outermost surface consists entirely of silanols, with the fraction of SiO_2 silicon increasing to 50% at maximum penetration depth (see Table 3). Methylsilane-treated glass shows no silanol silicon at the outer surface. However, it features a large carbon concentration at the outermost surface and a silicon concentration greatest at the innermost surface, as seen in Table 1. This is evidence for a relatively orderly mode of silane deposition with the methyl group outermost, and the silanols positioned for near complete condensation to siloxane (Table 3). Glass fibers treated with aminopropylsilane and chloropropyl silane both feature silanol silicon at the outermost surface. This may be additional evidence that a significant fraction of these molecules are deposited in an inverted conformation.

Heat treatment removes silanols from aminopropylsilane-treated fiber (Table 3), and the silicon profile (concentration vs. penetration depth, Table 1) resembles that of water-sized glass more closely than that of aminopropylsilane-treated glass. Diminished silicon concentration at the outermost surface of heat-treated fibers suggests that inverted molecules are thermally desorbed. Surface nitrogen concentration is completely eradicated, showing that the organic component of the silane is effectively removed.

3.2. ETA

Water desorbed from water-sized and silane-treated glass fibers is shown in Fig. 1. These curves result from equilibrating the samples in humidified nitrogen (2.5% RH) for several hours, measuring water desorbed upon heating at $8°C\ min^{-1}$, subtracting the appropriate blank measurement, and normalizing for the specific surface of the fiber. Figure 1 reveals that five states of water are associated with these samples: water desorbs between 20 to 100°C (peak 1), 100 to 160°C (peak 2), 160 to 220°C (peak 3), 220 to 400°C (peak 4) and 500 to 800°C (peak 5). Water desorption volumes computed from the area under each peak are displayed in Table 4.

Table 4.
Water desorbed from E glass fiber, molecules nm^{-2}

Peak	1	2	3	4	5
Water-sized	24.7	15.7	9.7	23.4	73.9
γ-Aminopropyltriethoxysilane	14.1	8.6	7.4	29.7	86.8
Methyltrimethoxysilane	16.5	17.3	11.1	34.8	68.6
γ-Chloropropyltrimethoxysilane	13.6	17.7	9.7	28.0	61.3

Peak 1 is caused by physically adsorbed water on the fiber surface, as was demonstrated previously [19, 30, 31]. The area under this peak varies linearly with the initial low equilibration humidity, indicating that the Henry's law region of the water adsorption isotherm is under investigation. Peak 5 arises from water dissolved within the bulk of the fiber; the area under peak 5 will correlate with independent measurements of dissolved water by infrared spectroscopy. Peak 4 is attributed to water evolving from the near-surface region of the fiber, and varies systematically with the ambient humidity around the forming cone [30]. Peaks 2 and 3 have not been observed in previous investigations of E glass fibers, and are attributed to water desorbed from organic contamination on the fiber surface.

It is apparent from Fig. 1 that the water evolution profile is qualitatively similar for water-sized and silane-treated glass fibers. Table 4 shows, however, that the desorption volume of physically adsorbed water (peak 1) is significantly larger for water-sized glass than for silane-treated specimens. This result is in qualitative accord with evidence from wetting experiments demonstrating that silane deposition diminishes the non-dispersive component of the work of adhesion with water [2–5]. When bare and silane-treated fibers were equilibrated with water for 6 months, as opposed to several hours in this study, the desorption volumes of

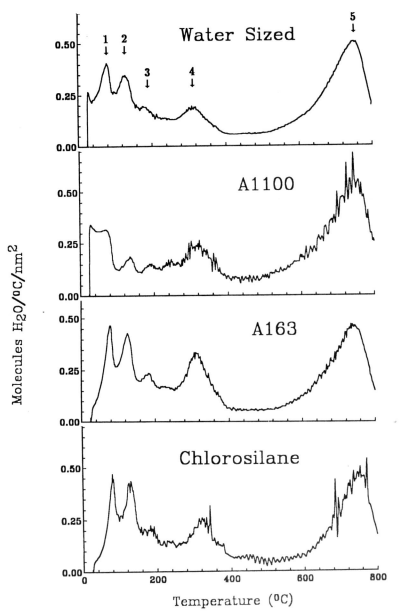

Figure 1. Water desorbed from E glass fiber. Arrows point to peak maxima in five temperature ranges.

physically adsorbed water were equivalent [31]. This suggests that silane deposition does not eliminate sites that attract water vapor to glass surfaces, but retards the kinetics of adsorption over the short term.

3.3. IGC

Isotherms for adsorption at 30°C of *t*-butylamine, a Lewis base, are shown in Fig. 2 (top); isotherms for adsorption of *t*-butanol, a Lewis acid, are also displayed

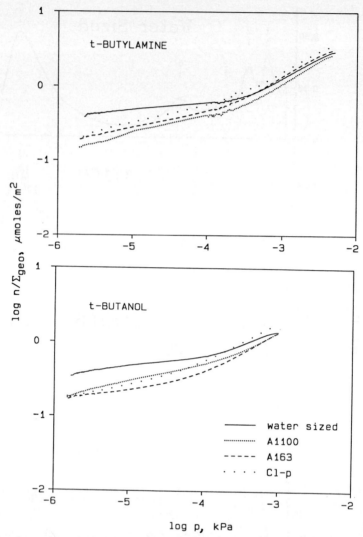

Figure 2. Monolayer adsorption isotherms on E glass fiber at 30°C. Substrate treatments are water size (solid lines); γ-aminopropyltriethoxysilane (closely spaced dots); methyltrimethoxysilane (dashes) and γ-chloropropyltrimethoxysilane (widely spaced dots).

(bottom). Isotherms depicted by solid lines show adsorption on the water-sized control, while dotted and dashed lines show adsorption on surfaces treated with A1100 (aminopropylsilane), A163 (methylsilane), and Cl-p (chloropropylsilane).

The main principle applied to interpreting monolayer adsorption is that high energy sites interact with probe vapors at low pressures. We may designate the pressure sector above 10^{-4} kPa as characterizing adsorption on low energy sites, and the region below 10^{-4} kPa as characteristic of adsorption on medium energy sites. Isotherms for both probes on all the substrates are essentially equivalent in the high pressure sector.

The adsorptive capacity of bare glass for both probes is substantially greater than that of any of the silane-treated substrates in the low pressure sector, however; this is the case for both the acidic and basic probes. The inference is that silane deposition partially passivates the glass surface, that is, it covers or reacts with polar surface sites such as silanols or silicate structures so as to render them inactive.

The isotherms on silane-treated surfaces are different from one another, considering the precision of the method, but these differences are minor, and are not worthy of detailed analysis. The main inference from analysis of physical adsorption is that silane deposition mitigates the sorptive properties of glass substrates, which is in accord with results of previous work [2–5]. This trend also correlates with results from water thermodesorption, which demonstrate that water-sized glass has the highest sorptive capacity for water, while silane-treated samples exhibit diminished sorptive properties.

3.4. PTD

The minute quantity of adsorbate remaining on the column after weakly bound probe has desorbed is chemisorbed to strongly acidic or basic sites on the substrate. The desorption profile obtained by ramping the column temperature is an index of the range of effective bond strength between the solid and adsorbed vapor. The flame ionization detector also registers desorption of adventitious organic contaminants; polytherms with no probe on the column must be obtained separately so that sorbate and contaminant desorption can be deconvoluted.

This process is shown in Fig. 3A, where the dashed curve denotes desorption of organic contaminants from a bare glass surface with no probe on the column. The dotted curve shows the desorption profile from a fresh load of bare glass obtained after injecting the Lewis base and waiting for the main body of the peak to pass through the column. The dashed curve is subtracted from the dotted curve to produce the envelope denoting *t*-butylamine desorption alone (solid curve). The subtraction curve is a complex composite of many peaks evolving between 70 to 240°C, evidence for a wide range of acidic site energies on water-sized glass fibers.

The subtraction curve in Fig. 3A (solid) is compared with that obtained by ramping the same column load twice with no probe, and then a third time after injecting the basic probe (dotted curve, Fig. 3B). The near congruence of the curves resulting from the first and third ramps demonstrates that acidic sites are part of the glass structure (silanols and silicates) and are not susceptible to desorption in this temperature range.

Polytherms from the same experiment performed with *t*-butanol, a Lewis acid, are shown in Fig. 3C. The subtraction curve from the first ramp shows essentially a single peak centered at 70°C. This peak is mitigated significantly by subsequent ramping, suggesting that it is caused in part by desorption of the probe from basic groups on adventitiously adsorbed contaminants. An alternative explanation is that basic sites on the glass are partially covered by contaminants that re-adsorb during the first ramp. It is therefore not possible to decide if the polytherm from the first ramp accurately characterizes the basicity of the glass surface.

The same sequence of desorption experiments was performed on silane-treated glass surfaces. Desorption of organic contaminants from glass fiber treated with methyltrimethoxysilane is shown by the dashed curve in Fig. 4A. The desorption envelope obtained from silane-treated glass with no probe on the column is

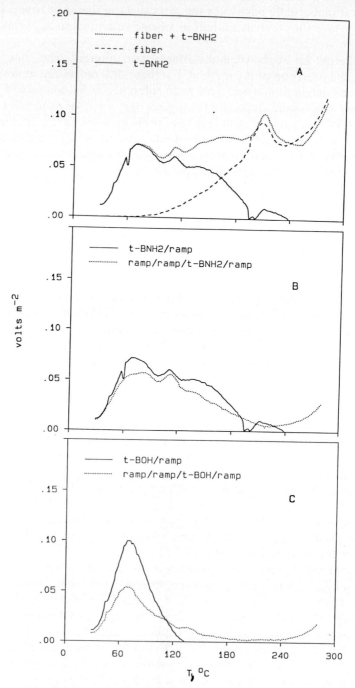

Figure 3. Thermal desorption polytherms for *t*-butylamine and *t*-butanol from water-sized E glass fiber. Polytherms in (B) and (C) are shown with desorption of adventitiously adsorbed organic material subtracted out.

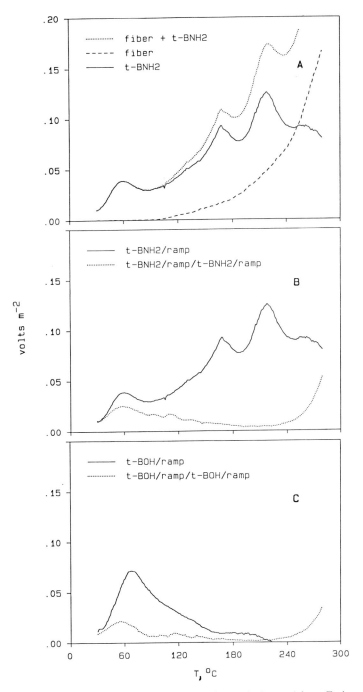

Figure 4. Thermal desorption polytherms for *t*-butylamine and *t*-butanol from E glass fiber treated with methyltrimethoxysilane. Polytherms in (B) and (C) are shown with desorption of adventitiously adsorbed organic material subtracted out.

substantially the same as that obtained from bare glass under the same conditions (see the dashed curve in Fig. 3A). This phenomenon, observed with all of the silane-treated specimens, demonstrates that desorption of organic substituents from silane molecules cannot be distinguished from the background of contaminant desorption.

The subtraction envelope obtained by ramping the t-butylamine from the silane-treated surface (Fig 4A, solid curve) has features above 120°C that are characteristic of methyltrimethoxysilane, and are probably dependent on all the details of glass surface chemistry, silane concentration in the size, and mode of deposition. These features are not thermally stable, as shown by the desorption envelope from a second ramp (Fig. 4B). The chemisorptive capacity of silanized glass for the Lewis acid is also severely mitigated by the first ramp (Fig. 4C).

Subtraction curves for the first ramp with t-butylamine from silane-treated surfaces are compared with those from water-sized glass in Fig. 3. The shape of the polytherm from aminosilane-treated glass is similar to that from bare glass from ambient temperature to 120°C, although the peak area in this region is considerably attenuated (Fig 5A). This is true also of the desorption curves from other silane-treated surfaces (Figs 5B and 5C). This suggests that, in each case, the low temperature region of the solid curve characterizes desorption from acid sites that were not covered by silane. The main inference is that silane deposition mitigates, but does not eliminate, the surface acidity of the glass substrate.

The high temperature sector of the polytherm from aminosilane-treated glass (Fig. 5A) features peaks that are not characteristic of desorption from bare glass. It is important to recognize that the peaks centered at 200 an 220°C are neither caused by organic contaminants (these have been subtracted out), nor by organic components of the silane (too dilute for detection), but by residual t-butylamine strongly bound to some acidic structure in the deposited coupling agent molecule. As indicated previously, this structure is not thermally stable: a second ramp shows no peaks in this temperature range.

The high temperature sector shows peaks of similar magnitude desorbing from methylsilane-treated glass, although the fine structure is different (Fig. 5B). No such feature is present in the envelope from chlorosilane treatment, which seems to passivate the glass partially without imparting any additional chemisorptive capacity (Fig. 5C). All three coupling agent molecules are tri-alkoxy silanes that co-polymerize in clumps featuring siloxane bonds and unreacted silanols (Table 3). These silanols, and the silicon atom itself, are acidic [32] and could give rise to chemisorptive attraction for a basic probe. It is not apparent why chlorosilane, the only one of the three, should impart no such additional acidity to the substrate.

Thermal desorption of t-butanol, the Lewis acid, from silane-treated surfaces is shown in Fig. 6. Aminosilane deposition mitigates the surface basicity of bare glass considerably, as evinced by the low temperature sector of polytherms in Fig. 6A. The peak in the high temperature sector of the solid curve could be attributed to acidic probe desorbing from the amine group, but it is not apparent why the shape and magnitude of these peaks should be so similar to those caused by basic probe desorbing from acid functionality (see Fig. 5A). We may infer from Figs 5A and 6A that aminosilane deposition imparts an amphoteric chemisorptive capacity that is not present on water-sized glass fibers.

Methylsilane deposition imparts no chemisorptive basicity to the substrate

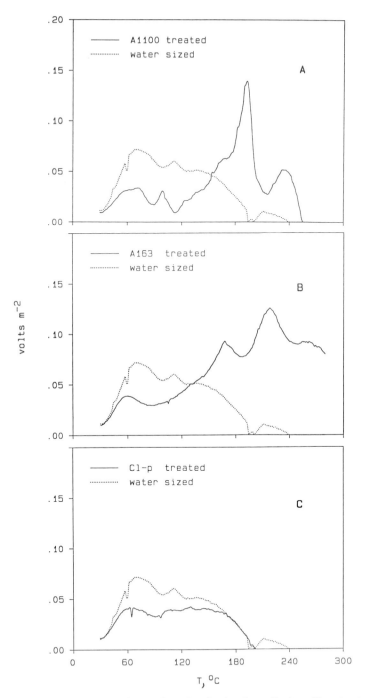

Figure 5. Thermal desorption polytherms for *t*-butylamine from E glass fiber treated with (A) γ-aminopropyltriethoxysilane; (B) methyltrimethoxysilane and (C) γ-chloropropyltrimethoxysilane. Polytherms are shown with desorption of adventitiously adsorbed organic material subtracted out.

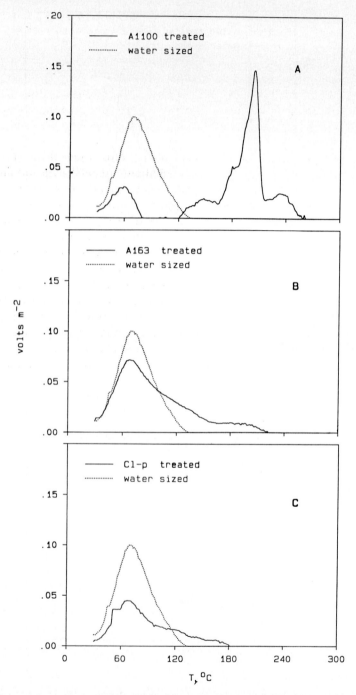

Figure 6. Thermal desorption polytherms for *t*-butanol from E glass fiber treated with (A) γ-amino-propyltriethoxysilane; (B) methyltrimethoxysilane and (C) γ-chloropropyltrimethoxysilane. Polytherms are shown with desorption of adventitiously adsorbed organic material subtracted out.

(Fig. 6B). The methyl group affords no possibility for backbonding or reversed orientation as shown by the carbon and silicon profiles in Table 1, so the only effect aside from partial surface passivation is the chemisorptive acidity shown in Fig. 5B.

Chlorosilane deposition imparts no basicity (Fig. 6C) and no acidity (Fig 5C); its only effect is to passivate the glass substrate partially. Carbon and chlorine concentration profiles that increase with penetration depth suggest an inverted orientation with the silicon atom uppermost (Table 1) bonded as siloxane (Table 3), seemingly a prospect for enhanced surface acidity, but not evinced by PTD measurements.

The high temperature features in polytherms from glass fibers treated with aminopropylsilane and methylsilane are qualitatively similar to the peak centered at 250°C when *t*-butylamine is thermally desorbed from silica–alumina catalysts [33, 34]. Figure 7 shows *t*-butylamine desorbed from a quartz sand (arbitrary detector units *DU* normalized for specific surface vs. temperature). The surface as received is so acidic that little of the probe desorbs below 200°C (solid line). Probe desorbs at low temperatures, and the high temperature peak is attenuated when the same sand is subjected to a second ramp. The first ramp partially passivates the surface, possibly by desorbing organic contaminants that re-adsorb on high energy sites. *t*-butylamine desorption from silane-treated fiberglass (Figs 5A and 5B, solid lines) appears similar to a composite of the two curves in Fig. 7. The polytherm from water-sized glass fibers has a shape analogous to that for desorption from contaminated sand. The presence of silane superimposes a high temperature component in the fiberglass polytherm for which desorption from pristine sand is a model. The main inference from this comparison is that the high temperature features in Figs 5A, 5B and 6A arise from probe interacting mainly with silicon from the silane rather than its organic functionality.

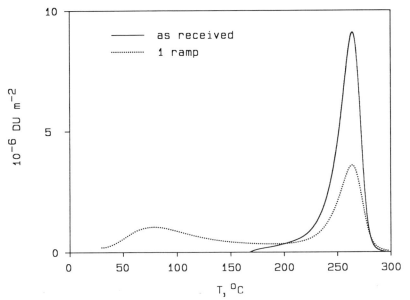

Figure 7. Thermal desorption polytherms for *t*-butylamine from sand as received (solid line) and after one ramp (dots). DU indicates detector units.

Substitution of aluminum for silicon in a silica lattice increases the concentration of both Lewis and Bronsted acidic sites over that of pure silica [35]. These sites, together with those created by other minor glass constituents, create the surface acidity of glass fibers that we observed in this work and has been noted by previous investigators [9–17]. We consider this acidity to be the primary effect that we observed in this study.

Metal ions that diffuse to the surface react with atmospheric gases to form hydroxides in a thin layer of high pH [36]. Fowkes improved adhesion of a basic polymer to glass with an acid wash [6, 7]; a similar treatment also improved adhesion to epoxy [37], an acidic polymer. The former effect was attributed to replacement of sodium silicate with silanols by ion exchange; the latter may result from removing a corrosive layer of metal hydroxides from the interface. IGC and PTD measurements in this study demonstrate a measurable basicity on bare fibers, which other investigators also noted [6, 14, 17].

Isotherms from IGC and moisture desorption polytherms from ETA show that silane deposition mitigates physical adsorption of organic vapors used as probes, and of water. This effect is not evident from other IGC studies [9–17]: it is difficult to compare our results directly with those, as we analyze sorption data from the entire diffuse branch of the chromatogram rather than from one point at low coverage.

4. SUMMARY

Aminopropylsilane and methylsilane imparted an acidic chemisorptive capacity that we attribute to the silicon portion of the coupling agent by analogy with the chemisorptive properties of sand. Aminopropylsilane treatment also showed considerable acidity by calorimetric pyridine titration [7]. Chloropropyl silane imparted no acid–base character to the glass substrate in this study, in contrast to the acidic character reported by Osmont and Schreiber [14].

We found no enhanced basicity attributable to the amine group on aminopropylsilane-treated glass, as reported by every other investigator cited above except Fowkes *et al.* [7]. Using calorimetric phenol titration, they found that an equivalent basicity was imparted to glass fiber by aminopropylsilane and two other silanes; all three were more basic than bare glass.

All of the silane treatments in this study diminish the physisorptive capacity of glass fiber substrates, as shown by the isotherms (Fig. 2) and the desorption volumes of physically adsorbed water (Table 4, peak 1). This is one reason for their efficacy at promoting wet strength retention and enhancing other composite properties that degrade when moisture adsorbs at the fiber–matrix interface. Chemisorptive properties for probe adsorbates that are imparted to the substrate by silane deposition may also influence fiber–matrix interaction.

The sorptive properties of silane-treated glass substrates will vary radically with minute changes in glass substrate composition and deposition conditions. The challenge is to choose from the huge matrix of possible experiments those that demonstrate the most general properties of silane-treated surfaces.

REFERENCES

1. E. P. Plueddemann, in: *Interfaces in Polymer, Ceramic, and Metal Matrix Composites*, H. Ishida (Ed.), p. 17. Elsevier Science Publishing Co. (1988).

2. G. M. Nishioka and S. P. Wesson, U. S. Patent #4455330 (1984).
3. S. P. Wesson and A. Tarantino, *J. Non-Crystalline Solids* **38–39**, 619 (1980).
4. S. P. Wesson and J. S. Jen, *Proc. 16th National SAMPE Tech. Conf.* **16**, 375 (1984).
5. G. M. Nishioka, *J. Non-Crystalline Solids* **120**, 102 (1990).
6. F. M. Fowkes, *J. Adhesion Sci. Technol.* **1**, 7 (1987).
7. F. M. Fowkes, D. W. Dwight and T. C. Huang, *J. Non-Crystalline Solids* **120**, 47 (1990).
8. D. W. Dwight, F. M. Fowkes, D. A. Cole, M. J. Kulp, P. J. Sabat, L. Salvati, Jr. and T. C. Huang, *J. Adhesion Sci. Technol.* **4**, 619 (1990).
9. C. Saint Flour and E. Papirer, *Ind. Eng. Chem. Prod. Res. Dev.* **21**, 337 (1982).
10. C. Saint Flour and E. Papirer, *Ind. Eng. Chem. Prod. Res. Dev.* **21**, 666 (1982).
11. C. Saint Flour and E. Papirer, *J. Colloid Interface Sci.* **91**, 69 (1983).
12. E. Papirer, in: *Composite Interfaces*, H. Ishida and J. L. Koenig (Eds), p. 203. Elsevier Science Publishing Co. (1986).
13. E. Papirer and H. Balard, *J. Adhesion Sci. Technol.* **4**, 357 (1990).
14. E. Osmont and H. P. Schreiber, in: *Inverse Gas Chromatography*, D. R. Lloyd, T. C. Ward and H. P. Schreiber (Eds), ACS Symposium Series No. 391, p. 230. American Chemical Society, Washington, DC (1989).
15. A. C. Tiburcio and J. A. Manson, *J. Appl. Polym. Sci.* **42**, 427 (1991).
16. A. C. Tiburcio and J. A. Manson, *J. Adhesion Sci. Technol.* **4**, 653 (1990).
17. K. Tsutsumi and T. Ohsuga, *Colloid Polym. Sci.* **236**, 38 (1990).
18. C. S. Fadley and S. A. L. Bergstrom, *Phys. Lett.* **A35**, 375 (1971).
19. G. M. Nishioka, *J. Non-Crystalline Solids* **120**, 34 (1990).
20. S. P. Wesson and R. E. Allred, in: *Inverse Gas Chromatography*, D. R. Lloyd, T. C. Ward and H. P. Schreiber (Eds), ACS Symposium Series No. 391, p. 204. American Chemical Society, Washington, DC (1989).
21. S. P. Wesson, J. J. Vajo and S. Ross, *J. Colloid Interface Sci.* **94**, 552 (1983).
22. S. P. Wesson and R. E. Allred, *J. Adhesion Sci. Technol.* **4**, 277 (1990).
23. F. J. Boerio and J. W. Williams, *Proc. 36th Annual Tech. Conf. Reinforced Plastics/Composites Div. SPI*, 23B (1980).
24. R. Bailey and J. E. Castle, *J. Mater. Sci.* **12**, 2049 (1977).
25. J. S. Jen, *Ceram. Eng. Sci. Proc.* **3**, 450 (1982).
26. Y. Eckstein, *J. Adhesion Sci. Technol.* **2**, 339, (1988).
27. Y. Eckstein, *J. Adhesion Sci. Technol.* **3**, 337, (1989).
28. H. Ishida, S. Naviroj and J. L. Koenig, in: *Physicochemical Aspects of Polymer Surfaces*, K. L. Mittal (Ed.), vol. 1, p. 91. Plenum Press, New York (1983).
29. G. M. Nishioka, unpublished data (1991).
30. G. M. Nishioka and J. A. Schramke, in: *Molecular Characterization of Composite Interfaces*, H. Ishida and G. Kumar (Eds), p. 387. Plenum Press, New York (1985).
31. G. M. Nishioka and J. A. Schramke, *J. Colloid Interface Sci.* **105**, 102 (1985).
32. S. Ross and N. Nguyen, *Langmuir* **4**, 1188 (1988).
33. R. L. Mieville and B. L. Meyers, *J. Catalysis* **74**, 196 (1982).
34. A. T. Aguayo, J. M. Arandes, M. Olazar and J. Bilbao, *Ind. Eng. Chem. Res.* **29**, 1621 (1990).
35. K. Tanabe, *Solid Acids and Bases*, pp. 58–66. Academic Press, New York (1970).
36. D. E. Clark, C. G. Pantano and L. L. Hench, *Corrosion of Glass*, Magazines for Industry, Inc., New York (1979).
37. R. Wong, *J. Adhesion* **4**, 171 (1972).

Part 3

Applications of Silane Coupling Agents

Silanes and Other Coupling Agents, pp. 401–409
Ed. K. L. Mittal
© VSP 1992.

Adhesion of polyimide to fluorine-contaminated SiO₂ surface. Effect of aminopropyltriethoxysilane on the adhesion

L. P. BUCHWALTER[1,*] and R. H. LACOMBE[2]

[1] *Dept. 466, Location 3-235, IBM T. J. Watson Research Center, Yorktown Heights, NY 10598, USA*
[2] *IBM East Fishkill, Hopewell Junction, NY 12590, USA*

Revised version received 5 February 1991

Abstract—The adhesion of pyromellitic dianhydride-oxydianiline (PMDA-ODA) polyimide to fluorine-contaminated silicon dioxide (F-SiO₂) with γ-aminopropyltriethoxysilane (APS) adhesion promoter has been studied as a function of the peel ambient humidity. The peel strength was not affected by the change in peel ambient relative humidity (RH) from 11–17% to 35–60% when APS was used at the interface. Without APS, the adhesion degraded significantly with this change in RH. It was found that although the dip application of APS caused the removal of about 80% of the initial atomic percentage of fluorine on the surface, it could not be totally removed even after several days in water at elevated temperature.

Keywords: Polyimide; adhesion; fluorine contamination; aminosilane; moisture effects.

1. INTRODUCTION

A recent review [1] on polyimide adhesion to metal and ceramic surfaces shows the relevance of this topic to many different technological areas. Of all the polyimides studied thus far, it is evident that the most popular one is PMDA-ODA. It has very good mechanical, thermal, and electrical properties, but it suffers from poor adhesion characteristics. This problem is often overcome by the application of an adhesion promoter to the surface of interest. The most popular adhesion promoter appears to be APS. An excellent review concerning APS has been written by Ishida [2]. A wealth of information concerning silane coupling agents can also be found in the book by Plueddemann [3].

This work continues our initial study concerning the effect of fluorine con-tamination on PMDA-ODA adhesion [4]. In this study we apply APS at the interface. We continue to use the peel test to monitor the adhesion and X-ray photoelectron spectroscopy (XPS) to study the surfaces and loci of failure after the peel test to elucidate the failure mechanisms.

2. EXPERIMENTAL

2.1. Sample preparation

Silicon wafers were oxygen plasma-cleaned in a parallel plate reactor at 400 W power, 300 mTorr pressure for 5 min. The gas flow was set to 100 sccm total,

---

*To whom correspondence should be addressed.

having 0–5% added CF_4 in it with oxygen. After the plasma cleaning of the surfaces, the wafers were dip-coated with APS in a 0.1% (v/v) solution in deionized water by submerging about three-quarters of the wafer into the solution. The wafer was held in the solution for 1 min and then drawn out slowly. This method of APS application was chosen *vis-à-vis* spin coating, since it allowed a convenient way of peel initiation in the wafer area not exposed to APS. One wafer after each surface treatment condition (0%, 2%, 5% added CF_4, and 2% added CF_4 + APS) was characterized using a Surface Science Laboratories SSX-100-05 small spot XPS unit. The analysis method used has been described in detail elsewhere [5]. The binding energy reference in the clean SiO_2 case is Si $2p$ at 99.8 eV [6] (the oxide is thin enough to allow access to the elemental Si). However, for all the locus-of-failure samples and the APS-coated samples, the XPS data are referred to C $1s$ at 285.0 ev.

PMDA-ODA polyamic acid was then spin-coated on the treated wafer surfaces from *N*-methylpyrrolidone (NMP) solution. The curing was done thermally to temperatures up to 400°C. The PMDA-ODA was applied in two coats, exposing the first one only up to 150°C to allow excellent PI/PI adhesion as shown by Brown *et al.* [7]. The resulting film thickness of PMDA-ODA was 40 μm.

After the PMDA-ODA films were fully cured, they were cut into 1.59 mm wide peel strips using a wafer dicer with deionized water coolant. The polyimide film was carefully cut down to the wafer, taking care that the wafer stayed intact.

2.2. Peel experiment

The peel test was done with an Instron Test using a constant 90° peel angle. The peel rate ranged from 0.05 to 500.0 mm/min for rate dependence analysis. A rate of 0.5 mm/min was used for the bulk of the adhesion studies. Three to five strips were peeled from each sample at low RH (11–17%) and at high RH (35–60%). The humidity was measured with a Cole-Parmer 9501 hygrometer. The peel force was determined as an average steady-state value evaluated from the strip chart by eye as shown in Fig. 1. The peel strength was then calculated by dividing the measured peel force by the peel strip width.

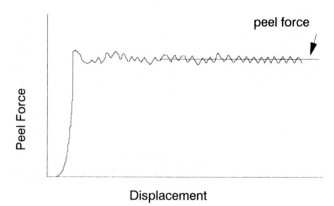

Figure 1. Schematic diagram of sample strip chart recording of the peel force vs. displacement.

2.3. Locus-of-failure analysis

The locus of failure after peel testing was determined using XPS analysis of both the silicon wafer and the polyimide failure surfaces. This analysis was also done as a function of the humidity.

3. RESULTS

3.1. Surface analysis

Table 1 shows the results for the 0%, 2%, and 5% added CF$_4$ in the oxygen plasma cleaning of the silicon wafers, as well as the effect of APS application on the surface composition.

Sample 1 in Table 1 shows that regardless of whether the fluorine is purpose-fully added to plasma gas or not, it will be deposited on the surface. This is due to the plasma system 'memory' as far as fluorine is concerned. Once the system has been exposed to fluorine-containing gas, it is not easy to get rid of it. Samples 2 and 3 give indications concerning the plasma cleaning run-to-run variation, which can be quite large (the F to Si atomic percentage ratio, F/Si, can range between 0.39 and 0.53 using the same plasma conditions). Samples 3 and 4 show the effect of APS dip-coating on the surface composition. About 80% of the fluorine is removed with this process. It should be noted, however, that F is still present in appreciable concentration. This amount of fluorine is capable of causing about a 90% drop in the peel strength when the humidity is increased from 11–22% to 40–55% RH [4].

Table 1.
Surface composition of the plasma-cleaned and APS-treated silicon surfaces

Sample	CF$_4$	APS	C%	O%	N%	F%	Si%	F/Si
1	0%	No	13	56	—	10	22	0.45
2	2%	No	7	61	—	9	23	0.39
3	2%	No	10	43	—	16	30	0.53
4	2%	Yes	25	43	2	3	27	0.11
5	5%	No	8	55	—	6	31	0.19

The F 1s binding energy (BE) is between 686.2 and 688.8 eV, which falls in the inorganic fluorine BE region [6, 8, 9]. In this work, most of the samples exhibited two F 1s peaks in the range given above, while this was not the case in our first study [4], where only a singlet F 1s peak was observed at about 686.9 ± 0.3 eV. This difference in fluorine chemistry may be primarily due to the differences in the plasma conditions (the system used in this study is different from the one used in the earlier study [4]). The higher BE F 1s is rather close to the organic F 1s BE, which ranges between 689.3 and 690.2 eV as reported by Clark *et al.* [10, 11]. If organic fluorine were present on the surface, then there would also be effects on the C 1s BE to corroborate it. The C 1s for F-containing organic compounds ranges from 288.0 to 292.1 eV [10, 11] for carbon atoms having 1–2 F atoms bonded to them. The C 1s peaks due to ambient contamination on a clean, out-of-the-box Si-wafer surface fall between 285.8 and 290.0 eV if the BEs are referred to Si $2p$ at 99.8 eV (elemental Si), as done by Chuang *et al.* [6]. When the surface is exposed to 100% oxygen plasma (with F contamination) or 2–5% added CF_4 in the oxygen plasma, the C $1s$ range is 285.9–290.8 eV. Based on the above information and the fact that the major component in the plama is O_2 (it is therefore unlikely that a polymer would be formed on the surface during plasma exposure), the fluorine found on these surfaces is entirely inorganic in nature. Table 2 shows the F $1s$ data for the different surface samples analysed.

Table 2.
Fluorine $1s$ peak binding energies and at.% concentrations of total fluorine on the surface

| Sample | Added CF_4 | F $1s$ peak BE and at.% of total peak area | | |
		688.6 eV	687.0 eV	685.1 eV
1	0%	82	18	
2	2%	88	12	
3	2%	88	12	
4	2% + APS		66	34
5	5%	75	25	

BE standard deviation is ± 0.4 eV for peaks at 688.6 and 687.0 eV.
No standard deviation calculated for peak at 685.1 eV (one data point).

The peak at about 687.0 eV is likely due to Si–F type fluorine [4, 6], while the assignments of the higher BE and the lower BE F $1s$ peaks are not clear at this time. However, it is likely that all the fluorine is bonded to Si [6, 8]. The higher BE peak at about 688.6 eV may be due to Si—O—F type bonding [12], although the BE is about 1.2 eV higher than the value reported by Vossen *et al.* [12]. It could also be due to some type of adsorbed F_2 species easily removed by the water solution of APS [8].

The N $1s$ region after exposure of the surface to APS has two peaks: one at about 400.0 eV and the other at about 402.1 eV, which may be assigned to free and protonated amine species [13, 14], respectively.

The amine bicarbonate salt of APS is commonly acknowledged to be present on surfaces treated with APS [2]. The higher BE N $1s$ peak may also be due to a protonated amine by the acidic surface hydroxyl groups, as shown in Fig. 2. This bonding scheme may not be such an unlikely one if one considered the acid–base nature of the two species; the SiO_2 surface is an acidic surface with an isoelectric

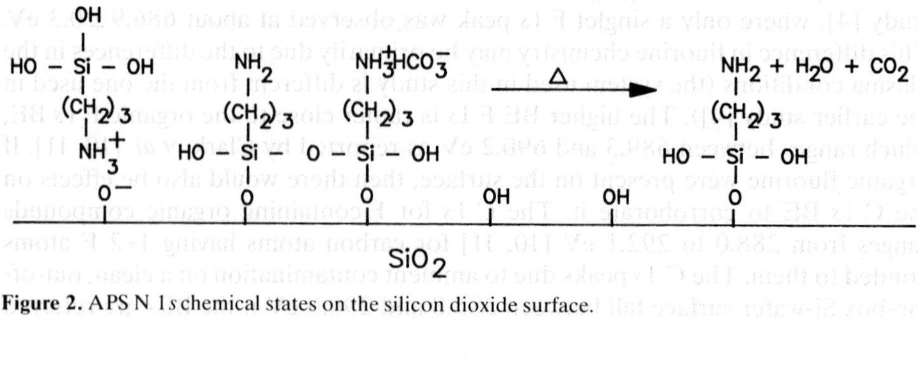

Figure 2. APS N $1s$ chemical states on the silicon dioxide surface.

point of the surface (IEPS) of about 2 [15], while the amine is basic, having a $pK_{A(B)}$ which can be approximated by that of ethyl amine, i.e. 10.6 [16]. If one calculated the degree of interaction between the amine and SiO_2 using Bolger's [16] approach, one would obtain

$$\Delta_B = pK_{A(B)} - \text{IEPS} = 10.6 - 2 = 8.6.$$

A large positive Δ_B value indicates that the ionic interactions at the interface dominate over the weaker dipole interactions [16]. It appears, then, that there is a high probability that the amine end will react with the silicon dioxide surface if this problem is approached purely from the acid–base point of view. It has been suggested by Linde [17] that indeed initially the bonding of APS would be with the amine group and the silicon dioxide surface. Upon heating, this would rearrange to bonding of the APS via the siloxane end of the molecule with the silicon dioxide surface as shown in Fig. 2.

As shown in Table 1, the application of APS to the F-contaminated SiO_2 surface causes a significant drop in the fluorine concentration. This result led us to study the removal of fluorine from the surface with water, which is the primary component in the APS solution. The results are shown in Table 3.

It is clear from the table that not all the fluorine can be removed from the silicon dioxide even after extended times in hot water. It should be noted here that the volume of water used was about 900 ml per experiment.

Table 3.
Elemental composition of F-SiO₂ surfaces after exposure to APS solution and water

Sample	C%	O%	N%	F%	Si%	F/Si
Clean out-of-the-box wafer	20	35	—	—	45	—
Oxygen plasma[a]	17	51	—	2	31	0.06
Oxygen plasma + APS spun	35	36	5	1	23	0.04
Oxygen plasma + 2 h in room temperature water	15	53	—	2	30	0.07
Oxygen plasma + 68 h in 76°C water	12	55	—	1	32	0.03
Oxygen plasma + 7.5 days in 92°C water	36	39	—	0.33	25	0.01

[a] As described in the Experimental section with no added CF₄.

3.2. Peel strength results

Table 4 lists the results of the peel test as a function of the added CF₄ and peel ambient RH.

A number of interesting observations can be made based on the data presented in Table 4. Storing even in nitrogen affects adhesion negatively in the case where no APS is used at the interface (samples 2 and 2a). In the 'no APS' case, the adhesion appears to be a function of the F% concentration on the surface prior to polyimide application, as is shown in Fig. 3.

It appears from the data that there may be a threshold value for the F concentration on the surface, where a very drastic degradation in the initial adhesion is observed. The data in Fig. 3 suggest that such a concentration would be at about

Table 4.
PMDA-ODA adhesion (J/m^2) to APS-treated SiO$_2$ as a function of the added CF$_4$ and peel ambient RH

Sample	%CF$_4$	Surface F%[a]	No APS		With APS	
			Low RH	High RH	Low RH	High RH
1	0	10	160	4	1070	1040
2	2	9	200	4	1120	1070
2a[b]	2	9	130	1	1110	1070
3	2	16	30	—	—	—
4	2	3	—	—	1260	1210
5	5	6	220	60	1020	930

[a] F% after plasma exposure, except in sample 4; which is sample 3 after APS exposure.
[b] The same as sample 2 but peeled after storing in nitrogen for 1 month.
Low RH = 11–17%; high RH = 35–60%.

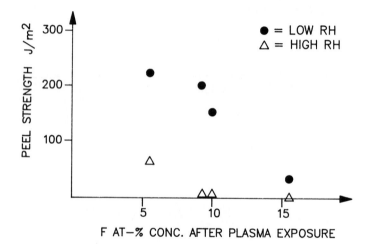

Figure 3. PMDA-ODA adhesion to F-SiO$_2$ without APS.

9–10 at.% of fluorine on the silicon dioxide surface (low RH case; the high RH case does not have enough data points to make such a conclusion).

The peel strength standard deviation in samples 1, 2, 2a, 3, and 4 is 3–8%, while in sample 5 it is 20–22%. If this is taken into consideration, one can see that the adhesion of PMDA-ODA to silicon dioxide with APS is not affected by the presence of fluorine, neither is it affected by the change in peel ambient RH as shown in Fig. 4.

All of the above peel data were collected using a 0.5 mm/min peel rate. A brief study was conducted on the effect of the peel rate on the measured peel strength in the PMDA-ODA/APS/SiO$_2$ system without any F contamination. These data are summarized in Table 5.

It is evident from Table 5 that an order of magnitude change in the peel rate when peeling PMDA-ODA polyimide does not make a significant difference in the measured peel strength. Actually, the difference between consecutive table entries may be considered to be within the experimental error.

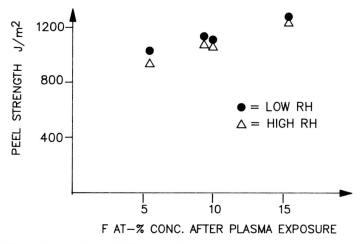

Figure 4. PMDA-ODA adhesion to F-SiO₂ with APS.

Table 5.
PMDA-ODA/SiO₂ peel strength as a function of
the peel rate

Rate (mm/min)	Peel strength (J/m²)	
	No APS	With APS
0.05	—	740
0.5	160	790
5.0	—	820
50.0	130	880
500.0	220	920

Peel strength standard deviation less than 8%.

3.3. Locus-of-failure analysis

The locus of failure was determined for samples 2 and 4 after peeling at low and high RH. The data are presented in Table 6.

The detailed oxidation state analysis of the locus-of-failure data shows that the O 1s peak area ratio in C=O/C—O—C is 3.0 ± 0.4 for all the failure surfaces studied, which is, within the experimental error, the same as that for virgin PMDA-ODA polyimide surface (i.e. 3.2) [18]. C 1s high resolution data show the presence of the C=O C 1s peak at about 288.9 eV as well as the π-to-π* peak due to the aromatic carbon atoms of the polyimide. This information combined with the data in Table 6 suggests that the failure locus in all cases is within the polyimide close to the PI/APS–SiO₂ interface, but not at it. This has also been observed for PMDA-ODA on SiO₂ and Si₃N₄ with APS and without F contamination by Anderson *et al.* [19]. Two hypotheses may be considered as the reasons for the lack of water sensitivity of this interface. First, siloxane films are known to be hydrophobic in nature [3], which would suggest a lack of water in the interfacial region. Without water at the interface, HF formation from the Si—F and H₂O reaction would not be possible. (HF formation and the following interaction with Si—O—Si bonds are remove the fluorine from the surface, but makes it ineffective in degrading adhesion.

(3) The peel strength of PMDA-ODA on SiO₂ is insensitive to an order of magnitude change in the peel rate.

REFERENCES

Silanes and Other Coupling Agents, pp. 411–421
Ed. K. L. Mittal
© VSP 1992.

Adhesion of polyimides to ceramics: Effects of aminopropyltriethoxysilane and temperature and humidity exposure on adhesion

L. P. BUCHWALTER,[1,*] T. S. OH[2] and J. KIM[1]

[1]*IBM T. J. Watson Research Center, P.O. Box 218, Yorktown Heights, NY 10598, USA*
[2]*Korea Institute of Science and Technology, P.O. Box 131, Cheongryang, Seoul, Korea*

Revised version received 20 December 1990

Abstract—The effects of both γ-aminopropyltriethoxysilane (APS) and elevated temperature and humidity ($T\&H$) exposure on the adhesion of pyromellitic dianhydride-oxydianiline polyimide to SiO_2, Al_2O_3, and MgO were studied using XPS, SEM, and peel test. Adhesion and $T\&H$ stability of PMDA-ODA on SiO_2 is significantly improved when APS is used at the interface, while no significant improvement is observed for Al_2O_3 or MgO. XPS analysis of the surfaces showed no retention of APS on Al_2O_3 or MgO, while SiO_2 did retain APS, as is expected. The APS retention is affected by surface treatment of the oxide prior to APS application.

Keywords: Polyimide; adhesion; ceramics; aminosilane; humidity.

1. INTRODUCTION

Adhesion of polyimides to inorganic substrates is of great importance to the microelectronics industry [1, 2]. The polyimide films are deposited most often by spin coating the polyamic acid (PAA) usually from a *N*-methylpyrrolidone (NMP) solution onto the substrate surface followed by thermal imidization at temperatures up to 400°C. The most studied polyimide is the pyromellitic dianhydride-oxydianiline (PMDA-ODA), which exhibits excellent mechanical and dielectric properties, but not so good adhesion characteristics. The latter has been generally overcome by application of an adhesion promoter, such as γ-aminopropyltriethoxysilane [3–7]. The reactions of APS (coated from water solution) with the silicon dioxide surface as well as with polyamic acid have been well characterized by Linde and Gleason [4]; however, we do not have such detailed information available on APS interaction with other ceramic surfaces.

Polyimide adhesion to the substrate surface is not only important initially after the interface preparation, but also after exposure for extended times to elevated temperature and humidity ($T\&H$) conditions. As Plueddemann [7] states: "Water molecules diffuse through any plastic and thus will reach the interface in composites exposed to humid environment. Individual water molecules, however, are relatively harmless at the interface unless they are capable of clustering into a liquid phase. The concentration of water at the interface is not determined by the rate of permeation of the water through the polymer matrix (silicones and

---

*To whom correspondence should be addressed.

hydrocarbon resins have highest permeability) but by the amount of moisture retained at the interface." In the microelectronics industry the various components experience exposure to water during the manufacture, as well as later during the operation of the computer (ambient humidity). Therefore, rigorous reliability testing, including extended exposures at *T&H* is of importance to the industry. It is necessary to have an assurance that the product will tolerate extended exposure to humidity.

In the present study we have examined the adhesion of PMDA-ODA (DuPont) to SiO_2, Al_2O_3, and MgO with APS (Union Carbide), and how that adhesion is affected by exposure to *T&H* conditions. This work is a continuation of our study published earlier [8].

2. SAMPLE PREPARATION AND EXPERIMENTAL PROCEDURE

The substrates used in this study were (0001) sapphire (Al_2O_3), (001) magnesia (MgO), and amorphous fused silica (SiO_2). All substrates were obtained with surface finish to 0.025 μm, and were cleaned with isopropylalcohol (IPA) prior to PAA or APS application. The surfaces and interfaces after peel test were characterized using X-ray photoelectron spectroscopy (XPS). The PMDA-ODA PAA was cast from NMP solution. Figure 1 shows the structure of the PAA and the thermally imidized PMDA-ODA polyimide.

APS was spin coated on the IPA cleaned substrates from 0.1% solution in deionized water after 1 h from mixing (fresh solution was made daily) and baked at 85°C in nitrogen. The PAA was then coated on the APS treated and non-treated substrates at 3000 rpm and exposed to 85°C bake for 30 min to remove most of the solvent. Multiple coatings were applied to obtain the desired PI film thickness by repeating the above PAA application process. Final cure was conducted at 150°C for 30 min, 200°C for 30 min, 300°C for 30 min, and 400°C for 40 min in nitrogen ambient. PI films were cut to 5 mm wide peel strips with a sharp scalpel.

The samples were exposed to *T&H* at 85°C and 81% RH for 100–700 h. The peel force was measured before and after *T&H* exposures using 5 mm wide peel

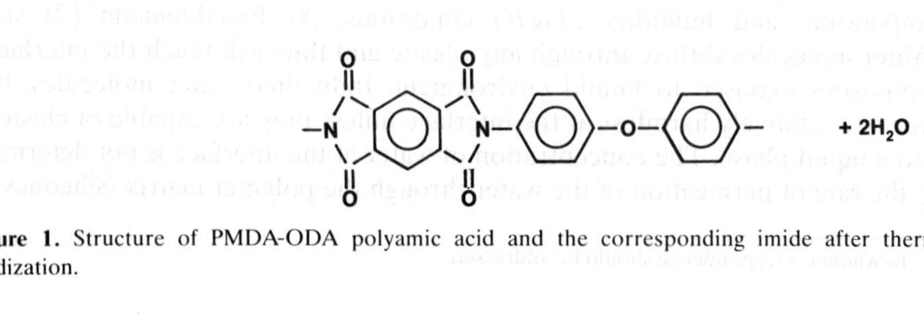

Figure 1. Structure of PMDA-ODA polyamic acid and the corresponding imide after thermal imidization.

strips, 90° peel angle, and 4 mm/min peel rate. The peel locus of failure was then determined using XPS and scanning electron microscopy (SEM). The XPS analysis was performed with a Surface Science Laboratories SSX-100-05 small spot unit. The details of the method are described elsewhere [9].

3. RESULTS AND DISCUSSION

3.1. Surface analysis of the substrates

XPS analysis results of the substrates after IPA cleaning and APS application and bake are shown in Table 1.

It is interesting to notice from the table that APS has only been retained on SiO_2 surface. This has in part to do with the surface cleaning process used in this study, which affects the retention of APS [10]. If these surfaces are cleaned in oxygen plasma (Tegal barrel reactor, 1 Torr oxygen pressure, 300 W power, 5 min) instead of IPA cleaning the APS is also retained on Al_2O_3, and on MgO. The nitrogen concentration on oxygen plasma cleaned surfaces after APS application (as analysed by XPS) is 4, 2, and 2 at.% for SiO_2, Al_2O_3, and MgO, respectively [11]. One reason for the difference in the APS retention, as compared to the IPA cleaned surfaces, may be possible IPA retention (a weak acid) on the more basic surfaces. This IPA is not replaced by the APS and thus can affect its bonding to these surfaces. The other reason may be that surface cleaning is only accomplished with SiO_2 surfaces using IPA (over 60% drop in carbon at.% concentration). In Al_2O_3 and MgO cases the carbon at.% concentration change is about 15 and 0% respectively (Table 1 in [8]). The APS retention can also be accomplished on both Al_2O_3 and MgO if the substrates are preheated to 100°C [12] (note: IPA b.p. is 82.4°C).

Based on the XPS N(1s) data the APS nitrogen is in two different chemical environments on SiO_2 surfaces. The two N(1s) peaks are at 400.1 and at 402.1 eV, which can be assigned to a free and protonated amine, respectively [10, 14]. The protonated amine is likely in a form of an amine bicarbonate salt of the APS as shown in Fig. 2, which is due to the amine reaction with moist CO_2 [6, 13]. This reaction is reversible thermally [13], leaving the amine available for reaction with the PAA.

These same N(1s) peaks have also been seen on F-contaminated SiO_2 surfaces, albeit at different relative concentrations (which may have to do with different surface cleaning processes [10]) [15].

Table 1.
XPS analysis of substrates after IPA cleaning and APS application

Substrate	Elemental composition					
	C%	O%	N%	Si%	Al%	Mg%
SiO_2 + IPA[a]	16	62	—	22	—	—
SiO_2 + IPA + APS	21	53	2	24	—	—
Al_2O_3 + IPA	19	51	—	—	30	—
Al_2O_3 + IPA + APS	19	43	—	—	40	—
MgO + IPA	37	31	—	—	—	32
MgO + IPA + APS	32	34	—	—	—	34

[a] Data from earlier work [8].

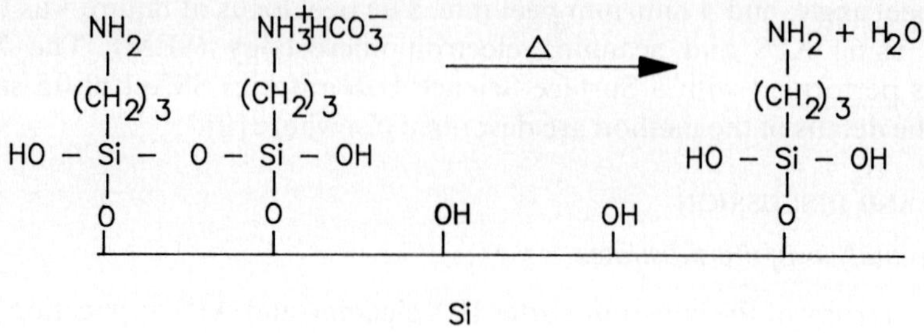

Figure 2. APS N(1s) chemical states on SiO₂ surface.

3.2. Adhesion of PMDA-ODA to the ceramic surfaces

3.2.1. PMDA-ODA on SiO₂. It is clear from Fig. 3 that the adhesion of PMDA-ODA to SiO₂ surface is significantly improved by the application of APS. This is not only seen initially but also after exposure to extended times at *T&H* conditions, i.e. the reliability of the interface has been improved. Notice the spontaneous delamination (zero peel strength) of the PMDA-ODA film from non-APS treated silica surface after only 100 h in *T&H*. It should be pointed out here that the 100 h exposure was the first point at which the samples were removed from the *T&H* test chamber. It is possible that the delamination may have occurred much earlier than the 100 h reported here. Table 2 shows the locus of failure analysis results for the interfaces after initial peel and after exposure to *T&H* for 100 (no APS only) and 700 h.

Generally speaking, the locus of failure is not changed after exposure to *T&H*, failing in the PI close to the failure surface. On occasion, as can be seen from the 700 h exposure in the non-APS case, a mixed mode failure is observed. As we had discussed in our previous paper [8], this can happen with very thick PI films due to the way in which the film was cut to peel strips creating weakness in the glass causing fracture in it also. Water molecules can also attack directly the Si—O bonds via a stress corrosion mechanism [16]. This has been established for fracture in bulk glass samples [17–19]. Therefore, it is quite probable that given enough time at *T&H*, glass fracture can play a role in the adhesion failure.

No significant difference is observed in the locus of failure data for the APS treated samples presented in Table 2. A closer look at the XPS high resolution scans as shown in Fig. 4 supports the interpretation that the failure is within the polyimide close to the PI-APS interface, but not at it. This is in particular verified by the presence of C=O C(1s) peak at about 288.9 eV, π-to-π^* transition peak, and the N(1s) peak at 400.8 eV due to imide nitrogens.

It is interesting to consider why the peel force is degraded significantly in the non-APS case after exposure to *T&H*, while the locus of failure remained the same before and after exposure to *T&H*, i.e. polyimide cohesive failure (apart from the sample exposed to 700 h with mixed mode failure). It has recently been suggested that the bonding between the surface bound polyimide chains and the bulk polyimide film is primarily via dispersion forces, and perhaps with some degree of chain entanglement (the latter component being much larger for the APS treated samples than the ones without it) [2]. Dispersion forces, as is well known, are severely affected by moisture (water) even to the point of spon-

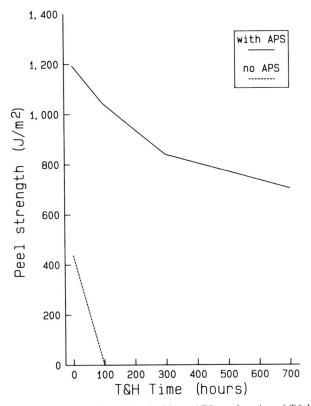

Figure 3. PMDA-ODA adhesion to SiO_2 with and without APS as a function of *T&H* exposure.

Table 2.
Locus of failure data for PMDA-ODA on SiO_2 with and without APS as a function of *T&H* exposure

Time at *T&H*	APS	Failure surface	C%	O%	N%	Si%	N/Si
None	no	PI	75	18	7	—	—
None	no	ceramic[a]	38	39	3	19	0.16
100 h	no	PI	76	18	7	—	—
100 h	no	ceramic	25	48	2	25	0.08
700 h	no	PI	78	15	7	1	7
700 h	no	ceramic	39	40	3	18	0.17
None	yes	PI	76	17	7	—	—
None	yes	ceramic	68	20	6	6	1.0
700 h	yes	PI	76	16	8	—	—
700 h	yes	ceramic	57	29	6	9	0.67

[a] Ceramic = SiO_2.

taneous delamination as observed in the non-APS treated samples. If polyimide hydrolysis [20] is considered as the reason for the loss in the peel strength after *T&H* exposure, then the same degree of peel strength loss should be seen with the non-APS and APS treated samples since they both, being the same polyimide, should hydrolyse in the same fashion. However, this is not observed. The non-APS treated samples have a 100% loss in peel strength after 100 h in *T&H*, while

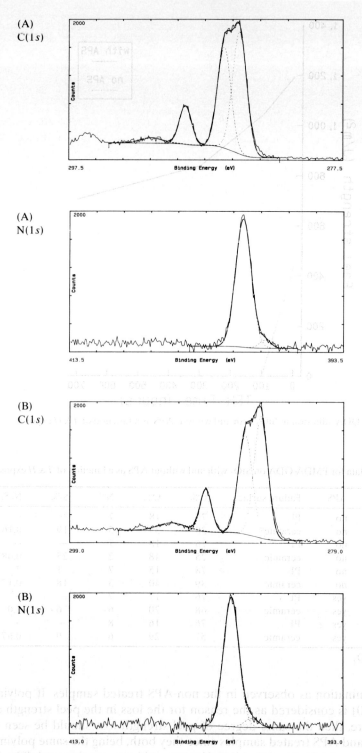

Figure 4. XPS locus of failure analysis of PMDA-ODA on SiO$_2$. (A) PI failure surface C(1s) and N(1s) data, (B) SiO$_2$ failure surface C(1s) and N(1s) data.

the APS treated samples show barely any degradation in the peel strength after such exposure (from initial value of 1193 to 1041 J/m^2 after 100 h of *T&H* exposure).

Figure 5 shows scanning electron micrographs of the failure surface of peeled polyimide films. It illustrates that peel crack propagates with a discontinuous stick-slip process, as reported previously [8]. The spacing of each striation becomes smaller as the exposure to *T&H* increases for APS treated surfaces. Since the peeling of a material from another is a series of discontinuous unstable crack propagations followed by arrest, blunting and propagation of crack, it is clear that larger spacing between the failure striations means more elastic stored energy for debonding and this corresponds nicely with higher peel force [8, 21].

3.2.2. PMDA-ODA on Al$_2$O$_3$. A minor improvement is noticed in the peel force of PMDA-ODA on Al$_2$O$_3$ when APS is applied to the surface. From the surface analysis results one can see that the APS was not retained on the IPA cleaned sapphire surface to any detectable level, which is likely the cause for no significant improvement in the peel force. The minor improvement in the results may have to do with a possible surface cleaning effect of the sapphire surface with APS solution. The data in Table 3 show that the failure locus has not changed significantly by the APS or *T&H* exposure, being essentially in the polyimide film close to the polyimide/ceramic interface.

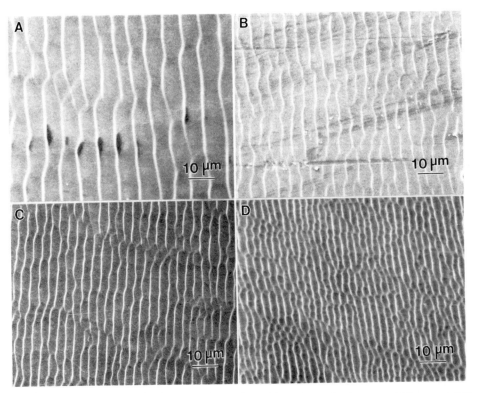

Figure 5. SEM micrographs of polyimide failure surface for PMDA-ODA on APS treated SiO$_2$ surface. (A) no *T&H*, (B) 100 h *T&H*, (C) 300 h *T&H*, and (D) 700 h *T&H*.

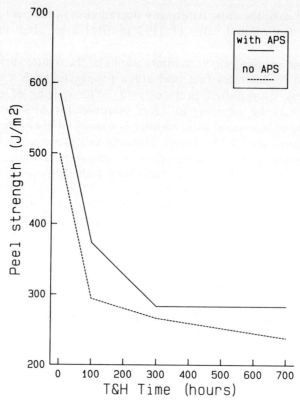

Figure 6. PMDA-ODA adhesion to Al_2O_3 with and without APS as a function of *T&H* exposure.

Table 3.
Locus of failure data for PMDA-ODA on Al_2O_3 with and without APS as a function of *T&H* exposure

Time at *T&H*	APS	Failure surface	C%	O%	N%	Al%	N/Al
None	no	PI	79	15	6	—	—
None	no	ceramic[a]	45	31	3	20	0.15
700 h	no	PI	77	15	7	—	—
700 h	no	ceramic	46	33	0.4	21	0.02
None	yes	PI	77	14	7	—	—
None	yes	ceramic	37	26	3	34	0.09
700 h	yes	PI	76	16	7	—	—
700 h	yes	ceramic	47	32	2	20	0.1

[a] Ceramic = Al_2O_3.

The N($1s$) peak on the sapphire failure surfaces indicates presence of imide (peak at 400.7–401.0 eV), which suggests presence of polyimide on the sapphire failure surface. This is supported by the C($1s$) data which show the presence of the carbonyl C($1s$) species as well as π-to-π^* transition peak due to aromatic species. Since the polyimide failure surface shows only C($1s$), N($1s$), and O($1s$) peaks in relative abundance, characteristic to the polyimide, it can be said that the failure in the PMDA-ODA on sapphire case is in the polyimide close to the polyimide/ceramic interface, but not at it.

3.2.3. PMDA-ODA on MgO. PMDA-ODA peel force data shown in Fig. 7 exhibit a very interesting phenomenon as a function of *T&H* exposure. The peel force is significantly increased as the time in *T&H* is increased. This is somewhat unusual, but apparently repeatable. The exposure to APS has not made much difference in the results, which is understandable from the initial surface analyses after IPA cleaning and APS exposure. The XPS data show no detectable amount of APS on the thus exposed MgO surface. The reasons for the peel force increase as a function of *T&H* exposure are not clear at this time. This is, however, due to increased interfacial strength, and not due to the polyimide mechanical properties (Young's modulus and yield stress) changes. If the latter were the case, then we should see similar effects also in the first two cases, which is not seen. However, more detailed analysis is essential to clarify the exact mechanism and this observation merits further study.

The locus of failure analysis results are shown in Table 4 below.

The locus of failure in all PMDA-ODA/MgO cases is consistently a mixed mode type leaving ceramic on the polyimide failure surface, and leaving polyimide on the ceramic failure surface. APS application prior to polyimide coating does not change the failure locus or the peel strength behavior of these interfaces.

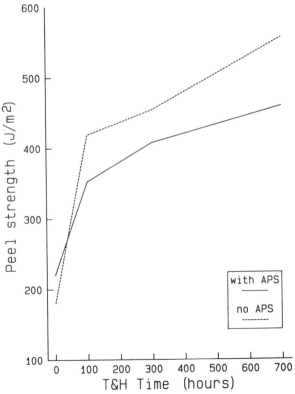

Figure 7. PMDA-ODA adhesion to MgO with and without APS as a function of *T&H* exposure.

Table 4.
Locus of failure data for PMDA-ODA on MgO with and without APS as a function of $T\&H$ exposure

Time at $T\&H$	APS	Failure surface	C%	O%	N%	Mg%	N/Mg
None	no	PI	73	19	7	2	3.5
None	no	ceramic[a]	41	32	3	25	0.12
700 h	no	PI	not analyzed				
700 h	no	ceramic	14	53	0.4	33	0.01
None	yes	PI	74	18	7	1	7
None	yes	ceramic	35	35	2	28	0.07
100 h	yes	PI	72	20	6	2	3
100 h	yes	ceramic	46	27	1	24	0.04
700 h	yes	not analyzed					

[a] Ceramic = MgO.

4. SUMMARY

The following statements can be made based on the above study:

(1) APS is retained only on SiO_2 surface cleaned with IPA, but not on Al_2O_3, or on MgO surfaces thus cleaned.

(2) Initial adhesion improvement and increased reliability is seen when APS is used in connection with SiO_2, but no significant change is observed with Al_2O_3.

(3) Adhesion is improved as a function of $T\&H$ exposure in the MgO. The reasons for this improvement are not clear at this time, and merits further study.

REFERENCES

1. R. R. Tummala and E. J. Rymaszewski (Eds), *Microelectronics Packaging Handbook*. Van Nostrand Reinhold (1989).
2. L. P. Buchwalter, *J. Adhesion Sci. Technol.* **4**, 697 (1990).
3. A. Saiki and S. Harada, *J. Electrochem. Soc.* **132**, 2700 (1982).
4. H. Linde and R. T. Gleason, *J. Polym. Sci. Polym. Chem. Ed.* **22**, 3043 (1984).
5. J. Greenblatt, C. J. Araps and H. R. Anderson, Jr., in: *Polyimides: Synthesis, Characterization, and Applications*, K. L. Mittal (Ed.), Vol. 1, p. 573. Plenum Press, New York (1984).
6. H. Ishida, in: *Adhesion Aspects of Polymeric Coatings*, K. L. Mittal (Ed.), p. 45. Plenum Press, New York (1983).
7. E. P. Plueddemann, *Silane Coupling Agents*. Plenum Press, New York (1982).
8. T. S. Oh, L. P. Buchwalter and J. Kim, *J. Adhesion Sci. Technol.* **4**, 303 (1990).
9. L. P. Buchwalter, B. D. Silverman, L. Witt and A. R. Rossi, *J. Vac. Sci. Technol.* **A5**, 226 (1987).
10. P. R. Moses, L. M. Wier, J. C. Lennox, H. O. Finklea, J. R. Lenhard and R. W. Murray, *Anal. Chem.* **50**, 576 (1978).
11. L. P. Buchwalter, T. S. Oh and J. Kim, unpublished results (1989).
12. S. P. Kowalczyk, private communication (1989).
13. H. Linde, private communication (1989).
14. R. Nordberg, R. G. Albridge, T. Bergmark, U. Ericson, J. Hedman, C. Nordling, K. Siegbahn and B. J. Lindberg, *Arkiv for Kemi* **28**, 257 (1967).
15. L. P. Buchwalter and R. H. Lacombe, *J. Adhesion Sci. Technol.* **5**, 449 (1991).
16. L. P. Buchwalter and R. H. Lacombe, *J. Adhesion Sci. Technol.* **2**, 463 (1988).
17. T. A. Michalske and B. C. Bunker, *Sci. Am.* **257**, 122 (1987).
18. T. A. Michalske and B. C. Bunker, *J. Appl. Phys.* **56**, 2686 (1984).

19. T. A. Michalske and C. R. Fuller, *J. Am. Ceram. Soc.* **68**, 596 (1985).
20. J. F. Heacock, paper presented at the Second Intl. Conference on Polyimides held in Ellenville, New York, Oct. 30–Nov. 1, 1985.
21. J. Kim, K. S. Kim and Y. H. Kim, *J. Adhesion Sci. Technol.* **3**, 175 (1989).

Silanes and Other Coupling Agents, pp. 423–437
Ed. K. L. Mittal
© VSP 1992.

Oligomerization of an aminosilane coupling agent and its effects on the adhesion of thin polyimide films to silica

C. J. LUND, P. D. MURPHY* and M. V. PLAT

IBM Corporation, East Fishkill Facility, Z/E40, Route 52, Hopewell Jct., NY 12533, USA

Revised version received 29 August 1991

Abstract—This work studies the effects of self-oligomerization of the aminosilane coupling agent 3-aminopropyltriethoxysilane—also called γ-aminopropyltriethoxysilane, 3-APS, γ-APS, APS or A1100 (Union Carbide)—on the adhesion of thin polyimide films to a native-oxide silica surface under no stress, i.e. T(0) conditions, and after standard 85°C/81% T&H (temperature and humidity) stress. Techniques have been developed using both silicon and hydrogen NMR to control and monitor the degree of oligomerization in aqueous solutions at low concentrations (0.1 vol %).

The results of these studies suggest that: (1) Highly self-oligomerized 0.1 vol % APS solutions promote adhesion as do those with very low amounts of self-oligomerization; (2) T&H conditions cause an overall decrease in adhesion. However, this loss is not well-correlated with the degree of oligomerization and the likely perturbations of the silicon–oxygen environment at the surface; (3) If the current ideas of APS/mineral surface interactions are to be supported, then the highly-oligomerized APS molecules must first revert to more reactive states such as monomers. Perhaps the surface itself catalyzes such a reversion. The newly formed APS silanols are then free to form surface bonds and eventually a two-dimensional surface network. Alternatively, the formation of Si—O covalent surface bonds may not even be required for APS-promoted adhesion.

Finally, studies of aqueous APS concentrations and their effects on adhesion under T&H stress suggest that aqueous APS solutions with concentrations above 0.01 vol % produce effective coupling at T(500) T&H conditions. Concentrations below this level produce lower adhesion that degrades quickly under T&H stress.

Keywords: Aminosilane; adhesion; polyimide; ceramic; PMDA–ODA; silica; interface; coupling.

1. INTRODUCTION

The aminosilane coupling agent 3-aminopropyltriethoxysilane or γ-amino-propyltriethoxy silane—also abbreviated as 3-APS, γ-APS, APS or A1100 (Union Carbide)—is widely used to promote adhesion between polyimide thin films and mineral surfaces such as native-oxide silica, alumina and various glass ceramics [1, 2]. The structure of APS and the hydrolysis reaction are shown in Fig. 1. Typically, dilute aqueous solutions of 0.1 vol % or approximately 0.080 wt % are employed to prime the mineral surface. The mechanism for the inter-action of the bifunctional aminosilane with the mineral surface is the subject of much speculation, although it is conjectured by Linde and Gleason [3] that the amine end initially forms an electrostatic bond with surface hydroxyls. Sub-sequently, possibly as the result of elevated temperatures, the silanol end of the molecule proceeds to form a siloxane-like bond with the surface and the amine

---

*To whom correspondence should be addressed.

end 'flips up' to point away from the surface. Ultimately, this strong alkyl amine interacts with the polyimide thin film precursor. An illustration of the proposed model [3] of APS coupling at PI thin film to a ceramic surface is shown in Fig. 2.

The APS molecule is quickly hydrolzyed from the triethoxy ester to the trisilanol in aqueous solution. (See below.) It is well known that the silanols are very reactive and will condense with each other to form oligomerized networks

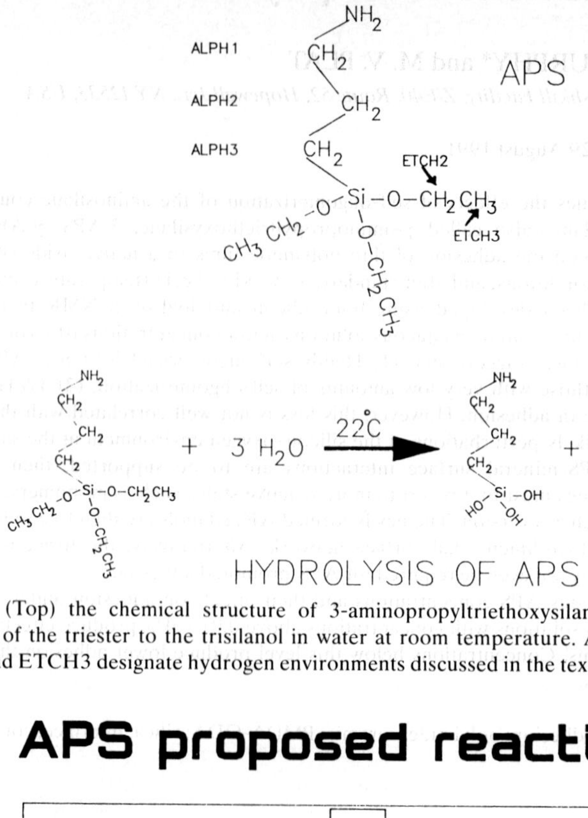

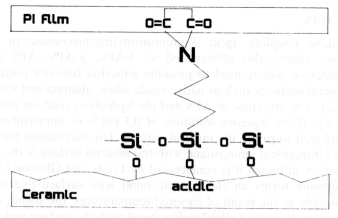

Figure 1. (Top) the chemical structure of 3-aminopropyltriethoxysilane or APS. (Bottom) the hydrolysis of the triester to the trisilanol in water at room temperature. ALPH1, ALPH2, ALPH3, ETCH2 and ETCH3 designate hydrogen environments discussed in the text.

APS proposed reactions

Figure 2. The proposed reaction of the bifunctional APS molecule when serving as an adhesion promoter or coupling agent for polyimide thin films on native-oxide ceramics such as silica and alumina.

[4, 5]. Because of this reactivity, primer solutions are formulated as very dilute (typically 0.1 vol%) solutions and used quickly. In industry, however, it is usually not practical to make a 'fresh' APS solution for each application. Therefore, there is some concern as to how oligomerization will affect the adhesion promoting efficiency of such primer solutions over a typical 8-hour shift.

The following studies attempt to characterize the degree of self-oligomerization of aqueous APS solutions and relate such data to effects on adhesion. Of particular interest are long-term effects that are usually evaluated by T&H acceleration.

2. EXPERIMENTAL

2.1. APS oligomerization

A-1100 silane primer from Ohio Valley Speciality Chemical (Marietta, Ohio) was used in all APS experiments. Heavy water or deuterium oxide (99.9% pure) was used in all experiments involving APS oligomerization. Heavy water is used to slow down the hydrolysis kinetics and as an NMR reference. Regular water (18 MΩ − cm resistivity at 25°C) was used in all dilutions and elsewhere.

Oligomerization was achieved in 50 vol% solutions of heavy water at room temperature. Proton NMR was used to calibrate oligomerization vs. time. The procedure is discussed in Section 3.

Adhesion strength between PI and Si in the presence of APS adhesion promoter was determined through the use of a 90° peel test. The test specimen was prepared by depositing a thin gold release layer on two edges of a (100) boron doped p-type polished Si wafer, leaving the middle section unaltered. Note, the active surface of this wafer is considered to be the native oxide. The APS solution is then puddled on the surface of the wafer and allowed to stand for 30 s. Next, the wafer is spun at 2000 rpm for 30 s and baked for 30 min at 90°C. The PI precursor solution is applied to the wafer and spun at 2200 rpm for 60 s. Du Pont PI-5878 (PMDA–ODA) was used for the PI coating. Two coats of the PI were necessary to achieve the desired thickness of 20–25 μm. The first coat of the PI was soft baked at 100°C for 60 min prior to application of the second coat. The second coat was soft baked identically as the first and the final curing was completed in a box oven under nitrogen purge. The cure schedule was 150°C/30 min, 230°C/30 min, 300°C/30 min and 400°C/40 min. The cured film was then diced, using a wafer dicer, into 3.18-mm strips perpendicular to the edge of the gold release layer.

An Instron 4204 uni-axial tension tester was used to perform the peel test experiments. The lower grip of the Instron was replaced by an X–Y stage that is free to translate in either direction to maintain, as much as possible, a 90° angle between the wafer and PI while being peeled. Peel testing was accomplished by securing the wafer to the X–Y stage, scribing along the edge of the wafer in the gold coated area to release the ends of the diced strips from the gold layer and inserting the released strip ends, one at a time, into the upper grip of the Instron and pulling at a rate of 5 mm min^{-1}. After an initial region of increasing load with displacement, the force required to sustain the peeling was constant. This constant force divided by the width of the peel strip is reported as the peel strength in units of g mm^{-1}

2.2. APS adhesion vs. concentration

In these experiments, fresh APS solutions at concentrations from 0.1 to 0.00001 vol % were prepared by sequential dilution. Initially, 200 ml of a 0.1 vol % solution was prepared. An aliquot of this solution was diluted to 0.01 vol % and then an aliquot of the 0.01 vol % solution was diluted to 0.001 vol %, and so on. This procedure ensured a progressive and accurate reduction of the APS concentration, especially at the very low vol %'s. Plastic (polypropylene) bottles used as hydrolyzed APS is known to react with glass.

Preparation of films on the silicon wafers, T&H exposure and adhesion testing were identically the same as discussed in the oligomerization experiments above.

2.3. NMR instrumentation

NMR analyses were done on an IBM Instruments NR-300 spectrometer and an Oxford 7 Tesla superconducting narrow-bore magnet. Silicon-29 (Si-29) NMR spectra were recorded at 59.6 MHz and hydrogen (also commonly called proton or H-1) NMR spectra at 300.13 MHz. Spectra were recorded using conventional single-pulse techniques with proton decoupling for Si-29 acquisitions. Si-29 experiments were structured so as to suppress nuclear Overhauser enhancement (NOE). For Si-29 acquisitions, spectral widths were 50 kHz and Fourier transform (FT) sizes were 4K points. For protons, spectral widths were 7.5 kHz and FT sizes were 16K points. Si-29 rf pulse widths were approximately 12 μs and proton rf pulse widths were 8 μs.

3. RESULTS AND DISCUSSION

A recent review of PI adhesion including discussion of APS has been published by Buchwalter [6]. It is conjectured that the bifunctional APS promotes adhesion or coupling by the formation of siloxane (Si—O—Si) bonds to the mineral surface and, eventually, imide bonds to the polyimide (PI) thin film [3]. Figure 2 illustrates this proposed coupling scheme. NMR studies in this laboratory of nitrogen-15-enriched APS have confirmed that the APS amine ultimately forms an imide bond with the PI thin film [7]. Figure 3 shows Si-29 NMR spectra of various concentrations of APS in heavy water. Heavy water or deuterium oxide is chosen for two reasons: (1) the deuteron is a common reference used in NMR tuning and locking and (2) the heavier hydrogen isotope will make studies of the rapid ester-to-alcohol hydrolysis somewhat easier because deuteron transfer is probably slower than proton transfer, possibly by as much as a factor of 2.

A fundamental discussion of NMR theory and technology has been published by Farrar and Becker [8] and, more recently, by Farrar [9] and Chandrakumar and Subramanian [10]. More advanced discussions appear in books by Abragam [11]; Slichter [12]; Carrington and McLachlan [13]; and Pople et al. [14].

NMR spectra are drawn such that the X-axis uses a dimensionless quantity called the chemical shift. The shift is the relative frequency variation divided by the Larmor frequency of the reference and expressed in ppm. For silicon NMR, the reference is usually neat tetramethylsilane (TMS) which is assigned 0 ppm. In the current work, the silicon reference is chosen as the N0 silicons in neat APS liquid which is assigned 0 ppm. Chemical shifts can be converted to the TMS

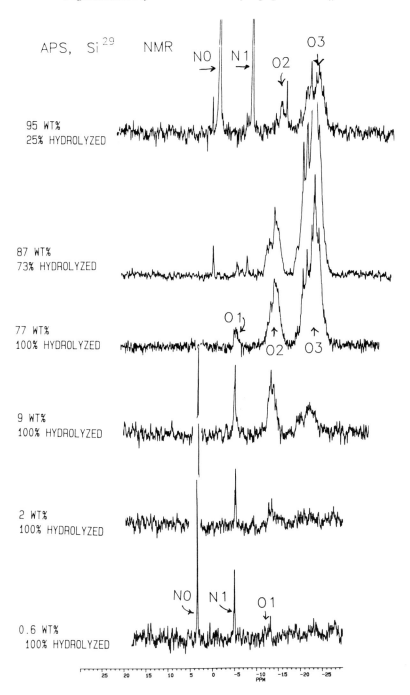

Figure 3. Si-29 NMR spectra as a function of the wt % APS in heavy water at room temperature. N0, N1, O1, O2 and O3 designate silicon environments discussed in the text. N0 is believed to be silicon from the monomer. The rest of the environments are those of APS with bridging Si—O—Si bonds. N1 is believed to be the dimer. O1, O2 and O3 are silicons in various stages of oligomerization. See text for discussion.

scale by subtracting approximately 45 ppm. The chemical shift axis is drawn with increasing shift from right to left. An increase in shift (more positive and towards the left) is termed 'downfield' and signifies less shielding of the silicon nucleus by its electron cloud. A decrease in shift (more negative and towards the right) is termed 'upfield' and signifies more shielding of the silicon nucleus by its electron cloud. A similar behavior can be attributed to the chemical shifts of the hydrogen (proton) nucleus.

The sharp lines (labeled as N0 and N1 in the 0.6 and 95 wt % spectra of Fig. 3 are believed to be the monomer (no additional Si—O bond) and dimer (1 additional Si—O bridging bond). Neat APS by itself shows two silicon environments (not shown here). The broad bands labeled 'O1', 'O2' and 'O3' are believed to be the Si—O—Si oligomer environments. O1 represents dimer environments (1 additional Si—O bridging bond), O2 is trimers (2 additional Si—O bonds), and O3 is tetramers (3 additional Si—O bonds). Such assignments are consistent with Si-29 studies of silanols/siloxanes which show an approximately 10 ppm decrement or upfield shift (usually to the right as spectra are plotted) for each silanol converted to a siloxane [15]. Recent Si-29 solution work of Besland *et al.* [16] is consistent with such interpretations.

For a 50/50 monomer/dimer solution above 83 wt %, there is insufficient heavy water available for complete hydrolysis of the ethyl ester and subsequent oligomerization. This fact is supported by the 87 and 95 wt % solutions shown in Fig. 3 in which the monomer N0 and dimer N1 reappear, but at different chemical shifts because supporting electrolyte is now essentially APS and not heavy water. This study supports the observation that ester hydrolysis must occur before self-oligomerization (loss of monomers).

Oligomerization in these solutions is most certainly complicated. Figure 4 shows the Si-29 NMR spectrum of 90 wt % APS in heavy water vs. time at room temperature. The Si-29 isotope is only about 4.7% of the silicons. Consequently, sensitivity suffers and recording 'real-time' Si-29 NMR spectra is difficult. The spectra of Fig. 4 are normalized and only the N0 band is prominent. The N0 band is seen to decay with increasing time as the APS monomers are 'used up' in self-oligomerizations. Because of the poor signal-to-noise, the N1 band and the

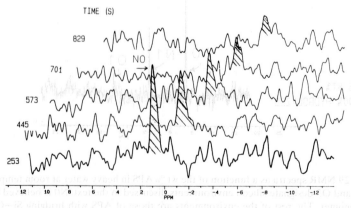

Figure 4. Sequential Si-29 NMR spectra vs. time of 90 wt % APS in heavy water. Because of low sensitivity, only the loss of the N0 monomer can be followed in real time.

products of oligomerization (the broad O1, O2 and O3 bands) are not visible. The decay appears to be well-approximated by first-order kinetics at room temperature. The half-life for the loss of the monomer in heavy water at room temperature is measured at 234 s (note: 200 s were required to transfer the sample into the magnet for analysis). The 'half-life' signifies the time for half of the current monomers to be 'used up' in the oligomerization process.

It is clear from Si-29 studies that Si-29 is a useful tool for measuring the steady-state oligomerizations but not for studying the actual oligomerization process in real time. In general, Si-29 sensitivities are not useful for analysis of solutions below 0.5 wt % or for measuring reactive half-lives for weight percents less than 90%.

Proton (hydrogen) NMR is more practical as the proton is greater than 99% abundant. Figure 5 shows the 300 MHz proton NMR spectrum of a 0.1 vol % APS solution in regular water. The reference is the proton line of TMS which is assigned 0 ppm and is not shown on the spectrum. The strong resonance from the water protons dominates the left part of the spectrum at approximately 4 ppm. There are two multiplet bands from the methylene (ETCH2) and methyl (ETCH3) groups of free ethanol at approximately 3.5 and 1 ppm, respectively. The three aliphatic protons of APS also show multiplet bands at 2.75, 1.55 and 0.45 ppm, respectively, and are labeled as ALPH1, ALPH2 and ALPH3. The ALPH3 band is believed to originate from the protons on the carbon which is adjacent to the silicon (also referred to as the 'alpha' carbon). Its lineshape appears to be particularly sensitive to the degree of oligomerization. The nitrogen protons are observed to resonate at approximately 1.2 ppm in the neat liquid. They 'disappear' from the aqueous spectra because of rapid exchange with the water protons which renders them invisible to NMR detection.

Figure 6 shows the three upfield lines of the Proton NMR of a 50 wt % solution of APS in heavy water at room temperature. As discussed above, the

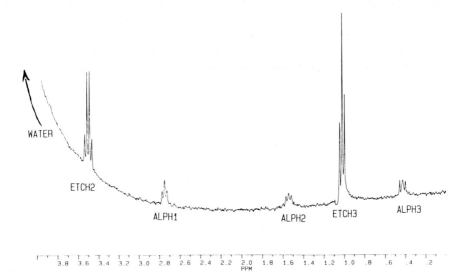

Figure 5. Proton (hydrogen) NMR spectrum of a 0.1 vol % APS in regular water. The dominant resonance of the water protons is at the left. The identities of the APS proton environments correspond to the structure and labels previously shown in Fig. 1.

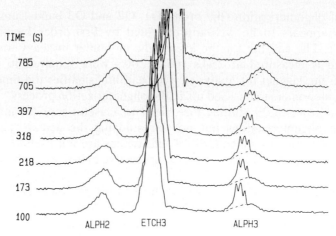

Figure 6. The upfield portion of the proton NMR spectrum of a 50 wt % APS in heavy water vs. time. Changes in the ALPH3 band are believed to be related to oligomerization.

bands labeled ALPH2 and ALPH3 originate from the aliphatic protons of the APS. The band labeled ETCH3 arises from the methyl protons of the ethyl ester/ethanol. Proton NMR is useful for at least three reasons: (1) The sensitivity is such that very low concentrations of APS can be analyzed; (2) The hydrolysis of the ethyl ester to ethanol can be studied; and (3) The effects of oligomerization appear to be present. The ETCH3 band moves upfield slightly as the ethyl ester is hydrolyzed to free ethanol. The changes in intensity of the ETCH3 bands or the ETCH2 bands can be used to measure the ethyl–ester hydrolysis constant. Figure 7 shows the actual hydrolysis of the APS triethoxyester to the trisilanol and free ethanol in a 50 wt % solution in heavy water. The downfield multiplet band is from the methylene protons of the ethyl ester and the emerging upfield band is from the same protons in free ethanol. The intensity of the upfield ETCH2 band can be used to study hydrolysis. The kinetics of the hydrolysis is certainly very complicated but seems to be well-fitted to first-order with a half-life of hydrolysis measured by NMR as 181 s for a 56 vol % (50 wt %) solution and as 284 s for a 50 vol % solution in heavy water at room temperature. In regular water, the reaction could be twice as fast and the half-life as little as 142 s.

Referring to Fig. 6, the ALPH3 band appears to contain two components: the first is three sharp lines that are believed to originate from the aliphatic protons of APS monomers and the second is a broad hump which is believed to originate from the same aliphatic protons but in oligomerized APS environments. The three sharp lines correspond to protons attached to silicons in environments previously labeled as N0 and N1 and shown in Fig. 3. The much broader and somewhat convoluted band is believed to represent the sum-total of all the protons attached to silicons in oligomer environments previously labeled as O1, O2 and O3, also shown in Fig. 3. It is consistent with the interpretation of NMR lineshape phenomena to expect oligomerization to result in a broadening of the sharp proton NMR lines associated with the smaller and more mobile APS monomers and dimers which are designated N0 and N1, respectively, in the Si-29 spectra. Such NMR phenomena are attributed to the relaxation effects

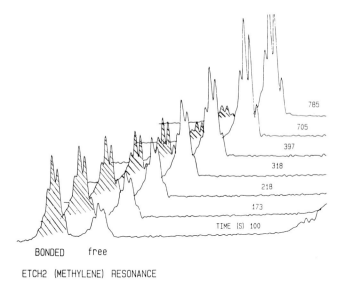

Figure 7. Sequential proton NMR spectra vs. time of a 50 wt % APS in heavy water solution which shows hydrolysis of the APS. The methylene resonance (labeled ETCH2) shifts upfield as the ester-bonded ethanol is released into the solution.

produced by residual dipolar-couplings and chemical-shift-anisotropy effects [8–14] as chain length grows.

Oligomerization-vs.-time data for the 50 vol % APS solution in heavy water at room temperature were derived from the increasing intensity of the ALPH3 'hump' vs. time which is seen in Fig. 6. Intensities were well-fitted to first-order process. The oligomerization half-life for a 50 vol % APS solution in heavy water at room temperature was well-fitted to a first-order growth and the half-life was measured as 349 s from the proton ALPH3 band. This is reasonably close to the 234 s half-life measured by silicon NMR of a 90 wt % solution. The proton NMR half-life is expected to be more accurate because of the overwhelmingly better signal-to-noise and increased number of data points. It should also be noted that half-lives of oligomerization seem to be longer than those of hydrolysis—a fact which is consistent with hydrolysis being required and preceding self-oligomerization.

An oligomerization-vs.-time curve has been generated and is shown in Fig. 8. This curve can be used to control the amount of oligomerization in a 50 vol % solution of APS in heavy water at room temperature. From a 10 ml 50 vol % APS master solution, 0.4 ml aliquots are removed progressively and quickly diluted into 200 ml of regular water to quench (or greatly reduce) the oligomerization. While studies in this laboratory show that oligomerization will occur even in very dilute solutions such as 0.1 vol %, the time frame is very long compared to that of very concentrated solutions. The procedure for the generation of oligomer-controlled APS primer solutions of 0.1 vol %, using the actual time response shown in Fig. 8, is diagrammed in Fig. 9. After preparation, the APS solutions were quickly used as adhesion promoters in the studies described below.

The ALPH3 proton band can also be used to indirectly monitor the Si—O—Si oligomerized state of the low concentration 0.1 vol % solutions. It is difficult and

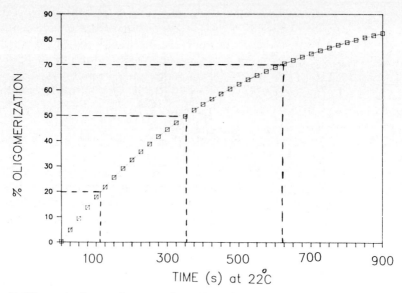

Figure 8. Oligomerization-vs.-time curve for a 50 vol % APS in heavy water solution. The curve is generated from time-dependent studies of the ALPH3 proton NMR band which is believed to reflect oligomerization.

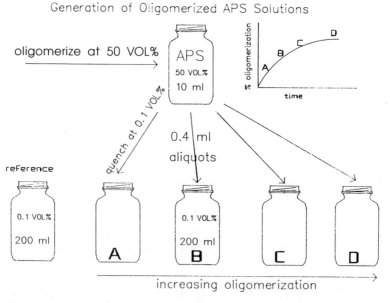

Figure 9. Diagram of the procedure used to generate 0.1 vol % solutions of APS with controlled percentages of oligomerization. Oligomerization is achieved in 50 vol % APS in heavy water. Then, 0.4 ml aliquots are diluted to 0.1 vol %. At this low vol %, oligomerization is believed to occur very slowly.

not practical to directly measure the silicon environments at these low concentrations as was mentioned above. Proton specta showing what is believed to be 'fresh' and 'stale' 0.1 vol % APS solutions are shown in Fig. 10. The 'fresh' solution was prepared with vigorous agitation of the solution after addition of APS. The 'stale' solution was made without agitation after APS addition. APS, being less dense than water, will float on its surface and mix slowly. It is believed that oligomerization can occur as it dilutes itself. Proton NMR analysis appears to support this assumption.

The procedures used to prepare and analyze the samples have been discussed in Section 2, above. Figure 11 illustrates the peel strength as a function of APS oligomerization after being exposed to 85°C/81% RH for varying times. The x-axis is not to scale. It should be noted that the same set of wafers was used in all experiments. By coating two edges with gold and dicing 3.18-mm strips, there was a sufficiently large number of samples to test the same wafer at T(0), T(200) and T(500).

To establish whether or not oligomerization affected adhesion, the adhesion data were least-squares-fitted to a line at each of the T&H conditions (not shown). The slopes of such lines can statistically infer if a correlation between adhesion and degree of oligomerization existed. The slopes (in % relative to a reference of 99 g mm^{-1} as 100%) are: -0.064, -0.031 and -0.043 for T(0), T(200) and T(500), respectively. The y-intercepts, or zero-oligomerized values of adhesions, of the T&H conditions, (in % relative to a reference of 99 g mm^{-1} as

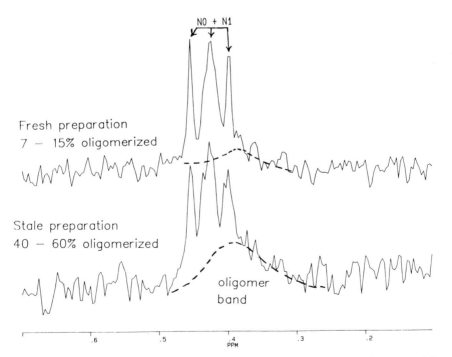

Figure 10. Proton NMR spectra of the ALPH3 band of 0.1 vol % APS in regular water. A large hump which is prominent in the stale solution is believed to result from oligomerization because of poor preparation techniques. See text for further discussion.

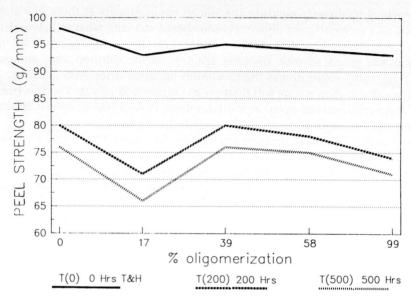

Figure 11. The peel strength in g mm^{-1} vs. the % oligomerization of the 0.1 vol % APS primer solutions used to couple PI 5878 (PMDA–ODA) to silica. The three lines represent from 0 to 500 h of 85°C/81% RH T&H stress. A definite correlation of adhesion with T&H stress, but little, if any, correlation with oligomerization is seen.

100%) are: 97.8, 78.5 and 75.2, respectively. For example, at T(200), there is a 0.031% decrease in adhesion for every 1% increase in the amount of oligomerization and an overall 19.3% drop in adhesion relative to T(0). Within estimated experimental error of a few percent, adhesion is certainly lost as a function of the length of T&H but not necessarily as a function of the degree of APS oligomerization! The negative slopes may suggest some degradation of adhesion with increasing oligomerization, but the magnitude is actually quite small and probably well within experimental error limits.

It was mildly suprising to find no apparent correlation of adhesion with APS oligomerization, to say the least. In order to confirm these findings, a second experiment was performed. A 50 vol % solution was sealed and allowed to age for 24 h at room temperature and then was quickly diluted to 0.1 vol % and used as an adhesion promoter within 15 min. A second 0.1 vol % reference solution was made from the neat APS reagent bottle. Films were spun and cured as described above. A Si-29 NMR spectrum was recorded on the 50 vol % solution and is shown in Fig. 12. From the areas of the N0, N1, O1, O2 and O3 bands, the amount of oligomerization was estimated at 98.7% with 1.3% of the N0 monomer detectable. O1 including N1 (dimers) was 9.3%; O2 (trimers) was 40.4%; and O3 (tetramers) was 49.0%. Proton NMR analysis of the ALPH3 band was consistent with the APS being highly oligomerized, but it cannot be ruled out that some oligomerized APS did not un-oligomerize when diluted from 50 to 0.1 vol %. Again, the new peel test results of films on silicon wafers showed virtually no loss of adhesion with respect to the reference film which used freshly-made APS, under T(0) and T(200) conditions. This second experiment confirms the earlier data.

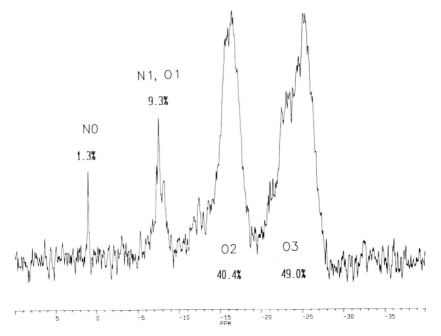

Figure 12. Si-29 NMR spectrum of a very highly oligomerized 50 vol % APS in heavy water solution which was subsequently diluted to 0.1 vol %. Adhesion measurements confirmed that little loss of adhesion with oligomerization seemed to have occurred. The symbols N0, N1, O1 and O2 designate Si-29 NMR absorption bands which are believed to correspond to various degrees of APS oligomerization.

It should be noted that approximately 1% of the APS used in the last experiment appears to be APS monomers before dilution. Could this low monomer concentration be responsible for adhesion? It should be noted that the residual monomers present in this highly-oligomerized solution correspond to a 0.001 vol % solution if one were to remove the oligomerized APS. As a result of this observation, the dependence of the adhesion of thin films to native-oxide silicon wafers as a function of the concentration of the APS under conditions of T&H stress was investigated. Adhesion studies were performed using APS solutions with concentrations that varied from the 'industry standard' of 0.1 vol % down to 0.00001 vol %. The test wafers were prepared and exposed to T(200) and T(500) conditions as discussed above. Adhesion was measured by 90° peel test, as discussed above. The results of this study are presented in Fig. 13. The '*x*-axis' is APS concentration which decreases from left to right. The *y*-axis is the adhesion in the units of g mm^{-1}. The three curves are the results at T(0), T(200) and T(500).

From the data, it can be concluded that an APS concentration of 0.01 vol % has virtually the same performance as the industry standard of 0.1 vol %. However, adhesion decreases significantly below 0.01 vol %, especially after T&H stress. It should be noted that the 0.001 vol % solution, which corresponds to the residual monomers in the highly-oligomerized 0.1 vol % solution, is very sensitive to T&H degradation. Such behavior was not observed for the oligomerized 0.1 vol % solution made from the 50 vol % APS aged for 24 h.

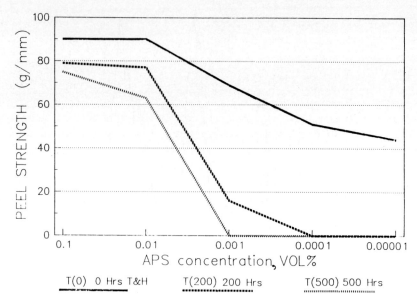

Figure 13. The peel strength of PI 5878 (PMDA–ODA) to silica vs. APS vol % and T & H stress from 0 to 500 h. Below 0.01 vol % APS, the coupling ability of APS suffers, especially after T & H stress.

4. SUMMARY

The following conclusions can be drawn from this study:
(1) Highly self-oligomerized 0.1 vol % APS solutions promote adhesion as do those with very low amounts of self-oligomerization. Formation of covalent Si—O bond with the mineral surface may not even be required to promote adhesion. Possibly, only hydrogen bonding or van der Waals interactions are sufficient. Such ideas have been discussed previously [6].
(2) Solutions with APS monomer concentrations below 0.01 vol % do not yield acceptable coupling, especially after T & H stress. Residual monomer concentration in highly-oligomerized solutions does not appear to be solely responsible for adhesion promotion.
(3) T & H conditions cause an overall decrease in adhesion. However, this loss is not well-correlated with perturbations of the silicon–oxygen environment of the APS molecule that one might expect when oligomerization occurs. Previous experiments in this laboratory have shown that overall adhesion losses can be partially reversed if parts are heated above 100°C after T & H stress. Previously, DeIasi reported such phenomena for the tensile properties of polyimide films [17]. In order to evaluate the possible beneficial effects of post T & H baking, the 0% and 99% APS-oligomerized parts (after T(500)) were heated at 100°C for 60 min and then at 230°C for an additional 60 min. The 0% part showed 0% and 12% adhesion recoveries (gain above T(500) adhesion), respectively. The 99% part showed 5% and 10%, respectively. With extended bake at 100°C for 24 h, the 57% oligomerized part did recover about 8% above its T(500) adhesion. Formation of surface-to-APS siloxane bonds are more likely at 100°C and above; however, formation of imide bonds from the APS to the PI film are also favored above 150°C. At

this time, it is not completely clear which end of the APS molecule is more susceptible to T&H stress. An illustration of the proposed coupling showing the siloxane bond to the surface and the imide bond to the PI has been shown in Fig. 2.

(4) If the current ideas of APS surface interactions are to be supported, then the highly-oligomerized APS molecules must first revert to more reactive species such as silanol monomers. Perhaps the surface acts as its own acid/base catalyst to break up the oligomerized fragments into monomers. The newly formed APS silanols are then free to form surface bonds and eventually a two-dimensional surface network [3]. Many experiments done in this laboratory and some discussed above are supportive of the labile nature of the siloxane bond. At or near the surface, this bond may be cleaved by water at elevated temperatures or at basic pH's, for instance, to form silanols or siloxanols. Subsequently, these molecules can react to form siloxanes and release the water molecule.

Further experiments to understand why 'stale' APS primer solutions work as well as freshly made ones are currently underway in our laboratory.

Acknowledgements

The authors would like to thank: Mike Lew, IBM EF Process Engineer, for his assistance with experiments on the APS Spin Apply Tool and R. Matos of IBM Product Quality Assurance for providing T&H facilities.

REFERENCES

1. E. P Plueddemann, in: *Interfaces in Polymer Matrix Composites, Vol. 6 of Composite Materials Series*, L. J. Broutman and R. H. Krock (Eds). Academic Press, New York (1974).
2. E. P. Plueddemann, *Silane Coupling Agents*. Plenum Press, New York (1982).
3. H. Linde and R. T. Gleason, *J. Polym. Sci., Polym. Chem. Ed.* **22**, 3043 (1984).
4. H. Schmidt, *J. Non-Cryst. Solids* **73**, 681 (1985).
5. G. L. Wilkes, B. Orler and H. H. Huang, *Polym. Preprints* **26**(2), 300 (1985).
6. L. P. Buchwalter, *J. Adhesion Sci. Technol.* **4**, 697 (1990).
7. P. D. Murphy and W. D. Weber, Unpublished results (1989).
8. T. C. Farrar and E. D. Becker, *Pulse and Fourier Transform NMR*. Academic Press, New York (1971).
9. T. C. Farrar, *Introduction to Pulsed NMR Spectroscopy*. Farragut Press, New York (1989).
10. N. Chandrakumar and S. Subramanian, *Modern Techniques in High Resolution FT-NMR*. Springer-Verlag, New York (1987).
11. A. Abragam, *The Principles of Nuclear Magnetic Resonance*. Harper and Row, New York (1963).
12. C. P. Slichter, *Principles of Magnetic Resonance*, 2nd edn. Springer-Verlag, New York (1980).
13. A. Carrington and A. D. McLachlan, *Introduction to Magnetic Resonance*. Harper and Row, New York (1967).
14. J. A. Pople, W. G. Schneider and H. J. Bernstein, *High-resolution Nuclear Magnetic Resonance*. McGraw-Hill Book Co., New York (1959).
15. E. Lippmaa, M. Magi, A. Samoson, G. Engelhardt and A. Grimmer, *J. Am. Chem. Soc.* **102**, 4889 (1980).
16. M. P. Besland, C. Guizard, N. Hovnanian, A. Larbot, L. Cot, J. Sanz, I. Sobrados and M. Gregorkiewitz, *J. Am. Chem. Soc.* **113**, 1982 (1991).
17. R. Delasi, *J. Appl. Polym. Sci.* **16**, 2902 (1972).

Silanes and Other Coupling Agents, pp. 439–459
Ed. K. L. Mittal
© VSP 1992.

Application of silanes for promoting resist patterning layer adhesion in semiconductor manufacturing

JOHN N. HELBERT* and NARESH SAHA

Advanced Technology Center, Motorola Semiconductor Sector, Mesa, AZ 85202, USA

Revised version received 20 May 1991

Abstract—The importance of adhesion science to semiconductor fabrication technology is established with emphasis on lithographic adhesion studies and advances. Phenomenological under-standings of silane adhesion promoter treatments developed to solve resist adhesion problems are provided from the correlations between surface analytical studies utilizing ESCA (XPS), contact angle measurements, and on-wafer empirical adhesion test results using special image adhesion test structures. Using the early work of Plueddemann and others as model systems, several silane adhesion promoter processes have been successfully developed to provide resist image adhesion to semiconductor manufacturing technology. In addition, surface studies have led to empirical measurement indicators that predict adhesion behavior well and these parameters are used to monitor adhesion performance even in volume semiconductor manufacturing facilities.

Keywords: Adhesion; photoresist; silanes; photoresist image adhesion; semiconductor process adhesion.

1. INTRODUCTION

Adhesion is an important ingredient in the fabrication of integrated circuits (ICs). These circuits are fabricated by placing layer upon layer vertically from the silicon crystal wafer surface until the entire circuit or device is completed (see Fig. 1). These thin layers (i.e. 0.2–3.0 μm) must all adhere intimately for the device to function as designed, even though some of these layers are applied under a significant amount of stress.

Device layers can range from single crystal and doped polycrystalline silicon, silicon nitride, thermally grown oxides of silicon to plasma-enhanced deposited dielectric layers to sputtered or CVD deposited dielectrics, and metal or metal silicide layers. Layers that do not adhere can usually be made to adhere through special adhesion promoting processes or treatments. For example, sputter-deposited TiW is routinely used to promote adhesion and lower contact resistance for nonreducing metals such as tungsten. Similarly, silane adhesion promoters are extensively used to promote polyimide (PI) dielectric adhesion to underlying silicon and metal layered patterns [1]. Saiki and Harada [1a] demonstrated improved polyimide peel strength for N-(β-aminoethyl)-amino-propylmethyldimethoxysilane-treated substrates and also showed a dependence of PI adhesion on cure temperature. Narechania *et al.* [1b] have further confirmed these effects but on silicon nitride substrates and with another adhesion promoter, 3-aminopropyltriethoxysilane. From these references, it is

---

*To whom correspondence should be addressed.

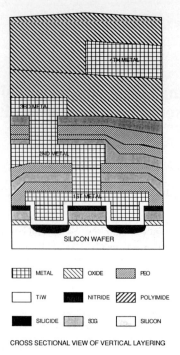

	METAL		OXIDE		PEO
	TiW		NITRIDE		POLYIMIDE
	SILICIDE		SOG		SILICON

CROSS SECTIONAL VIEW OF VERTICAL LAYERING

Figure 1. Cross-sectional view of a vertically layered IC wafer showing one transistor and four layers of metal circuit interconnections.

clear that organosilanes are effective in promoting structural adhesion for semi-conductor device fabrication.

Individual circuit layers, layers not directly patterned economically, must be patterned indirectly with the required specific pattern. This is accomplished with lithographic resist materials and processes (see Fig. 2), and again adhesion plays a critical role. The integrity of each lithographic patterning layer is essential for successful IC fabrication. Resist image adhesion is of paramount importance, because if IC patterns are missing, have jagged pattern edges, or are not the right dimension, the device will simply not function as designed.

Before spinning or applying a lithographic resist onto the IC wafer in process, a resist adhesion promotion process is typically carried out. Resist adhesion can be accomplished by several methods ranging from ion bombardment [2] to silane surface treatments [3] to polymeric coating treatments [4] (see Table 1). Most semiconductor photoresist adhesion processes involve either liquid or vapor phase treatment of the wafer before resist coating with hexamethyldisilizane (HMDS), which has become an industry standard. HMDS is typically used at every lithographic step in the IC fabrication process whether it accomplishes surface modification or not. HMDS processing can be carried out on automatic wafer tracks with liquid or vapor HMDS modules (e.g. Silicone Valley Group, SVG) or in stand alone microprocessor-controlled all stainless-steel commercial reactor chambers (e.g. Yield Engineering Systems, YES); these processes are proven for high volume production. Recently, however, new IC materials or layers have appeared, and resist adhesion problems are continuing and will

RESIST AND ETCH PROCESSING SEQUENCE

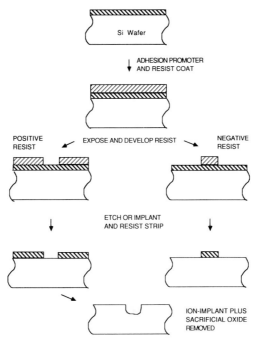

Figure 2. Schematic diagram illustrating lithographic resist imaging technology.

Table 1.
Successful treatments exhibiting semiconductor resist adhesion promotion

Treatment	Mechanism	Reference
HMDS	Surface reaction	3,5
Al chelates	Complex formation	1
Chlorosilanes	Thin polymer layer	6
Alkoxysilanes	Thin polymer layer	6
Negative resist	Thin polymer layer	4
Inert gas reactive-ion etching	Carbon/oxide layer	2
Oxygen plasma/HMDS	Oxidation of nitride	7
Argon ion-milling	SiC layer	2
Ti/Zr/Hf oxides	Thin oxide layers	8

continue to be common to advanced developmental IC pilot lines employing new advanced fabrication substrate materials.

Other classes of silanes, namely alkoxy, halogenated, and other silanes [3, 9], are known to react with —OH containing compounds, and, therefore, should also function as adhesion promoters or surface modifiers for —OH containing substrates. Texas Instruments, for example, employed a 2% xylene solution of phenyltrichlorosilane to provide resist image adhesion to various oxide wafer substrates [10]. References 3, 9, and 10 describe many of these materials applied to silicone dioxide substrates. As for HMDS treatment, ESCA evidence of reactions to verify covalent bonding to surface silanol groups will be provided in this paper; in some cases, surface coverage or thicknesses will also be provided by angle-resolved ESCA analysis.

Organosilanes can be used to treat hydrated surfaces and reduce problems associated with poor polymer adhesion due to moisture or silanol formation [11–15]. They can act in several ways. One is the simple reaction with adsorbed water to remove it. They can also form a chemical bond by reacting with surface hydroxyl groups to form a molecular surface film to which the photoresist adheres better chemically or physically depending on the organosilane and its functional groups [11]. Most of the widely used organosilanes have the structural formula $R_nSiX_{(4-n)}$, where n is at least 1 and less than 4. X is the reactive, labile group. The organic substituent R can be chosen, in principle, to match the functionality of the photoresist resin employed [11, 12]. According to Plueddemann [11], only a monolayer of organosilane is theoretically needed to achieve adhesion for glass systems being bonded to polymer resins like resists.

The type of adhesion dealt with in the examples in the second paragraph above and Fig. 1 is mechanical or structural; while for the lithographic resist adhesion requirements described in this paper a more practical definition of adhesion, one first proposed by Mittal [16], is being referenced and used. Resist patterning layer–substrate adhesion is required only to process or pattern a particular device layer. After the circuit layer is patterned, the resist layer is removed and does not become an integral part of the circuit, as opposed to a PI interlevel metal dielectric layer which does. As such, it is not required to possess high mechanical adhesion strength. In fact, the resist layer must be quantitatively removed after the circuit required layer has been patterned. If the resist layer adheres too well and becomes difficult to remove, it actually interferes with successful circuit fabrication.

Problems have also been observed at resist coat due to substrate nonwetting or oleophobicity. This problem is independent of the more commonly occurring 'pattern lifting' (see Fig. 3 for examples of missing or displaced images) adhesion failure problem, which is observed after resist exposure and development. We will also focus on work addressing these latter two specific problem areas in this paper.

2. SUBSTRATE CHEMISTRY

There is a direct relationship between the condition of the SiO_2 or Si substrate and photoresist adhesion [2]. After high temperature oxidation and diffusion processes, oxide and silicon surfaces are known to consist mainly of Si—O—Si and Si—Si bonds. However, such surfaces, when exposed to the ambient, will adsorb water and slowly rehydrolyze to reform Si—OH bonds [6]. Such substrates are polar, and photoresists, especially negative resists, do not adhere well to them. The rate of hydrolysis is thought to increase with relative humidity, proceed more rapidly in liquid water, and be accelerated by acids or bases. All other conditions being equal, wafers that have been stored for some time period are observed to give poorer resist adhesion than those fresh from a high temperature treatment, such as those from an oxidation furnace. The effect becomes more serious as the relative humidity increases. For older semiconductor fabrication areas with poor humidity control, rework rates due to high

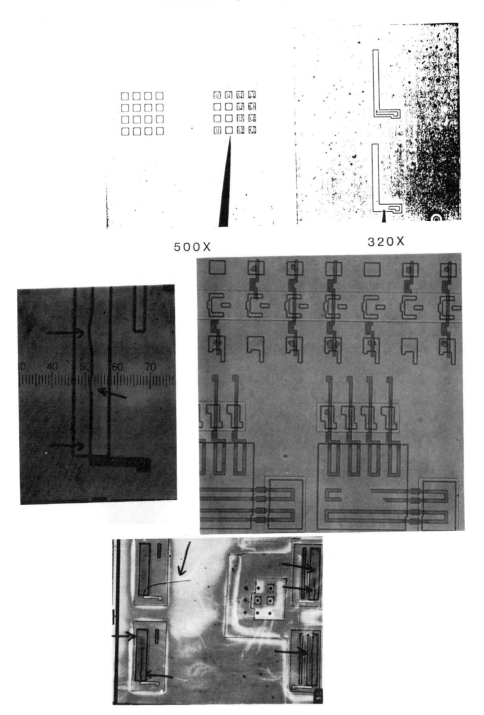

500X   320X

Figure 3. Optical micrographs ($\times 500$) of positive photoresist island test patterns for resist 'lifting' tests. The upper photograph shows missing islands of the blank wafer (i.e. no promoter) where adhesion failure has occurred. The lower photographs demonstrate varying degrees of resist image lifting from partial displacement to break off and loss.

humidity and poor HMDS or other silane processing can be as high as 30% of the total rework (calculated from months worth of rework data; rework is defined in a later section), while values below 6% are more typical during drier or lower relative humidity or for semiconductor factories with low and controlled humidity conditions.

Silanes have also been employed to improve adhesion to metal substrates. For example, Toray Chemicals of Japan has observed adhesion improvement for Mo by employing an organosilane (isopropanol solution) and for Al, Ti, Ni, and Pd by employing an organotitanate spun from toluene. Results for other silanes to Au and Al are reviewed later. Similarly, Dynamit Nobel recommended employing glycidoxypropyltrimethoxysilane to improve positive photoresist image adhesion to Ta, W, Al, and Nb.

3. RESIST BACKGROUND

Resist materials are usually polymeric in nature and are spun from organic solvents of moderate to low volatility. As such, these bi- or tricomponent liquid spinning systems also usually possess low surface tension and will wet most solid surfaces involved in the technology discussed above, except those with a very low energy and/or contaminated surfaces. This type of substrate presents a problem at resist coat due to nonwetting [2, 9]. In cases where it is observed, the lithographic process cannot be completed without further substrate surface process modification. Needless to say, added process steps are economically unattractive, but sometimes required.

After the resist film is successfully coated onto the wafer substrate and the film exposed to the electromagnetic energy of appropriate wavelength, the film is rendered relief image-developable. It is absolutely essential that each resist pattern adheres to the substrate or the device layer will be flawed with missing circuit elements. When resist patterns lift at development and are missing, the resist has undergone 'adhesion failure' (Fig. 3). Adhesion failure for this processing step has been well documented in the literature [3, 5, 6]. Treatment of the circuit layers before resist coat with silanes, or any surface treatment, to prevent resist image 'lifting' is referred to as adhesion promotion or priming. Results for several kinds of adhesion promoting schemes are provided in this paper, but the main emphasis will be placed on silane promoter processes developed using the early work of Plueddemann [11, 12] on nonsemiconductor substrate systems as a guide.

The resist systems most susceptible to resist image lifting at the development stage of processing are positive e-beam and conventional photoresists [2, 6]. Historically, negative photoresists have also required silane adhesion promotion, but in that case the goal was reduced undercut from wet etch processing and not the prevention of simple image lifting at development [6, 17]. Successfully employed silanes include vinyltrichlorosilane [6], and dimethyldichlorosilane [18]; HMDS also worked well [3]. Undercutting is less of a consideration for positive photoresists, because they are used primarily in conjunction with anisotropic dry etching processes. Examples of negative resist wet etch undercut and positive lifting at development are shown in Figs 4 and 3, respectively.

WET ETCHING ADHESION TEST ON Si(240X)

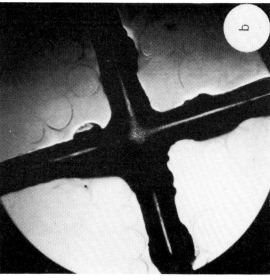

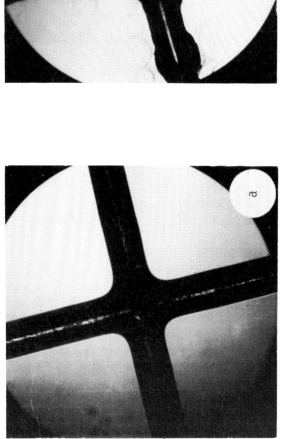

VTS   NO PROMOTER

Figure 4. Si wet-etched samples with resist removed. In (a) the substrate was treated with VTS, while in (b) the control wafer had no organosilane treatment. Severe undercutting defects occur in (b) resulting from adhesion loss at the edges of the large cross pattern (×240).

4. REWORK DEFINITION

After silicon wafers have been fabricated to a certain level, say at the 20th level of a 25-step process, they become too valuable to terminate fabrication when misprocessed. Sometimes in lithography either the wrong mask is used, the resist exposure is adjusted improperly, lithographic equipment setting drifts occurred, or the resist is improperly developed. Since the resist pattern is just a thin polymer layer and not an integral part of the circuit at that point, the wafer can be saved and completed by simply removing the misprocessed layer. A good rework process, sometimes referred to as recycle [17] or redo, is required to accomplish this task. For device implantation, etch, or deposited layers, this flexibility is lost because the resulting process structures of these processes become a permanent part of the device and cannot be reprocessed. As a result, lithographic rework or redo is quite common in fabrication facilities. Unfortunately, Deckert and Peters [17] found that silicon dioxide wafer surfaces exhibit random photoresist adhesion variation, which is very much affected by previous chemical treatments, such as those for rework. These effects can cause defects, and the additional wafer processing required is known to increase defect levels statistically. This, in turn, decreases circuit electrical yield. Obviously, original and rework processes which prevent adhesion failure, and are clean from a surface point of view, are very important economically. The best case situation occurs when the rework process is also the resist removal process, the one utilized to remove the resist following layer patterning completion.

The surface characteristics of reworked substrates and several rework processes are described later in several tables and the discussion. The substrate condition following the rework process is characterized by the results of three tests: ESCA or XPS [19] surface chemical composition spectra, actual lithographic resist pattern lift testing [2, 6], and water droplet contact angle measurement, θ_{H_2O} [3].

5. EXPERIMENTAL

The resists employed for 'wet etching' adhesion experiments were typical commercially available azide initiated/cyclized isoprene type negative photoresists.

With many of the Union Carbide organosilanes investigated, it was difficult to distinguish which materials worked best on freshly 1050°C steam-oxidized (1 μm) wafers. Wafers were therefore intentionally 'soaked' in water for 48 h, spun 'dry', followed by the addition of the organosilane (100%) to the substrate at 3000 rpm for 10 s. The negative resist (1 μm) coated wafers were patterned and processed in a conventional manner, and the oxide was etched in 4:1 buffered NH_4F–HF at room temperature until the oxide was removed. Examination (by microscope) of the resist surface for resist lift, followed by resist strip to observe undercutting was performed.

Additional adhesion promoters were tested on a silicon substrate. A 5.5 μm thick negative resist was employed and 75 μm deep moats were etched in the silicon using HNO_3–HF–acetic acid (8:4:3) at 10–15°C. Again, edge definition was examined by optical microscopy.

Development resist image 'lifting' tests were carried out using silane promoters obtained from Petrarch Systems, Inc. and conventional naphthoquinone diazide inhibited/Novolak resin type positive photoresists. They are all proprietary formulations, but generically they are composed of a mixture of (1) Novolak resins, (2) photoactive components of the diazoquinone type, (3) leveling agents and/or surfactants, and (4) glycol-based spinning solvents. The commercially available adhesion promoters (Petrarch Systems, Inc.) tested were applied as dilute solutions 0.3–7% by weight in acetone or xylene, or as for HMDS, either as a liquid or vapor (YES, Imtec Star 2000, or SVG wafer track).

A practical experimental approach for testing positive photoresist adhesion during development was adopted, where the test pattern (or mask) is one that yields a series of unexposed islands of varying dimensions from one to several micrometers, surrounded by larger exposed areas. Pior to resist spinning by conventional spin techniques, the substrates were treated with adhesion promoters, again usually by conventional spinning. Following cure of the promoter, the resist was applied and prebaked. Substrates not treated with silane promoter usually result in >90% island 'lifting' during development (see Fig. 3). Tests were carried out using both e-beam and photolithographic exposure techniques.

All XPS or ESCA measurements were performed using a Perkin Elmer 5300 ESCA spectrometer equipped with a dual anode (Mg, Al) X-ray source, differentially pumped Ar^+ sputter gun, and the variable angle measurement set-up for angle-resolved photoelectron spectroscopic measurements. The data collection and treatment, e.g. smoothing, curve-fitting, intensity measurements, were accomplished by a Perkin Elmer 7500 dedicated computer system using PHI software package.

Typically, a broad energy (0–1000 eV) survey spectrum was acquired from each sample for elemental detection and then high resolution data for each element were collected to determine the surface chemistry and compositions of different samples. The elemental compositions of different samples were determined from the integrated area intensities of respective photoelectron peaks after normalizing for their relative sensitivity factors [20].

The depth of analysis (d) in XPS is approximately given by $3l \sin \theta$ [21] where l is the mean escape depth and θ is the take-off angle of the photoelectron with respect to the sample surface plane. Thicknesses of surface coverage layers on different samples were estimated from the attenuation of the XPS signal from the substrate by the overlayer using the relation [22] $\ln [R/R^\infty + 1] = d/l \sin \theta$, where d is the overlayer thickness, R is the ratio of photoelectron signal intensities from the overlayer to substrate of any particular element, and R^∞ is the photoelectron intensity from the same element of infinite thickness.

6. HMDS AND REWORK RESULTS

Substrate nonwetting at resist coat has been observed most frequently with reworked and mistakenly overpromoted wafers. It occurs after repeated treatments, when the wrong liquid silane treatment has been applied, or when vapor times exceed the optimum time for that respective substrate. It can also occur in selected circuit pattern areas and not for the whole layer. Nonwetting is also characterized by a higher water droplet contact angle, θ_{H_2O} [23]. Although it is not generally well understood, it can be prevented by reducing priming times for vapor treatments, corrected by an ion bombardment treatment of the wafer [2], and can be prevented by using resist containing a solvent of lower surface tension or by double resist application under dynamic spin conditions.

The silicon-based substrate layers, which represent nearly 80% of all the layers encountered in device production (i.e. nonmetal layers), can usually be successfully promoted against lifting by treatment with liquid silanes or silane vapor treatments at reduced pressures [2, 3]. Figures 5 and 6 demonstrate what is actually accomplished when a Si wafer is treated with HMDS. The ESCA spectra shown clearly illustrate the removal of carbon-containing adsorption species at higher binding energies in favor of the monolayer of trimethylsilyl surface reaction product (see Fig. 5b). In addition, the surface is dehydrated *in situ* as verified by an increased θ_{H_2O} [3] and a lower O/Si ratio as measured by ESCA [5]. Furthermore, this converted surface is stabilized for days against recontamination; therefore, the HMDS process provides a very stable surface for resist adhesion. The Si 2p ESCA spectrum of Fig. 6b verifies the appearance of the trimethylsilyl, $-Si(CH)_3$, silanol surface reaction species.

In Fig. 7, the difference in C 1s ESCA spectra for vapor phase HMDS vs. liquid phase silane treatments is provided. The C 1s XPS spectral changes are

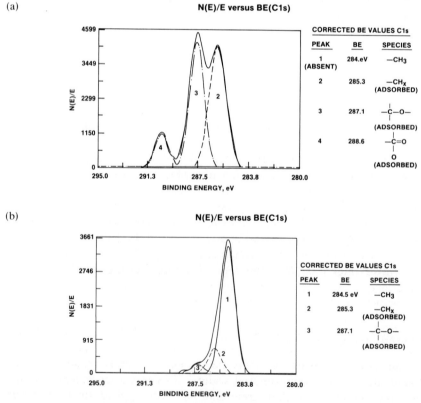

Figure 5. (a) C 1s ESCA spectrum for the blank silicon wafer (Y58). (b) C 1s spectrum for IMTEC Star 2000 vapor HMDS-treated silicon wafer.

(a)

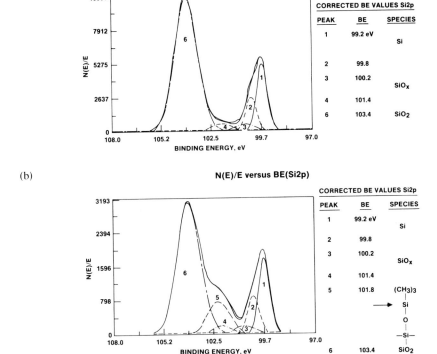

(b)

Figure 6. (a) Si 2*p* ESCA spectrum from the untreated blank Y58 wafer. (b) Si 2*p* ESCA spectrum from Star 2000 vapor HMDS-primed Y58 wafer.

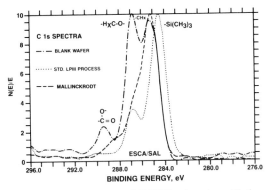

Figure 7. C 1*s* ESCA spectra of differently treated Si(100) substrates with the native oxide.

shown for a Si(100) substrate with native oxide (<50 Å). The LP III process, a vapor HMDS process, efficiently removes the carboxylic, etheral, and hydrocarbon impurities from the surface and replaces them with a blanket of trimethylsilyl groups comprising a monolayer. The Mallinckrodt system, a model liquid silane primer with both aminc and alkoxysilane molecular moieties, replaces the carbon surface species with CH_x species from the condensation polymerization reaction on the surface, which produces a 20–50 Å thick adhesion-promoting layer and a substrate with a θ_{H_2O} value of 60°. When the Si $2p$ spectrum is observed, a new signal appears at 101.8 eV from the Si(CH)$_3$ groups for the vapor-treated HMDS substrates, while no such signal appears for the Mallinckrodt system. Hence, the two comparison systems differ in the basic method of adhesion lifting prevention mechanism, even though they are both successful 'lift preventing' processes. In Table 2, the ESCA results and water droplet contact angle (θ_{H_2O}) measurements are listed for a range of representative processes. Total surface C/Si ratios from ESCA are also listed because lifting has been shown to occur when this parameter is found to be 30–100% larger than that for primed wafers [5].

Table 2.
Comparison of adhesion priming methods

| Method | Type | ESCA | | | θ_{H_2O} |
		Carboxylic	Trimethylsilyl	C/Si	
Star 2002 (5 min)	Vapor (100°C)	No	Yes	0.64/NA	75
Star 2002 (90 s)	Vapor (100°C)	No	Yes	0.58/NA	69
Star 2000	Vapor (150°C)	No	Yes	NA/0.8	77
YES LP-3	Vapor (120°C)	No	Yes	0.54/NA	76
SVG	Vapor (RT; 760 mm)	Yes; CA< 70 No; CA> 70	Yes	0.46/NA	75
HMDS SVG	Liquid	Yes	Smaller	NA/1.1	58
Mallinckrodt	Liquid	Small	No	NA/1.5	60
Virgin control	NA	Yes	No	1.0/1.1	24

The θ_{H_2O} measurements of Table 2 indicate that these treatments are also very successful at removing wafer surface water contamination, as has been verified by others (see [3] and [10] and references cited therein). It is notable, however, that there is a correlation between θ_{H_2O}, resist image lifting results, and ESCA surface condition. If θ_{H_2O} is less than 60° (or <50° [20]), the ESCA C $1s$ carboxylic peak is present at 290 eV, and there is a high relative total C/Si ratio, lifting or poor resist image adhesion is very likely to occur, either intermittently or quite frequently. For semiconductor manufacturing or any manufacturing process this kind of process uncertainty is unacceptable, and the vapor and some liquid adhesion promotion treatments have created better process reproducibility. HMDS SVG, a wafer track applied liquid HMDS process, is an example of a process that worked most of the time, but provided only marginal resist image lift prevention reproducibility. Vapor temperature, at vacuum pump pressure levels, is seen in Table 2 (i.e. the top four systems) to make little difference to the measured parameters, while priming time does provide more attractive, i.e. higher, θ_{H_2O} values. Multiple priming and long prime times can also lead to overpromoted or nonwetting (i.e. a coat) wafers; therefore, the priming time should be optimized for each substrate. Since θ_{H_2O} values level off at long prime times, an optimized time is really the minimum prime time usable while maintaining a value large enough to prevent poor adhesion. For wafer track vapor silane systems, systems that operate at house vacuum (i.e. higher pressure levels), contact angles are usually observed to be lower (see Fig. 8); again as for low pressure systems, treatment time is found to be a primary process variable statistically, as well as temperature. Contact angles for wafers treated in high-pressure SVG-type systems range from 50 to 66° for silicon nitride, from 60 to 65° for oxide, and from 52 to 71° for silicon. For these systems, multiple processing can sometimes lead to improved adhesion and larger contact angles (see Fig. 9 and [23]).

Turning to rework wafers, we must first look at the effect on substrate surface chemistry of a representative group of resist strip or removal processes. This is done in Table 3. The same wafer parameters as those for Table 2 are used. Oxygen plasma and sulfuric acid/peroxide are both oxidizing carbon removal techniques. Carbitol is a commercial mildly alkaline organic solvent stripper and acetone is a representative organic resist solvent stripper. All but the acetone treatment restore the wafer to a state close to the original state before priming and coat (see Table 3 and Fig. 10). However, the simple acetone dissolution strip leaves the substrate in the promoted and ready for recoat state, thus saving a repriming processing step (see Fig. 11). Importantly, the other processes tend to leave the substrate less clean and with larger θ_{H_2O} values than the virgin wafers.

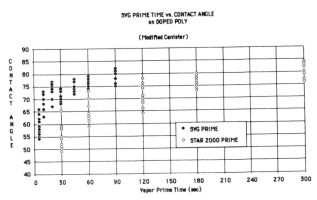

Figure 8. Water droplet θ vs. silane prime time for SVG, a high pressure system, and Star 2000, a low pressure system. The closed diamonds represent wafer data obtained from an SVG wafer track HMDS vapor prime system operating near 1 atm, and the open diamonds represent wafer data obtained from a low-pressure HMDS vapor prime system, a Star 2000 system (Stauffer Marketing or IMTEC).

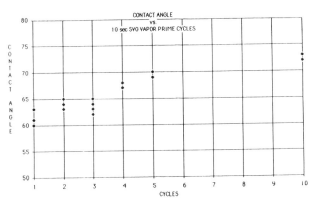

Figure 9. Contact angle vs. the number of prime cycles for an SVG track silane process at 1 atm.

Table 3.
Rework method wafer characterization data for silicon wafers

	ESCA			
Rework process	Carboxylic	Trimethylsilyl	C/Si	θ_{H_2O}
Acetone dissolution	–	+	NA/0.48	70
Oxygen plasma	+	–	NA/0.7	39
Acetone/plasma	+	–	NA/0.8	26
Sulfuric acid/peroxide	+	–	0.6–0.8/NA	25
Carbitol strip	+	–	0.6–0.8/NA	48
Virgin control	+	–	0.6/1.0	25

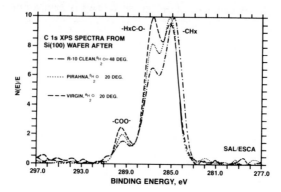

Figure 10. C 1*s* ESCA spectra and θ values for two reworked wafers vs. the virgin control.

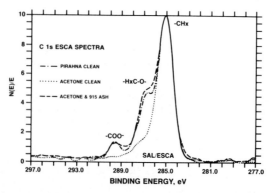

Figure 11. C 1*s* ESCA spectra for three reworked wafers. Note that the best results are observed for the acetone stripped wafer in the sense that it is ready to be recoated without silane repriming.

These observations are consistent with those reported by Deckert and Peters [9], where greater lifting or poorer adhesion was reported to occur for sulfuric acid/peroxide reworked wafers. Obviously, reworking wafers is not desirable; however, when economically necessary, simple solvent treatments on a wafer track like the acetone treatment are attractive and can be effective. Actually, other organics work as well as acetone; cellosolve acetate and other cellosolves also work well.

Rework/reprimed wafer results are found in Table 4. Here it is seen that all the rework processes return the substrate to a primed condition, but they are all a little less properly conditioned than the HMDS-treated virgin wafer control (see Fig. 12). Both the θ_{H_2O} and ESCA data are a little worse than those values for the virgin controls (i.e. they have values where resist image lifting becomes more probable). Most importantly, these processes must all be concluded to provide a wafer with unwanted circuit yield decreasing particles, just simply due to the increased handling and processing involved.

Table 4.
Surface characterization data for reworked and reprimed silicon wafers

Rework process	ESCA			
	Carboxylic	Trimethylsilyl	C/Si	θ_{H_2O}
Acetone dissolution only	No	Yes	0.48	70
Sulfuric acid/peroxide/SVG	No	Yes	0.48	68
Sulfuric acid/peroxide/Star 2000	Very small	Yes	0.59	77
Carbitol/SVG	Very small	Yes	0.56	77
Carbitol/Star 2000	No	Yes	0.60	70
Virgin/SVG	No	Yes	0.43	73
Virgin/Star 2000	No	Yes	0.46	78

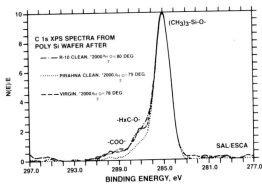

Figure 12. C 1s ESCA spectra and θ values for two reworked/reprimed wafers vs. the Star 2000 primed virgin control.

7. SPECIAL LIQUID-PHASE SILANE PROCESSES

Since SiO_2 substrates appear frequently during IC fabrication, the adhesion test results for this substrate are important. Four types of oxides have been extensively tested. They are (1) thermal oxide grown at $T > 1000°C$, (2) softer oxide processed by conventional spin-on-glass technology, (3) phosphorus-doped LPCVD oxide, and (4) low-temperature (200°C) plasma deposited oxide.

Adhesion has been achieved on these oxides through a variety of silane precoat treatment processes. Solutions of γ-aminopropyltriethoxysilane, γ-methacryloxytrimethoxysilane, and γ-glycidoxytrimethoxysilane applied to thermal oxide substrates all improved resist image adhesion for conventioanl positive photoresists; Mallinckrodt Multisurf also works well. Conventional liquid-phase application of HMDS, however, was not adequate for the latter three tougher substrates listed above; it did provide adequate photoresist adhesion for thermal oxides, however. For the last three substrates, a double silane adhesion promoter process was needed and developed. This process has been incorporated into actual device fabrication processing.

The double promoter process involves the successive application of liquid promoter solutions of vinyltrichlorosilane (VTS) and 3-chloropropyltrimethoxy-silane followed by successive cure cycles in dry N_2 at 90°C after each application and before photoresist application. The double promoter process evolved because it was felt that the silane reaction with the SiOH surface groups of low temperature oxides was incomplete for a single promoter application, and because vapor silane equipment did not exist at that time. Interestingly, a double HMDS liquid promoter process failed to yield adequate adhesion as well. Later in time, the successful but somewhat complex double promoter process was replaced by the 'vapor phase' HMDS process in the Star 1000 (or 2000); then superior resist image adhesion was obtained on all four oxide substrates with all the photoresists tested. Before the advent of the HMDS vapor priming in stand-alone or wafer track equipment module chambers, liquid priming solutions were widely used, especially in development areas.

The double promoter process is felt to approach the same quality of surface preparation as that for the vapor HMDS treatment through improved surface cleaning capability. Furthermore, it must be concluded that the Star 1000 *in situ* dehydration baking and vapor phase processing is of great importance. If this were not so, then suitable adhesion should have been achieved on SiO_2 substrates by a simple double HMDS (liquid) treatment.

What are the mechanisms of these two successful promoter processes? To answer this question, ESCA analysis was employed.

ESCA survey scans recorded for thermal oxide surfaces indicated the presence of Si, C, and O. The elemental compositions for these oxides are similar to those for the Y58 blank system of Table 5. Again, the simple elemental compositions do not show any dramatic changes due to the surface chemical modifications by

Table 5.

XPS elemental compositions from different Y58 samples

Samples	Take-off angle	Composition (at%)			C/S	O/Si
		Si	C	O		
Blank Y58 (1)	5°	25.28	21.24	53.48	0.84	2.11
Blank Y58 (2)	5°	21.20	30.49	48.31	1.43	2.28
Blank Y58 (3)	5°	25.21	23.11	51.68	0.92	2.05
Blank 58 (cured)	5°	19.78	28.25	51.97	1.43	2.62
HMDS (SVG)/Y58						
Cured (1)	5°	24.65	29.27	46.08	1.18	1.87
Cured (2)	5°	27.88	23.02	49.15	0.82	1.76
No cure	5°	22.49	31.58	45.93	1.40	2.04
HMDS (Star 2000)/Y58						
Cured (1)	5°	28.53	27.42	44.05	0.96	1.54
Cured (2)	5°	27.30	23.43	49.26	0.85	1.80
Cured (3) [a]	5°	31.16	21.88	46.96	0.70	1.50

[a] Sample stored in air for 7 days.

Samples numbered 1, 2, and 3 in each section of the table are replicates to show wafer-to-wafer reproducibility of the ESCA results.

the HMDS treatment, except for the decrease in O/Si ratio. But the ESCA data do show that the chemical nature of these elements changes significantly, i.e. C/O-containing molecular impurities are being replaced by a covalently bonded Si/O/C-containing stabilization layer. As seen for the Y58 (Monsanto manufacturer's ID) blank silicon wafer, XPS spectra resulting from Si $2p$ transitions for HMDS-treated oxide wafers also depict the evolution of a new peak at 101.8 ± 0.1 eV. Again this peak is most prominent in the Imtec Star 2000 HMDS (or YES) treated oxide sample, and is assigned to the Si $2p$ peak present on the oxide surface due to $(CH_3)_3Si{-}O{-}$ surface species.

C $1s$ XPS spectra for the treated surfaces are not well resolved. From the deconvoluted spectra, the decreases in the main contamination peak at 284.8 eV, and the other two peaks at 1.7 and 4.0 eV higher binding energy (BE) can be followed. The intensities of these peaks are notably much lower in the oxide samples as compared with those of Y58 wafers, consistent with the lower density of surface silanols or contamination adsorption sites between the two surfaces. After vapor-phase HMDS treatment, the contribution of these peaks is greatly reduced and a new main C $1s$ peak centered at 284.6 eV appears, as for the Y58 samples, which is assigned to the $-CH_3$ group, due to the HMDS stabilization reaction.

When conventionally applied VTS (dilute xylene solution) is used, adhesion is increased to SiO_2 substrates and no trace of chlorine from the VTS reaction remains at the surface. ESCA results for VTS-treated PSG oxides show reproducible increases in carbon concentration as expected from $-SiO{-}Si(CH{=}CH_2)$ surface reaction products. Consistent with this hypothesis, broadening of the ESCA Si $2p$ peak at lower BE is also observed. The broadened Si $2p$ spectrum can be simulated by two Gaussian curves with 1.8 eV FWHM centered at 103.0 and 103.7 BE values. The 103.7 eV peak is the same as that observed for the 'no HMDS' blank (see Fig. 6a), with the second peak at 103.0 eV attributed to the Si product of the VTS surface reaction.

ESCA results for double promoted oxides are less definitive than those previously described, but definitely show differences in the Si $2p$ spectra, hence indicating that again surface changes are occurring due to the irreversible surface chemical reactions. At the same time, these oxide surfaces are being at least partially cleaned of C/O-containing atmospheric contaminants.

Polysilicon subtrates are usually encountered during MOS device fabrication for defining gate structures. For submicrometer CMOS processing, adhesion problems encountered with this substrate are even more severe.

Here again, standard liquid HMDS promoter processing fails to be adequate. Historically, double promoter processes were needed as found for the troublesome low-temperature oxides. Great care must be exercised here, however, because oleophobicity is often created if the wrong two promoters are selected, resulting in dewetting of the treated substrate by the photoresist spinning solution. This often occurs, even though photoresist solutions contain surfactants to aid in wetting. This more 'black art' processing involved the use of VTS again, but followed this time by the use of nitrogen-containing silanes, such as 1,3-divinyltetramethyldisilazane (6208) or HMDS, and in this order.

Again, double liquid treatments of HMDS did not prove adequate for polysilicon gate substrates, substrates comprised of amorphous deposited silicon as opposed to single-crystal silicon wafers. As for the troublesome oxide surfaces, vapor-phase HMDS also works. Thus, the double promoter process utilizing somewhat 'poor safety handling' promoters for polysilicon can be replaced with the vapor-phase HMDS process.

ESCA results of treated polysilicon substrates indicate that the polysilicon surface, like that of the blank single-crystal Si wafer, is covered with a thin layer of SiO_2 (<20 Å). For this substrate, surface layer changes are detected by observing the SiO_x/Si ratio vs. that of the blank. VTS/HMDS and VTS/6208 treated surfaces have increased SiO_x/Si ratios of 6.8–7.0, while the blank exhibits values of 2.5–2.6. Concomitantly, increases also occur in the carbon concentration. The increases in SiO_x/Si ratio and carbon content after surface treatment are interpreted as due to the formation of an additional surface layer through the reaction of the organosilanes and surface —SiOH groups. If a thicker layer had resulted, no Si^0 would have been detected, because it would have been out of the ESCA detection depth; therefore, the reactions are restricted primarily to the first few surface layers.

It is not too surprising that vapor-phase HMDS also improves adhesion to this substrate. The native oxide has been shown on Y58 wafers to be easily treated and/or passivated. Actual resist image 'lift' testing on vapor promoted polysilicon wafers produced superior results and no image lifting occurred even for first generation resists known to be susceptible to 'lifting'.

Gold and aluminum substrates are used in conventional metallization fabrication processing as device element metal interconnects. For aluminum and gold, adhesion failure is somewhat less frequent than for the other substrates studied; actually, adhesion failure could not be observed for test resist images on aluminum substrates unless intentional substrate rework occurred. In fact, no promoter is usually required for these layers even though they are frequently used. 'Lifting' of resist test images for these substrates usually occurs when the device wafers are reworked due to photoresist processing errors or other lithographic problems. Reworked wafers exhibit statistically poorer adhesion [17], and a promoter is usually needed to ensure good adhesion confidence.

Good adhesion has been achieved for gold substrates [24], using chelating type adhesion promoters. Chemisorption of carbon monoxide (CO), adsorption with stronger and more directional bonding than that involving van der Waals attraction forces, is well known for metal surfaces like gold [25]. Sellwood has shown that the chemisorbed CO aligns on metal surfaces with the carbon towards the substrate and the oxygen pointing outward [26]. Therefore, trimethylsilylacetamide (Huls T3250), a carbonyl-containing silane, is proposed to very likely adsorb and align preferentially on the Au surface similar to that observed for CO, the model compound; however, no direct spectroscopic evidence for this hypothesis has been obtained. The possibility of electrostatic adhesion forces existing between the C–O dipole and polar groups of the Novolak resin is also thought to be very likely.

Although no adhesion failures were observed for virgin aluminum substrates as found for gold with a variety of liquid silanes, VTS and vapor-phase HMDS-treated wafers were analyzed via ESCA to determine whether surface changes occurred with these treatments. For VTS-treated substrates, a dramatic 20-fold increase in the Si/Al ratio occurred with a concurrent decrease in carbon surface concentration. These findings support the hypothesis of a covalently bonded $-O-Si-CH=CH_2$ species $5-10\,\text{Å}$ in thickness across the wafer. Nistler [23] reported HMDS-treated substrates to have $5-10°$ contact angle increases vs. untreated surfaces; thus, either adsorption or surface bonding had occurred. ESCA results for HMDS-treated Al wafers were void of any trace of surface reaction product, however. Therefore, HMDS is a less effective surface modifier than VTS. When Al metal substrates are reworked by flood exposure followed by aqueous base development or O_2 plasma ashing, the aluminum oxide layer becomes thicker, indicating that further oxidation is occurring (see Table 6 and Fig. 13); adhesion tests on these reworked substrates usually result in severe pattern lifting. Nistler [23] observed the same results, even when HMDS was employed after a post-metal deposition cleaning or an oxygen plasma ashing process. To successfully rework oxidized Al substrates, either a bifunctional silane such as the alkoxy-organosilane Mallinckrodt Multisurf must be used or the oxidized Al surface must be physically sputter-cleaned to provide a new fresh Al surface; both methods have been successfully employed at Motorola.

Gold-coated substrates treated with three silanes were adhesion-tested and ESCA-analyzed. First of all, the three promoters yielded differing layer thicknesses on the wafers. For T3250, the promoted layer was $30\,\text{Å}$ or $1-2$ layers in

Table 6.
Al wafer rework ESCA results

Sample	Take-off angle	Composition (at%)					
		Al	O	C	Si	F	S
Virgin Al	5°	26.9	46.3	26.8	—	—	—
Virgin Al	45°	36.0	50.7	13.3	—	—	—
Reworked Al, flood exposure	5°	21.6	46.3	26.8	3.9	—	—
	45°	27.5	49.5	20.3	2.7	—	—
Reworked Al, plasma ash	5°	21.2	43.8	30.5	—	1.1	3.5
	45°	28.6	52.5	14.3	—	1.5	3.1

ESCA X-ray take-off angles were set as described in the text.

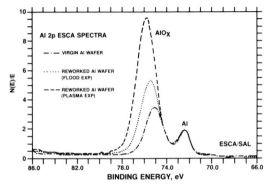

Figure 13. Al $2p$ spectra illustrating the effect of rework processes on the aluminum oxide spectral peak.

thickness. For 2-(diphenylphosphino)ethyltriethoxysilane (6110) and 3-mer-captopropyltriethoxysilane (8502), layers of thicknesses $\gg 50\,\text{Å}$ to $<1000\,\text{Å}$ were observed. All three promoters were successful in promoting resist adhesion, but the latter promoters are not recommended unless reactive-ion etching or ion-milling etch processes are used. Owing to their thickness, these layers may create a masking problem for plating or wet etching processing applications. Two of these processes have been extensive applications as adhesion promoters for the labeling of ceramic packages.

For T3250 processed wafers, ESCA shows an increase in Si and oxygen surface concentrations with a concurrent loss in carbon concentration. There-fore, carbon surface contamination on the gold is being replaced with the chemisorbed or physisorbed primer species. HMDS-treated Au surfaces exhibited no such change in surface chemical nature, consistent with the 'no improvement' adhesion 'lifting' test results for that promoter.

The latter thicker adhesion layers are most likely polysiloxane condensation polymer layers. Photoresist adhesion to these layers may be electrostatic in nature, while the adhesion of the promoter layers to the gold substrate may be the result of chemisorption forces between the chelating moieties of the promoter layer and the gold surface atoms.

8. CONCLUSIONS

What is accomplished by adhesion promotion treatments in IC manufacturing should actually be referred to as wafer substrate preparation, and not adhesion. Adhesion in the structural sense, as experienced in airplane composite material parts attachment, is not accomplished by silane wafer processing treatments except for the PI applications discussed early in this paper. The term adhesion, as it is used here, refers to a more practical definition—that is, resist image adhesion. Nevertheless, this type of adhesion is essential to the huge inter-national semiconductor business, and the early silane work of Plueddemann and others was essential to early wafer adhesion process development.

Furthermore, this work provides insightful engineering knowledge about surface conditions relevant to good lithographic processing of silicon integrated circuit wafers. Even though HMDS is extensively employed as an adhesion promotion process, sometimes it is not totally satisfactory and special silane treatments are required [27]. Experimental parameters have been established which can provide quality controls for predicting and monitoring resist image adhesion on both virgin and reworked wafers. An understanding of why resist adhesion is achieved for promoted and reworked wafers is also provided from a wafer surface chemistry or condition view.

REFERENCES

1. (a) A. Saiki and S. Harada, *J. Electrochem. Soc.* **129**, 2278 (1982); (b) R. G. Narechania, J. Bruce and S. Fridmann, *Ibid.* **132**, 2700 (1985).
2. J. N. Helbert, F. Y. Robb, B. R. Svechovsky and N. C. Saha, in: *Surface and Colloid Science in Computer Technology*, K. L. Mittal (Ed.), p. 121. Plenum Press, New York (1987).
3. K. L. Mittal, *Solid State Technol.* 89 (May 1979).
4. C. Johnson, Jr., M. S. Pak, B. Patnaik and R. Wilbarg, *IBM Tech. Disclosure Bull.* **26**, 1045 (1983); M. Levinson and C. W. Wilkins, Jr., *J. Electrochem. Soc.* **133**, 619 (1986).

5. J. N. Helbert and N. C. Saha, *ACS Polym. Mater. Sci. Eng.* **55**, 91 (1986).
6. J. N. Helbert and H. G. Hughes, in: *Adhesion Aspects of Polymeric Coatings*, K. L. Mittal (Ed.), p. 499. Plenum Press, New York (1983).
7. M. R. Gulett, M. L. Trudel and K. Stewart, Jr., US Patent 4,330,569 (1982).
8. J. C. Marinace and R. C. McGibbon, *J. Electrochem. Soc.* **129**, 2389 (1982).
9. C. A. Deckert and D. A. Peters, in: *Adhesion Aspects of Polymeric Coatings*, K. L. Mittal (Ed.), p. 469. Plenum Press, New York (1983).
10. G. L. Varnell, R. A. Williamson, T. L. Brewer, R. A. Robbins and C. D. Winborn, *US Army ERADCOM Rep.* No. 03-78-07, 17 (Nov. 1978).
11. E. P. Plueddemann, *J. Adhesion* **2**, 184 (1970).
12. E. P. Plueddemann, in: *Adhesion Aspects of Polymeric Coatings*, K. L. Mittal (Ed.), p. 363. Plenum Press, New York (1983).
13. R. L. Kaas and J. L. Kardos, *Polym. Eng. Sci.* **11**, 11 (1971).
14. R. G. Bortfield, *Proc. 1968 Kodak Microelectronics Semin.* **2**, 30 (1968).
15. S. Brelant, in: *Treatise on Adhesion and Adhesives*, R. L. Patrick (Ed.), p. 363. Marcel Dekker, New York (1969).
16. K. L. Mittal, *Polym. Eng. Sci.* **17**, 467 (1977).
17. C. A. Deckert and D. A. Peters, *Proc. 1977 Kodak Microelectronics Semin.* 13 (1977); *Circuits Manuf.* p. 1 (April 1979).
18. R. Steel, Motorola, private communication (1971).
19. T. A. Carlson, *Photoelectron and Auger Spectroscopy.* Plenum Press, New York (1975).
20. C. D. Wagner, L. E. Davis, M. V. Zeller, J. A. Taylor, R. H. Raymond and L. H. Gale, *Surf. Interface Anal.* **3**, 211 (1981).
21. C. S. Fadley, R. J. Baird, W. Siekhaus, T. Novakov and S. A. L. Bergstrom, *J. Electron Spectrosc.* **4**, 93 (1974).
22. M. Pijolat and G. Hollinger, *Surf. Sci.* **105**, 114 (1981).
23. J. L. Nistler, *Proc. 1988 Kodak Microelectronics Semin.* 233 (1988).
24. J. N. Helbert, *J. Electrochem Soc.* **131**, 451 (1984); US Patent 4,497,890 (1985).
25. N. A. DeBruyne, in: *Adhesion and Adhesives*, R. Houwink and G. Salomon (Eds), Elsevier, Amsterdam (1979).
26. P. W. Sellwood, *Adsorption and Collective Paramagnetism.* Academic Press, New York (1962).
27. Editorial, *Semicond. Int.* (April 1987).

Silanes and Other Coupling Agents, pp. 461–472
Ed. K. L. Mittal
© VSP 1992.

Plasma polymerized organosilanes as interfacial modifiers in polymer–metal systems

T. J. LIN,[1,*] B. H. CHUN,[1] H. K. YASUDA,[1] D. J. YANG[2] and
J. A. ANTONELLI[2]

[1] *Department of Chemical Engineering, University of Missouri-Columbia, Columbia, MO 65211, USA*
[2] *E. I. Du Pont de Nemours and Company, 3500 Grays Ferry Avenue, Philadelphia, PA 19146, USA*

Revised version received 25 June 1991

Abstract—Plasma polymers deposited under vacuum conditions have the advantage of creating strong interfacial bonds (possibly covalent bonds) at polymer–metal interfaces. This is particularly advantageous when silane chemistry is involved. Polymer–metal oxane bonds (M—O—Si) can be created in the form of tight networks by plasma polymerization directly on the metal substrates, e.g. steel. Using this tightly-networked silane plasma polymer as a priming layer on steel followed by an organofunctional silane polymer coating (i.e. polymer resin containing organofunctional silane) will form an effective corrosion protective layer as a whole. The structure and properties of the plasma-deposited organosilane polymers, the corrosion performance of such coating systems on steel substrates, and the interfacial chemistry were investigated in this study.

Keywords: Plasma polymer; polymer-metal interface; organosilane; coupling agent; corrosion resistance.

1. INTRODUCTION

Plasma polymers of organosilane compounds have been widely studied [1–10]. A large number of these studies deal mainly with the characterization, structure, and properties of polymer films, and reaction mechanisms of plasma polymerization of organosilicon compounds [11]. Efforts have also been expended in exploiting possible applications of these organosilane plasma polymers. The electrical properties of plasma-polymerized organosilane thin films which may be applied as protective coatings or as dielectrics in microelectronics were thoroughly studied [2, 3, 12]. The biocompatibility of plasma organosilane polymer was studied for potential applications as biomedical materials [4]. Plasma silane polymers were found to be good moisture barriers [7]. Plasma organosilane polymers have also been used as intermediate oxygen etch resistance layers in resist lithography [9]. Gas permeability studies and applications of plasma-polymerized organosilane films as a gas separation membrane have been of much interest to researchers [13–15].

We have long been interested in applying organosilane plasma polymers as a protective coating on metal and plastic substrates [16, 17]. This work focuses on the study of depositing on steel substrates a thin organosilane plasma film which serves as a priming layer (or coupling agent as used in the composite arena) for

---

*To whom correspondence should be addressed. Present address: E. I. Du Pont de Nemours and Company, 3500 Grays Ferry Avenue, Philadelphia, PA 19146, USA.

subsequent coating. The whole coating provides good corrosion protection for a steel substrate.

Silane chemical reactions can proceed under dry conditions in which a silane coupling effect can be easily accomplished. 'Contaminant-free' interfaces between the plasma polymer and metal substrate can easily be obtained by plasma cleaning in vacuum. This reduces the possibility of weak interfacial bonding caused by contaminants, e.g. adsorbed hydrocarbons, as well as air entrapment. An organo-functional silane polymer is then applied on the silane plasma-coated substrate. With the reactivities of the organofunctional silane polymer and the silane plasma polymer surface, possible reactions such as co-polymerization in the interfacial region can occur during the resin curing. The tightly-networked coating and strong interfacial bond result in an effective corrosion-resistant coating.

2. EXPERIMENTAL

The plasma polymerization was carried out in the set up shown in Fig. 1. The vacuum chamber was a typical bell-jar (18″ in diameter and 30″ height) containing two internal parallel electrodes (7″ × 7″ × ⅛″) made of stainless steel plates spaced 4″ apart. The substrate (4″ × 6″ × 0.03″) sat in the middle between the two parallel electrodes. The metal substrate can either be connected as an electrode or can be floating in the plasma environment. The chamber was pumped down to a background pressure of less than 1 mTorr by a rotary pump.

Gas or vapor, such as oxygen or organosilane, was fed into the chamber through a feeding tube with its outlet located at the top portion of the chamber so that uniform expansion and flow of the gas were ensured. The flow rates of gases were controlled by MKS 1259B mass flow controllers. Flow rates of gases ranged from 1 to 10 sccm (standard cubic centimeter per minute). The system pressure was controlled by an MKS 253A throttle valve which was installed at the outlet of the chamber. The typical pressure range for plasma cleaning and polymerization was 20–200 mTorr. Power supply units included Advanced Energy MDX-1K DC, PE-1000 55 kHz AC plasma generator, and RFX-600 13.56 MHz RF generators with automatic matching network. The power duration was normally 30 s to 2 min for oxygen–argon plasma cleaning and 1–10 minutes for silane plasma polymeriz-ation, which provides a film of thickness between 50 and 500 nm. Film thickness

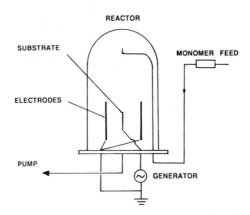

Figure 1. Schematic diagram of the experimental set-up for plasma polymerization.

was estimated by measuring the weight gain of a quartz sensor, with 1 cm^2 deposition area, which was placed on the substrate during the deposition.

X-ray photoelectron spectroscopy (XPS) was used for elemental analysis of plasma-deposited polymer films. The photoelectron spectrometer (Physical Electronics, Model 548) was used with an X-ray source of Mg K_α (1253.6 eV). Fourier transform infrared (FTIR) spectra of plasma polymers deposited on the steel substrate were recorded on a Perkin-Elmer Model 1750 spectrophotometer using the attenuated total reflection (ATR) technique. The silane plasma-deposited steel sample was cut to match precisely the surface of the reflection element, which was a high refractive index KRS-5 crystal.

Cold-rolled steel panels were purchased from Advanced Coating Technologies, Inc. (Hillsdale, Michigan). Silane chemicals (methylsilane, trimethylsilane, and tetramethylsilane) were purchased from Petrarch Systems, Inc. The silane plasma-deposited steel was then dip-coated with a polymer film 10–25 μm thick. The polymer coating resins used were silane-modified polymers with functionalities such as hydroxyl, acrylate, or amine.

Corrosion performance was evaluated by the scab corrosion test. The coated panels were scribed and subjected to 25 cycles as follows: 15 min immersion in 5% NaCl solution, 75 min air-dry at room temperature, followed by 22.5 h exposure to 85% relative humidity (RH) and 60°C environment. The tested samples were examined visually for failure such as corrosion, film lifting, peeling, adhesion loss, or blistering. The distance between the scribe line and the unaffected coating was measured as the corrosion creepage.

3. RESULTS AND DISCUSSION

3.1. Plasma surface cleaning

Plasma surface cleaning is widely used for removing trace surface contaminants, such as adsorbed hydrocarbons and surface oxides [18, 19]. As-received 'pre-cleaned' steel substrates were subjected to plasma of oxygen, argon, or oxygen–argon mixture prior to the deposition. Samples without plasma surface cleaning showed delamination of coating after the cyclic corrosion test, while samples which were cleaned with plasma showed good adhesion. This indicates that the plasma cleaning step is essential to ensure a cleaned surface for strong interfacial bonding.

Plasma surface cleaning can be a very reactive process if a gas such as oxygen is used. Surface hydrocarbons react with the oxygen plasma and form volatile compounds such as CO and CO_2. The effectiveness of surface carbon removal is due to a combined process of ion sputtering and chemical reactions [18]. We performed *in-situ* XPS analysis of steel surfaces after argon or oxygen plasma cleaning. The carbon content of substrate surfaces subjected to plasma environments was found to be significantly reduced. Little or no carbon signal was detected by XPS. To take advantage of the plasma-cleaned substrate surface, the deposition was carried out in the same vacuum chamber.

3.2. Characterization of silane plasma polymers

Elemental analysis of surfaces of plasma-deposited polymers on steel substrates was done by XPS. The XPS spectra of a representative plasma-polymerized trimethylsilane film are shown in Fig. 2. Oxygen was found to be incorporated into the polymer film and surface. A typical composition of trimethylsilane plasma film deposited at 2 W is C 52 at.%, Si 34 at.%, and O 14 at.%. The source of oxygen was probably from the residual water or air in the vacuum chamber and possibly from adsorbed oxygen after the deposited film was exposed to air. There is a high probability that such oxygen takes part in the oxane (Fe—O—Si) bond formation at the initial stage of plasma polymerization. The oxygen also contributes to the formation of siloxane chains in the plasma polymerization of the alkylsilane and oxygen mixture. No iron was detected in the polymer film. This indicates that the electrode material or substrate was subjected to insignificant sputter deposition.

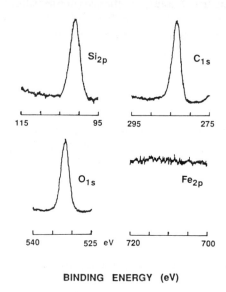

BINDING ENERGY (eV)

Figure 2. XPS spectra of a plasma-polymerized trimethylsilane film on steel. Plasma deposition conditions: flow rate 2 sccm; pressure 50 mTorr, power 2 W, deposition time 2 min.

The XPS spectra of plasma-polymerized silane polymer showed peaks that were a combination of two or more component peaks. In Fig. 3, the dashed lines show the Gaussian component peaks which have a sum almost equal to the total peak. The C_{1s} and Si_{2p} peaks can be deconvoluted into several peaks characteristic of different possible bonding environments such as aliphatic carbon and C—Si, C—O, C=O, Si—C, Si—O, and Si—O$_2$ [6, 17, 20]. For the C_{1s} core-level spectra, the peaks centered at 284.5, 286.2 and 287.6 eV are assigned to aliphatic carbon and C—Si, C—O, and —C=O groups; while for the Si_{2p} core-level spectra, the peaks centred at 100.8, 101.8 and 102.8 eV are assigned to Si—C, Si—O, and Si—O$_2$ groups, respectively. The results of peak deconvolution show that the compositions of peak components corresponding to the functional groups are Si—C:Si—O:SiO$_2$ = 47%:38%:15% and C—C (or C—Si):C—O:C=O = 78%:16%:6%. This suggests that the film consists of a combination of carbon chains, siloxane chains, and some hydroxylated groups of carbon and silicon. This agrees with the results obtained from FTIR analysis described in the following.

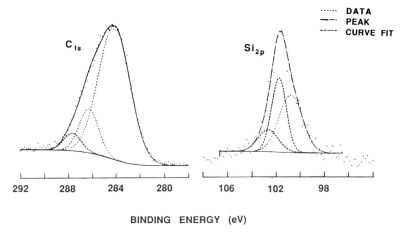

Figure 3. Curve fits of C_{1s} and Si_{2p} peaks in XPS spectra of a plasma-polymerized trimethylsilane film. Plasma conditions are the same as those in Fig. 2.

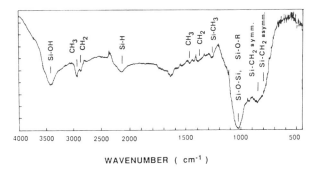

Figure 4. FTIR spectrum of trimethylsilane plasma polymer. Deposition conditions are the same as those in Fig. 2.

The chemical structure and possible bonds in the plasma polymer films were also characterized by means of IR spectroscopy. An FTIR spectrum of a typical trimethylsilane plasma polymer is shown in Fig. 4. According to the literature [8, 21–25], the peaks can be assigned as follows: 800 cm^{-1} (Si—CH$_2$ asymmetric rocking), 847 cm^{-1} (Si—CH$_2$ symmetric rocking), 1024 cm^{-1} (Si—O—Si, Si—O—alkyl stretching), 1269 cm^{-1} (Si—CH$_3$ symmetric deformation), 1350 cm^{-1} (CH$_2$ deformation), 1460 cm^{-1} (CH$_3$ antisymmetric deformation), 2140 cm^{-1} (Si—H stretching), 2865 cm^{-1} (CH$_2$ stretching), 2930 cm^{-1} (CH$_3$ stretching) and 3426 cm^{-1} (Si—OH stretching). The bonding environments surrounding Si atoms are essentially Si—O—Si, Si—O—alkyl, Si—CH$_3$, Si—CH$_2$, Si—H, and Si—OH.

3.3. Silane polymer structure and deposition kinetics

When alkylsilanes such as methylsilane, dimethylsilane, trimethylsilane, tetramethylsilane, and hexamethyldisilane polymerize in a glow discharge, the stoichiometry, chemical structure, and properties of the resulting polymers generally depend on the discharge conditions [2, 8, 21, 22]. Elemental compositions from XPS for plasma trimethylsilane polymers deposited under different power levels are illustrated in Fig. 5. Silane plasma polymers were deposited on steel substrates which were connected as a cathode in a DC glow discharge. Two other parallel electrodes were connected to the anode. For alkylsilanes, such as trimethylsilane, the discharges were fairly stable during the deposition. As shown in Fig. 5, both atomic ratios C/Si and O/Si increase with the discharge power. Polymers formed within this power range (2–50 W) have C/Si ratios (1.5–2.3) which are lower than that in the starting monomers (3.0). The carbon content of films increases with the deposition power. From a mechanistic viewpoint, this can be explained as follows: under the plasma conditions used, the silicon-containing segments contribute to the polymer film deposition to a larger degree than the hydrocarbon segments. As a result, the plasma-deposited polymers have a smaller C/Si ratio than the monomers. Kokai *et al.* [21] suggested that the cleavage of the Si—C bond (89.4 kcal/mol) is the dominant reaction in tetramethylsilane plasma polymerization. The stronger C—H bond (99.2 kcal/mol) makes the alkyl portion of the silane molecule difficult to decompose under plasma conditions. Higher energy causes more C—H bond breakage and will result in a higher degree of hydrocarbon crosslinking in the process of plasma polymerization. At a lower power level, however, less C—H bond cleavage in the alkyl portion takes place and the formation of small, stable hydrocarbon molecules such as CH_4 and C_2H_4 [21] results in a smaller contribution of hydrocarbon during the polymer deposition. As a result, this polymer contained relatively large amounts of silicon-containing moieties.

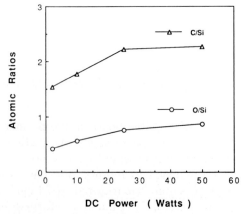

Figure 5. Atomic ratios (C/Si and O/Si) from XPS for plasma-deposited trimethylsilane polymer as a function of the DC power.

Similar results were obtained in the polymerization of monomethylsilane and tetramethylsilane. Instead of controlling the deposition in a constant power mode, these polymerizations were carried out in a constant voltage mode to give a better correlation between the deposited polymer structure and the applied electric field. Elemental compositions for films deposited from methylsilane (MS) and tetramethylsilane (TMS) under different voltages are shown in Fig. 6. As expected, within the operating voltage range (400–1000 V), the C/Si ratio for polymers deposited from TMS increased from 2.3 to 4.8, while that for the MS plasma polymer varied slightly from 1.5 to 1.7. Plasma polymers of tetramethylsilane and

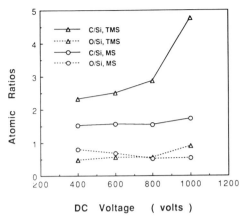

Figure 6. Atomic ratios of C/Si and O/Si for plasma polymers of methylsilane (MS) and tetramethyl-silane (TMS) as a function of the applied DC voltage.

trimethylsilane showed C/Si ratios (2.3–2.86, except for that at 1000 V) smaller than those of their starting monomers (C/Si = 4 and 3, respectively). The C/Si ratios of polymers from methylsilane (1.5–1.7) and dimethylsilane (2.2–2.5, not shown in Fig. 6) were greater than those of their monomers (C/Si = 1 and 2, respectively). The bond strengths [26] of $H-SiH_2CH_3$ (89.6 kcal/mol) and $H-SiH(CH_3)_2$ (89.4 kcal/mol) are close to that of the Si—C bond (89.4 kcal/mol). Therefore, under the energetic plasma conditions, the probability of Si—H bond cleavage was about the same as that of the Si—C bond. In the process of plasma polymerization, instead of bonding with H, the Si provided its bonding sites to either C or O for the formation of crosslinked polymer, resulting in bond linkage of Si with more than one carbon atom. In the case of trimethylsilane or tetramethyl-silane, incorporation of silicon in the polymer network was enhanced. This, together with the incorporation of oxygen, which may bond with Si through Si—O—C and Si—O—Si linkages, resulted in C/Si ratios smaller than those of their starting monomers.

3.4. Corrosion resistance of coatings on metal substrates

When a metal substrate was cleaned with oxygen–argon plasma followed by a thin layer (100–500 nm) of plasma silane polymer deposit, the coated substrates showed good humidity and corrosion resistance. Samples were prepared and placed either in a humidity chamber (85% RH and 60°C) or immersed in a salt solution (5% NaCl) for 5 days. The plasma-coated samples showed little or no pitting on the surfaces, while severe corrosion appeared on the uncoated sample.

For the cyclic corrosion test, a layer of acrylosilane polymer coating (10–25 μm thick) was dip-coated onto the plasma-deposited substrates. The coated samples were then subjected to 25 scab cycles. The test results are plotted in Fig. 7. Corrosion performance (as described by the length of scribe creep) was correlated to the wattage used for plasma film deposition. As discussed in the previous section, the chemical structure and properties correlated with the deposition conditions, especially the power level applied. Therefore, atomic compositions for plasma polymers deposited at different power levels were also plotted in Fig. 7. A

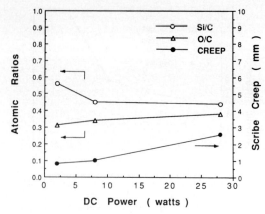

Figure 7. Corrosion performance (scribe creep distance after corrosion test) of coated cold-rolled steel and atomic ratios (Si/C and O/C) for plasma-deposited TMS films as a function of the applied DC power.

better performance was found for samples with plasma polymers deposited at lower wattages (e.g. 2–10 W). The lower power-deposited plasma polymers showed a higher silicon content and a lower oxygen content than those deposited at higher power (28 W). This suggested that the silane polymer coating had better compatibility with plasma polymers (higher silicon content) deposited at lower powers.

The well-known interfacial phenomenon effected by silane coupling have been well elucidated by Plueddemann [27]. Silane coupling agents are capable of reacting with surface moisture to generate silanol groups, which may also form hydrogen bonds with the hydroxylated surfaces. They are also capable of reacting with surface hydroxyl groups to form covalent oxane bonds with the mineral surface. Through this chemical bonding process, the strength of the interface between a mineral surface and an organic polymer coating can be significantly enhanced.

Plueddemann suggested [27] that conditions favorable for silane interfacial bonding include (1) maximum initial formation of oxane linkages (M—O—Si), (2) minimum penetration of water molecules to the interface, and (3) a polymer structure that retains silanols at the interface. In terms of these favorable conditions, plasma-deposited silane polymers are considered to be very good interfacial modifiers. In the initial stage of plasma film deposition, the reactive silane plasma (e.g. trimethylsilane plasma) reacts with the substrate surface. The initial formation of oxane bonds will be more effective if the surface is pretreated with oxygen-containing plasmas. Extensive Fe—O—Si oxane bonds will be created at the interface between the plasma polymer and the steel surface. As the deposition occurs under dry conditions (under vacuum), the oxane reaction can possibly be driven to the maximum extent without the formation of intermediate silanols due to the presence of moisture. *In-situ* XPS monitoring of the interfacial chemical reactions during the plasma deposition is now in progress to obtain a better understanding of this argument about the interfacial chemistry. Polymers formed by plasma polymerization are normally found to have a high degree of crosslinking [28, 29]. This tightly-networked polymer film reveals low moisture or gas (O_2, N_2) permeability [29]. Because of this highly crosslinked nature and the strong interfacial bonding, the water penetration through the interface region is minimized during the humidity exposure. The water molecules, permeable to the polymers, reach the interface and possibly cluster into a liquid phase. This will cause interface weakening via the water plasticization of the interphase region [27]. Normally, plasma polymers are relatively rigid due to their highly cross-linked nature. A smaller plasticizing effect caused by water at the interphase region is expected in the plasma polymer case.

Reactions of plasma polymers with an organofunctional silane polymer coating are not yet well understood. However, an analogy to the conventional chemistry of silane coupling agents can be assumed. In the case of a liquid phase coupling agent, the hydrolyzed silane condenses to oligomeric silanols which are initially soluble and fusible, but ultimately can condense to rigid crosslinked structures. Plasma polymers could also be hydrolyzed by the adsorbed moisture during exposure to the ambient after removal from the deposition chamber. The plasma polymer surfaces can virtually become silanol in nature as characterized by XPS and IR spectroscopy. It is also possible that the oligomeric phase exists in the plasma polymers [3, 21]. These low molecular weight oligomeric components may have a higher solubility with the organofunctional silane polymer and co-polymerize with it during the curing.

A plasma silane polymer was also deposited on a zinc-phosphated, cold-rolled steel substrate. Organofunctional silane polymer was applied on the plasma-deposited phosphated steel. The sample thus prepared showed excellent corrosion performance. This can be attributed to (1) the micro-mechanical inter-locking mechanism by the porous phosphate layer, (2) interfacial and oxide stability, and (3) the formation of primary interfacial bonds, as suggested by Kinloch [30]. The coating system including a phosphate layer, a plasma polymer, and an organofunctional silane polymer is illustrated in Fig. 8. The phosphate layer is essentially a layer of zinc (or iron) phosphate crystals. It is highly porous. When the phosphated substrates were 'immersed' in the silane plasma gas for film deposition, the plasma polymer deposited on all surfaces exposed to the plasma gases. This included the pore surfaces. Each individual crystal in the near surface region was uniformly coated with a thin polymer film, as is shown in Fig. 9. After 2 min deposition (Fig. 9b), the crystal surfaces were covered with a thin polymer

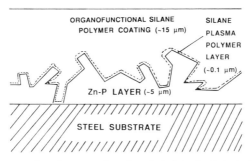

Figure 8. Schematic diagram of coating on a zinc-phosphated steel substrate. Thin plasma polymer is first deposited on the phosphate crystal surface, followed by an organofunctional silane polymer coating.

Figure 9. SEM micrographs of surfaces of phosphated steel with and without deposition of plasma polymer of trimethylsilane. (a) No plasma film deposition; (b) 2 min deposition; (c) 8 min deposition.

film, compared with Fig. 9a. This can be seen more clearly in the SEM micrograph of 8 min deposition (Fig. 9c). XPS spectra of the phosphated steel surface without plasma deposition and of surfaces with 2 min and 8 min trimethylsilane plasma deposition are shown in Fig. 10. For the steel surface without plasma deposition (0 min), no silicon was detected (Si_{2p} shows only background noise), but the phosphate surface was clearly seen with peaks such as C_{1s}, O_{1s}, P_{2p}, Zn_{2p}, Fe_{2p}, and Cr_{2p}. In the case of 2 min deposition, the spectra showed no peaks of P_{2p}, Zn_{2p}, Fe_{2p}, and Cr_{2p}, but a strong Si_{2p} peak was detected. Similar XPS spectra were obtained for 8 min deposition. This indicates that there is a high percentage of coverage of plasma film on the phosphate crystal surfaces. The type of bonding between the silane plasma polymer and the zinc phosphate surface is not known. However, the micro-mechanical interlocking and bonding between the plasma polymer and organofunctional silane polymer provided an effective corrosion protection coating.

4. CONCLUSIONS

Plasma surface cleaning ($Ar-O_2$) of the metal substrate is critical for good interfacial bonding with the subsequent coating. XPS and IR analyses show that the plasma-deposited silane polymers are essentially polyorganosiloxane in nature.

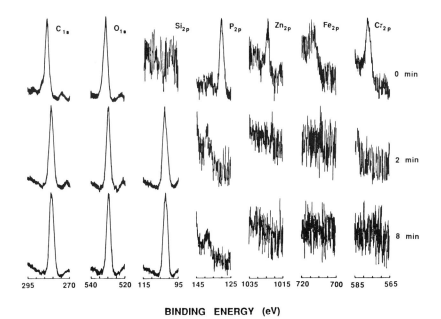

BINDING ENERGY (eV)

Figure 10. XPS spectra of surfaces of phosphated steels without plasma deposition (0 min), and with 2 and 8 min plasma trimethylsilane polymer deposition. SEM micrographs of these samples are shown in Fig. 9.

The bonding environments surrounding Si atoms are basically Si—O—Si, Si—O—alkyl, Si—CH$_3$, Si—CH$_2$, Si—H, and Si—OH. The polymer–metal interface bonding is believed to be a typical oxane linkage (M—O—Si). The elemental compositions of deposited silane plasma polymers vary with the deposition power. Both C/Si and O/Si atomic ratios increase with the discharge wattage (or voltage). Organofunctional silane polymer coatings show good compatibility with the plasma silane polymers. Corrosion performance of thus-coated steel samples were found to be excellent. The results were particularly encouraging when plasma-deposited silane polymer and organofunctional silane polymer coating were applied on phosphated steel substrates.

REFERENCES

1. G. Smolinsky and M. J. Vasile, *Int. J. Mass Spectrom. Ion Phys.* **12**, 147 (1974).
2. M. Kryszewski, A. M. Wrobel and J. Tyczkowski, in: *Plasma Polymerization*, M. Shen and A. T. Bell (eds), ACS Symposium Series No. 108 American Chemical Society, Washington, DC (1979).
3. A. M. Wrobel, M. Kryszewski and M. Gazicki, *J. Macromol. Sci. Chem.* **A20**, 583 (1983).
4. A. K. Sharma and H. Yasuda, *Thin Solid Films* **110**, 171 (1983).
5. K. G. Sachdev and H. S. Sachdev, *Thin Solid Films* **107**, 245 (1983).
6. N. Inagaki and A. Kishi, *J. Polym. Sci. Polym. Chem. Ed.* **21**, 1847 (1983).
7. E. Sacher, J. E. Klemberg-Sapieha, H. P. Schreiber and M. R. Wertheimer, *J. Appl. Polym. Sci. Appl. Polym. Symp.* **38**, 163 (1984).
8. Y. Catherine and A. Zamouche, *Plasma Chem. Plasma Proc.* **5**, 353 (1985).
9. V. S. Nguyen, J. Underhill, S. Fridmann and P. Pan, *J. Electrochem. Soc.* **132**, 1925 (1985).
10. I. Tajima and M. Yamamoto, *J. Polym. Sci., Part A* **25**, 1737 (1987).
11. A. M. Wrobel and M. R. Wertheimer, in: *Plasma Deposition, Treatment, and Etching of Polymers*, R. d'Agostino (Ed.). Academic Press, Orlando, Florida (1990) and references cited therein, p. 163.

12. P. Kazimierski and J. Tyczkowski, *J. Macromol. Sci. Phys.* **B27**, 233 (1988).
13. M. Sanchez Urrutia, H. P. Schreiber and M. R. Wertheimer, *J. Appl. Polym. Sci., Appl. Polym. Symp.* **42**, 305 (1988).
14. J. Sakato and M. Yamamoto, *J. Appl. Polym. Sci., Appl. Polym. Symp.* **42**, 339 (1988).
15. N. Inagaki and H. Katsuoka, *J. Membrane Sci.* **34**, 297 (1987).
16. D. L. Cho and H. Yasuda, *J. Appl. Polym. Sci., Appl. Polym. Symp.* **42**, 233 (1988).
17. P. Laoharojanaphand, T. J. Lin and J. O. Stoffer, *J. Appl. Polym. Sci.* **40**, 369 (1990).
18. D. M. Mattox, in: *Deposition Technologies for Films and Coatings*, R. F. Bunshah, J. M. Blocher, D. M. Mattox, T. D. Bonifield, G. E. McGuire, J. G. Fish, M. Schwartz, P. B. Ghate, J. A. Thornton, B. E. Jacobson, and R. C. Tucker, Jr. (Eds), p. 63. Noyes Publications, New Jersey (1982).
19. D. F. O'Kane and K. L. Mittal, *J. Vac. Sci. Technol.* **11**, 567 (1974).
20. K. Miyoshi and D. H. Buckley, *Appl. Surf. Sci.* **10**, 357 (1982).
21. F. Kokai, T. Kubota, M. Ichjyo and K. Wakai, *J. App. Polym. Sci., Appl. Polym. Symp.* **42**, 197 (1988).
22. A. Wrobel, J. E. Klemberg, M. R. Wertheimer and H. P. Schreiber, *J. Macromol. Sci. Chem.* **A15**, 197 (1981).
23. A. L. Smith, *Spectrochim. Acta.* **16**, 87 (1960).
24. D. R. Anderson, in: *Analysis of Silicones*, A. L. Smith (Ed.). Wiley, New York (1974).
25. N. Inagaki and M. Taki, *J. Appl. Polym. Sci.* **27**, 4337 (1982).
26. D. R. Linde (Ed.), *CRC Handbook of Chemistry and Physics*, 71st edn. CRC Press, Boca Raton (1990).
27. E. P. Plueddemann, *Silane Coupling Agents*. Plenum Press, New York (1982).
28. H. V. Boenig, *Plasma Science and Technology*. Cornell University Press, Ithaca (1982).
29. H. Yasuda, *Plasma Polymerization*. Academic Press, Orlando, Florida (1985).
30. A. J. Kinloch, in: *Polymer Surfaces and Interfaces*, W. J. Feast and H. S. Munro (Eds), p. 75. Wiley, New York (1987).

Silanes and Other Coupling Agents, pp. 473–491
Ed. K. L. Mittal
© VSP 1992.

Interfacial shear strength and durability improvement by monomeric and polymeric silanes in basalt fiber/epoxy single-filament composite specimens

J. M. PARK and R. V. SUBRAMANIAN*

Department of Mechanical and Materials Engineering, Washington State University, Pullman, WA 99164-2920, USA

Revised version received 27 February 1991

Abstract—Silane coupling agent effects in basalt fiber-epoxy systems have been investigated through measurement of the interfacial shear strength (IFSS) in single-fiber composite (SFC) specimens. Three silane were studied: 3-aminopropyltriethoxysilane (APS) and two polymeric silanes in which dimethoxy- or trimethoxysilane groups are attached via side chains to a polyethyleneimine backbone. Optimal conditions for silane application were standardized. Crosslinking of the deposited silanes is shown to result in decreased interpenetration by the matrix epoxy resin and lower values of IFSS. The polymeric silane with trimethoxy groups was found, as expected, to be inferior to the other two. APS and the polymeric dimethoxysilane gave similar results in improvement of IFSS and its retention after 1 h boiling in water. Monitoring of acoustic emission (AE) during straining of SFC specimens established a one-to-one correspondence between the number of AE events and fiber breaks. Measurement of AE pulse energies provided evidence for the sizing effect of polymeric silane coatings, through healing of surface flaws, as well as for moisture attack at severe surface flaws on the fiber during silane treatment.

Keywords: Polymeric silanes; basalt fiber; epoxy composite; acoustic emission.

1. INTRODUCTION

Bonding at the fiber–matrix interface plays an important role in controlling composite properties. A strong interfacial bond is needed for efficient transfer of the applied load, whereas fracture toughness can be improved by debonding and crack blunting mechanisms occurring at the interface. In glass fiber composites, silane coupling agents are used to improve composite performance generally under ambient and especially under high humidity conditions.

As reviewed extensively by Plueddemann [1], silane coupling agents are known to improve the wettability of fiber by resin; to increase the compatibility of the two materials; to enhance the physical, mechanical, and chemical bonding between the fibers and the polymer matrix; and to give greater durability and resistance to water attack at the interface. Several mechanisms have been proposed to explain the interfacial reinforcement by silane coupling agents resulting in improvement in mechanical strength and hygrothermal stability of composites. Among them, the most widely accepted is chemical bonding [2]; others invoke preferential adsorption of silanes or correlate the interfacial properties to wettability of the

---

*To whom correspondence should be addressed.

components and surface energy effects. All of the proposed theories have their merit because many complex factors contribute to interfacial mechanical properties. It should also be recognized that hydrogen bonding between silanols in the coupling agent and hydroxyl groups on the glass surface is a significant factor for the stronger interfacial interaction causing enhanced mechanical strength, compared with the weaker interactions by van der Waals forces at the interface which may also contribute to the interfacial bonding.

It is known that more than a monolayer of silane is needed to attain the optimum strength in a composite material [1, 3]. This may be due to the necessity for interfacial chemical bonds, interpenetrating network formation in the chemi-sorbed silane layers, and proper orientation of the functional groups. The silanes can form a bridge-like structure in the first monolayer where the silanol groups covalently bond with the surface-active sites. The silane coupling agents carry an organic functional group that can react with the matrix, and alkoxy groups which, on hydrolysis to silanols, condense with silanols on the fiber surfaces to form hydrogen bonds as well as Si—O—Si bonds. A molecular bridge is thus formed between the fiber and the matrix.

In recent work by Arkles *et al.* [4, 5], it has been proposed that, in comparison with monomeric silanes, polymeric silanes may react with substrates more efficiently. A typical polymeric silane is shown in Fig. 1a, in which pendant chains of siloxanes are attached through methylene chain spacers to a polyethyleneimine backbone. The film-forming polymeric silane thus provides a more continuous reactive surface to the polymer matrix in the composite. In this case, the recurring amino groups on the polymeric silane backbone can react with an epoxy resin matrix through chemical bond formation.

Another noteworthy feature of the polymeric silane depicted in Fig. 1a is that the silane chains on the polymer backbone carry *dialkoxy* groups. In comparison with monomeric silanes which carry *trialkoxy* groups, these polymeric silanes are claimed to offer superior substrate reactivity and to provide further performance improvements [4, 5]. A probable explanation seems to be the difference in the number of siloxane bonds that can possibly be formed with the glass surface by the two types.

When monomeric trialkoxysilanes are employed, it seems likely that only two siloxane bonds are formed with the glass surface, and one silanol group (Si—OH) is left dangling without reacting at the surface [4]. This is because distortions of surface silicon oxide on glass make it highly improbable for trialkoxysilanes to react to form three oxane bonds (Si—O—Si). While silanol condensation to form silane networks occurs to a significant extent, the free dangling Si—OH is a site for

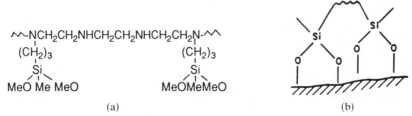

(a)   (b)

Figure 1. (a) Polymeric aminofunctional silane-carrying dialkoxy silane groups. (b) Multiple 'sawhorse'-type siloxane bonding to the fiber surface.

water absorption, which adversely affects the hydrolytic stability of the bond and detracts from the integrity of the fiber–matrix interphase. It is particularly important for reinforced composites to minimize this vulnerability to hydrolytic attack.

It is reported [4, 5] that the molecular architecture of the polymeric silanes is designed to eliminate this susceptibility. Since they carry only *dialkoxy*silanes, the excess, dangling Si—OH group, which is the site of hydrophilicity, is absent from the interfacial region. The polymeric backbone, from which the dialkoxysilanes hang, is also capable of forming large ring structures with enough flexibility to adjust to the steric constraints imposed by the surface. This bonding scheme, shown in Fig. 1b, is described as a series of connected sawhorses.

Another noteworthy advantage of polymeric silanes is their better film-forming ability because they are polymeric. Interpenetration of the silane polymer segments into adhesive/matrix polymer segments should be expected to provide improved bond strengths. The critical role of such interdiffusion in the interfacial region in controlling interfacial bond strength is well documented [6, 7].

Schmidt and Bell have also pointed out the potential advantages of polymeric coupling agents in their studies of ethylene mercaptoester (EME) co-polymers for adhesion promotion in steel/epoxy systems [8, 9]. In addition to their increased load-bearing capacity and toughness compared with low molecular weight mercaptoesters, the EME co-polymers enabled the investigation of the effects of increasing coupling agent functionality on interfacial properties and adhesion durability. Their results have shown, as would be expected, that increased hydrophilicity through increase in the concentration of the polar functional groups decreased the durabilty of the joints. The authors rightly emphasize the need for optimization of polar group interactions to achieve the desired levels of dry strength and durability in designing adhesion systems.

In view of the interesting features described above, the objective of the study reported here was the comparison of the effects of different polymeric silanes with that of a typical well-studied monomeric silane, namely 3-aminopropyltriethoxy-silane (APS). As candidate polymeric silanes, the polymeric dialkoxysilane shown in Fig. 1a and its trialkoxy analog were chosen for our experiments. In both of these cases, the polymeric backbone carries aminofunctional groups, as in the case of monomeric APS, which can co-react with the curing epoxy resin. Thus, it was possible to assess the effects arising from the presence in the silane structure of a polymeric C—C backbone not susceptible to hydrolysis. Furthermore, between the two polymeric silanes, it should be possible to compare the effects of attaching dialkoxy vs. trialkoxysilanes to the backbone on durability under high humidity conditions.

The composite system chosen was basalt fiber-reinforced epoxy resin. The study of basalt fibers as reinforcement for polymers has been a topic of continuing interest from our early investigations [10–12].

The method adopted for the comparison of silane treatment effects on fiber–matrix bond strength was the determination of interfacial shear strength (IFSS) in single-filament composite (SFC) specimens [13, 14], which we have used extensively in investigating fiber–matrix interactions. The conditions of silane treatment of single fibers as well as the corresponding effects on IFSS could thus be carefully controlled, measured, and compared.

In testing the SFC specimens, acoustic emission (AE) was monitored while the specimens were strained, by counting AE events and recording the energy associated with the events. In combination with optical microscopy, it was of interest to identify the correspondence between AE events and fiber fragmentation. The first results along these lines are also reported.

2. EXPERIMENTAL

2.1. Materials

2.1.1. Basalt fiber. Basalt fiber was made from naturally occurring basalt rock in the Pullman, WA area, which has the main composition SiO_2, 49%; Al_2O_3, 14%; FeO, 12%; CaO, 9.5%; and MgO, 5%. The fiber was produced in a prototype device built in our laboratory. The diameters of the fibers used were mostly 15 μm, and also 35 and 92 μm. The basalt rock was melted in a platinum–rhodium crucible at 1250–1350°C; the fiber was drawn from the melt through an orifice in the crucible and continuously wound on a rotating drum as described in detail elsewhere [15]. The virgin fiber strength, varying in the range 2–4 GPa depending on the drawing conditions, is comparable to that of E-glass fibers which are used widely in reinforced plastics; the Young's modulus is 85 GPa [15]. The strengths of the experimental fibers were measured before use as described below.

2.1.2. Silane coupling agents. Monomeric and polymeric silane coupling agents were purchased from Petrarch Systems Co. (Bristol, PA) and used without further purification. Their chemical structures are given in Table 1. APS was obtained as the neat compound, and the polymeric silanes were supplied as 50% solution in isopropanol. The silanes were diluted to the required concentration in chosen solvents for application on fibers.

2.1.3. Epoxy matrix. The SFC specimens were prepared using Epon 828, which was purchased from Miller-Stephenson Chemical Co. and is based on diglycidyl

Table 1.
Chemical structures of the monomeric and polymeric silanes

Item no.[a]	Chemical name	Chemical structure
A0750	3-Aminopropyltri-ethoxysilane	$H_2NCH_2CH_2CH_2Si(OEt)_3$
PS076.5[b]	Dimethoxymethyl-silylpropyl-substituted polyethyleneimine (aminofunctional)	$-NCH_2CH_2NHCH_2CH_2NHCH_2CH_2N-$ $(CH_2)_3$ ... $(CH_2)_3$ Si ... Si MeO Me MeO ... MeO Me MeO
PS076[b]	Trimethoxysilyl-propyl-substituted polyethyleneimine (aminofunctional)	$-NCH_2CH_2NHCH_2CH_2NHCH_2CH_2N-$ $(CH_2)_3$ ... $(CH_2)_3$ Si ... Si MeO MeO MeO ... MeO MeO MeO

[a] Petrarch Systems designation.
[b] Molecular weight = 1500–2000.

ether of bisphenol-A. Curing agents Jeffamine D400 and D2000 (polyoxy-propylenediamine) were purchased from Texaco Chemical Co. (Austin, TX). The flexibility of the specimens was controlled by changing the relative proportions of D400 vs. D2000 in the curing mixture. Their chemical structure is

$$H_2NCH_2-CH-(OCH_2CH)_x-NH_2$$
$$\underset{CH_3}{|} \qquad \underset{CH_3}{|}$$

where x is, on average, 5.6 for D400 and 33.1 for D2000. Specimen casting was done in silicone molds prepared using Dow Corning 3112RTV encapsulant and catalyst.

2.2. Methods

2.2.1. Fiber strength measurement. Fiber diameter was measured using an American Optical Co. type 1-60 microscope and a Vickers image-splitting eyepiece. Tensile strength of the fiber was measured with a mini-tensile tester constructed in-house. The fiber was fixed with tape on the center line of a paper frame across an 8 mm diameter hole. The fiber was then glued with 5 min epoxy glue applied at the two ends of the fiber. A strain-gauge type load cell, which was mounted on the crosshead, moved at a constant rate of 1 mm/min. The load cell was calibrated with standard weights. At least 40 fiber specimens were tested in each set to calculate mean strengths.

2.2.2. Preparation of silane-treated specimens. The main experimental procedure was to dip-coat fibers with silane coupling agents and to evaluate the IFSS of coated fibers embedded in a resin matrix. The basalt fibers were coated individually in a steel frame, to ensure uniform coating and to avoid the complications of neighboring fiber interactions in a tow. Optimized conditions for silane application were arrived at after experimenting with different solvents, concentrations, drying conditions, and coating times. Treated fibers were embedded with epoxy matrix in a dog-bone-shaped silicone mold, precured at 80°C for 3 h, and post-cured at 130°C for 2 h in order to obtain SFC specimens having a 1 inch gauge length. The specimens were tested after ageing for 3 days at room temperature and humidity.

2.2.3. Interfacial shear strength (IFSS) measurements. In the method used here for IFSS measurements, a single fiber cast in a polymer resin to produce a dog-bone-shaped specimen, with the fiber aligned along the longitudinal axis as described above. During testing, the specimen is incrementally stressed, and the fiber undergoes multiple fracture within the matrix at flaws of decreasing severity. The stress is increased until no further fiber breaks are observed (while the specimen is still intact), and the ultimate fragment lengths within the matrix are measured *in situ* with the aid of a polarized-light microscope.

The ultimate fragment lengths are approximated to equal the critical length for stress transfer, l_c, which is related to the fiber diameter, d, strength, σ_f, and the IFSS, τ, through the expression $l_c/d = \sigma_f/2\tau$. This expression is derived from a simple force balance and predicts unique values for these quantities [16]. In practice, a wide distribution of values is obtained—a result of random distribution of flaws and heterogeneities in the fibers. The data for both fragment length and fiber strength may be approximated by lognormal distributions, and these have been combined by a theory recently developed in our laboratory to yield a lognormal distribution for IFSS [13, 14].

In order to ensure that the critical length of the fiber has been reached without fracture of the dog-bone specimen, a high elongation-to-break (about 35%) is needed in the epoxy resin. This was achieved by adjusting the proportions of the flexible curing agents D400 and D2000, as mentioned earlier. The fragment lengths were measured from at least six specimens from each set of experimental conditions to obtain a few hundred fragment lengths adequate for statistical analysis.

In order to evaluate the effect of moisture exposure on IFSS, the SFC specimens were immersed in distilled water at boiling temperature for 1 h. IFSS was then measured after equilibration of the specimens for 1 h at room temperature.

2.2.4. Acoustic emission measurement. By using an AE analyzer (Physical Acoustic Co., Lawrenceville, NJ), AE tests were conducted during tensile testing using a specially designed tensile device. A piezoelectric transducer was positioned directly on the center of the SFC specimens. As soon as an AE test was finished, the number of fiber breakages in the SFC specimens was measured via polarized-light microscopy.

2.2.5. Infrared spectroscopy. IR spectra were recorded on a Nicolet 5DX FT-IR spectrometer. A thin coating of the condensed silanes was applied on KBr pellets made from a very high purity KBr powder (J. T. Baker Chemical Co.). The spectra of these pellets were recorded using the transmission method. Liquid silanes were analyzed using the internal reflection attachment (IRA). The number of scans was chosen to be 20 for both background and samples.

3. RESULTS AND DISCUSSION

3.1. Silane treatment

In order to realize, in practice, the coupling effects of silanes, it is necessary to standardize the conditions of application such that the silane is able to provide a molecular bridge without itself becoming a weak interphase between the fiber and matrix. As already mentioned, the principal experimental parameters to be controlled in the application of silane coupling agents on the fibers are the concentration, the solvent, and the treatment time. In order to arrive at the best conditions for each parameter, the IFSS was determined for each set of treatment conditions. Optimization of the conditions of silane application was thus progressively achieved.

The uniformity of silane coatings was examined by scanning electron microscopy (SEM). SEM examination of the coated silanes showed a thin uniform coating to be present when a 0.5% silane aqueous solution was used. At higher silane concentrations, the coating was formed in lumps, which could clearly form a weak interphase when the fiber is embedded in the epoxy matrix. These observations are illustrated in Fig. 2. The formation of lumps of APS from a 5%

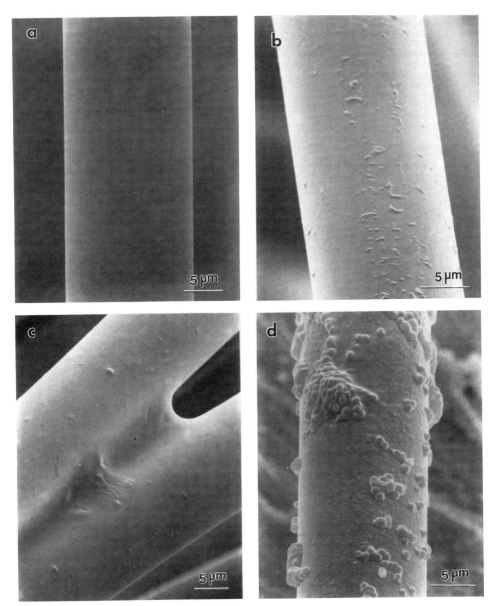

Figure 2. SEM of basalt fiber: (a) untreated; (b) monomeric APS (A0750) treated (0.5 wt%, 1 min); (c) polymeric aminosilane (PS076.5) treated (0.5 wt%, 5 min); (d) monomeric APS (A0750) treated (5 wt%, 1 min).

aqueous solution is evident in Fig. 2d in comparison with the smooth coating formed from 0.5% APS solution, (Fig. 2a) which shows very little surface roughness. The excellent film-forming capability of polymeric silanes is seen in Fig. 2c, which shows two unseparated fibers bridged by the polymeric dialkoxy-silane PS076.5.

When the treatment times for silane application were varied up to 30 min immersion of the fibers in 0.5% aqueous silane solution, it was found that the maximum values of IFSS were observed after 1 min dip time for APS and after a 5 min dip time in the case of the two polymeric silanes. For example, after residence times of 0, 0.25, 1.0, 5, 15, and 30 min, the IFSS values were 41.2, 43.3, 48.3, 46.7, 46.8, and 46.7 MPa respectively in the case of APS application, and 41.2, 45.9, 46.1, 50.1, 44.9, and 43.0 MPa respectively in the case of the polymeric dialkoxysilane, thus, establishing the desired treatment times for the best IFSS values as 1 min in the case of APS and 5 min in the case of the polymeric silane. Further prolonged immersion did not increase the measured IFSS.

By conducting further experiments in different solvents, such as acetone, isopropanol, ethanol, and toluene, it was found that application of APS from an aqueous solution gave the best results. Extensive experimentation with the polymeric dimethoxy- and trimethoxysilanes, PS076.5 and PS076 respectively, gave similar results. Representative data for PS076.5 are shown in Table 2. Application from 0.5 wt% solutions in water produced the maximum IFSS and retention after exposure to 1 h boiling in water.

Similar experiments using silane concentration as the variable showed that the highest IFSS values were observed for application from 0.5 wt% concentration for all three silanes studied. The typical results shown in Table 3 for the polymeric dialkoxysilane illustrate this trend. It can be seen that the IFSS of the specimens under dry conditions is highest when the silane concentration is 0.5 wt% in the application solution. Furthermore, it should be noted that after 1 h boiling in water, retention of IFSS is also highest for this set.

3.2. The silane interlayer

The dependence of IFSS on the silane concentration in solution, increasing up to 0.5 wt% and decreasing at higher concentrations (Table 3), illustrates the point made earlier that the optimal silane interphase requires a certain thickness of the silane, while a silane layer too thick can result in a weak boundary layer. Certainly, in practice, more than a monolayer of silane is needed to attain the optimum strength of a composite material, largely due to the necessity of forming interfacial chemical bonds, an interpenetrating network in the chemisorbed silane layers, and proper orientation of the functional groups [1].

Table 2.
Effects of solvents for polymeric silane PS076.5[a] application on dry/wet IFSS in basalt fiber–epoxy SFC specimens

Solvent	Dry IFSS (MPa)	Wet IFSS[b] (MPa)
Untreated	41.2	24.4
Water	50.1	38.6
Acetone	47.8	36.4
Ethanol	47.4	32.4
Isopropanol	47.3	30.4
Toluene	39.4	21.7

[a] Treatment time = 5 min; 0.5 wt% silane concentration.
[b] After 1 h boiling in water.

Table 3.

Dependence of the interfacial shear strength (IFSS) in SFC specimens on the polymeric dialkoxysilane PS076.5 concentration in aqueous solution[a]

PS076.5 solution conc. (wt%)	Dry IFSS (MPa)	Wet IFSS[b] (MPa)
0	41.2	24.4
0.1	46.3	37.5
0.5	50.1	38.6
1.0	47.2	34.3
5.0	45.6	33.9
10.0	45.2	33.6

[a] Treatment time: 5 min.
[b] After 1 h boiling in water.

It may be that the ideal condition is where a monolayer of silane reacts through its bifunctional groups with the glass and matrix resin, forming a stable, strong bridging interface. However, many layers of the coupling agent seem to be necessary to perform the expected role.

As the thickness of the deposited silanes increases, another factor, having an opposite effect on the interfacial properties, comes into play—that is, the non-bonded oligomeric silane interphase between the two main substrates of the composites. This interphase, having low mechanical strength and resistance to environmental attack, is the major factor for the decrease in IFSS. The thicker this oligomeric region, the greater is this effect. An interphase of low molecular weight silane oligomers, called the physisorbed fraction, has few or no bonds between the molecules, and consists of weak oligomeric silanols. Since they have low mechanical strength and hygrothermal stability, these oligomers can be hydro-lyzed easily, and stress failure and environmental attack can also occur easily, whereas the chemisorbed fraction should have a better resistance to hydrolysis than the physisorbed portion.

The initial build-up of the silane layer during fiber surface treatment involves the hydrolysis and condensation of alkoxysilanes, the relative rates of which determine the oligomeric nature and other characteristics of the deposited silanes. In addition, further reactions can occur during drying; for example in the present case, reactions involving the amine groups of the silane. Crosslinking can result from such reactions, leading to the formation of a tighter network which is not well penetrated by the matrix resin.

The data in Fig. 3 illustrate these points in the case of the aminosilane APS. In this figure, the spectrum of APS (a), is compared with the spectra obtained after three different drying conditions, at 25°C (b), 125°C (c), and 170°C (d) for 3 h. The disappearance of the bands due to ethoxy groups at 1072, 1100, and 1166 cm^{-1} is accompanied by the appearance of new ones. An indication of the condensation of silanes is seen in the appearance of two very strong bands at 1028 and 1129 cm^{-1} due to the asymmetric Si—O—Si stretching vibrations. The corresponding disappearance of the silanol group is seen in the significant

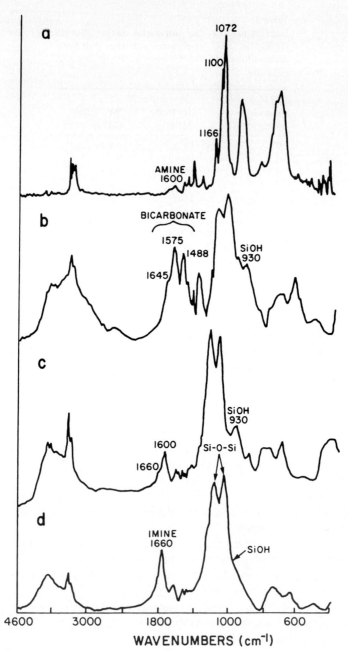

Figure 3. FT-IR spectra of (a) neat APS, and APS films after different drying temperatures, (b) 25°C, (c) 125°C, and (d) 170°C, for 3 h.

decrease of Si—OH peaks at about 950 cm^{-1}, with increasing drying temperature. Therefore, the ratio of the intensities of the Si—O—Si to Si—OH peaks increases.

It is also seen that after drying in air at 25°C, bicarbonate peaks appear (spectrum b). But after drying at 125°C the bicarbonate salt peaks have disappeared (spectrum c) and the free amine peaks appear at about 1600 cm^{-1}. In the case of drying at 170°C (spectrum d), the imine peak at 1660 cm⁻¹ is observed due to oxidation of APS at the higher temperature of drying [17, 18].

The effect of these chemical changes on the reaction of the amine functional groups of the silane with the epoxy resin matrix is seen in Fig. 4. The expected reaction is the opening of the epoxide by amine hydrogens to form hydroxy groups, as shown below:

$$R-NH_2 + 2\ -CH-CH_2 \rightarrow R-N \begin{cases} CH_2-CH- \quad OH \\ \\ CH_2-CH- \quad OH \end{cases}$$

The depletion of epoxide groups can therefore be followed by observing the intensities of the epoxide peak at 912 cm⁻¹ and the hydroxyl band at 3500 cm⁻¹.

On comparing the spectrum of the epoxy resin in Fig. 4a with that obtained after reaction with APS dried at 25°C (Fig. 4b), one can see the disappearance of the epoxide peak at 912 cm⁻¹ and the appearance of a strong band at 3500 cm⁻¹ due to —OH groups, as expected from the above reaction. However, after reaction for the same duration with APS dried at 170°C, the disappearance of the epoxide peak at 912 cm⁻¹ and the appearance of the hydroxyl band at 3500 cm⁻¹ are both less significant. The ratio of peak intensities, 912/3500 cm⁻¹, remains high, indicating inhibition of the amine–epoxy reaction when APS is dried at 170°C.

When considered in conjunction with the results of Fig. 3, it would appear that amine oxidation has caused crosslinking to occur in the silane layer. This is in addition to network formation by silanol condensation already present. Interpenetration of this crosslinked layer by epoxy is consequently hindered, which thereby reduces the accessibility of amine groups for reaction with the epoxy resin. When the amine–epoxy reaction is inhibited, the expected consequence is a reduction in interfacial bonding in the composite, which is actually what is observed. The IFSS of basalt fibers treated with APS and dried at 25°C shows better than 15% improvement over that of the untreated fibers. However, if the silane-treated fibers are dried at 170°C, there is negligible improvement over that of untreated fibers. These results emphasize the importance of interdiffusion and interpenetration of the matrix resin into the interphase region of composites in order for chemical bond formation between the coupling agent and the matrix polymer to occur, and are consistent with observations made in other reports [6], including those with graphite fiber composites [7]. In the latter case, when an electrodeposited polymer interlayer of butadiene–maleic anhydride became crosslinked, the interlaminar shear strength of the epoxy composite prepared from the electrocoated fiber was significantly reduced.

3.3. IFSS in SFC specimens

The measured IFSS in SFC specimens can now be discussed in light of the above considerations. Qualitative aspects of the observation of strained SFC specimens provide some interesting clues, as seen in the photomicrographs in Fig. 5. In the

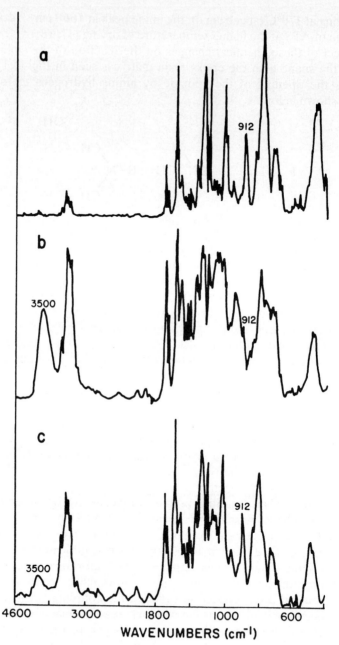

Figure 4. FT-IR spectra of (a) neat Epon 828 resin, (b) the reaction product of APS film dried at 25°C and Epon 828, and (c) the reaction product of APS film dried at 170°C and Epon 828.

case of the untreated fiber, fiber fracture is seen to result in some debonding along the fragments on either side of the break (Fig. 5a). When the bonding is good, as in the silane-treated fibers, matrix crack is initiated which grows as it is pulled by well-bonded fiber fragments on either side of the fracture (Fig. 4b); interfacial debonding is not a significant failure mode in this case.

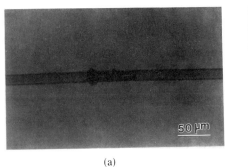 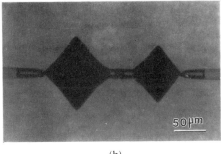

(a)   (b)

Figure 5. Optical micrographs of fiber fracture in SFC specimens under the wet condition: (a) untreated basalt fiber: (b) APS silane-treated basalt fiber showing matrix crack.

The improvements in fiber–matrix bonding brought about by the application of silane coupling agents should be reflected in the results of IFSS measurement using the SFC specimen. The relative merits of the three silanes studied here can thus be compared through the IFSS values obtained under optimal conditions of silane application in each case. These results are presented in Figs 6 and 7.

The primary data obtained from the IFSS test is the distribution of ultimate fiber fragment lengths. When the fiber–matrix bond is improved, the critical fiber length for stress transfer is reduced, and the ultimate fiber fragment length in the SFC test is correspondingly shortened. The fiber fragment aspect ratio plotted in Fig. 6 illustrates this trend; it is smaller for the silane-treated specimens, compared with the untreated fiber, as a result of improvement in fiber adhesion to the matrix. The distribution of IFSS, plotted in Fig. 7, shows a corresponding shift to higher values for the silane-treated specimens.

The data shown in Fig. 7 are for the dry specimens. When the specimens had been exposed to boiling water for 1 h and then tested after equilibration at room temperature, the results shown in Fig. 8 were obtained. Here, the differences in

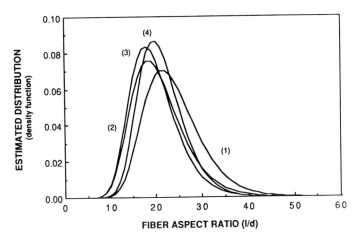

Figure 6. Comparison of the ultimate fiber fragment aspect ratio in SFC specimens prepared from treated and untreated fibers under the dry condition: (1) untreated; (2) APS: 3-aminopropyltriethoxy-silane; (3) PS076.5: dimethoxymethylsilylpropyl-substituted polyethyleneimine (aminofunctional); (4) PS076: trimethoxysilylpropyl-substituted polyethyleneimine (aminofunctional).

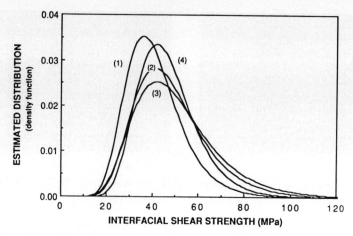

Figure 7. Comparison of IFSS for the treated and untreated fibers under the dry condition: (1) un-treated, and (2) APS, (3) PS076.5, (4) PS076 silane treated.

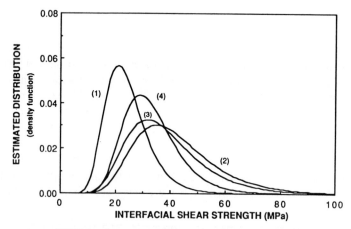

Figure 8. Comparison of IFSS for the treated and untreated basalt fibers after 1 h boiling in water: (1) untreated, and (2) APS, (3) PS076.5, (4) PS076 treated fiber.

IFSS are much more striking, emphasizing the fact that the effectiveness of silane treatment is more important in retaining fiber–matrix adhesion under conditions of exposure to moisture.

While the shift of the IFSS distribution curves to higher values is clear enough, the calculated mean values presented in Table 4 and the wet strength improvements shown in Fig. 9 provide better comparisons of the effectiveness of individual silane treatments. Aminopropylsilane is the best, followed closely by PS076.5 and PS076. This is not surprising since the polymeric silanes also carry aminofunctional groups capable of co-reacting with the epoxy matrix. The dramatic differences in improvement in IFSS after exposure to water require some comment.

As expected, the polymeric dimethoxysilane PS076.5 confers better moisture resistance than the polymeric trimethoxysilane PS076 (Fig. 9). However, the improvement obtained with APS, the monomeric silane, is somewhat better. This

Table 4.
Mean fiber strength and IFSS in SFC specimens of untreated and silane-treated basalt fiber–epoxy resin[a]

Silane treatment	Fiber strength (MPa)	IFSS (dry) (MPa)	IFSS (wet)[b] (MPa)
Untreated	1831 (269)	41.2 (13.0)	24.4 (8.0)
APS	1849 (287)	48.3 (15.7)	42.4 (12.4)
PS076.5	1862 (393)	50.1 (17.8)	38.6 (14.4)
PS076	1885 (275)	46.7 (12.8)	33.2 (10.2)

[a] Standard deviation in parentheses.
[b] After 1 h boiling in water.

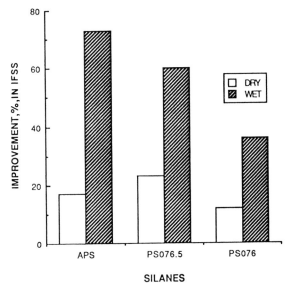

Figure 9. Improvements in IFSS under dry and wet conditions for the silane-treated fibers compared with IFSS of untreated fibers.

is probably related to an observation that we made of the behavior of the polymeric silane coating during drying after application on the fiber. The coating tends to become crosslinked. For this reason the coating was dried under mild conditions, at room temperature for 3 h. As described earlier, the effect of cross-linking would be to reduce the degree of interpenetration of the silane layer and matrix resin and, consequently, the measured value of IFSS. In fact, it was easy to see the effect of crosslinking by conducting the drying step at a higher temperature; the IFSS values obtained with the polymeric silane were reduced by this treatment.

It was shown above that the crosslinking observed with APS through self-condensation and oxidation could be followed by changes in the IR spectra (Figs 3 and 4). A similar reaction with the polymeric silane would lead more easily to a tightly bound network because of the polymeric backbone to which the silanol groups are attached. These aspects of the silane reactions need further careful investigation.

As shown in Table 1, the polymeric silanes on average have a molecular weight in the range 1500–2000. They are obtained as a 50% solution in isopropanol solvent and the IR spectra are complicated by the presence of traces of the solvent, which is difficult to remove. The SFC test provides reliable evidence for trends in fiber–matrix bonding efficiency under different conditions of surface treatment. At the same time, a degree of caution is required in quantitative interpretations. The matrix resin, in our case, was formulated to be very flexible, with about 35% elongation to failure. This was necessary since the basalt fiber has about 6% elongation to failure, and the SFC test succeeds in producing ultimate fiber fragments when the matrix has at least 5 times the elongation to failure as the fiber, so that the specimen can be strained enough to produce fiber fragmentation without fracturing the specimen itself. Matrix cracking was observed at the points of fiber fragmentation during the test, a phenomenon which we have not observed before in tests with carbon fibers. Since the carbon fibers fail at about 1% elongation, the epoxy resin used in that case was much higher in modulus.

Matrix cracking in the SFC tests in the case of low modulus resins has been observed recently by Netravali *et al.* [19], who have concluded that the SFC technique is experimentally simple but analytically complex. It can be expected that composite strength properties can show different degrees of improvement on silane treatment than indicated by IFSS measurements by the SFC test.

3.4. Acoustic emission

In addition to the direct observation of fiber fragmentation through the optical microscope, acoustic emission accompanying fiber fracture was also monitored by a piezoelectric transducer positioned at the center of the SFC specimen. Both the number of acoustic events and the energy of the pulses were measured. Thus, the correspondence between fiber fracture, observed microscopically, and the AE events accompanying it could be established as shown in Fig. 10.

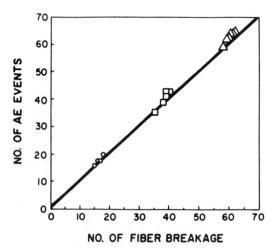

Figure 10. Correspondence between fiber breakage and the number of AE events for SFC specimens with three different diameter fibers: 15 μm ($\triangle$), 35 μm ($\square$), and 92 μm ($\bigcirc$).

It can be observed that the number of fiber breakages is reduced as the diameter of the fibers is increased. This follows directly from the relationship discussed under IFSS measurements, the ultimate aspect ratio l_c/d being equal to $\sigma_f/2\tau$. Thus, for the same fiber and surface treatment (σ_f and τ the same), the ultimate fragment length l_c will be longer for larger diameter fibers, in order to maintain constant the ultimate aspect ratio. This means that the number of breaks within the fixed gauge length of the SFC specimen will be smaller for larger diameter fibers. These observations lend strong support to the deduction made above that the AE events have their source in fiber fracture within the specimen. Similar conclusions have been drawn by Netravali *et al.* in a recent report [19].

The acoustic energies of the pulses are shown in Fig. 11 and indicate the stress levels at which fracture has occurred in fibers of the same diameter. It is interesting to note here that fibers treated by the polymeric silane produce failure pulses at a higher energy than the untreated fibers. The protective effect of the polymeric silane coating in healing some surface flaws can be deduced from this observation. This effect is not significant when the monomeric silane is used.

One cannot also fail to observe the increase in the number of counts at the lowest energy level corresponding to the most severe flaws. It is likely that these flaws are a result of corrosion by water used as solvent during the silane treatment. This feature was observed for all three silanes.

It might be recalled that the occurrence of matrix cracking for well-bonded silane-treated fibers was discussed earlier (Fig. 5). However, it is not likely that the changes in energy pulse distribution are caused by matrix cracking since we have observed that AE events from matrix cracking and debonding occur at much lower energies than does fiber fracture.

It is clear that the acoustic emission technique is capable of providing valuable information on the *in situ* strength of surface treated fibers, especially at very small gauge lengths approaching the critical aspect ratio of the fiber in the composite. The AE method should also be useful in extending the SFC test to opaque matrix resins, through which fiber fracture cannot be observed by the optical microscope.

4. CONCLUSIONS

The results presented here suggest that the polymeric dimethoxysilane can provide significant improvement under both wet and dry conditions, comparable to that obtained with monomeric APS, of epoxy bonding to basalt fiber. The results should be applicable to E-glass fibers also, in view of the similarity in the compositions of E-glass and basalt fibers. While the film-forming capability of the polymeric aminosilane is a decided advantage, careful control is needed of the crosslinking reactions that can occur in the silane layer, in order to provide for optimum interdiffusion with the matrix resin. Thus, the relative merits of monomeric APS and the polymeric dimethoxysilane are not clearly established by the SFC tests, even though the polymeric trimethoxysilane is clearly shown to be less effective. Strength tests of composites prepared from the treated fibers may be needed to obtain more definitive results for comparative evaluation of the two effective silanes. Acoustic emission studies of SFC specimens have the potential to provide valuable information on the effects of fiber surface treatment on fiber–matrix bonding and *in situ* fiber strength in composites.

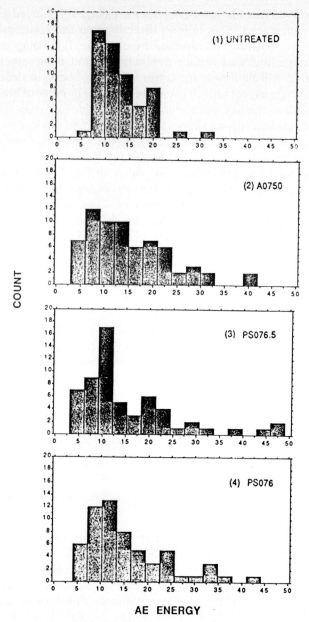

Figure 11. AE energy distribution of fiber fracture in SFC specimens: (1) untreated, and (2) A0750 (APS), (3) PS076.5, and (4) PS076 treated fibers.

Acknowledgements

An initiation grant from the Weyerhaeuser Foundation, and further funding from the M. J. Murdock Charitable Trust and a Consortium of Wood Industries are gratefully acknowledged.

REFERENCES

1. E. P. Plueddemann, *Silane Coupling Agents.* Plenum Press, New York (1982).
2. C. H. Chiang, H. Ishida and J. L. Koenig, *J. Colloid Interface Sci.* **74**, 396 (1980).
3. J. L. Koenig and C. H. Chiang, in: *The Role of the Polymeric Matrix in the Processing and Structural Properties of Composite Materials.* J. C. Seferies and L. Nicolais (Eds), p. 503. Plenum Press, New York (1983).
4. B. Arkles, J. Steinmetz and J. Hogan, *SPI Reinforced Plastics/Composites Annual Tech. Conf.* **42**, 21-C (1987).
5. B. Arkles, J. Steinmetz and J. Hogan, *Mod. Plast.* 138 (May 1987).
6. M. K. Chaudhury, T. M. Gentle and E. P. Plueddemann, *SPI Reinforced Plastics/Composites Annual Tech. Conf.* **42**, 21-B (1987).
7. R. V. Subramanian and A. S. Crasto, *Polym. Composites* **7**, 201 (1986).
8. R. G. Schmidt and J. P. Bell, *J. Adhesion* **25**, 85 (1988).
9. R. G. Schmidt and J. P. Bell, *J. Adhesion* **27**, 135 (1989).
10. R. V. Subramanian, T. J. Wang and H. F. Austin, *SAMPE Q.* **8**, 1 (1977).
11. R. V. Subramanian and K. H. Shu, in: *Molecular Characterization of Composite Interfaces,* H. Ishida and G. Kumar (Eds), p. 205. Plenum Press, New York (1985).
12. R. V. Subramanian and H. F. Austin, *Int. J. Adhesion Adhesives* **1**, 50 (1980).
13. R. V. Subramanian, A. S. Crasto and S. H. Own, in: *Composite interfaces,* H. Ishida and J. Koenig (Eds), p. 113. Academic Press, New York (1986).
14. R. V. Subramanian, A. R. Sanadi and A. Crasto, *J. Adhesion Sci. Technol.* **4**, 829 (1990).
15. R. V. Subramanian, in: *Handbook of Reinforcements for Plastics,* J. Milewski and H. Katz (eds), p. 287. Van Nostrand Reinhold, New York (1987).
16. A. Kelly, *Proc. R. Soc. London, Ser. A.* **319**, 95 (1970).
17. C. H. Chiang, H. Ishida and J. L. Koenig, *J. Colloid Interface Sci.* **74**, 396 (1980).
18. S. R. Culler, H. Ishida and J. L. Koenig, *J. Colloid Interface Sci.* **109**, 1 (1986).
19. A. N. Netravali, R. B. Henstenburg, S. L. Phoenix and P. Schwartz, *Polym. Composites* **10**, 226 (1989).

Silanes and Other Coupling Agents, pp. 493–511
Ed. K. L. Mittal
© VSP 1992.

The effect of a multicomponent silane primer on the interphase structure of aluminum/epoxy adhesive joints

R. G. DILLINGHAM[1,*] and F. J. BOERIO[2]

[1]*Central Research-Advanced Composites Laboratory, 1702 Building, The Dow Chemical Company, Midland, MI 48674, USA*
[2]*Department of Materials Science and Engineering, University of Cincinnati, Cincinnati, OH 45221, USA*

Revised version received 19 September 1991

Abstract—The structure of films formed by a multicomponent silane primer applied to an aluminum adherend and the interactions of this primer with an amine-cured epoxy adhesive were studied using X-ray photoelectron spectroscopy, reflection–absorption infrared spectroscopy, and attenuated total reflectance infrared spectroscopy. The failure in joints prepared from primed adherends occurred extremely close to the adherend surface in a region that contained much interpenetrated primer and epoxy. IR spectra showed evidence of oxidation in the primer. Fracture occurred in a region of inter-penetrated primer and adhesive with higher than normal crosslink density. The primer films have a stratified structure that is retained even after curing of the adhesive.

Keywords: Silane; interphase; adhesive joint; epoxy.

1. INTRODUCTION

The performance of coatings and adhesives depends to a large extent on the nature of the interactions occurring at the *interface* between the adhesive or coating and the substrate material. These interactions may be physical, as in the mechanical interlocking that occurs when a liquid wets a rough surface and then solidifies. They may also be chemical, resulting, for example, from acid–base or van der Waals attractions. Owing to the short-range nature of these intermolecular interactions, this region of chemical interaction may extend no more than a few molecular layers into either phase.

Many investigations of adhesion phenomena have focused on the structure and properties of interfaces. Less has been accomplished in determining the manner in which the structure and properties of an organic polymer are altered by curing it in contact with a substrate surface. The region of the polymer whose structure and/or chemistry has been influenced by the presence of an interface may extend many molecular layers into the organic phase and is referred to as the *interphase region* [1] to distinguish it from either the bulk phase or interface.

In some instances, interphase regions are intentionally created to enhance the strength and durability of a composite structure. An example of this is the application of silane coupling agents to adhesive joints, fiber-reinforced composites, and coating formulations [2]. Coupling agents are compounds with a high degree of functionality, capable of strong interaction with both the adhsive or coating and

---

*To whom correspondence should be addressed.

the substrate. Typically, thin films only a few molecular layers thick give the best improvements in performance. These compounds appear to be capable of improving composite performance by several mechanisms. They can function to improve the *mechanical strength* of an interphase by forming interpenetrating polymer networks with the organic phase [3, 4]. Primary chemical bonds may also be formed with the organic phase if the adhesive or coating and primer are chemically compatible [2]. Some investigators have also found evidence for the formation of primary chemical bonds between the silane and the inorganic phase [2]. Silanes have also been shown to function as *corrosion inhibitors* on substrates such as aluminum [5, 6].

This work discusses the structure of films formed by a multicomponent silane primer as applied to an aluminum oxide surface as well as the interactions of this primer with the adhesive and oxide to form an interphase region with a distinct composition and properties. The mecanical properties and durability of adhesive joints prepared using this primer system have yet to be evaluated.

When applied to adherend surfaces from dilute solutions, silane primer films, even if only a few molecular layers thick, contain functionality much in excess of that necessary to chemically bond to the adhesive. It has been postulated [7] that the cured system that results in these situations may have an interphase region that consists of overcured (and perhaps brittle) matrix resin, while still containing free, hydrophilic functional groups from unreacted silane molecules. The dilution of the reactive silane with another silane which is incapable of co-reacting with the resin could potentially improve the mechanical properties of the interphase region by limiting (and allowing optimization of) the degree of cure. When chosen with regards to solubility considerations, this additional silane could also improve interfacial strength by promoting interpenetration of the silane film and the adhesive. Moreover, if the functional group is non-polar, the filler could conceivably impart hydrophobicity to the interface, thereby improving the environmental resistance of the system.

The primer chosen for this investigation consisted of an equimolar mixture of phenyl- and amino-functional silanes, suggested as a potential superior primer for aluminum/epoxy adhesive joints [7]. The amino-functional silane is known to be effective as an adhesion promoter for fiber-reinforced composite materials [1, 2] as well as for epoxy/metal adhesive joints [8, 9] and provides for strong chemical interaction between the adhesive and primer, while the phenyl functional silane should reduce the overall concentration of polar, hydrophilic functional groups in the interphase region and at the same time maintain or improve the ability of the resin and primer to interpenetrate due to its structural similarity to the adhesive resin.

The primary techniques used in this study include X-ray photoelectron spectroscopy (XPS), reflection–absorption infrared spectroscopy (RAIR), and attenuated total reflectance infrared spectroscopy (ATR). XPS is the most surface-sensitive technique of the three. It provides quantitative information about the elemental composition of near-surface regions ($< ca.$ 50 Å sampling depth), but gives the least specific information about chemical structure. RAIR is restricted to the study of thin films on reflective substrates and is ideal for film thicknesses of the order of a few tens of angstroms. As a vibrational spectroscopy, it provides the type of structure-specific information that is difficult to obtain from XPS. The sample geometry utilized in this study allowed its application for analysis of the adherend side of the fracture surface. ATR is the least surface-sensitive of the three techniques, with effective sampling depths of the order of micrometers, but is ideal for obtaining structural information about the near-interface regions of the adhesive side of the fracture surface. These techniques overlap enough in the types of information that they provide so as to allow internal checks on the validity of the results.

One major problem encountered in the study of interfaces and interphases is to obtain the region of the composite of interest in a physical form that is possible to study. Both of the reflection IR techniques as well as XPS require relatively planar surfaces for analysis, but fracture surfaces of adhesive joints and composite structures are typically too rough to allow analysis by these techniques. As a result, investigators frequently must resort to modelling the desired interface or interphase in a form more suitable for study by these techniques. In this study, the problem of obtaining a planar surface for study was solved by constructing samples consisting of beams of adhesive cast onto polished aluminum substrates with a geometry that allows almost ideal interfacial debonding. The adhesive (being organic) has a much greater coefficient of thermal expansion than the aluminum substrate and contracts much more during cooling from the elevated temperature cure. Residual stresses that develop at the interface are calculable through measurement of substrate deflection using techniques similar to those developed for analysis of stresses in oxides, paint films, or coatings [10, 11]. These stresses, acting primarily in shear, produce strains that allow cracks to be easily propagated parallel to the interface. Because these cracks pass through the weakest part of the interfacial region, this technique provides the analyst two quite planar surfaces for study which represent the weakest link in the composite system.

2. EXPERIMENTAL

The specimen substrates were sectioned from 2024-T3 aluminum barstock using a band saw and then cut to final size on a vertical end mill. After cutting, the samples were first degreased by soaking in acetone and then washed with detergent (Alconox®, Fisher Scientific Co.), followed by a 5 min soak in a sulfuric acid–sodium dichromate cleaning solution (Chromerge®, Fisher Scientific Co.). This was followed by a final rinse with reverse-osmosis, deionized carbon-filtered water. This water was used during all phases of sample preparation to ensure surfaces with a minimum of contamination. All handling of the samples after the initial degreasing was done either with acid-cleaned haemostats or with rubber gloves that had been washed thoroughly with detergent. Next, metallographic polishing techniques were used to obtain a clean, specularly reflecting surface. The initial step consisted of dry grinding of the surfaces with successively finer grades of SiC paper, followed by wet grinding using 14.5 μm alumina powder on Microcloth® (Beuhler, Inc.). Specimens were then polished on Microcloth using MgO as the polishing medium and finally blown dry with nitrogen. The resulting surfaces were bright mirrors that were initially extremely hydrophilic. Freshly polished surfaces had a water contact angle of only 1° or 2°. After exposure of these surfaces to the laboratory atomosphere for a few minutes, however, the contact angle of the surfaces with water increased to between 15° and 20°, suggesting that contaminants were being adsorbed. Visual evaluation of the contact angle of water with these surfaces served as a qualitative check on the quality of the surface preparation and relative cleanliness.

The adhesive used in this study consisted of Epon 828 (Shell Chemical Co.) cured with 10 phr (parts per hundred resin by weight) of triethylenetetraamine (TETA) (Fisher Scientific Co.). The resin was degassed in a vacuum oven at 50°C, combined with the curing agent, and then mixed for about 1 min using a magnetic stirrer.

The primer mixture consisted of a 50:50 molar ratio of aminosilane (Dow Corning Z-6020) and phenylsilane (phenyltrimethoxysilane, Petrarch Systems). This combination was diluted with 50% by weight of monomethyl ether of propylene glycol (Dowanol PM®, Dow Chemical Co.) as a solvent and then combined with 5 parts by weight of distilled, deionized water to partially hydrolyze the methoxy groups of the silanes. Before use, this mixture was further diluted to 1% V/V in methanol. This primer contains only about two-fifths of the water necessary for complete hydrolysis. The primer was applied to the freshly polished adherend surfaces from the methanol solution using a Kimwipe® (Kimberly-Clark Corp.) and then allowed to air-dry for about 10 min before casting the adhesive–curing agent mixture. The Kimwipe® technique was recommended by Dow Corning as part of the procedure for utilizing this particular primer mixture, although it was recognized as being a less than ideal method from the viewpoint of the surface analyst.

Casting was accomplished by clamping a fluorocarbon-lined aluminum mold to the freshly prepared substrate and filling with the warm adhesive–curing agent mixture. The resulting sample geometry is shown in Fig. 1. The samples were allowed to cure overnight at room temperature, removed from the mold, and then postcured at 100°C for 1 h in a convection oven.

Interfacial cracks were easily propagated by applying a light restoring force to the beams. Alternatively, cracks may be propagated by immersion of the specimen in liquid nitrogen. Specimens for XPS, ATR, and RAIR analyses were sectioned from both the adhesive and the adherend using a band saw. Spectra were obtained from these surfaces as soon after fracture as possible, usually within minutes. XPS spectra were obtained at four different exit angles to assist in depth profiling. To allow quantitative interpretation of the XPS spectra of the cured resin and fracture surfaces, spectra were first obtained from model materials of known composition. These standards included polyethylene and Epon 1007® (Shell Chemical Co.), a

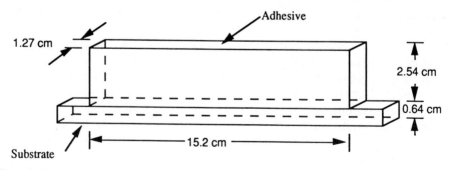

Figure 1. Sample geometry used in the preparation of specimens for interfacial analysis.

high molecular weight diglycidyl ether of bisphenol A resin (DGEBA) which could be analyzed under UHV conditions without outgassing problems. XPS spectra were obtained on a PHI 5400 using Mg K_a at 300 W. All binding energies were referenced to the aliphatic C(1s) peak at 284.6 eV.

The RAIR and ATR spectra were obtained using a Perkin-Elmer Model 1800 FTIR equipped with Harrick Scientific Co. external reflection and internal reflection accessories. Reflection spectra of the aluminum side of the fracture surfaces were referenced to spectra obtained from freshly polished aluminum substrates to obtain difference spectra of material left on these surfaces.

3. RESULTS AND DISCUSSION

3.1. Structure of primer films

This section discusses analysis of the structure of the bulk primer and of the films formed by the primer mixture as cast onto polished aluminum substrates using transmission IR, RAIR, and XPS techniques. To assist in analysis of the RAIR spectra, transmission IR spectra were obtained of the bulk, unhydrolyzed primer components from films of these materials deposited onto KBr pellets (Figs 2A and 2B). Figure 3A shows the RAIR spectrum obtained from the primer mixture as deposited onto polished 2024-T3 aluminum from a 1% solution in methanol, and Table 1 is a list of the bands seen in these spectra as well as tentative band assignments. The small amount of water added to this mixture (5 parts by weight) has resulted in incomplete hydrolysis to the silanol form as evidenced by the band near 807 cm^{-1}, assigned to the stretching vibrations of the Si—O—C species. It has polymerized to some extent, however, to form a low molecular weight polysiloxane, as shown by the intense band due to Si—O—Si stretching near 1145 cm^{-1}. The amines present in the film exist as bicarbonate salts due to reaction with

Table 1.
Observed bands (in cm^{-1}) and tentative assignments for the IR spectra of the primer components and primer mixture

Z-6020	Phenyl silane	Partially hydrolyzed mixture	Assignment
	656		Si(Cl)$_n$
691			
	698	698	Phenyl C=C
	745	737	Phenyl
815	812	809	Si—O—C
929		948	Si—O—Et or Si—C$_2$H$_5$
1071	1088	1097, 1055	Si—O—C or Si—O—Si
1191	1188	1193, 1197	CH$_3$ rock of Si—O—Me
1285		1285	Si—C$_2$H$_5$
1312		1312	CH$_2$ wag of Si—CH$_2$—R
	1308, 1325		Various bands of monosubstituted phenyls
1343			
		1381, 1375	
1411		1411	CH$_2$ def. of Si—CH$_2$
	1429	1431	Si—O—Phenyl

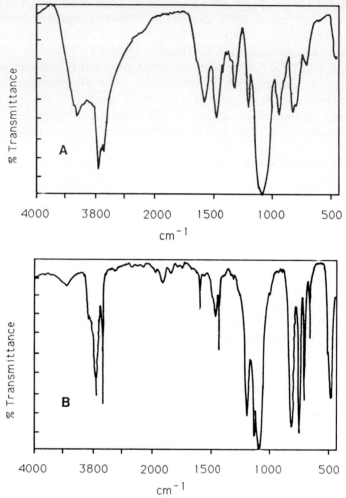

Figure 2. Transmission IR spectra of the silanes used in the primer. (A) Phenyltrimethoxysilane; (B) 2-aminoethyl-3-aminopropyltrimethoxysilane.

water and atmospheric carbon dioxide, as indicated by the presence of bands near 1570 and 1470 cm^{-1} which have been shown to be due to the asymmetric and symmetric stretching vibrations, respectively, of the bicarbonate ions [9].

Figure 3B shows the spectrum of this film after heating at 100°C for 1 h, and Fig 3C is the difference spectrum between Figs 3B and 3A, showing the spectral changes that result from gentle heating of this film. These changes include a decrease in the relative intensity of the bands near 1570 and 1470 cm^{-1} due to decomposition of the bicarbonate salt to produce free amines, as well as an increase in the degree of polymerization due to condensation of the silanols, indicated by an increase in the frequency of the Si—O—Si stretching vibration to about 1152 cm^{-1}. The decrease in intensity of the band near 807 cm^{-1} suggests that a reduction in the concentration of methoxy groups has occurred. This may be due to further hydrolysis of methoxysilane, or possibly to the evaporation of unhydrolyzed (and therefore unpolymerized) material. There is also a band near

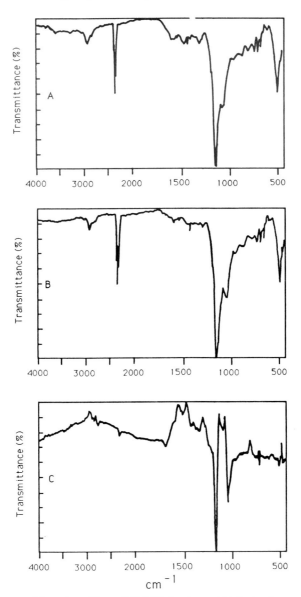

Figure 3. RAIR spectra of the primer film on 2024-T3 aluminum. (A) 5 min after application; (B) after 1 h at 100°C; (C) difference between (A) and (B).

1669 cm^{-1} which appears after heating and is due to the formation of imines through decomposition of the amines during heating in the presence of aluminum [12, 13].

The XPS spectra of the as-cast primer films show some differences as a function of the exit angle that give further clues as to film structure. Figure 4A shows the C(1s) spectrum obtained at a 15° exit angle, while Fig. 4B shows the same film analyzed at a 30° exit angle. The component near 285.6 eV, shifted 1 eV higher in binding energy than the main peak, is characteristic of carbon atoms bonded to nitrogen atoms in the aminosilane, while the component near 286.3 eV represents

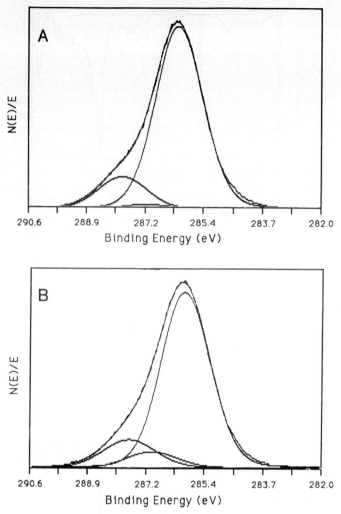

Figure 4. C(1s) XPS spectra, primer film. (A) 15° exit angle; (B) 30° exit angle.

carbon atoms singly bonded to oxygen atoms in the methoxy groups that are present due to incomplete hydrolysis. The peak due to the C—N species increases from about 2% of the total envelope area at 15° to almost 8% at 30°, indicating that the uppermost few angstroms of the primer film contain much less C—N type material than the material located a few atomic layers below the film surface. This may be due to the tendency of the system to minimize its surface free energy during casting, and may represent segregation or orientation of the low energy functionalities towards the film surface. This phenomenon has been demonstrated in films cast from block co-polymers [14] in which the lower energy co-polymer segregates to the surface during solidification.

   The N(1s) spectra of these films show that the amines present in the regions of the film close to the aluminum oxide surface were protonated, an effect similar to that seen in the N(1s) spectra of the adhesive near the fracture surface of unprimed adhesive joints [15]. Figure 5 shows the spectra obtained from two

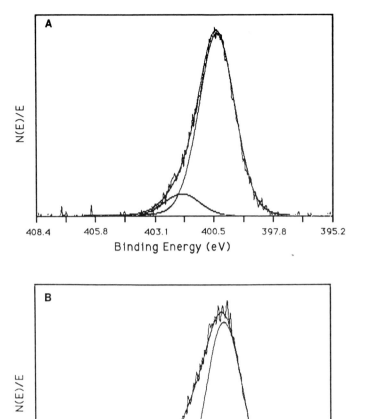

Figure 5. N(1*s*) XPS spectra of the primer film as applied onto polished 2024-T3 aluminum. (A) 15° exit angle; (B) 75° exit angle.

different exit angles and indicates that protonation of the amine nitrogens in the primer is occurring due to interaction with the hydroxyls of the oxide.

3.2. *Interactions between primer films and epoxy resins*

The interaction between these films and bulk epoxy resin was assessed by immersing an aluminum mirror coated with an air-dried primer film in a Petri dish filled with the epoxy resin, heating the dish in an oven at 100°C for 1 h, allowing the dish to cool overnight, and then extracting any unreacted material from the surface of the mirror by MEK extraction. Figure 6A is the reflection spectrum of a relatively thick film (ca. 3 μm) of neat DGEBA resin (cast onto polished aluminum from a 3% solution in toluene), and Fig. 6B shows the RAIR spectrum obtained from the mirror that was primed, heated in resin, and extracted. The

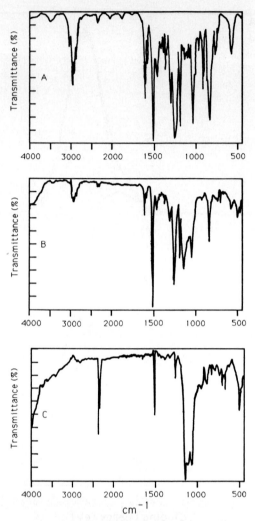

Figure 6. (A) RAIR spectrum of neat epoxy film; (B) RAIR spectrum of primed mirror that was heated in resin, extracted with MEK; (C) difference spectrum between (A) and (B).

relatively low frequency of the Si—O—Si stretching vibration in the spectrum shown in Fig. 6B (near 1036 cm^{-1}) suggests that the degree of polymerization of the primer in this composite film is lower than that of the neat oven-dried films, and the presence of a band near 917 cm^{-1} due to C—O—C asymmetric stretching of the epoxide rings shows that the retained resin is not completely reacted. The retained resin is partially crosslinked, however. Figure 6C results from subtraction of the neat epoxy resin spectrum (Fig. 6A) from the primer + resin spectrum (Fig. 6B) using the epoxide ring band near 917 cm^{-1} as a scaling reference. The bands in this difference spectrum characteristic of the epoxy resin (e.g. 1510 cm^{-1} ring mode) that remain represent epoxy molecules that have crosslinked with the primer through opening of the epoxide ring. The intensity of these bands relative to the Si—O—Si stretching mode near 1080 cm^{-1} indicates that a large amount of epoxy has been retained.

3.3. The interphase region in primed aluminum/epoxy adhesive joints

3.3.1. IR data. Spectroscopic techniques were then employed to determine more about the composition of the fracture surfaces. ATR of the adhesive side of the fracture surfaces showed only slight differences in the composition of the organic phase near the interface as a result of applying primer to the adherend surface before applying the adhesive. The spectrum shown in Fig. 7C (obtained with the 45° KRS-5 reflection element) is the difference between a sample prepared from an unprimed adherend (Fig. 7B) and one prepared from a primed

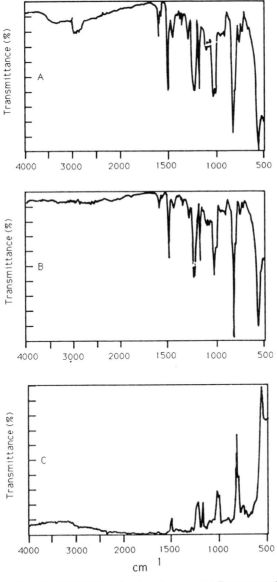

Figure 7. ATR spectra using the 45° KRS-5 reflection element. (A) Fracture surface, primed sample; (B) fracture surface, unprimed sample; (C) difference between (A) and (B).

adherend (Fig. 7A). The epoxy ring mode near 1610 cm^{-1} was used as the scaling reference. This spectrum shows only bands due to the epoxy resin and indicates that the thickness of the interphase region (consisting of material spectrally different from the bulk, cured epoxy resin) is much less than the sampling depth of the KRS-5 element (about 1–10 μm in the epoxy, depending on wavelength). Very little evidence of residual epoxide rings was present near 917 cm^{-1}, suggesting an essentially complete cure.

The spectra obtained with the 45° germanium ATR element showed some differences between the primed specimen (Fig. 8A) and the unprimed specimen (Fig. 8B), however. The difference spectrum is shown in Fig. 8C. The shallower sampling depth with this element (about 0.1–1 μm) results in greater sensitivity to the structure of the material close to the fracture surface, and the spectrum in Fig. 8C shows bands which may be due to both polymerized aminosilane and cured epoxy resin. In addition to the band due to the Si—O—Si stretching vibration of the polymerized silane near 1130 cm^{-1}, there is a band near 1650 cm^{-1} which may be due to C=N stretching. This band is characteristic of the spectra obtained from the primer films, particularly after heating (cf. Fig. 3C), and has been shown to be due to the oxidation of the amines in the aminosilane to form imines [11, 12]. The relative intensity of this band and the lack of bands near 1470 and 1570 cm^{-1} suggest that the amine bicarbonate species present in the freshly deposited films is not stable during subsequent curing, and that some of the primer in this region has been oxidized to the imine during the curing process. The broad band near 3500 cm^{-1} is due to O—H stretching of the hydroxyls formed during the epoxy ring opening. (The possibility also exists that these bands near 1650 and 3500 cm^{-1} are due to water adsorbed onto the germanium ATR crystal. If this were the case, then the bands near 3500 cm^{-1} would result from the O—H stretching modes while the bands near 1640 cm^{-1} would be due to the O—H bending modes of molecular water.)

The matching RAIR spectrum of the adherend side from the primed specimens (Fig. 9) was weak because of the small amount of organic material left on the surface. This spectrum shows a trace of epoxy, as suggested by the bands near 1510 and 1600 cm^{-1}. The presence of primer is also indicated by the band near 1080 cm^{-1}, but the weakness of this band in comparison with the bands due to the epoxy indicates that only a small amount of primer is present relative to the amount of epoxy present.

3.3.2. XPS data. XPS analysis provided more information about the structure of the fracture surfaces. Figures 10A and 10B show the Al($2p$) spectra obtained from the aluminum side of the fracture surface at 15° and 75° exit angles, respectively. The appearance of the peak due to metallic aluminum near 72 eV even at a grazing angle, as well as the strong angular dependence of the intensity of this component, shows that very little organic material is left on the primed surfaces. The atomic percentages calculated from the peak areas as a function of the exit angle are shown in Table 2. This type of angle resolved data, combined with the fact that SEM did not show discrete island type structures on the adherend fracture surface, leads to the conclusion that the organic is present on the oxide surface as a thin, continuous film with a thickness that is of the order of a few molecular layers.

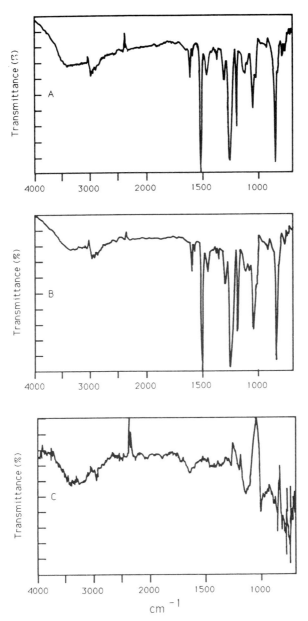

Figure 8. ATR spectra using the 45° germanium reflection element. (A) Fracture surface, primed; (B) fracture surface, unprimed; (C) difference between (A) and (B).

Table 2 also shows that the amount of nitrogen relative to the amount of silicon on the aluminum side of the fracture surface is almost 2 to 1, while the relative amount present in the XPS spectra of the neat primer films is about 1 to 2. The large excess of nitrogen relative to the silicon present on the fracture surfaces most likely originates from the curing agent and confirms the conclusion drawn from the RAIR data that much epoxy relative to primer is present on these surfaces. Table 2 shows that at steeper exit angles (analyzing material located farther

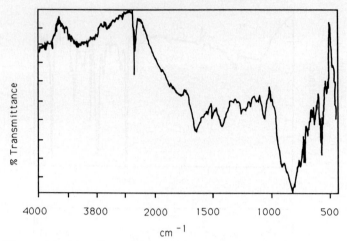

Figure 9. RAIR spectrum of the adherend side of the fracture surface.

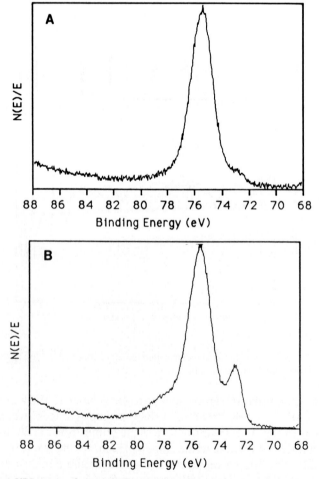

Figure 10. Al(2*p*) XPS spectra from the adherend side of the fracture surface. (A) 15° exit angle; (B) 75° exit angle.

Table 2.
Atomic concentrations (in %) as a function of the exit angle for elements detected on the aluminum side of fracture surfaces of primed samples

Element	Exit angle 15°	30°	45°	75°
Aluminum	15	23	29	35
Carbon	50	31	22	16
Oxygen	32	42	47	47
Nitrogen	2.2	1.6	1.2	0.7
Silicon	1.0	0.7	0.8	0.5
Copper	0.3	0.4	0.2	0.4

beneath the surface) the amount of silicon relative to the other elements decreases while the ratio of silicon to nitrogen increases greatly. This is consistent with the film stratification seen in the C(1s) curve fits of the neat primer films, which showed less nitrogen near the free surface of the film. Another possible explanation for the presence of a higher nitrogen to silicon ratio at the oxide surface is preferential adsorption of curing agent from the adhesive onto the oxide, which has been observed in epoxy/aluminum adhesive joints in other situations [16].

The C(1s) spectra obtained from these surfaces at 15° and 75° exit angles (Fig. 11) again show a larger amount of high binding energy carbon material in the regions closer to the oxide surface. This is similar to what was observed on the fracture surfaces of the specimens prepared from unprimed adherends [15]. Determination of the origin of this material would require considerably more analysis, although it may be due either to adsorbed hydrocarbons present on the polished surfaces before bonding, or to more highly oxidized organic molecules adsorbed from the resin and/or primer, or possibly to material that degraded during curing in close proximity to the oxide.

The N(1s) spectrum from this surface obtained at an exit angle of 15° (Fig. 12) shows the high binding energy component characteristic of amines that have been protonated by the oxide hydroxyls. The low intensity of the N(1s) spectra obtained at steeper exit angles prevented accurate curve fitting of these spectra.

More information about the locus of failure was obtained through curve fitting of the C(1s) XPS spectra obtained from the corresponding epoxy side of the fracture surfaces, shown in Fig. 13. While the bulk, cured epoxy showed about 8.5% C—N and the epoxy side of the fracture surface of the unprimed sample contained about 11.4%, roughly 13% C—N type material is present in the regions near the fracture surface of the primed specimens. This may originate from two sources: the presence of amine functional primer that has interdiffused with the adhesive and/or from the ability of the amine functionality in the primer to catalyze a more complete crosslinking of the resin in the interphase region [15]. The relative peak areas for the different carbon species, listed in Table 3, show that the relative amounts of C—N decrease while the amount of C—O increases slightly with the exit angle. This effect does not result from a lower concentration of primer, as the concentrations of silicon and nitrogen do not change as a function of the distance from the interface. It most likely results from a lower crosslink density farther away from the interface, due to less proximity to the oxide.

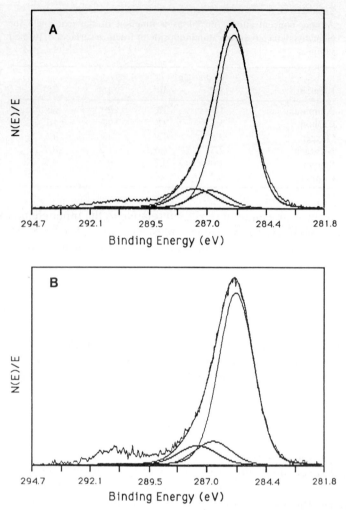

Figure 11. C(1s) XPS spectra from the adherend side of the fracture surface. (A) 15° exit angle. (B) 75° exit angle.

Table 4 shows the relative atomic concentrations as a function of the exit angle for the epoxy side of the fracture surface. The precision of these calculations is not good when the peaks are of weak intensity, but the angle dependence of the relative concentration of aluminum suggests that it is present in a very thin film in the uppermost regions of the fracture surface. This film is probably discontinuous because the other strong peaks (i.e. carbon and oxygen) do not show a significant angle dependence and hence should be exposed. This is in contrast to the behavior of the unprimed samples [15], where the XPS data were consistent with aluminum oxide present on the adhesive side of the fracture surface as island structures which were thicker than the photoelectron escape depth.

These data show that fracture in the primed specimens occurred very close to the oxide surface, leaving a small amount of primer and epoxy on the oxide and a thin, discontinuous film of oxide on the adhesive surface. Significant interdiffusion

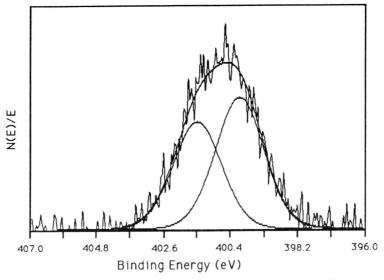

Figure 12. N(1s) XPS spectrum from the adherend side of the fracture surface, 15° exit angle.

Table 3.
Relative peak areas (in %) for curve fitting of the C(1s) envelope from the epoxy side of the fracture surface of a primed specimen

	Exit angle			
eV shift	15°	30°	45°	75°
0.0 (C—C)	60	58	59	59
1.0 (C—N)	15	15	13	12
1.7 (C—O)	20	23	24	25
3.0 (C=O)	5.3	4.3	4.4	4.4

Table 4.
Atomic concentrations (in %) as a function of the exit angle for elements detected on the epoxy side of the fracture surfaces of primed samples

	Exit angle			
Element	15°	30°	45°	75°
Aluminum	1.2	0.8	0.9	0.6
Carbon	70	70	70	71
Oxygen	21	21	21	20
Nitrogen	4.2	4.4	4.7	4.4
Silicon	3.4	3.6	3.4	3.1
Copper	0.05	0.06	0.13	0.12

of the primer and adhesive occurred and much C—N type material was present in the near-interfacial regions, possibly due to both amine functional primer and a higher degree of resin crosslinking. There is also evidence that the primer film retained some of the stratification introduced during its application even after interdiffusion and crosslinking with the adhesive.

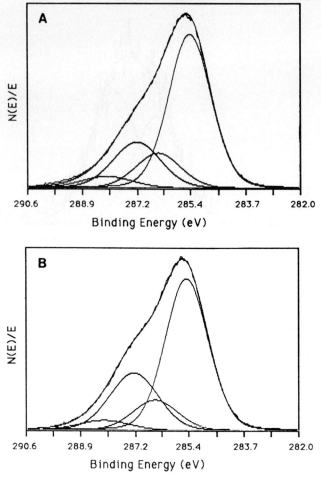

Figure 13. C(1s) XPS spectra from the adhesive side of the fracture surface, primed sample. (A) 15° exit angle; (B) 75° exit angle.

4. CONCLUSIONS

Prior analysis of the oxide surface [17], model adhesive [15] and primer compounds, and films of primer and of primer plus adhesive allowed data obtained from the fracture surfaces to be analyzed and interpreted. This produced a relatively complete picture of the strata which formed in the adhesive joints during cure due to interaction of the individual components. The failure in the joints prepared from adherends primed with a thin layer of silane primer occurred extremely close to the adherend surface in a region that contained much inter-penetrated primer and epoxy. The IR spectra showed evidence of possible oxidation reactions occurring in the primer during the curing process. The presence of bands due to the silane in the ATR spectra from the adhesive side and the lack of an angle dependence in the XPS spectra suggest that a region of inter-diffused primer and adhesive extending perhaps hundreds of angstroms away from the interface had been formed during curing. The high nitrogen to silicon ratio found in the XPS data of both the adherend and the adhesive sides combined with the large percentage of C—N type species in the C(1s) spectrum of the adhesive side suggests that the fracture occurred in a region of interpenetrated primer and adhesive which may have had a higher than normal crosslink density. Moreover, the primer film appears to have a stratified structure as cast onto the adherend surface that is retained, to some degree, even after application of the adhesive.

REFERENCES

1. L. T. Drzal, in: *Advances in Polymer Science*, K. Dusek (Ed.), Vol. 75. Springer-Verlag, New York (1986).
2. E. P. Plueddemann, *Silane Coupling Agents*. Plenum Press, New York (1982).
3. N. H. Sung, A. Kaul, I. Chin and C. S. P. Sung, *Polym. Eng. Sci.* **22**, 637 (1982).
4. E. P. Plueddemann and G. L. Stark, *Proc. 35th Annu. Tech. Conf.*, SPI Reinforced Plastics/ Composites Inst., Sect. 20-B (1980).
5. F. J. Boerio, R. G. Dillingham and R. C. Bozian, *Proc. 39th Annu. Tech. Conf.*, SPI Reinforced Plastics/Composites Inst., Sect. 1-A (1984).
6. F. J. Boerio and C. H. Ho, *J. Adhesion* **21**, 25 (1987).
7. E. P. Plueddemann, Private communication (1986).
8. F. J. Boerio and R. G. Dillingham, in: *Adhesive Joints: Formation, Characteristics, and Testing*, K. L. Mittal (Ed.), p. 541. Plenum Press, New York (1984).
9. F. J. Boerio and J. W. Williams, *Proc. 36th Annu. Tech. Conf.*, Reinforced Plastics/Composites Inst., Sect. 2-F (1981).
10. R. Pawel, J. V. Cathcart and J. Campbell, *J. Electrochem. Soc.* **110**, 551 (1963).
11. P. H. Townsend, D. M. Barnett and T. A. Brunner, *J. Appl. Phys.* **62**, 4438 (1987).
12. D. J. Ondrus and F. J. Boerio, *J. Colloid Interface Sci.* **124**, 349 (1988).
13. S. R. Culler, H. Ishida and J. L. Koenig, *Polym. Composites* **7**, 321 (1986).
14. H. P. Thomas and J. M. O'Malley, *Macromolecules* **12**, 323 (1979).
15. R. G. Dillingham and F. J. Boerio, *J. Adhesion* **24**, 315 (1987).
16. S. G. Hong, N. G. Cave and F. J. Boerio, *J. Adhesion* (submitted).
17. R. G. Dillingham and F. J. Boerio, *J. Adhesion* **21**, 95 (1987).

Silanes and Other Coupling Agents, pp. 513–529
Ed. K. L. Mittal
© VSP 1992.

Glass fiber 'sizings' and their role in fiber–matrix adhesion

E. K. DROWN, H. AL MOUSSAWI and L. T. DRZAL*

Department of Chemical Engineering, Composite Materials and Structures Center, Michigan State University, East Lansing, MI 48824-1326, USA

Revised version received 5 June 1991

Abstract—Silane coupling agents are but one of the many ingredients in commercial sizings that are applied to glass fibers. The action of epoxy-compatible silane coupling agents alone is to increase the fiber–matrix adhesion; however, the action of a silane coupling agent-containing sizing system is not well understood. Research has been conducted in order to determine to what degree an epoxy-compatible glass fiber sizing alters the adhesion between fiber and matrix, as well as to what degree it changes the mechanical properties of the resulting composite. By using blends of epoxy-compatible sizing with bulk matrix, it has been possible to model the properties of the fiber–matrix interphase formed when the sizing interacts with the matrix during composite processing and fabrication. It has been shown in this case that the sizing's interaction with the matrix produces a material with a higher modulus, a greater tensile strength, but a lower toughness. The level of fiber–matrix adhesion increases along with a change in failure mode of the composite caused by the presence of the lower toughness interphase. The results from this study show that a chemical interaction theory of adhesion is not sufficient to explain the effect of fiber–matrix adhesion on composite properties. An interphase-based theory in which the mechanical properties of the interphase are considered along with the chemical interactions between the fiber surface and the sizing offers the best approach for developing these relationships.

Keywords: Sizing; finish; adhesion; glass fiber; silane coupling agent; epoxy; composite.

1. INTRODUCTION

Glass fiber-reinforced polymer matrix composites have been in continuous use for over 30 years largely because of the development of silane coupling agents, which are used both as protective treatments for glass surfaces and as coupling agents to the matrix polymer. Commercial practice has resulted in the development of fiber 'sizes', which contain silanes along with other components and are applied to the glass fiber surfaces shortly after fiber manufacture and before composite fabrication. While the subject of silane chemistry and its interaction with both the glass surface and the polymer matrix have been extensively studied, little fundamental information of a predictive nature attesting to the relationships between sizing application and composite mechanical properties has been published. Contemporary art views the application of 'sizings' as a necessary condition for acceptable levels of mechanical and durability performance of glass fiber-reinforced polymer matrix composites. This study was undertaken to investigate the glass fiber–sizing–polymer matrix interphase encountered in commercial fiber 'sizing'

---

*To whom correspondence should be addressed.

systems in order to better understand the role of the silanes in the performance of sizing systems and to provide the basis for the development of a framework upon which a predictive methodology can be developed. Single fiber studies of fiber–matrix adhesion were combined with mechanical property measurements on sizing/matrix compositions which were expected to model interphase compositions encountered with these materials. Continuous fiber composites were also made from these same materials, and mechanical properties and failure modes were measured.

The 'sizings' which are applied to glass fibers are multicomponent systems. The composition of a silane coupling agent-based size is complex, with the coupling agent comprising a relatively small portion of the material applied to the fibers. The interaction of silane coupling agents with glass and silica surfaces at the chemical and molecular levels has been extensively studied. The structure of physisorbed and chemisorbed silane coupling agents at silica and glass surfaces has been investigated via FTIR [1], NMR [2], and XPS [3]. The coupling agents have been found to form covalent oxane bonds with the hydroxyl groups present on the glass surface. This interaction also extends beyond the monolayer range and evidence has accumulated which shows that the silane coupling agent adsorbs in multilayers when possible. The presence of covalent bonding has bolstered the chemical bonding concept of the coupling agents acting as a bridge between the glass surface and the polymeric matrix.

However, the chemical bonding theory cannot account for the increase in adhesion experienced between non-reactive matrices such as polyolefins and inorganic reinforcements in which chemical bonds will not be formed [4]. This observation, among others, leads to an alternative proposal that an 'interphase' composed of various constituents forms surrounding the reinforcement. This third phase in the composite is possibly formed through interdiffusion of physisorbed silane and matrix molecules in the interphase and perhaps via preferential adsorption of both matrix components as well as silane coupling agents on the reinforcement surface [5].

Alteration of the interphase mechanical properties alone near the reinforcement surface [6] can explain many of the phenomena associated with silane coupling agents. The increased hydrolytic stability of coupling agent-treated composites likewise has been proposed to occur through hydrolyzable bonds between the silane and reinforcement surface which can break and reform easily enough to relieve stress at the interface and preserve the integrity of the composite. This type of argument is supported by an interphase model [7] in which the molecular structure of the interphase is most often considered to be an interpenetrating network between the coupling agent and matrix resin.

The problem of using molecular approaches alone for understanding sizing behavior is complicated because silanes are rarely used alone as a sizing agent. Most commercial sizes, especially those used in glass/epoxy composites, are proprietary complex compositions. For example, Table 1 lists the general proportion of components in a commercial size, excluding the solvent or carrier used, to apply the mixture.

These constituents of typical sizing systems are compatible and can interact with each other. The resulting distribution of these components in the interphase region is not well understood, with the formulation of effective sizing systems

Table 1.
Typical components of a glass fiber size [8]

Component	Percent
Film-forming resin	1–5
Antistatic agent	0.1–0.2
Lubricant	0.1–0.2
Coupling agent	0.1–0.5

being an empirical undertaking for the most part. The processing aids present in the size may change the interphase properties in a manner similar or counter to that of the silane coupling agent acting alone.

Here we have conducted experiments to develop an understanding of how the commercial size interacts with the matrix in the glass fiber–matrix interphase. Careful characterization of the mechanical response of the fiber–matrix interphase (interfacial shear strength and failure mode) with measurements of the relevant materials properties (tensile modulus, tensile strength, Poisson's ratio, and toughness) of size/matrix compositions typical of expected interphases has been used to develop a materials perspective of the fiber–sizing–matrix interphase which can be used to explain composite mechanical behavior and which can aid in the formulation of new sizing systems.

2. EXPERIMENTAL

2.1. Materials

A typical epoxy matrix system, a commercial size, and glass reinforcing fibers were selected as the materials to be used in this study for both single-fiber adhesion measurements and composite mechanical property measurements. The epoxy resin used in this study was DER 383 (Dow Chemical Company). DER 383 is a diglycidyl ether of bisphenol A-based resin. 1,2-Diaminocyclohexane (DACH) (Aldrich Chemical Company) was used as the hardener at 15.6 parts per hundred of resin. E-glass tows (1600 filament/tow, 13 μm filament diameter) were supplied by the Pittsburgh Plate Glass Company (PPG). The fibers were supplied with two surface treatments, the first being deionized water only and the second a commercial epoxy-compatible size. The glass fiber sizing was also supplied in bulk form for the preparation of epoxy/size blend samples which were used for mechanical property measurements.

2.2. Composite fabrication

Unidirectional laminates were prepared for single-fiber microindentation measurements with an interfacial testing system (ITS) (Schares Instrument Corporation), flexure, and short beam shear testing. A Research Tool prepregger was used to prepare glass/epoxy prepreg tapes 6 feet long by 11 in. wide. Tapes were cut into 12 in. lengths and stacked to produce unidirectional laminates of 12 and 18 plies. The uncured laminates were placed between $\frac{1}{2}$ in. thick aluminum caul plates that were covered with 2 plies of bleeder cloth and 1 ply of porous Teflon release film. Self-adhesive cork dams surrounding the prepreg prevented the matrix from escaping from between the caul plates.

The temperature cycle used to cure the lamintes was 16 h at room temperature, 2 h at 80°C, and 2 h at 175°C with 5°C/min ramps between soaks. Directly after lay-up, the caul plate/laminate assembly was placed in a Carver Laboratory Press, and sufficient load was applied to produce 0.7 MPa (100 psig) in the laminate. This load was maintained for 30 min in order to bleed-out excess resin. The assembly was then placed in a vacuum bag and loaded into a United-McGill auto-clave for the elevated temperature portion of the cure cycle. A pressure of 0.9 MPa (125 psig) was applied to outside of the vacuum bag assembly for the duration of the autoclaving. Vacuum of -0.09 MPa (-28 in. Hg) was applied for the first 30 min of the 80°C soak period and vented to atmosphere for the remainder. Laminate quality was monitored by optic numeric volume fraction analysis [9] (ONVFA), a Michigan State University developed technique which allows quantitative assessment of the fiber volume fraction and void content using optical methods. Using this technique, the void and fiber volume fraction of the composites were found to be less than 1% and 60–65%, respectively. ITS, short beam shear, and flexure specimens were cut from the laminates using a water-cooled diamond saw.

Dynamic mechanical testing of the epoxy matrix/epoxy sizing blends was done on a DuPont model 983 DMA interfaced to a 9900 model controller. Stoichio-metric mixtures of DER 383 and DACH were first prepared, and then sufficient size was added to produce the desired concentration on a weight percent basis. The mixture was degassed and poured into silicone RTV-664 [10] molds that contained four cavities, $3.2 \times 12.5 \times 60$ mm. The specimens were cured at room temperature for 16 h in a desiccator, placed in a forced convection oven, and ramped to 80°C at 5°C/min and held at that temperature for 2 h. The samples were allowed to cool to room temperature and then removed from the mold. The specimens were replaced in the oven on a metal sheet and postcured at 175°C for 2 h. The free surface of the specimens was ground on a Struers Abramin polisher using 320 grit SiC paper and water to produce parallel faces on the specimens.

The DMA was operated in the fixed frequency mode at 1 Hz and a peak-to-peak amplitude of 0.06 mm. A DuPont Liquid Nitrogen Cooling Apparatus (LNCA-II) was used to achieve sub-ambient temperatures. The temperature profile used was a ramp to $-20°C$, soak for 5 min, and then ramp at 5°C/min to 220°C. Three specimens were tested for each blend.

Tensile properties of the blends were measured according to ASTM standard D638-84 [11] using type I specimens which were cast in a manner identical to that used to prepare DMA test specimens. The specimens were polished to a 1 μm finish on all surfaces and stored in a desiccator until tested.

The single-fiber embedded shear strength test uses a model composite made of a single fiber encapsulated in an epoxy tensile coupon [12]. A silicone RTV-664 mold with eight cavities was used for the fabrication of the coupons. The cavities were ASTM standard tensile coupons, 60 mm long with a gage section 25.4 mm long, 4 mm wide, and 1.5 mm thick. Sprue slots were molded into the end of each cavity. A single filament was removed from a fiber tow and secured in the sprue slots with hot-melt adhesive. Care was taken to ensure that no tension was applied to the filaments during separation or while they were secured in the slots.

The components of the matrix were thoroughly mixed and degassed in a vacuum oven for 5 min at room temperature to remove air entrained during the mixing process. The molds with the single fibers were also treated in the vacuum oven. The mold was filled at one end of the cavity, allowing the resin to flow to the other end, thereby preventing the entrapment of air bubbles in the coupon. The mold was then transferred to a desiccator for a 16 h cure at room temperature followed by a 5°C/min ramp to 80°C for a 2 h soak. After the molds had cooled, the test coupons were removed, placed on an aluminum sheet, and ramped to 175°C at 5°C/min for a 2 h postcure. The coupons were allowed to cool at 5°C/min to room temperature.

2.3. Mechanical property measurements

The free surfaces of the coupons were polished to a 1 μm finish on Struers Abramin polisher to allow clear visualization of the fracture process under the microscope. Specimens were mounted in a tensile testing apparatus that permits direct observation of the fracture process via a transmitted light microscope. The specimen was strained in increments indicated on a dial gauge; at each increment, the strain and the number of fiber breaks were recorded. After the number of breaks had reached a plateau, the fiber fragment lengths and fiber diameter were measured with a filar eyepiece. The measured lengths do not give the interfacial shear strength directly. They are first fit to a Weibull distribution to account for the fiber strength distribution and fiber inhomogeneties [13].

The interfacial testing system (ITS) shown in Fig. 1 measures the interfacial shear stress at debond of a fiber in an actual composite [14–16]. The system consists of a Mitutoyo Metallographic FS100 microscope with the stage modified to house Klinger linear motion stages for *x*, *y*, and *z* translation, with 1 μm resolution in the *x* and *y* directions and 0.04 μm resolution in the *z* direction. Mounted on top of the stages is a Sartorius L610 weighing mechanism and the specimen holder. Mounted on the objective lens is a collar which holds a diamond-tipped indenter, the profile of which is a 90° cone with an included tip radius of

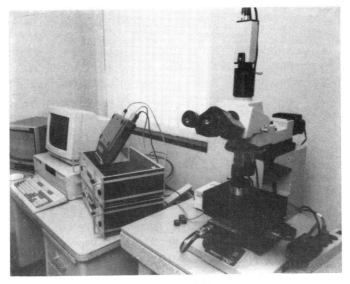

Figure 1. Photograph of the interfacial testing system (ITS),

10 μm. Stage motion controllers and the balance readout are interfaced to a Zenith 386 20 MHz microcomputer for control through the software. Observation and selection of the fiber ends for testing are done via a video camera and monitor.

The specimen used for ITS testing can be any section of the composite, provided that after mounting and polishing, fibers normal to the polished surface can be found. A section of composite approximately 1 cm × 1 cm was cut from the laminates previously prepared. The sections were embedded in 1 in. diameter phenolic ring mounts with room temperature-cured epoxy mounting media. The sections were oriented so that the fiber ends were normal to the face of mount to be polished. The initial steps of the polishing were accomplished on a Struers Abramin polisher and the final step was a on Beuhler Vibromet I vibrating lap polisher. The steps used are listed in Table 2.

Table 2.
ITS specimen polishing protocol

Abrasive	Lubricant	Duration (min)
320 grit silicon carbide	Water	0.5
1000 mesh silicon carbide	Water	2
2400 mesh silicon carbide	Water	3
4000 mesh silicon carbide	Water	4
1 μm diamond	Oil	30

After polishing, the specimens were stored in a desiccator until testing. Once mounted in the specimen holder, the specimen was scanned for fibers that were spaced a minimum of 2 μm and a maximum of one half of their diameter from the nearest neighbor. Matrix conditions near the fiber of interest were also considered. Fibers near voids, cracks, or surface damage or those fibers that showed a pre-existing interfacial crack were rejected. Once a suitable fiber end was located, its diameter and distance to the nearest neighbor were entered for data reduction purposes.

The fiber modulus and matrix shear modulus are also required for the analysis. The fiber's coordinates are recorded directly from the stage controllers to the computer. The operator begins the test from the keyboard. The x and y stages move the fiber end to a position directly under the debonder tip; the z stage then moves the sample surface to within 4 μm of the tip. The z-stage approach is slowed down to 0.04 μm/step at a rate of 6 steps/s. The balance readout is monitored, at a load of 2 g the loading is stopped, and the fiber end returned to the field of view of the camera. The location of the indent is noted and corrections are made, if necessary, to center the point of contact. Loading is then continued from 4 g in approximately 1 g increments. Debond is determined to have occurred when an interfacial crack is visible for 90–120° on the fiber perimeter. The load at which this occurs is used to calculate the interfacial shear stress at debond.

Two types of composite physical property tests were conducted to measure properties which are sensitive to the degree of adhesion and failure mode of the fiber–matrix interphase. Short beam shear tests (ASTM D2344-84) were conducted on 18 ply unidirectional laminates. The support span-to-thickness ratio used was 5:1 with a corresponding specimen length to thickness ratio of 7:1. The specimen width was fixed at 6.4 mm. The specimen was three point loaded, two end supports were set to the support span previously determined, and the load was applied to the midpoint via a loading nose. The crosshead speed was 1.3 mm/min. The load to break was recorded along with the failure mode, i.e. shear or tensile failure.

Flexure testing (ASTM D790-76) was performed on 12 ply unidirectional laminates with the fibers oriented at 0° and 90° to the support span. Specimens were 25 mm wide, nominally 100 mm long and tested at a span-to-thickness ratio of 32:1. The specimens were three point loaded and the rate of strain in the outer fibers was kept constant at 0.01 (mm/min) per min. Load vs. deflection curves were recorded for determination of the modulus and yield properties.

3. RESULTS AND DISCUSSION

3.1. Interphase mechanical properties

Dynamic mechanical testing provides information on the elastic and viscoelastic properties of the epoxy/sizing blends as a function of temperature. The T_g of a polymer is qualitatively the point at which it undergoes a transition from the glassy state to the rubber state. For these measurements, the T_g was taken to occur at the inflection point in a plot of the storage modulus (E') vs. temperature for comparison purposes. Figure 2 is a plot of T_g determined in this manner vs. the amount of epoxy-compatible size added to the stoichiometric mixture in weight percent. The properties of these blends would then represent the possible range of properties that could exist in the fiber–matrix interphase.

It is obvious that the polymer created by the addition of the commercial size exhibits a monotonic decrease in T_g. This indicates that the silanes and other ingredients in the sizing are acting to reduce the crosslink density of mixtures. This is not surprising since the silanes present in commercial size usually contain a chemical group which is reactive with the matrix constituents.

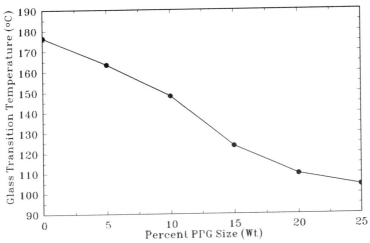

Figure 2. Glass transition temperature T_g measured by dynamic mechanical analysis as a function of the weight percent epoxy size added to the stoichiometric mixture of epoxy matrix.

The tensile properties of the blends—Young's modulus, Poisson's ratio, and tensile strength—were determined via mechanical testing. The storage or 0% tangent Young's modulus determined via DMA or by standard ASTM tensile coupon and extensometer methods, respectively, was relatively the same. Toughness, a measure of the material's ability to absorb energy before fracturing can be calculated by integrating the area under the stress vs. strain curve. Table 3 lists the values of Young's modulus (E), storage modulus (E'), tensile strength (σ), crosshead failure strain (ε), toughness, and Poisson's ratio (v) for the blends.

Using the epoxy at the stoichiometric ratio without any additive, as representing bulk matrix properties, it is readily apparent that adding the sizing has significant effects on the properties of the matrix in the interphase region. Comparing either DMA or extensometer mechanical tests shows that the modulus of the blends increases with increasing commercial sizing content.

For comparison purposes, the tensile modulus, tensile strength, and toughness are plotted against percent of sizing added to the stoichiometric mixture in Fig. 3. The room temperature modulus increases with increasing sizing content. The tensile strength and failure strain also change with additive concentration. Since all samples were prepared in an identical manner, the relative changes in the stress and strain at failure must reflect material properties, although the fracture properties are certainly sensitive to flaws. For the commercial sizing, the tensile strength actually increases with increasing sizing content. This may be a similar phenomenon [17] to that reported previously in which difunctional additives with differing reactivities when added to an epoxy matrix produced increases in tensile strength. Addition of the commercial size changes the material behavior of the blends so that they have a lower value of toughness than the stoichiometrix epoxy matrix.

Taking the 25% sizing additive composition as a model for the interphase near the fiber surface, the interphase consisting of this epoxy with this commercial sizing agent would have a lower T_g than the bulk by about 70°C (176°C vs. 100°C),

Table 3.
Mechanical properties of DER 383/DACH blends

Additive	Extensometer tensile modulus, E (GPa)	DMA[a] storage modulus, E' (GPa)	Tensile strength, σ (MPa)	Failure strain, (ε) (%)	Toughness (MPa)	Poisson ratio (v)
None	3.09 ±0.17	3.46 ±0.02	72.9 ±5.0	4.11 ±0.40	1.58 ±0.34	0.41 ±0.03
5% epoxy size	3.38 ±0.04	3.30 ±0.02	71.8 ±0.7	3.39 ±0.05	1.19 ±0.20	0.45 ±0.01
10% epoxy size	3.69 ±0.06	3.39 ±0.04	73.4 ±5.2	3.25 ±0.35	1.16 ±0.22	0.45 ±0.01
15% epoxy size	3.71 ±0.10	3.62 ±0.06	72.3 ±7.3	3.39 ±0.29	1.25 ±0.28	0.43 ±0.01
20% epoxy size	4.01 ±0.07	3.76 ±0.03	78.0 ±4.8	3.33 ±0.61	1.19 ±0.21	0.44 ±0.01
25% epoxy size	3.97 ±0.12	3.82 ±0.09	81.6 ±3.5	3.39 ±0.28	1.36 ±0.12	0.45 ±0.02

[a] DMA = Dynamic mechanical analysis.

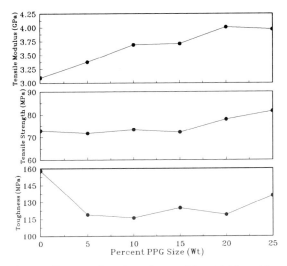

Figure 3. Tensile modulus, tensile strength, and toughness of the stoichiometric mixture of epoxy with increasing weight percent of the epoxy sizing as determined by tensile testing.

a larger tensile modulus by 28% (3.97 GPa vs. 3.09 GPa), a greater tensile strength by 12% (81.6 MPa vs. 72.9 MPa), a reduced strain at failure by 18% (3.30% vs. 4.11%), and decreased toughness by 14% (1.36 MPa vs. 1.58 MPa). Overall, the interphase created by the interaction between the commercial size and the epoxy would be characterized as a stiffer, stronger, lower toughness interphase.

In summary, the interphase formed when the sized fiber is brought into contact with the matrix during formation of the composite can have significantly different mechanical properties than those of the bulk matrix. In a real situation, the solubility of the size and the matrix would determine the interaction and the resultant composition. There would undoubtedly be a gradient of properties where the interphase material near the fiber surface would have a high concentration of sizing and a low bulk matrix. As the distance from the fiber surface increases, the sizing content decreases until the bulk matrix properties are reached at some distance away from the fiber surface. Since we have no way of quantitatively determining this gradient and distribution in a real interphase, the values obtained here serve to provide some limits that 'bracket' the expected material behavior of the fiber–matrix interphase.

3.2 Fiber–matrix adhesion and the interphase

Determination of fiber interfacial shear strengths and failure modes provides insight into how the interphase is affected by the sizing and how the sizing affects fiber–matrix adhesion. Figure 4 is a polarized transmitted light photomicrograph of a typical single bare (unsized) E-glass fiber embedded in a coupon of DER 383/DACH epoxy matrix. The specimen was strained in a direction parallel to the fiber axis until the fiber started to fracture in the coupon. The figure shows the region around the break taken directly after the fiber fractured. There is a small matrix crack that extends outward from the fiber into the matrix. The stress birefringence patterns show up as light regions extending approximately 3 fiber diameters into the matrix.

Table 4.
ITS results for DER 383/DACH/E-glass composites

Fiber size	Interfacial shear strength (MPa)	SD (MPa)	No. of specimens tested
Bare			
(water–sized)	44.6	± 3.0	30
Epoxy-sized	60.1	± 3.3	28

IFSS of the sized fibers vs. the bare. These results indicate that the application of the sizing and the consequent formation of the interphase have resulted in an increase in the level of fiber–matrix adhesion. It could reasonably be expected that under the multiaxial state of stress at the fiber–matrix interphase, stronger fiber–matrix adhesion would reduce the tendency to grow an interfacial crack, thereby placing more energy into driving the matrix crack as observed for the sized fiber.

The increase in fiber–matrix interfacial shear strength can be predicted from a purely mechanistic viewpoint. Rosen [20], Cox [21], and Whitney and Drzal [22] have shown that the square root of the shear modulus of the matrix appears explicitly in any model of the interfacial shear strength. It has been demonstrated experimentally [23, 24] that the fiber–matrix interfacial shear strength has a dependence on both the product of the strain-to-failure of the matrix times the square root of the shear modulus and on the difference between the test temperature and T_g when the interfacial chemistry is held constant.

3.3. Effect of fiber–matrix adhesion and the interphase on composite mechanical properties

Both fiber–matrix interphase-sensitive mechanical tests (interlaminar shear strength, 90° flexure) and interphase-insensitive tests (0° flexure) were conducted on high volume composite samples fabricated from the same materials and in the same manner as discussed above to see if the interphase and its properties altered the composite mechanical properties and in what manner. A summary of the data is plotted as a bar graph in Fig. 7. The first set of bars represents the difference in fiber–matrix adhesion measured between the bare fibers and the sized fibers by the ITS. The composite properties plotted on the figure also show increased values for the epoxy-sized material over the bare fiber composite.

Short beam tests provided information on the apparent interlaminar shear strength (ILSS) of laminates made with bare and sized E-glass fibers. In all cases, the specimens failed in shear at or near the midplane, allowing comparisons between fiber types. The ILSS, SD, and number of specimens tested are given in Table 5.

The increase in ILSS for the epoxy-sized fibers over the bare fibers is 12.4%, approximately 50% of the increase observed in the interfacial shear strength as measured by ITS testing. Changes in the failure mode at the fiber–matrix interface may account for the differences. The sized fibers produced large matrix cracks that grew quickly to catastrophic size under load. This would tend to limit the increase in composite shear properties if at every fiber break in the tensile surface of the coupon a matrix crack was created. The presence of these matrix cracks would lead to premature composite failure from coalescence of the microcracks into critical sized flaws on the tensile surface.

The 90° flexural properties were determined using the procedure described above. Table 6 lists the properties determined for laminates with bare and sized E-glass fibers.

In the 90° direction, the fiber–matrix interphase is a controlling factor in the flexural strength. Here a 35% increase in flexural strength is measured in direct proportion to the increase in interfacial shear strength measured with the ITS tests. The transverse tensile strength of the fiber–matrix interphase is greater for the finished fiber than for the bare fiber. Apparently the lower 'toughness' of the

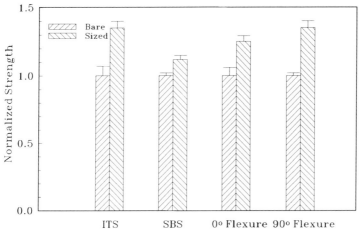

Figure 7. Normalized values for the interfacial shear strength as measured by the ITS plotted against the short beam shear strength (SBS), 0° flexure strength, and 90° flexure strength for the 'bare' and 'Epoxy-sized' fibers.

Table 5.
ILSS results for DER 383/DACH/E-glass composites

Fiber type	Interlaminar shear strength (MPa)	SD (MPa)	No. of specimens tested
Bare (water-sized)	71.3	1.6	9
Epoxy-sized	80.2	2.2	12

Table 6.
90° flexural properties of DER 383/DACH/E-glass unidirectional laminates

Fiber type	90° flexural strength (MPa)	90° flexural strain (%)	Tangent modulus of elasticity (GPa)
Bare (water-sized)	75.6 ± 1.8	0.48 ± 0.01	15.7 ± 0.26
Epoxy-sized	102 ± 5.2	0.69 ± 0.05	15.6 ± 0.23

interphase and its tendency to create matrix cracks and not to debond under shear loading do not affect the interphase when it is loaded in tranverse tension.

Flexural properties in the 0° direction are largely determined by the fiber properties; however, the matrix/interphase may also control the failure mode or path which may in turn affect the strength properties of the composite. The 0° flexural properties of the composites made with bare and sized E-glass are listed in Table 7.

Laminates with either bare or sized E-glass fibers failed first on the tensile surface of the test specimen but the nature of the event differed. Figures 8 and 9 are photographs of the tensile surfaces and edges of failed test coupons that were fabricated with bare E-glass. Figure 8 illustrates the breakage of fiber bundles which then propagates delaminations along the surface of the specimen. This process occurred over the last 10–15 s of the test, and failure occurred gradually rather than catastrophically. The lack of penetration of the failure zone into the interior of the specimens, as shown in Fig. 9, indicates that strain energy was being absorbed by delamination rather than by matrix cracks into the specimen.

The failure behaviour of the laminates made with sized E-glass fibers was different. Figures 10 and 11 are photographs of the tensile surfaces and edges of failed specimens. The specimens failed catastrophically through the thickness, and the delaminations shown in Fig. 10 occurred during the failure, not before. Regions of tensile failure (fiber pull-out) and compressive failure (crushing) can be seen in

Table 7.
0° flexural properties of DER 383/DACH/E-glass unidirectional laminates

Fiber type	Failure stress (MPa)	Failure strain (%)	Tangent modulus of elasticity (GPa)
Bare			
(water-sized)	1010 ± 60	3.0 ± 0.2	39.5 ± 1.3
Epoxy-sized	1260 ± 50	3.5 ± 0.2	44.2 ± 3.3

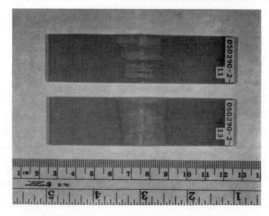

Figure 8. Photographs of 0° bare (water-sized) E-glass fiber laminates tested in flexure showing the tensile side of the specimen. The tensile side fibers failed first, then propagated delaminations along the tensile surface of the specimen.

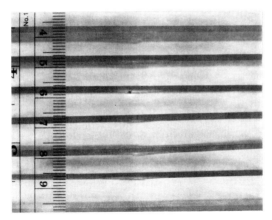

Figure 9. Photographs of 0° bare (water-sized) E-glass fiber laminates tested in flexure showing the edge of the specimen. The tensile side fibers failed first, then propagated delaminations along the tensile surface of the specimen. The strain energy is being used to delaminate the specimen.

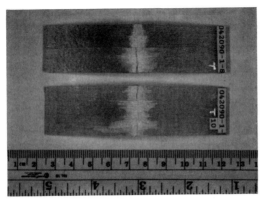

Figure 10. Photographs of 0° epoxy-sized E-glass fiber laminates tested in flexure showing the tensile side of the specimen. The tensile side fibers failed first but the fracture then propagated as both delaminations and cracks that extend through the specimen thickness.

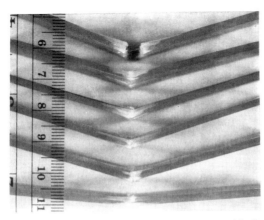

Figure 11. Photographs of 0° epoxy-sized E-glass fiber laminates tested in flexure showing the edge of the specimen. The failure locus extends through the sample thickness as well as by propagating delaminations along the tensile surface of the specimen.

Fig. 11. The change in failure at the single-fiber level as shown by the embedded single-fiber test is mirrored by the failure behavior at the macroscopic level. The increase in adhesion also tracks the same between the ITS results and 0° flexure data.

It should also be noted that the failure surfaces of the specimens also reflect changes in the degree of adhesion. The bare glass fibers always appear devoid of any matrix on the examined composite fracture surfaces, while the epoxy-compatible sized fibers always show substantial matrix adhering to the fiber surface. The remaining matrix also has undergone a high degree of deformation. Considering that the adhesion level has increased and that an interphase has been created with a gradient of properties extending out into the bulk away from the fiber surface, these differences are not unexpected. The higher level of adhesion places the interphase under a higher state of stress and the gradient distributes the stress, allowing the strain energy to be distributed over a larger volume.

The results reported here are similar to a study recently completed in which the adhesion of carbon fibers to epoxy matrices was varied [25, 26]. Over three different levels of adhesion and three different failure modes, composite properties both in the fiber direction and perpendicular to the fiber–matrix interface were shown to be dependent both on the level of adhesion and on the interphase properties and failure modes.

4. CONCLUSIONS

An alternative approach to understanding the role which silane-based fiber sizes play in fiber–matrix adhesion has been proposed in which the three-dimensional nature of the interphase formed by the interaction of the sizing agent with the bulk matrix has been shown to be a potential key factor in understanding the role that fiber sizings have in affecting fiber–matrix adhesion and composite mechanical properties.

The mechanical properties of a fiber–matrix interphase composed of high concentrations of sizing, exclusive of the presence or absence of specific chemical interactions between the fiber surface and the surrounding matrix, have been demonstrated to be potentially responsible for the level of fiber–matrix adhesion.

The modulus and toughness of this interphase combined with the increased fiber–matrix adhesion can be used to explain the resulting mechanical properties of these composite materials. For both interphase-sensitive and -insensitive mechanical properties, it has been concluded that the strength of the interphase and the failure mode initiated by the interphase properties can be responsible for composite mechanical properties.

Acknowledgements

Support for this work was provided by the Dow Chemical Company. The glass fibers and commerical sizing were graciously provided by the PPG Company. A portion of this work was supported by the Research Excellence Fund of the State of Michigan.

REFERENCES

1. C. H. Chiang, H. Ishida and J. L. Koenig, *J. Colloid Interface Sci.* **74**, 396 (1980).
2. K. P. Hoh, H. Ishida and J. L. Koenig, *Polym. Compos.* **11**, 121 (1990).
3. D. J. Pawson and F. R. Jones, in: *Controlled Interphases in Composite Materials*, H. Ishida (Ed.), p. 407. Elsevier, Amsterdam (1990).
4. E. P. Plueddemann, *Silane Coupling Agents*, p. 20. Plenum Press, New York (1982).
5. E. P. Plueddemann, *Silane Coupling Agents*, Ch. 5. Plenum Press, New York (1982).
6. L. T. Drzal, *SAMPE J.* **19**, 7–13 (1983).
7. L. T. Drzal, in: *Advances in Polymer Science II*, K. Dusek (Ed.), Vol. 75, Springer-Verlag, New York (1985).
8. Dow Corning Corporation, *A Guide to Dow Corning Silane Coupling Agents*, p. 15 (1985).
9. M. C. Waterbury and L. T. Drzal, *J. Reinf. Plast. Compos.* **8**, 627–636 (1989).
10. RTV-664, High-Temperature Silicone Molding Materials, General Electric Company, Silicone Products Division, Schenectady, NY.
11. American Society for Testing Materials, ASTM Standard D638-84, Philadelphia, PA.
12. L. T. Drzal, M. J. Rich, J. D. Camping and W. J. Park, 35th Annual Technical Conference, Reinforced Plastics/Composites Institute, The Society of the Plastics Industry, Section 20-C, p. 1 (1980).
13. L. T. Drzal and P. Herrera-Franco, in: *The Engineered Materials Handbook: Adhesives and Sealants*, Vol. 3, pp. 391–405. ASM Int., Materials Park, OH (1990).
14. D. L. Caldwell, D. A. Babbington and C. F. Johnson, in: *Interfacial Phenomena in Composite Materials '89*, F. R. Jones (Ed.), p. 44. Butterworth, London (1989).
15. D. L. Caldwell and F. M. Cortez, *Mod. Plast.* p. 132 (Sept. 1988).
16. D. H. Grande, M.S. Thesis, M.I.T., Cambridge, MA (1983).
17. A. Garton and G. S. Haldankar, *J. Adhesion* **29**, 13–26 (1989).
18. A. N. Netravali, P. Schwartz and S. L. Phoenix, *Polym. Compos.* **10**, 385 (1989).
19. D. Hull, *An Introduction to Composite Materials*, p. 14. Cambridge University Press, Cambridge (1981).
20. B. Rosen, in: *Fiber Composite Materials*, pp. 37–75. American Society for Metals, Metals Park, Ohio (1964).
21. H. L. Cox, *Br. J. Appl. Phys.* **3**, 72–79 (1952).
22. J. M. Whitney and L. T. Drzal, in: *Toughened Composites*, STP 937, pp. 179–196. ASTM Philadelphia, PA (1987).
23. V. Rao and L. T. Drzal, *Polym. Compos.* **12**, 48 (1991).
24. L. T. Drzal, *Mater. Sci. Eng. Trans. ASME* **A126**, 289–293 (1990).
25. M. Madhukar and L. T. Drzal, *J. Compos. Mater.* (in press).
26. M. Madhukar and L. T. Drzal, *J. Compos. Mater.* (in press).

Silanes and Other Coupling Agents, pp. 531–539
Ed. K. L. Mittal
© VSP 1992.

Use of coupling agents for internally reinforced rayon fibers

B. J. COLLIER[1,*] and J. R. COLLIER[2]

[1] *School of Human Ecology, Louisiana State University, Baton Rouge, LA 70803, USA*
[2] *Department of Chemical Engineering, Louisiana State University, Baton Rouge, LA 70803, USA*

Revised version received 18 October 1991

Abstract—Two difunctional, sterically hindered compounds, 1,4-bis(3-aminopropyldimethylsilyl)-benzene and fumaric acid, were evaluated as possible coupling agents for composite skin–core textile fibers. Other silane coupling agents and surface treatments had been tried previously. The fibers were formed using a process similar to wire coating in which a commercially produced nylon 66 monofilament core fiber, pretreated with coupling agent, contacted a viscose rayon solution in the die. Subsequently, the coated fiber was passed through a sulfuric acid coagulation bath, water wash, and dryer. The coupling agent concentration and coating line speed were evaluated for their effect on enhancement of adhesion. The composite fibers were stitched into a carrier fabric which was subsequently abraded in an Accelerotor to test the effect of the variables on adhesion of the skin to the core. Certain combinations of line speed, concentration, and type of coupling agent were found to provide significant enhancement of adhesion compared with control fibers formed without the use of either coupling agent.

Keywords: Coupling agents; textile fibers; bigeneric fibers; rayon; nylon.

1. INTRODUCTION

Composite fibers have been produced for a number of years to create structures with enhanced properties or combinations of properties. Different configurations of the materials selected for composite fibers are matrix–fibril, side-by-side, and skin–core. In the latter configuration, the core component can be made to dominate the mechanical properties of the fiber, while the skin controls surface properties. This permits decoupling of the two types of properties and closer control of the overall characteristics of the final product.

In previous work, a model skin–core composite fiber with a nylon core and a rayon skin was produced by a coating process [1, 2]. The composite fibers exhibited the mechanical properties of the nylon core, while the moisture regain was proportional to the thickness of the rayon skin.

One important aspect of these composite fibers is the adhesion between the skin and core. The fibers must maintain their integrity during processing into yarns and fabrics, and during subsequent consumer use. In the coating process, the skin and core are not formed at the same time, so sufficient interdiffusion across the interface does not occur, as in the melt forming process reported by Southern *et al.* [3]. Etching of the core with acid and corona discharge improved

---

*To whom correspondence should be addressed.

the adhesion between the skin and core; however, this improvement was not considered sufficient [4, 5].

Another method of enhancing adhesion is the use of coupling agents. These compounds have been used to improve the bonding between fibers and matrix in fiber-reinforced composites [6]. Silane coupling agents designed for coupling glass fiber in various polymeric matrices were examined for use in the reinforced rayon fibers described above [7]. A vinyl silane was partially effective in enhancing the adhesion between polypropylene core fibers and the cellulosic rayon skin, decreasing the skin loss during abrasion testing from 45% to less than 5%. Another, epoxy functional silane, was used for rayon skin–nylon core composite fibers. The improvement in adhesion was not as pronounced, as these fibers had a better initial adhesion between the skin and core. In this research, it was found through energy dispersive X-ray analysis that the silane coupling agent remained with the core when the skin was deliberately scraped off. Therefore, while the bonding of the coupling agent to the synthetic core fibers appears to be sufficient, the silane moiety is not adequately bonded to the cellulosic skin. In the present study, other coupling agents selected for greater compatibility between the rayon skin and nylon core fibers were studied.

2. EXPERIMENTAL

Commercially produced nylon 66 core fibers were washed with water to remove the spin finish and then pretreated with two different coupling agents: 1,4-bis(3-aminopropyldimethylsilyl)-benzene (APDMSB) obtained from Petrarch, and fumaric acid (FA):

APDMSB   FA

The major criteria for the choice of coupling agent were difunctionality of the reactive groups and steric hindrance of the remainder of the molecule. Solutions of the coupling agents, APDMSB in toluene and FA in ethanol, were prepared at 0.5% and 1.0% concentrations. The core fiber was passed through the solution in a pan with the speed adjusted to give a retention time of 30 s, and wound on a cone.

The pretreated fibers were coated with a solution of viscose (sodium cellulose xanthate in sodium hydroxide), using a specially designed fiber coating die [1, 2]. The pressure on the viscose reservoir feeding the die was 10 kPa. Two line speeds were used for coating fibers pretreated with each coupling agent and concentration: 21 and 41 m/min. The coated fibers were passed through a regeneration bath containing 9 wt % sulfuric acid and 13 wt % sodium sulfate. The acid bath was followed by a water rinse and passage through a drying tube before take-up.

The adhesion of the skin on the experimental fibers was determined using a carrier fabric into which the fibers were woven [4, 8]. A 100% polyester leno weave was used as the carrier fabric. A 2 m length of fiber was stitched into a

6.4×6.4 cm square of carrier fabric following the weave. The squares were conditioned at $21 \pm 1°C$ and $65 \pm 2\%$ relative humidity and then abraded individually in an Accelerotor for 2 min at 3000 rpm [9]. The Accelerotor abrasion method had been previously used to determine weight loss attributable to removal of the rayon coating [7]. However, the former study used small skeins of fiber which could occasionally become entangled in the impeller of the Accelerotor, affecting the results. Protecting the experimental fibers by incorporating them into an already formed fabric structure made the specimens more appropriate for testing in this instrument developed for fabrics. Weight loss during abrasion was determined, and abraded specimens were also qualitatively assessed for the amount of rayon skin remaining. Each of five abraded specimens for each treatment (coupling agent, concentration, and line speed combination) were examined microscopically and the numbers of fibers with coating remaining, coating peeling, and no coating remaining were recorded.

Analysis of variance (ANOVA) was used to determine the effects of coupling agent, concentration, and line speed on adhesion. Significant differences among means were determined by Duncan's multiple range test, using a 95% confidence level.

3. RESULTS AND DISCUSSION

The initial intent was to analyze the results of Accelerotor testing by determining the weight loss from the samples. This was not entirely successful since the fabric samples with test fibers in them were affected so severely that the weight data had little reliability within groups and even exhibited weight gains rather than losses on a rather random basis (Fig. 1). The use of this abrasion test on a leno weave carrier fabric stitched with a test fiber is an excessively severe test. Besides abrading away the coating on some of the test fibers, 6% of the carrier fabric samples were torn apart in the Accelerotor. Apparently, the leno weave carrier fabric is either disrupted sufficiently during stitching the test fibers or the test fibers serve as 'hooks' for the impeller of the Accelerotor to snag. When unstitched samples of the carrier fabric were used, the severe effects of the Accelerotor test were not observed.

Determination of the number of fibers remaining coated after the severe abrasion test was more successful. The results of an ANOVA on these data to determine the effect of coupling agent, concentration, and line speed are given in Table 1. As indicated in Section 2, two coupling agents were used, APDMSB and FA, and a control without a coupling agent. Two concentrations (0.5 and 1.0%) were used for each coupling agent and two line speeds (21 and 41 m/min) for each and for the control. The coupling agent was the only factor that was significant by itself at the 95% confidence level. *Post-hoc* analysis revealed that there was a significant difference between the use of APDMSB coupling agent compared with the control but the two coupling agents were not significantly different.

Amine and carboxyl functional coupling agents are recommended for both nylon and cellulosic materials. The amine groups of APDMSB can react with the carbonyl functionality and the carboxyl end groups of nylon, and with the hydroxyl groups of cellulose. The carboxyl groups of FA react with the nylon

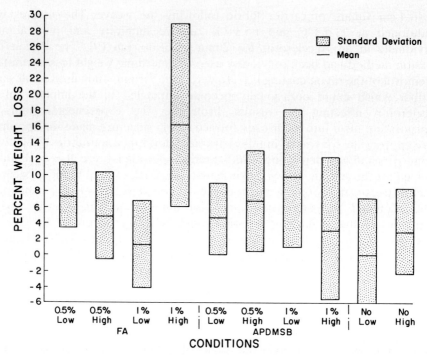

Figure 1. Weight loss of abraded fibers as a function of the coupling agent, line speed, and concentration.

Table 1.
ANOVA for percent of fibers remaining coated after the abrasion test

Variable(s)	DF[a]	F value[b]	Probability
Coupling agent	2	3.54	0.038
Concentration	1	0.01	0.932
Line speed	1	0.59	0.448
Coupling agent–concentration[c]	1	1.65	0.206
Coupling agent–line speed	2	2.46	0.099
Concentration–line speed[c]	1	6.17	0.017
Coupling agent–concentration–line speed[c]	1	7.05	0.011
Error[d]	40		

[a] Degrees of freedom. For single factors DF equals the number of factor levels minus 1. For interactions, DF is the product of the DF for the factors involved.

[b] The *F* value is the ratio of the mean square for a factor to the error mean square. The higher this ratio, the more significant is the variance among levels of a factor relative to the variance within each level of this factor.

[c] These interactions do not include the control groups, as there was only one level of concentration for the controls.

[d] Includes the variance not accounted for by the variables listed.

amine end and amide groups and the hydroxyl groups of cellulose. It is possible that the amine functional APDMSB and the carboxylic FA have different responses to the alkaline viscose solution, the acidic cellulose regeneration bath, and the residence time in each environment, accounting for the interaction between coupling agent and treatment conditions. Subsequent studies to this work suggest that a longer exposure to the acidic regeneration conditions improves the performance of the FA coupling agent.

As shown in Table 1, the interaction of concentration and line speed, as well as the three-way interaction of coupling agent, concentration, and line speed, was also significant. The APDMSB coupling agent is contributing to the significant interaction. For this coupling agent, the effect of line speed depends on the concentration. At the higher concentration of 1.0%, the higher line speed gave better adhesion results, while a lower line speed was better with the 0.5% concentration of coupling agent solution. For the FA coupling agent, however, the effect of line speed on adhesion was not dependent on concentration.

To determine the best coupling agent and treatment conditions, ANOVAs were performed using all the treatment combinations as independent groups (Table 2) and combinations of concentration and line speed as separate treatment groups (Table 3). The results of the first analysis indicate that two APDMSB coupling agent treatment conditions—low concentration and low line speed, and high concentration and high line speed—and FA at high concentration and low line speed are significantly different from the treatments which resulted in the lowest percent of coated fibers remaining after abrasion. The two best APDMSB coupling agent conditions had 80 and 78%, respectively, of the fibers surviving as coated in the hostile abrasion test. It should also be noted that FA at high concentration and low line speed had nearly as high a percent of coated fibers surviving, i.e. 70%. The four worst treatment combinations—no coupling agent at high and low line speeds, APDMSB at high concentration and low line speed, and FA at low concentration and high line speed—retained coating on 50% or fewer of the fibers.

Table 3 contains the results of an analysis of the variables of concentration and line speed across all coupling agents. As shown by this test, combinations of low concentration and low line speed, and high concentration and high line speed

Table 2.
Effect of the coupling agent, concentration, and line speed on percent of fibers remaining coated after the abrasion test

Coupling agent	Concentration (%)	Line speed (m/min)	% Coated fibers	Statistical grouping[a]
APDMSB	1.0	41	80	A
APDMSB	0.5	21	78	A
FA	1.0	21	70	A B
FA	0.5	21	62	A B C
APDMSB	0.5	41	58	A B C
FA	1.0	41	54	A B C
None	0.0	21	50	B C
FA	0.5	41	48	B C
APDMSB	1.0	21	40	C
None	0.0	41	40	C

[a] The statistical groupings are determined from significant differences between pairs of treatment combinations. Means of % coated fibers with the same letter are not significantly different at $P < 0.05$, i.e. all of the levels having a statistical grouping label of A are not significantly different from each other, but are significantly different from those labelled B, etc.

Table 3.
Effect of the line speed and concentration, independent of
the coupling agent, on percent of fibers remaining coated
after the abrasion test

Concentration (%)	Line speed (m/min)	% Coated fibers	Statistical grouping[a]
0.5	21	70	A
1.0	41	67	A
1.0	21	55	A B
0.5	41	53	A B
0.0	21	50	A B
0.0	41	40	B

[a] The statistical groupings are determined from significant
differences between pairs of treatment combinations. Means
of % coated fibers with the same letter are not significantly
different at $P < 0.05$, i.e. all of the levels having a statistical
grouping label of A are not significantly different from each
other, but are significantly different from those labelled B.

were significantly different from the control at high line speed. The other groups
were not significantly different.

The effect of line speed on the composite fibers may be due to some
combination of three effects. As discussed in an earlier paper [2], the coating
thickness tends to be periodically variable when a lower line speed is used. When
this occurs, it should be easier to abrade off the coating. The effect of line speed
on the coated fibers pretreated with 1.0% APDMSB was not consistent with that
pertaining in other instances. Additional work indicates that the instabilities due
to line speed may be related to the surface tension between the viscose solution
and the core fiber. With poorer wetting the instability should onset at a lower
speed and its amplitude should be greater at a given speed below the onset value.
In a subsequent series of multiple fiber coatings, it was determined that the best
wetting characteristics occurred with FA-coated fibers, followed in order of
decreasing wetting by polyamide fibers, polyester fibers, and polyolefin (very
hydrophobic) fibers. Coating the polyamide core fibers with the APDMSB used
in this study would make them increasingly more hydrophobic as the APDMSB
concentration increased. This hydrophobicity should develop when the
APDMSB is applied to the core fiber prior to coating with viscose. The
APDMSB has sufficient flexibility to have both ends form hydrogen bonds with
the polyamide surface, leaving non-polar groups on the surface. The shifting of
some of the hydrogen bonds of the core with APDMSB to the cellulose would
develop predominantly after the cellulose was precipitated and after the
instability had developed.

Another effect of line speed is that since the process is a combination of drag
and pressure flow [2], increasing the line speed decreases the fraction of total
flow due to pressure and results in a thinner coating. In this work, fibers coated
at the higher line speed had a lower linear density, also indicating a thinner
coating. Since the viscose coagulation bath is of fixed length and is in line with
coating, increasing the line speed also decreases the residence time in the acid
bath. The change and variation in coating thickness and degree of regeneration

of cellulose may be responsible for the significance of the combination of coupling agent, concentration, and line speed. Although not part of this study, ongoing research has shown that the residence time may be important. A series of runs was made using the FA coupling agent in which the coated fiber was not washed in line after the acid bath, but after holding for 30 min on the take-up package. Initial indications are that the adhesion is definitely much better after the longer regeneration time provided by delaying the afterwashing. This may also be related to FA being an acidic coupling agent.

As mentioned earlier, statistical analysis indicated no significant differences in the weight loss data. However, the trend in the weight loss data, as shown in Fig. 1, agrees with the analysis of percent coated fibers for the three-way groupings presented in Table 2 and depicted in Fig. 2. The standard deviation indicated on the bar graph in Fig. 1 demonstrates why the statistical analysis of the weight loss data did not show a significant difference.

The two APDMSB groupings with the highest percent of coated samples after abrasion testing were distinctive in another fashion. Both had a significant number of coated fibers that exhibited a fuzzy appearance in the microscope, as shown in Fig. 3b, and contrasted to a fiber with an even coating in Fig. 3a, an uncoated fiber in Fig. 3c, and a fiber in which the skin appears to be peeling in Fig. 3d. The APDMSB group with the highest percent of coated fibers after abrasion exhibited a fuzzy surface on seven out of every ten of the coated fibers after abrasion. In the next best group of fibers treated with APDMSB, about one

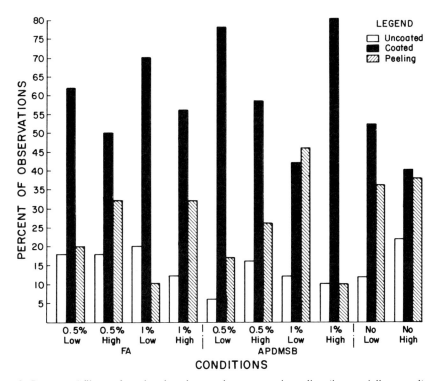

Figure 2. Percent of fibers after abrasion that are intact coated, peeling (i.e. partially coated), and uncoated.

Figure 3. Abraded fibers. (a) Intact coated fiber; (b) fuzzy intact coated fiber; (c) fiber with no coating remaining; (d) fiber with coating peeling.

out of every four coated fibers exhibited a fuzzy appearance after abrasion. None of the FA-treated fibers appeared fuzzy after abrasion.

4. CONCLUSIONS

Coupling agents can enhance the adhesion in internally reinforced rayon fibers. This improvement over fibers coated without a coupling agent is dependent on the concentration of the coupling agent, and especially on the line speed used for the coating operation. The APDMSB coupling agent at two treatment and coating conditions—high concentration and high line speed, and low concentration and low line speed—was the most successful. Apparently nearly as good as the FA coupling agent if used at high concentration and low line speed. It should be noted, however, that APDMSB may be sensitive to some conditions such as static build-up, which results in the attraction of small fibrous material giving the fuzzy appearance that high percentages of the two best groups exhibited. Qualitative measurements of the ability to scrape the coating off with a sharp object suggest that the best group is the FA coupling agent at high concentration and low line speed.

The higher concentration pretreatment with APDMSB did not show a decrease in adhesion with increasing line speed, as did the other coupling agent–concentration combinations. This could be due to the difunctional coupling agent initially forming hydrogen bonds only with the core fiber during pretreatment, leaving a more hydrophobic surface for the viscose coating when subsequently applied. At higher concentrations of APDMSB, it is more probable that this phenomenon could occur.

This study is continuing with changes being made in the operating procedures for coupling agent application, processing, and testing. Improvements in the method for analyzing the coating effectiveness are also being pursued.

Acknowledgements

National Science Foundation grant MSM 8896Q33 provided partial support for this research effort.

REFERENCES

1. J. R. Collier, B. J. Collier, R. L. Rabe and D. L. Wright, in: *Wood and Cellulosics*, J. F. Kennedy, G. O. Phillips and P. A. Williams (Eds), pp. 581–586. Ellis Horwood, Chichester (1987).
2. R. L. Rabe, B. J. Collier and J. R. Collier, *Text. Res. J.* **58**, 735–742 (1988).
3. J. H. Southern, D. H. Martin and D. G. Baird, *Text. Res. J.* **50**, 411–416 (1980).
4. B. J. Collier, J. R. Collier and J. E. McDonald, *Home Economics Res. J.* **18**, 126–132 (1989).
5. R. L. Rabe, Ph.D. Dissertation, Ohio University, Athens (1989).
6. E. P. Plueddemann, *Silane Coupling Agents*. Plenum Press, New York (1982).
7. B. J. Collier, J. X. Li and J. R. Collier, *Text. Chem. Colorist* **22**, 31–34 (1990).
8. H. M. Elder, T. S. Ellis and F. Yahya, *J. Text. Inst.* **78**, 72–79 (1987).
9. American Association of Textile Chemists and Colorists, *Technical Manual*, pp. 133–135. AATCC, Research Triangle Park, NC (1988).

Silanes and Other Coupling Agents, pp. 541–544
Ed. K. L. Mittal
© VSP 1992.

A study of silane coupling for commercially bonding dissimilar thermoset materials

PAUL E. KOCH* and D. JAY SCHENCK

Plastics Technical Center, Pennsylvania State University at Erie, The Behrend College, Erie, PA 16563-0203, USA

Revised version received 11 October 1991

Abstract—Application of a silane coupling agent as a means of bonding melamine foil to an unsaturated, mineral-filled polyester substrate in the form of a molded dinner plate, 25.4 cm in diameter, was investigated. Bonding was achieved which was satisfactory from industry standards, but the adhesion promoter, a glycidoxypropyltrimethoxysilane, migrated through the foil thickness and deposited an unacceptable residue which caused topical cosmetic failure of the molded plate.

Keywords: Coupling; dinnerware; foil; melamine; silane; polyester.

1. INTRODUCTION

Alpha-cellulose-filled melamine–formaldehyde has long been used for unbreakable, scratch-resistant, dishwasher-safe dinnerware. Its thermosetting nature has made it the logical choice for the rapidly-expanding microwaveable tray market. Unfortunately, owing to the disposability requirement of the product, the cost has made melamine prohibitive.

Mineral-filled polyester continues to represent a very attractive alternative, but it lacks the required scratch resistance and decorating capabilities. This project examined the potential of using a melamine foil bonded during the conventional molding process to the more cost-effective polyester substrate. Melamine foil is an overlay produced from a thin paper sheet printed with a decorative design and then impregnated with melamine–formaldehyde resin. The finished thickness of the foils is typically 0.17 mm. Dispatch Printing of Erie, Pennsylvania provided the melamine foil and Premix, Inc. of North Kingsville, Ohio supplied the food grade glass-reinforced polyester used throughout this study. Three commercially applicable tests were used to evaluate the viability of different silane coupling agents that were used for bonding these two dissimilar materials.

The intent in this project was to work only with commercial processes so that a successful solution could be readily incorporated into production. Therefore, the finished product being evaluated was a 25.4 cm dinnerware plate which was molded on a standard compression molding machine. Plates were prepared following processing parameters and production techniques typical of commercial molding standards.

---

*To whom correspondence should be addressed.

2. EXPERIMENTAL

2.1. Foil preparation

The melamine foil was prepared for bonding by diluting four parts silane coupling agent with one part anhydrous isopropyl alcohol. The alcohol reduced the viscosity of the silane and thus aided in flowing the solution over the foil surface. Two methods of applying the silane to the plate contact surface of the foil were implemented. One technique, suggested by Dr. Plueddemann [1], consisted of wetting a lint-free disposable tissue in the silane solution and wiping the tissue once in a single direction across the foil. The second approach used was intended to distribute a denser layer of silane; in this case, a fine-hair artist's brush was used to coat the foils with a single stroke.

Table 1 lists the silane coupling agents that were used and their methods of application to the melamine foil. Considerable difficulty was encountered in trying to obtain a thin uniform layer on the highly porous melamine foil; the methods listed in Table 1 represent the final iteration in a series of tests.

Table 1.
Coupling agents and their methods of application to the melamine foil

Coupling agent	Chemical name	Application method
Dow Corning Z-6020	Diaminopentatrimethoxysilane	Dilute and hand wipe
Dow Corning 25 Additive[a]	Glycidoxypropyltrimethoxysilane	Dilute and hand wipe
Dow Corning X1-6146[b]	Glycidoxypropyltrimethoxysilane	Dilute and brush

[a] Ingredients: hexamethoxymethyl melamine, 90 wt%; glycidoxypropyltrimethoxysilane, 10 wt%; formaldehyde, 0.4 wt%.
[b] Ingredients: methylated melamine–formaldehyde resin, 72 wt%; isobutyl alcohol, 18 wt%; formaldehyde, < 0.9 wt%; glycidoxypropyltrimethoxysilane, 10 wt%.

2.2. Molding method

The experimental plates were formed through compression molding in the following sequence:
(1) the mold was electrically heated to 157°C.
(2) a preweighed polyester charge was inserted between the two mold halves;
(3) the mold was then closed, forcing the resin into the mold cavity shape;
(4) the mold was opened briefly for evacuation of trapped volatiles;
(5) a silane-coated melamine foil was inserted with the coated surface in contact with the polyester substrate; the mold was closed and the devolatilization cycle was repeated;
(6) the resin was allowed to crosslink and cure for 2 min, while compressed within the mold cavity at temperatures of 149–142°C; and
(7) the press was opened and the plate was ejected and allowed to air cool.

2.3. Test methods

The test methods chosen were designed to evaluate the relative strength and durability of the bond between the melamine foil and polyester base.

The first test was an abrasion test in which a diamond stylus was used to scratch and rupture the foil. This was done repeatedly, forming a crosshatch pattern in which the surface area of each separate circumscribed patch became progressively smaller. The surface was vigorously wiped by hand to determine which interior squares failed to remain bonded. A uniformly-coated, well bonded foil will adhere within as small as a 1.5 mm × 1.5 mm patch. This test was extremely effective as a screening tool and since it can be done at the molding press work station, it allowed for rapid process assessment.

The second test, developed by the Society of the Plastics Industry, involved preparing an amount of bacon that covered at least 50% of the dinner plate surface area [2]. The plate with the bacon layer was heated in a microwave oven set at full power until the temperature at the center of the plate area reached 121°C ($\pm$ 5°C). The time period required to reach this temperature within the bacon fat was 5 min on the microwave unit used for this test. Immediately after the cooking cycle, the plate was cleaned on the lower rack of an automatic dishwasher. The dishwasher cycle was set at 'normal' and included a heated dry cycle. The water temperature reached 66°C ($\pm$ 5°C). Once the cycle was complete, the plate was inspected for surface blemishes such as cracks or blisters. This test is considered the most severe of the three due to the extreme localized heat generated, which causes the difference between the thermal coefficient of expansion for the two materials to shear the interface resulting in delamination.

The third test was a boil-and-soak test recommended by Dr. Plueddemann [3]. The plates were cut into 2.54 cm wide strips and immersed in boiling water to stimulate an enhanced sanitizing cycle in a typical dishwasher. The plate strips were pulled from the water once an hour and the interface between the polyester and the melamine foil was inspected. This was performed by probing the exposed interface with a stainless steel stylus, attempting to initiate delamination of the two layers.

3. RESULTS

Table 2 summarizes the findings of the three tests performed.

Table 2.
Quantification of test results

Coupling agent	Abrasion scratch[a]	'Microwave bacon' (cycles)[b]	Boil-and-soak time (h)[c]
Control, no silane	Fail	0	0
Z-6020	Fail	0	0
25 Additive	Pass	10	8
X1-6146	Pass	10	8

[a] A passing qualification required a scribed perimeter of 1.5 × 1.5 mm remaining intact and bonded to the plate.

[b] Cycles signify the number of times a plate endured cooking bacon in the microwave followed by cleaning in a dishwasher without showing evidence of delamination.

[c] The time shown is the number of hours during which a 2.54 cm wide section of the plate soaked in boiling water until loss of adhesion was evidenced.

4. DISCUSSION

The results show that two of the silane coupling agents, Dow Corning 25 Additive and X1-6146, performed well enough to be considered for further study, although both appeared to migrate through the foil during the crosslink cycle and caused the plate to stick to the top half of the mold. In addition, when the plate was stripped, the mold cavity had a substantial residue built up on the surface which eliminated the process from being commercially viable. This problem is probably caused by the high mobility and low viscosity of the silane molecules in the coupling agents.

Owing to the excessive deposits of residue experienced, an additional step was incorporated in the foil preparation process. The melamine foil coated with the silane was given a 'B-stage' which created a barrier to the migration of the silane through the foil during the final cure, thereby preserving the glossy finish on the molded plate. Additional silane application developments will be required to guarantee a thin uniform coating of coupling agent and make this a viable approach to part manufacture.

5. CONCLUSION

The viability of incorporating a silane coupling agent to bond dissimilar materials was confirmed. The process, with some improvement in silane application onto the foils, can be a commercial success. Enhanced methods of silane coating being considered include a spray mist mechanism and a screen print technique. Either the Dow Corning 25 Additive or X1-6146 will perform, although our best results in terms of surface quality were achieved with Additive 25. Further investigation needs to be focused upon the silane migration through the foils when subjected to the thermal pressures of compression molding.

Acknowledgements

We wish to express our gratitude to Dr. Edwin Plueddemann for his encouragement and technical support; Mr. Nathaniel Reyburn, who first conceived the idea and contributed his decades of experience in thermoset molding; and Dr. K. L. Mittal for editorial assistance in the preparation of the manuscript.

REFERENCES

1. E. P. Plueddemann, *Silane Coupling Agents.* Plenum Press, New York (1982).
2. The Society of the Plastics Industry, *Recommended Test Methods—Plastics Cookware for Microwave Ovens* (1982).
3. Telephone conversations with E. P. Plueddemann, Dow Corning Company, Midland MI (1989–1990).

Part 4

Non - Silane Coupling Agents

Silanes and Other Coupling Agents, pp. 547–557
Ed. K. L. Mittal
© VSP 1992.

Zirconium-based coupling agents and adhesion promoters

PETER J. MOLES

Magnesium Elektron Ltd, P.O. Box 6, Swinton, Manchester M27 2LS, UK

Revised version received 21 October 1991

Abstract—The use of zirconium compounds as coupling agents is reviewed, with particular emphasis on their use in flexographic and gravure inks. The modes of action favoured by zirconium compounds are, firstly, reaction with carboxyl groups and, secondly, reaction with hydroxyl groups. Usage of zirconium compounds is increasing in application areas such as inks, adhesives, and paints.

Keywords: Zirconium compounds; coupling agents; adhesion promoters.

1. INTRODUCTION

When people talk of coupling agents, most think of silicon-based compounds. However, there are other elements whose compounds exhibit suitable chemistry for adhesion promotion effects. In recent years, zirconium and titanium compounds have been studied as coupling agents/adhesion promoters. A recent review [1] on coupling agents discusses the recent work on silanes and briefly discusses the usage of titanium and zirconium compounds as coupling agents/ adhesion promoters. This paper concentrates on reviewing the published work on zirconium compounds and attempts to highlight similarities and differences between the chemistries of zirconium, titanium, and silicon.

It is worth noting at the start of this paper that a literature search revealed that Dr. Plueddemann had worked with zirconium compounds in the past, albeit using silane coupling agents for zirconium silicate minerals [2], rather than zirconium coupling agents in their own right.

2. THE CHEMISTRY OF ZIRCONIUM

The chemistry of zirconium has some similarities to that of silicon and titanium, since it is in Group IV of the periodic table. Zirconium has a normal oxidation state of 4 with limited redox chemistry and a coordination number of up to 8. Zirconium compounds are normally colourless.

The solution chemistry of zirconium, both aqueous and solvent-based, is dominated by the tendency to form polymeric species. This has been reviewed in several articles [3], but the basic facts bear repeating here once again.

Zirconium demonstrates a marked preference for the formation of bonds with oxygenated species. In aqueous solution, zirconium polymers based on hydroxyl bridged species can be found. These polymeric zirconium compounds vary in size depending on conditions such as pH, concentration, and temperature. Figure 1 illustrates the crystal structure of zirconium oxychloride, which shows it to be based on cationic tetramers. The chloride ions do not bond to the zirconium, and hence are not shown in the diagram.

All aqueous zirconium compounds have polymeric structures, sometimes with ligands bonded to the zirconium-based polymer. Whether ligands are bonded to the zirconium depends on the counterion. In very general terms, oxygenated ions such as carbonate and sulphate bond to the zirconium whilst halides such as chloride do not.

It has been usual to classify zirconium compounds as either anionic, cationic, or neutral on the basis of the structures shown in Fig. 2 (which are schematic representations in which water ligands have been omitted for the sake of simplicity).

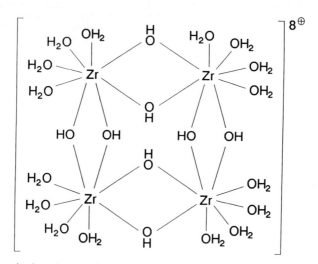

Figure 1. Tetrameric zirconium unit found in zirconium oxychloride crystals.

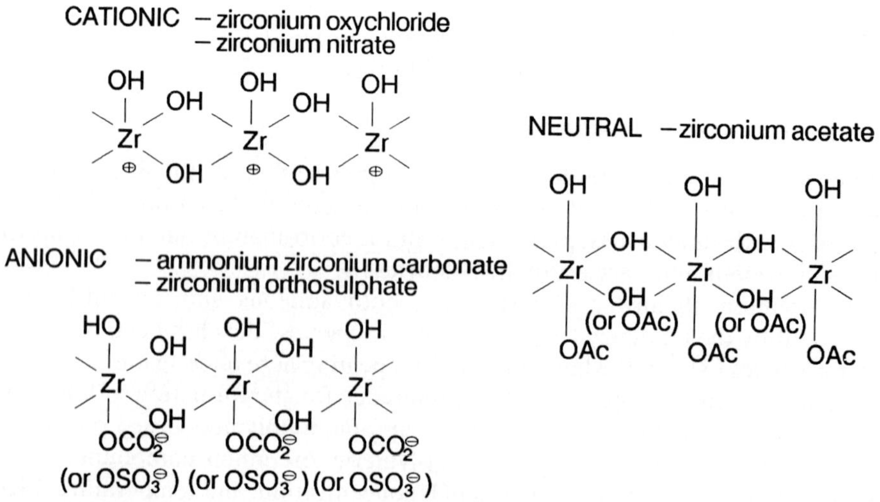

CATIONIC – zirconium oxychloride
– zirconium nitrate

NEUTRAL – zirconium acetate

ANIONIC – ammonium zirconium carbonate
– zirconium orthosulphate

Figure 2. Schematic polymeric structures for aqueous zirconium compounds.

In the case of organic derivatives such as zirconium acetate, direct bonding of the carboxylate to the zirconium is found. Similar structures are also found in solvent-soluble, water-insoluble carboxylates such as zirconium propionate. Zirconium alkoxide derivatives tend to be monomeric in solvent-based systems but hydrolyse rapidly with ambient water to give polymeric species.

The polymeric nature of zirconium in aqueous systems is similar to that of titanium compounds. However, zirconium compounds tend to be significantly more stable towards hydrolytic polymerization.

These zirconium polymers tend to display two basic reactions. First, there is the reaction with carboxyl groups, where strong covalent bonds are formed (Fig. 3); second, they can undergo hydrogen bonding interactions with hydroxyl groups, as shown in Fig. 4.

It is generally observed that zirconium reacts most strongly with carboxyl groups and less strongly with other oxygenated functional groups such as ester, ether, etc. Understanding this chemistry is of great importance in the determination of how zirconium compounds function as coupling agents or adhesion promoters.

Finally, the chemistry of zirconium fluorides should be mentioned. This is a rather special case but does have some relevance to industrial usage, as discussed later. Zirconium fluorides exhibit a marked reluctance to hydrolyse at acid pHs. These zirconium fluoride derivatives tend to be less polymeric, although some polymeric nature is believed to exist [3].

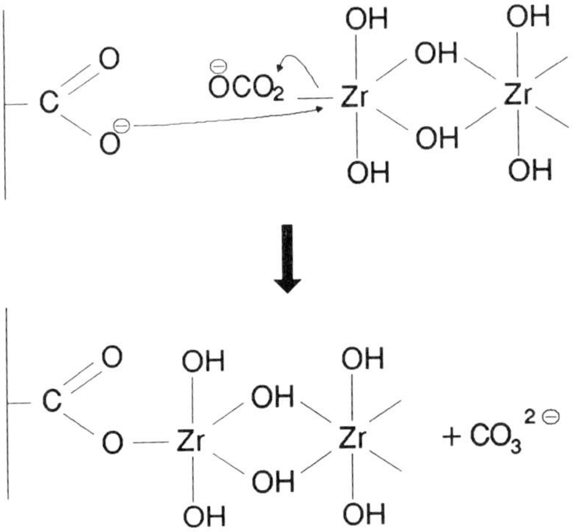

Figure 3. Interaction between zirconium polymers and carboxyl groups.

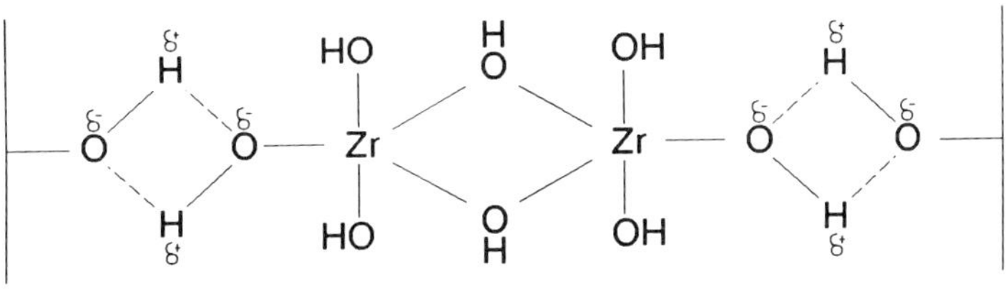

Figure 4. Interaction between zirconium polymers and hydroxyl groups.

3.1.1.3. Aluminium drinking cans. There are many patents [8] referring to the use of fluorozirconic acid (H$_2$ZrF$_6$)-based systems to treat the surface of aluminium cans to improve the corrosion resistance of the metal and the adhesion of the applied coatings. Typically, the zirconium fluoride will be used in conjunction with polyacrylic acid, presumably to form a complex *in situ* which acts as an adhesion promoter. Such surface treatment of aluminium is not restricted to zirconium fluorides, as ammonium zirconium carbonate displays similar properties in such application areas.

3.1.1.4. Pigment coating. It is well known that the surface treatment of titania with zirconium compounds (in essence zirconium hydroxide) gives improved dispersibility in paints and related systems. Although not normally regarded as a coupling agent effect, this type of surface treatment obviously modifies the pigment–polymer matrix interaction and, as such, is worth mentioning here.

3.2. Fundamental studies on zirconium-based coupling agents

Published work on the use of zirconium compounds as coupling agents/adhesion promoters is limited. In this section, the work of major researchers in the field will be described followed by a more in-depth review of the use of zirconium compounds in the ink industry.

3.2.1. Literature review. As noted previously, Plueddemann investigated the use of silanes for surface treating zirconium silicates for use as a filler in thermosetting resins [2].

As for actual zirconium coupling agents, one of the earliest reports concerned the use of zirconium compounds as a coupling agent between dental enamel and polymeric filler [9]; this was taken up further by Misra [10], who concluded that zirconium carboxylates and, in particular, zirconium methacrylate had merit as a coupling agent for dental polymer composites. A composite of apatite treated with zirconium methacrylate and methyl methacrylate had a tensile strength 50% greater than a composite using untreated apatite. Misra suggested that the zirconium was hydrogen bonded to the surface of the apatite.

Miedaner [11] in several patents has disclosed that zirconium and titanium complexes of orthofunctional aromatic compounds (monomeric or polymeric) could be used as coupling agents for phenolic or resorcinol resins and glass.

Calvert *et al.* [12] studied the interfacial coupling of alkoxy titanium and zirconium tricarboxyls with metal oxides. They showed that isopropoxytitanium and zirconium tristearates interact with silica and alumina by exclusive loss of isopropanol, and that the titanate and zirconate were resistant to desorption by treatment with hot water.

Cohen has presented a series of papers on the extremely interesting and versatile zircoaluminates [13]. The structures of these compounds are not known with certainty, but various structures have been put forward and one such suggestion is shown in Fig. 7. Typically, these zircoaluminates are solvent-soluble, but can be used in aqueous solution in combination with various aqueous zirconium compounds.

Cohen has shown that zircoaluminates can be used as coupling agents for bonding polymers to metals and inorganic pigments. He proposes that with metals the zirconium and aluminium can bond to the surface hydroxyls on the metal substrate, leaving the organofunctional group available for reaction with the resin. This is shown in Fig. 8.

An analogous chemistry is proposed by Cohen for the use of zircoaluminates with various pigments. This coupling agent chemistry can be utilized to give commercially useful effects such as improved salt spray and humidity resistance in coatings and reduced viscosities in pigment dispersions which leads to improved colour strength.

Sugerman and Monte have reported the use of neoalkoxy organotitanates and zirconates [14] as coupling and polymer processing agents, in particular, in polyurethane systems; they have reported that good corrosion resistance properties are found by utilizing neoalkoxy pyrophosphato-organozirconates.

In particular, they found enhanced bonding between metal surfaces and resins such as acrylics (solvent- and water-based), epoxy chlorinated rubbers, silicones, and polysulphides. It was noted that titanium complexes caused colouration with phenolics, whilst zirconium complexes did not.

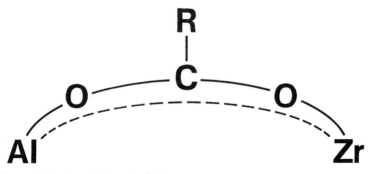

Figure 7. Proposed structure of zircoaluminates.

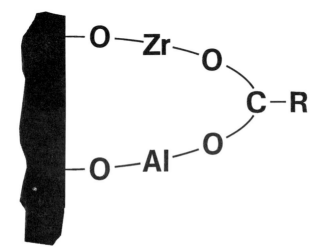

Figure 8. Proposed mode of interaction for zircoaluminates and oxygenated substrates.

A recent patent by Thomason [15] has revealed that ammonium zirconium carbonate when applied to a substrate such as glass, aluminium, or polypropylene can improve the adhesion of microsphere polymer-based adhesives. It is proposed that the zirconium after reacting with the substrate surface reacts with carboxyl groups at the surface of the polymer microspheres.

Merck in Japan has recently patented [16] a process for the production of water and weather-resistant pearlescent pigments produced by coating mica with hydrous zirconia. This is in many ways similar to processes operated in the titanium oxide industry and mentioned previously. The zirconium hydroxide aids dispersion and gives better compatibility with the polymer matrix that it is incorporated in.

Hunt *et al.* [17] have investigated the role of various coupling agents, such as zirconates, titanates, and zircoaluminates, in zirconia–polypropylene suspensions for use in the production of ceramics by injection moulding. All the coupling agents were observed to reduce the melt viscosity.

Brewis [5] has reported that tack reduction observed with various adhesives after treatment with aqueous zirconium compounds is due to zirconium reacting with carboxyl groups on the surface of the polymer particles. This has been confirmed by XPS surface analysis, which showed that the zirconium compounds reacted with carboxyl groups and that the effectiveness of the zirconium compounds was directly related to the carboxyl concentration.

Archer and Wang, in a series of recent papers [18], have investigated some novel zirconium compounds and their usage as adhesion promoters. Zirconium [tetrakis (salicylidene) diaminobenzidine] co-polymers (Fig. 9) are rather interesting zirconium polymer systems in their own right, but they have also been shown to act as useful adhesion promoters or coupling agents for glass and metal substrates.

These zirconium compounds are interesting, as they represent rather inert, highly chelated zirconium polymers. All previous work on zirconium coupling agents used zirconium compounds where the ligands were unidentate or bidentate. Surface coverage of silica has been examined and Zr—O—Si bonding has been postulated as the mode of interaction. Adhesion tests such as scribe stripping, indentation debonding, and peel have shown that these zirconium compounds demonstrate marked adhesion promotion properties when deposited onto the surface of a glass or aluminium–polymer laminate as shown in Table 2. These zirconium co-polymers also demonstrate marked hot water resistance.

Overall, there has not been a great deal of detailed published work on zirconium adhesion promoters or coupling agents. It is known that they will react with various metal, metal oxide, and oxidized plastic surfaces, and given a

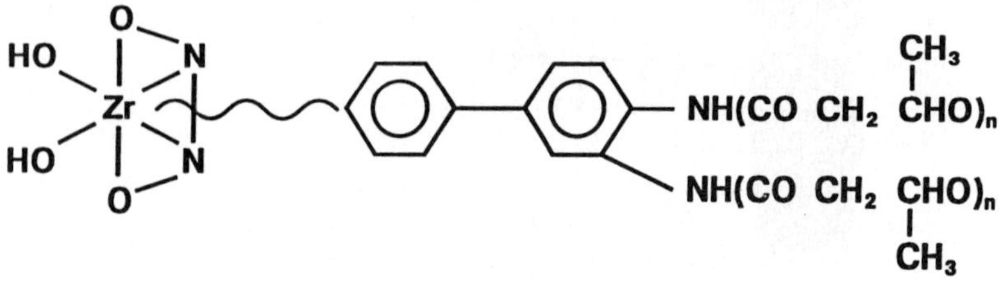

Figure 9. Structure of zirconium [tetrakis (salicylidene) diaminobenzidine] co-polymer.

Table 2.
Adhesion strength of polymeric coatings using 180° peel test

Polymer	Substrate	Peel strength (J/m^2)
PMMA[a]	Glass only	3.5×10^2
PMMA	Glass/ZrPL[b]	1.2×10^3
PMMA	Al only	2.4×10^2
PMMA	Al/ZrPL[b]	9.0×10^2

[a] Poly(methyl methacrylate).
[b] Zirconium co-polymer.
Taken from ref. 18c.

suitable functionality on the organic polymer coating, they can act as coupling agents.

To conclude this discussion on zirconium, it is appropriate to look at the adhesion promotion effects of various zirconium compounds in flexographic and gravure ink printed on corona discharge-treated polyolefins and polyester. Flexographic and gravure inks are basically a pigment (often titanium dioxide) suspended in a polymer (normally called the binder) dissolved in a solvent. Actual commercial ink formulations are rather more complicated. These inks are either water-based when acrylic polymers and co-polymers are typically the binders, or solvent-based (usually ethanol–ethyl acetate mixtures) when the binder is typically nitrocellulose or cellulose acetate propionate.

A basic nitrocellulose ink formulation is given in Table 3.

Table 3.
Typical simple ink formulation

Nitrocellulose	20 parts
Solvent[a]	100 parts
Pigments	1–5 parts
Adhesion promoter	0–2 parts

[a] Typically a 1:1 mixture of low boiling esters and alcohols.

Zirconium propionate is a polymeric zirconium carboxylate; its structure is illustrated in Fig. 10. Use of zirconium propionate markedly increases the adhesion of an ink applied to treated polypropylene film. Figure 11 compares zirconium propionate with titanium acetylacetonate, which is commonly regarded as the industry standard. The standard test method used in the ink industry is the so-called 'tape test'. Sticky tape is placed on the printed film and pressure is applied by the operator's thumb. The tape is then pulled off, by hand, and the amount of ink removed is visually assessed. Although extremely crude, it can be, and is, used for control in the ink industry.

Zirconium acetylacetonate can react in a similar way and as noted previously, surface analysis of zirconium acetylacetonate derivatives on corona-discharged polypropylene has shown bonding to the surface carboxyl groups [6].

Further work by Comyn (private communication) has attempted to use a peel test to measure the actual adhesion improvements. Zirconium propionate was

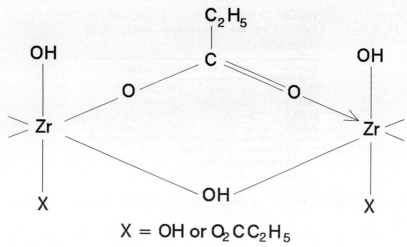

Figure 10. Schematic structure of zirconium propionate.

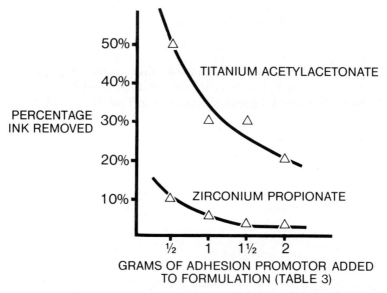

Figure 11. Comparison between zirconium propionate and titanium acetylacetonate as adhesion promoters in a simple ink formulation.

added to nitrocellulose solution (in ethyl acetate–ethanol) and this solution was used to bond a cotton fabric to corona discharge-treated polypropylene. Preliminary work has shown that zirconium propionate markedly improves the adhesion over a non-zirconium-containing system. Quantitative data are being gathered.

4. CONCLUSIONS

Although not nearly as well researched as silicon, zirconium compounds have managed to find utilization as adhesion promoters. They may never displace silanes as the major class of coupling agents, but they have a growing part to play in the field of adhesion in the years to come.

REFERENCES

1. P. Walker, in: *Surface Coatings 1*, A. D. Wilson, J. W. Nicholson and H. J. Prosser (Eds), pp. 189–232. Applied Science Publishers, New York (1988).
2. E. P. Plueddemann, *Adv. Chem. Ser.* **124**, 86–94 (1974).
3. P. J. Moles, *Proc. Water Borne Higher Solids Coatings Symp.* **14**, 314, University of Southern Mississippi, Hattiesburg, MS (1987).
4. G. E. Ealer, *Tappi*, 145–150 (1990).
5. D. Brewis, *J. Adhesion* **26**, 215 (1988).
6. D. Palmer, Ph.D. Thesis, University of Newcastle, UK (1990).
7. G. B. Patent 1128506 (Polychrome Corp.).
8. US Patents 2,795,518, 3,695,942, 3,966,502 and 3,912,548 (all Amchem Products Inc.).
9. US Patent 350 7041 (US Dept. of Health).
10. D. N. Misra, *J. Dent. Res.* **64**, 1405–1408 (1985).
11. (a) US Patent 399 3835 (PPG Corp.); (b) GP Patent 1566446 (PPG Corp.).
12. P. D. Calvert, R. R. Lalanandham and D. R. M. Walton, in: *Adhesion Aspects of Polymeric Coatings*, K. L. Mittal (Ed.), pp. 457–466. Plenum Press, New York (1983).
13. (a) L. B. Cohen, *High Solids Coatings* **9**, 2–8 (1984); (b) L. B. Cohen, World Patent Application 840 3093; (c) L. B. Cohen, *Proc. Water Borne Higher Solids Coatings Symp.* **13**, 216–236, University of Southern Mississippi, Hattiesburg, MS (1986); (d) L. B. Cohen, *High Solids Coatings* **12**, 24–36 (1987); (e) L. B. Cohen, *Polym. Paint Colour J.* **177**, 686 (1987); (f) L. B. Cohen, *J. Adhesion Sci. Technol.* **5**, 439–448 (1991); (g) L. B. Cohen, in: *Surface and Colloid Science in Computer Technology*, K. L. Mittal (Ed.), pp. 171–178. Plenum Press, New York (1987).
14. (a) S. J. Monte and G. Sugerman, *Proc. SPI Annu. Tech. Marketing Conf.* **28**, 322–330 (1984); (b) S. J. Monte and G. Sugerman, European Patent 164227; (c) G. Sugerman and S. J. Monte, *Proc. Water Borne Higher Solids Coating Symp.* **15**, 462–477, University of Southern Mississippi, Hattiesburg, MS (1988); (d) G. Sugerman and S. J. Monte, *Mod. Paint Coat.* **78**, 50 (1988); (e) G. Sugerman and S. J. Monte, in: *Interfaces in Polymer Ceramic and Metal Matrix Composites*, H. Ishida (Ed.), pp. 657–667. Elsevier, New York (1988); (f) G. Sugerman and S. J. Monte, *Int. SAMPE Tech. Conf.* **20**, 201, 423–437 (1988); (g) S. J. Monte and G. Sugerman, in: *Adhesion Aspects of Polymeric Coatings*, K. L. Mittal (Ed.), pp. 421–441. Plenum Press, New York (1983).
15. European Patent 86305367 (Smith and McLawin Ltd).
16. (a) European Patent 268918 (Merck GmbH); (b) European Patent 342533 (Merck GmbH).
17. K. N. Hunt, J. R. Evans and J. Woodthorpe, *Polym. Eng. Sci.* **28**, 1572–1577 (1988).
18. (a) B. Wang and R. D. Archer, *Polym. Mater. Sci. Eng.* **59**, 120–144 (1988); (b) B. Wang and R. D. Archer, *ibid.* **60**, 710–714 (1989); (c) R. D. Archer and B. Wang, *ibid.* **61**, 101–105 (1989); (d) R. D. Archer and B. Wang, *Inorg. Chem.* **29**, 39 (1990).

Silanes and Other Coupling Agents, pp. 559–568
Ed. K. L. Mittal
© VSP 1992.

Zircoaluminate adhesion promoters

LAWRENCE B. COHEN

Rhone-Poulenc Chemicals, Inc., 32 Condor Road, Sharon, MA 02067, USA

Revised version received 10 January 1991

Abstract—Zircoaluminates constitute a unique class of adhesion promoters which are highly reactive with a diverse array of substrates and do not suffer from the hydrolytic sensitivity of the silanes. The zircoaluminate compositions are useful in prolonging the service life of solvent- and water-borne coatings by improving wet adhesion and reducing concomitantly corrosion on many metal surfaces when incorporated into coatings applied to such surfaces. Similarly, the bond strength to difficult substrates is improved when incorporated into adhesives or rubber compositions. Successful use requires incorporation by high shear mixing, predissolution in hydroxy-containing solvents, or the selective use of co-dispersants.

Keywords: Zircoaluminates; adhesives; polyester coating; alkyd coating; rubber high shear mixer.

1. INTRODUCTION

The definition of an adhesion promoter in its most literal sense may be stated as any substance which when placed between two adherends results in a measurable increase in the force required to separate the two materials. This definition does not address the basic mechanism responsible for the increased adhesion nor does it concern itself with the mode by which the promoter is contacted with the adherend surface, i.e. as primer, by *in situ* incorporation into an adhesive or coating, or by other means.

The adhesion to substrates may be increased by (i) mechanical alterations of the substrate, (ii) polar interactions with the bonding agent, or (iii) chemical bonding facilitated by an adhesion promoter [1]. Although both mechanisms (i) and (ii) are effective in improving adhesion, this paper will focus on the use of chemical bonding (iii).

The mechanism (iii) has been largely the domain of the silanes over the last 30 years. The class of compounds has demonstrated outstanding performance advantages, most notably in the fiberglass industry, but also in other varied applications including mineral filled/reinforced rubber, plastics, and high performance adhesives. Nonetheless, the silanes are not without their deficiencies: specifically, high cost, severe hydrolytic instability, and nonoptimal performance on many nonsiliceous substrates.

In 1983 a new class of metal organic adhesion promoters—the zircoaluminates—emerged. Evaluations by all of the major glass manufacturers ultimately provided compelling evidence that zircoaluminate application to fiberglass did not result in the *sine qua non* for success, that is, the 'irreversible attachment' to the glass surface in the presence of moisture. (Note: Irreversible is not intended to mean the absence of a dynamic equilibrium, but rather as seen in reaction (1) it refers to a reaction in which $K_{eq} \geqslant 1$.)

$$\text{Substrate (S) + Adhesion promoter (AP)} \rightarrow \text{S—AP} \qquad (1)$$

Fortunately, the absence of such irreversible attachment to E-glass did not extend to other important substrates. Early work conducted with ATH (alumina trihydrate), calcium carbonate, and titanium dioxide demonstrated 'irreversible attachment' to such substrates [2, 3], thereby laying the foundation for use in a diverse array of non-fiberglass applications.

2. ZIRCOALUMINATE CHEMISTRY

Zircoaluminates (ZAs) constitute a unique class of bifunctional additives which are synthesized in accordance with a highly specific procedure [4–6] to produce stable covalent molecules. As shown below, the essential chemistry involves the reaction of two solid inorganic raw materials with a number of different functionalized carboxylic acids in the presence of selected solvents [reaction (2)].

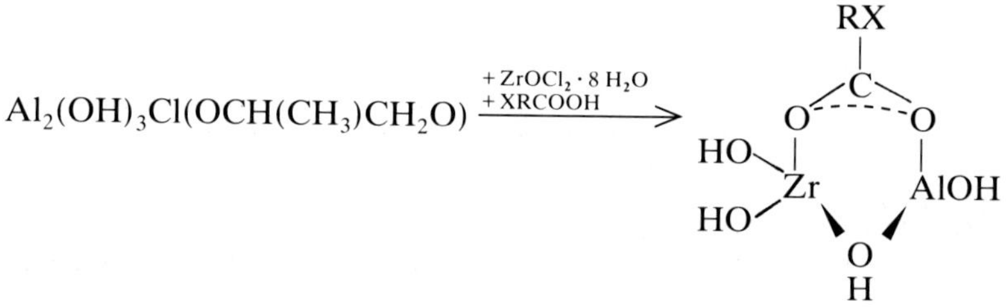

$$Al_2(OH)_3Cl(OCH(CH_3)CH_2O) \xrightarrow[\text{+ XRCOOH}]{\text{+ ZrOCl}_2 \cdot 8\,H_2O} \qquad (2)$$

Such products are found to undergo continued inorganic polymerization resulting from pendant hydroxy groups forming bridges between adjacent metal centers. Kinetic studies of this reaction mechanism indicate that a general rate law may be written as

$$r = k[I]^x,$$

where $x \geqslant 1$, $[I]$ is the concentration of zircoaluminate, and k is the empirical rate constant.

Products synthesized wherein $[I] > 30\%$ by weight result in gross inorganic polymerization occurring sufficiently rapidly so as to preclude any possible usefulness. Therefore, all zircoaluminates are furnished at a concentration of 20–26 wt% active matter. Note: There is one anomalous product, zircoaluminate APG-X, having an active matter content of 42%. Ishida has previously hypothesized the formation of an N-heterocyclic ring with aminopropylsilane [7]. In like fashion, it is thought that reactive ZA metal sites are occupied by a similar ring formation, thereby resulting in reduced inorganic polymerization of the zircoaluminate.

The product mix (Table 1) is observed to vary as a function of both the solvent in which the product is synthesized and the organofunctional group. Appropriate selection will depend on the resin component of a system and also on the zircoaluminate solvent which will determine the dispersibility of the active component and thereby the performance (see Section 4.2.2).

Table 1.
Zircoaluminate products

Commercial designation	Organofunctionality	Solvent
A	Amino	Alcohols
APG	Amino	Propylene glycol
APG-X	Amino	Propylene glycol
C	Carboxy	Alcohols
CPM	Carboxy	Propylene glycol, Meth ether
CPG	Carboxy	Propylene glycol
F	Oleophilic	Alcohols
FPM	Oleophilic	Propylene glycol, Meth ether
M	Methacryloxy	Alcohols
MPM	Methacryloxy	Propylene glycol, Meth ether
MPG	Methacryloxy	Propylene glycol
M-1	Methacryloxy/oleo.	Alcohols
S	Mercapto	Alcohols
APG-1	sec-Amino, pri-Hydroxy	Propylene glycol
APG-2	pri-Amino, sec-Amino	Propylene glycol
APG-3	sec-Amino, pri-Mercapto	Propylene glycol

3. EXPERIMENTAL

3.1. T-Peel strengths of EPDM bonded with adhesive containing ZA or silane

ZA (APG-1, APG-2, APG-3) or silane, 0.90 g, was added to 100.00 g of adhesive (formulation in Table 2) and thereupon mixed with a three-prong impeller for 30 min. The adhesive mix was then brush applied to two 1″ × 6″ strips of EPDM at a thickness of 0.86–0.96 mm. The adhesive was allowed to dry partially for 15 min and the adherends were then bonded. After 168 h of curing (at 20°C, 50% relative humidity), adhesion was determined in accordance with ASTM D-1876.

3.2. Coating evaluation

The ZA was added to the resin at 1.5% on resin solids and thereupon dispersed using a Cowles mixer at high speed for 10 min. All coatings were applied by spraying at 0.051 mm dry film thickness and cured at 77°F, 50% relative humidity for 168 h (alkyd) or 375°F for 12 min (polyester).

Blistering and rusting evaluations were conducted in accordance with ASTM D-2247 and B-117, respectively.

4. ZIRCOALUMINATE PERFORMANCE IN END APPLICATIONS

4.1. Adhesives

The single ply roof membrane industry uses various rubber-based adhesives to bond the seams at which adjacent rolls of membrane overlap. Since such roofs frequently carry 20-year warranties, maximizing adhesion is always desirable. The adhesion problem is exacerbated in that the membranes used are based on highly durable (nonreactive) rubbers like EPDM. Such rubbers contain no polar or organoreactive sites, and are thus a challenging substrate for an adhesive [8].

Zircoaluminates were evaluated in a Kraton (Shell SBS block co-polymer)-based adhesive. The data (Table 2) clearly show that the *in situ* introduction of 3-aminopropyltriethoxysilane (3-APS) increases the T-peel strength to 16.0 kg/cm when compared with the control (2.3–4.6 kg/cm). Three zircoaluminates with different organofunctional groups were evaluated at level equal to or less than that of the of silane (based on material as supplied). All three yielded values greater than or equal to that of the silane-containing specimen. The adhesive containing Manchem APG-3 (primary mercapto, secondary amine) showed a 52% increase in T-peel strength when compared with the same adhesive with the aminosilane.

Table 2.
T-peel strengths or EPDM bonded with zircoaluminate or silane-containing adhesive[a]

Adhesion promoter	T-peel strength (kg/cm)	SD
None	2.3–4.6	—
3-Aminopropyltriethoxysilane	16.0	18.1
APG	16.2	—
APG-2	19.0	20.1
APG-3	24.5	19.5

[a] Formulation: component: Kraton G1650, 65 parts; Kraton G 1657, 35 parts; Neville LX-685, 60 parts; Piccofyn A-135, 60 parts; Picovar AB165, 65 parts; zircoaluminate, 4–6 parts.

4.2. Coatings

Beyond the appearance, a coating may impart corrosion protection to a bare piece of metal. Adhesion is fundamental to this process since it will act to preclude the agents of corrosion—water, oxygen, and salts—from having access to the metal substrate [9]. Zircoaluminates have proven to be outstanding agents for the adhesion promotion of a wide range of both solvent-borne and water-borne coatings to metallic surfaces.

It is envisioned (see Fig. 1 for adhesion mechanism) that upon application of a zircoaluminate-containing adhesive or coating, the zircoaluminate is characterized by a net migration toward the high energy metallic substrate. Upon colliding with the substrate, hydroxy bridging linkages are formed between the ZA metal centers and surface metal sites. Subsequently upon curing, the ZA carboxy group undergoes condensation with the resin terminal hydroxy groups, the result being chemical linkage of the resin to the substrate through a zircoaluminate 'chemical bridge'.

An evaluation was conducted to identify the benefits that might accrue when ZAs were used in high solid polyesters, high solid epoxies, and conventional alkyds, in each instance the coating being applied to two or more different surfaces. Recognizing that alkyds and polyesters are both characterized by available carboxy and hydroxy groups, the decision was made to evaluate amino and carboxy ZAs in these systems (Table 3).

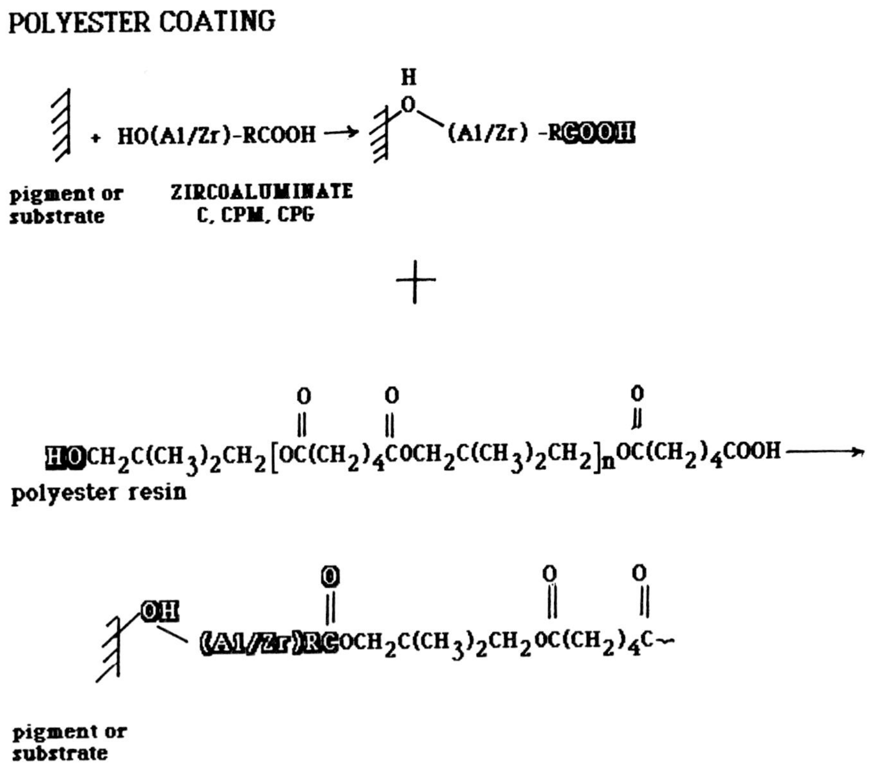

POLYESTER COATING

Figure 1. Postulated mechanism for enhanced adhesion by zircoaluminates.

Table 3.
Zircoaluminate type selection as a function of the resin

| Resin type | Zircoaluminate | | | | | |
	A	APG	APG-2	C	CPM	CPG
Alkyd		X		X		X
Polyester				X	X	X
Epoxy	X	X	X			X

Evaluation of salt fog (ASTM B-117) and humidity resistance (ASTM D-2247) after 300 h of exposure was conducted on scribed panels of cold rolled steel (CRS), phosphatized steel, oily CRS, and aluminum. In accordance with test protocols, performance was rated from 2 to 10, 2 representing the most extensive rusting and largest blisters and 10 representing the absence of any rust or blisters. Additionally, the frequency of the blisters was assessed by indicating dense (D), medium (M), or few (F).

4.2.1. Polyester coating. A high solids polyester baking enamel (Cargill 5776) was prepared and spray-applied to yield a 50 μm dry film thickness (Table 4).

The performance of the CRS control was not surprising (Table 5). Blistering was substantial both at the scribe and overall and rusting was similarly extensive.

Table 4.
High solids polyester enamel formulation

Cargill 5776 (85%)	278.4
MIAK	77.5
Manchem C	6.3 (1.5 phr)
Manchem CPM	
Manchem CPG	
Manchem D/C	
Zopaque RCL-9	422.8
Cargill 5776 (85%)	188.5
Cymel 303	132.1
Byk 303	0.9
Byk VP-451	10.6
MIAK	77.5
Total weight (lbs)	1194.6
Total yield (gals)	100.0

Table 5.
Rusting and blistering of high solids polyester coatings on metal substrates. Salt fog resistance (300 h)

Substrate	Zircoaluminate			
	None	C	CPM	CPG
1. CRS				
Blistering				
Overall	4F	8F	8F	8F
Scribe	2F	4F	4F	2F
Rusting	6	8	8	10
2. CRS (oily)				
Blistering				
Overall	4MD	4M	6M	9F
Scribe	2M	2M	2F	2M
Rusting	6	6	6	8
3. Phosphatized steel				
Blistering				
Overall	10	10	10	10
Scribe	6MD	6M	6M	8M
Rusting	10	10	10	10
4. Aluminum				
Blistering				
Overall	10	10	10	10
Scribe	4F	10	10	6F

The use of ZA C, CPM, and CPG all resulted in substantial improvements, with CPG showing the optimal tool performance. Likewise, on oily steel, CPG provided optimal performance. Further specific adhesion data, although desirable, are not available.

Adhesion may also be improved by mechanical means which is dependent on maximizing the surface area exposed. The deposition of zinc phosphate serves to increase the surface area, and thus the data observed for the phosphatized steel are uniformly superior to the corresponding data for the CRS. Moreover, at 300 h of salt fog exposure, little deterioration of the control occurred.

4.2.2. Alkyd coatings. A medium oil conventional solids alkyd (52.5% solids by weight, Reichhold Beckosol 11-070) was prepared and spray-applied as already described to both CRS and phosphatized steel (Table 6).

The coating was cured at 25°C and 50% relative humidity for 7 days; hence, all the improvements observed were achieved by enhancement of adhesion which took place in the absence of a thermal input.

The data (Table 7) clearly indicate the appearance of large, dense blisters both at the scribe and overall on CRS. Moreover, the rusting was observed to be extreme, covering more than 33% of the total surface.

Each of the zircoalumintes C, CPG, and APG shows some improvement on CRS. The performance of the CPG and APG is outstanding, manifesting no

Table 6.
Medium oil conventional solids alkyd formulation

Beckosol 11-070 (50%)	403
Mineral spirits	50
Thixogel UP	2
Zopaque RCL-9	300
Manchem C	9 (1.5 phr)
Manchem APG	
Manchem CPG	
Beckosol 11-070 (50%)	200
Mineral spirits	50
Nuxtra Co 6%	0.6
Ca 5%	0.6
Zr 6%	2.5
Exkin No. 2	1.0
Total weight (lbs)	1009.7
Total yield (gals)	100.0

Table 7.
Rusting and blistering of an alkyd enamel coating on metal substrates. Salt fog resistance (300 h)

	Zircoaluminate			
Substrate	None	C	APG	CPG
1. CRS				
Blistering				
Overall	2D	2M	10	10
Scribe	2D	2D	4D	4D
Rusting	2	4	8	8
2. Phosphatized steel				
Blistering				
Overall	2F	10	10	10
Scribe	2D	4M	4M	4M
Rusting	4	10	10	10

blistering overall, only a few isolated rust spots, and even at the scribe fewer and smaller blisters. Both ZA C and CPG contain the same active chemistry. It is therefore postulated that the dramatic difference in performance is attributable to the difference in the respective carrier solvents, propylene glycol in the instance of the CPG and lower alcohols in the case of C. The greater solubility of the propylene glycol as compared with the alcohol solubility in the alkyd resin medium facilitates the preparation of a much finer and a more homogeneous dispersion of the active ZA contained therein.

On phosphatized steel, this alkyd, although exhibiting improved performance relative to the same coating on a CRS substrate, does not compare favorably with the performance of the high solid polyester on the same substrate. It is noteworthy that all three ZAs impart excellent performance to the alkyd on this substrate.

4.2.3. Epoxy coatings. A high solids two-component epoxy coating was prepared (Ciba Geigy Araldiate 6010 epoxy/Polyamide 840 hardener) and spray-applied to CRS and aluminum (Table 8). The coating was cured at 25°C and 50% relative humidity for 7 days.

Epoxy coatings are known to have inherently superior adhesion to metals which may be attributed to polar interactions occurring between sites in the resin backbone and the metal substrate. Therefore, failure is often difficult to observe in less than 1000 h of exposure testing. Since the time available allowed the coatings to undergo only 500 h of exposure testing, the differences are not as dramatic as those observed in the earlier alkyd coating. Nonetheless, ZA APG, in particular, was shown to contribute to reduced blistering on CRS and aluminum.

Table 8.
High solids two-component epoxy formulation

Part A	
Araldite 6010	543.2
Manchem A	7.8 (1.5 phr)
Manchem APG	
Manchem APG-2	
Manchem CPG	
Zopaque RCL-9	350.3
Total weight (lb)	901.3
Part B	
Polyamide 840	233.6
Cavco Mod A	2.3 (1.0 phr)
Cavco Mod APG	
Cavco Mod APG-2	
Cavco Mod CPG	
Total weight (lb)	235.9
Total yield (gals)	100.0

4.3. Rubber

Mercaptosilane has long been recommended for use in crosslinkable rubber compositions [10], i.e. NBR, SBR, etc., both as an adhesion promoter to aid bonding between rubber adherends (or rubber to metal) and to enhance bonding between the rubber backbone and a variety of fillers, i.e. clay, silica, thereby enhancing the physical properties of the filled elastomer. More recently, ZA S (mercaptofunctional) has found application in NBR and SBR-molded articles (Rhone-Poulenc Chemicals Ltd., Manchester, private communication). Compounding of the ZA S into the rubber composition prior to molding enables the manufacturer to achieve enhanced adhesion when the rubber article is bonded to a variety of other polymers. The ZA S affords a 33% increase in adhesion to PVC when compared with the same composition containing a similar amount of mercaptosilane.

5. ZIRCOALUMINATE USE PROCEDURE AND USE LEVEL

Zircoaluminates are predominantly inorganic (50–75 wt%) metal organic adhesion promoters in contrast to the silanes ($< 50\%$). As such, they are highly polar and do not dissolve appreciably or disperse easily in resin/solvent-based media. Thus, the user must employ a high shear mixer (Cowles in paint manufacture, Banbury in rubber manufacture) to achieve as fine and homogeneous a dispersion of ZA as possible within a given system.

Alternatively, the user may elect to combine a specific (surfactant) co-dispersant with the ZA (4:1, ZA:co-dispersant) prior to introduction to the resin. The co-dispersant will frequently aid in the solubilization of the active component in the resin. Although generally not a concern at the recommended levels, caution must be exercised in following this approach since the same anionic co-dispersants will have a tendency to compromise wet adhesion properties.

Zircoaluminate starting point use level is typically 1.5 phr (parts ZA solution per hundred resin). Since the ZA is not specific in terms of the surfaces with which it will react it is recommended that this level be increased by 3–5 wt% on pigment in end products containing high surface area pigments and fillers, i.e carbon black, fumed silica, other organic pigments. Moreover, it may be useful to investigate the performance of ZA at 3.0 phr in systems containing other metal additives with which the ZA might react. A notable example is in alkyd coatings where the ZA will react with classical cobalt driers although with resultant diminution in ZA performance (although no apparent effect on dry time) (Rhone-Poulenc Chemicals Ltd., Manchester, private communication).

6. CONCLUSION

Zircoaluminate adhesion promoters constitute a novel class of compounds which have proven themselves useful in arresting corrosion on coated metals, enhancing adhesion of adhesives to rubber and metal, and improving bonding of formed rubber articles containing ZA to other substrates.

Thus, in alkyd, polyester, and epoxy coatings applied to CRS, phosphatized steel, and aluminum, the use of ZAs APG (aminofunctional) and CPG (carboxyfunctional) has allowed for the virtual elimination of blister formation and corrosion after 300 h of salt fog exposure. The use of multifunctional ZAs in a Kraton base adhesive has allowed for a 52% increase in T-peel strength on EPDM rubber when compared with the same adhesive containing aminofunctional silane. Incorporation of mercaptofunctional ZA into crosslinkable elastomers has yielded a 33% increase in the adhesion of such elastomers to PVC relative to the same elastomer containing mercaptosilane.

High shear mixing of the ZA into the resin is a critical process step toward achieving a fine, homogeneous dispersion of ZA owing to the high polarity of the ZA molecule. Normal use levels of 1.5 phr must be augmented (by further addition of 3–5% on pigment) in the presence of high surface area pigments or other reactive metal additives (3.0 phr recommended).

REFERENCES

1. C. G. Munger, *Corrosion Protection by Protective Coatings*, pp. 206–212. National Association of Corrosion Engineers, Houston, 1984.
2. L. B. Cohen, *Plastics Eng.* **39**, (1983).
3. L. B. Cohen, in: *Surface and Colloid Science in Computer Technology*, K. L. Mittal (Ed.), pp. 171–178. Plenum Press, New York (1987).
4. Cavedon Chemical Co., Inc., US Patent No. 4,539,048 (1985).
5. Cavedon Chemical Co., Inc., US Patent No. 4,539,049 (1985).
6. Cavedon Chemical Co., Inc., US Patent No. 4,764,632 (1988).
7. H. Ishida, in: *Adhesion Aspects of Polymeric Coatings*, K. L. Mittal (Ed.), pp. 45–106. Plenum Press, New York (1983).
8. S. A. Westley, in: *Single Ply Roofing Technology*, W. H. Gumpertz (Ed.), pp. 90–113. American Society for Testing and Materials, Philadelphia, 1982.
9. H. Leidheiser, Jr. and W. Funke, *J. Oil Colour Chem. Assoc.* **70**, 121 (1987).
10. E. P. Plueddemann, *Silane Coupling Agents*, pp. 43–44. Plenum Press, New York (1982).

Silanes and Other Coupling Agents, pp. 569–578
Ed. K. L. Mittal
© VSP 1992.

Metal alkoxide primers in the adhesive bonding of mild steel*

BEENA MENON,† R. A. PIKE‡ and J. P. WIGHTMAN§

Department of Chemistry, Center for Adhesive and Sealant Science, Virginia Polytechnic Institute and State University, Blacksburg, VA 24061, USA

Revised version received 20 June 1991

Abstract—The effects of metal alkoxide type and relative humidity on the durability of alkoxide-primed, adhesively bonded steel wedge crack specimens have been determined. Aluminum tri-sec-butoxide, aluminum tri-tert-butoxide, tetrabutyl orthosilicate, and titanium(IV) butoxide were used as alkoxide primers. Grit-blasted, acetone-rinsed mild steel adherends were the substrates bonded with epoxy and polyethersulfone. The two aluminum alkoxides significantly enhanced the durability of the adhesively bonded steel, while the titanium alkoxide showed no improvement in durability over a nonprimed control. The silicon alkoxide-primed samples gave an intermediate response. The failure plane in the adhesively bonded samples varied with the relative humidity during the priming process.

Keywords: Aluminum alkoxide; titanium alkoxide; silicon alkoxide; primers; durability; steel; relative humidity; adhesion.

1. INTRODUCTION

Coupling agents have been used extensively in producing tailored interphases in a number of technologically important areas including fiber sizing and adhesive bonding. Plueddemann [1] pioneered the development of silane coupling agents. Metal alkoxides have been used in sol–gel chemistry and in the formation of glasses for many years [2–6]. A new application for alkoxides is their use as primers on metal adherend surfaces to enhance polymer–metal bond durability [7–13].

Organic primers formulated with corrosion inhibitors are typically applied to pretreated metal surfaces to protect the surfaces prior to adhesive bonding and during environmental exposure. Pike [7–11] found that inorganic primers, such as sec-butyl aluminum alkoxide, improved the durability of aluminum–epoxy bonds when applied to both porous and nonporous aluminum oxide surfaces. It was shown that the effective thickness of the inorganic primer was directly related to the degree of oxide porosity and the depth of the porous oxide layer resulting from the normally used pretreatments for aluminum [10, 11].

Metal alkoxides react rapidly with water. This is due to the presence of electronegative alkoxy groups, which make the metal atoms highly prone to

---

*Paper based on the M.S. thesis of Beena Menon at Virginia Polytechnic Institute and State University.

†Present address: Hoechst Celanese, 3340 West Norfolk Road, Portsmouth, VA 23707, USA.

‡Present address: United Technologies Research Center, East Hartford, CT 06108, USA.

§To whom correspondence should be addressed.

3. RESULTS AND DISCUSSION

3.1. *Crack propagation*

The effect of alkoxide primers on crack propagation in the adhesively bonded joints is shown in Figs 1–5. The effect of alkoxide primer on crack propagation for PES-bonded steel primed at 34% RH and tested by immersion in DI water at 100°C for 200 h is illustrated in Fig. 1, plotted as crack length (in cm) vs. time (in h). The initial crack length for both the AlTSB and AlTTB primed samples was 2.5 cm and the initial crack length for the TBOSi primed sample was 3.5 cm. The initial crack length for both the control, grit-blasted only, and Ti(IV)B primed samples was 4.0 cm. The standard deviation for the initial crack length was ± 0.005 cm and the maximum standard deviation for all other data points on the primed surfaces was ± 0.125 cm. The curves representing the crack propagation in AlTSB and AlTTB bonded systems indicate that the crack propagated fairly slowly and the maximum crack length in both of these systems was 3.5 cm after the pre-crack was initiated. The TBOSi primed system gave an intermediate response and propagated an additional 3 cm after pre-crack initiation. By contrast, the Ti(IV)B and control samples showed rapid crack propagation to 5–6 cm beyond the pre-crack length. Thus, the control and Ti(IV)B primed samples showed the poorest durability. After a period of 200 h, the AlTSB and AlTTB primed samples had not completely delaminated and were failed in an Instron. The TI(IV)B and control samples delaminated completely in the same environment, i.e. immersion in DI water at 100°C.

The effect of alkoxide primer on crack propagation for PES-bonded steel primed at 51% RH and tested by immersion in DI water at 100°C is shown in Fig. 2. The initial crack length for both the AlTSB and AlTTB primed samples and the control was 2.5 cm. The initial crack length of the TBOSi and Ti(IV)B primed samples was 4.5 cm. The AlTSB and AlTTB primed samples exhibited the least crack propagation. After 200 h, the cracks propagated to approximately 5.5 cm. The effect of TBOSi on crack propagation was unlike the case reported in

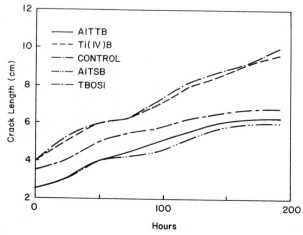

Figure 1. Crack length as a function of time for mild steel primed at 34% RH and bonded with PES and placed in a 100°C water bath.

Fig. 1. The crack propagated rapidly and erratically through the adhesive as observed visually. Although the initial crack length in the control samples was 2.5 cm, the crack propagated an additional 6.5 cm through the bondline under the test environment. The results for the Ti(IV)B samples under these conditions were comparable to the results reported in Fig. 1.

The effect of alkoxide primer on crack propagation for PES-bonded steel primed at 18% RH and tested by immersion in DI water at 100°C is shown in Fig. 3. The initial crack length for both the AlTSB and AlTTB primed samples was 2.7 cm and the same crack length was also observed from the control sample. The initial crack length for both the TBOSi and Ti(IV)B primed systems was approximately 4.5–5.0 cm. As a result, AlTSB and AlTTB primed systems demonstrated the best durability and had to be failed in an Instron after 200 h.

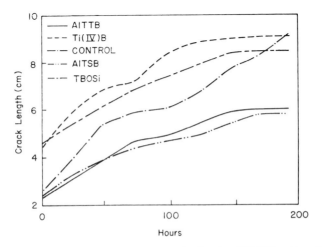

Figure 2. Crack length as a function of time for mild steel primed at 51% RH and bonded with PES and placed in a 100°C water bath.

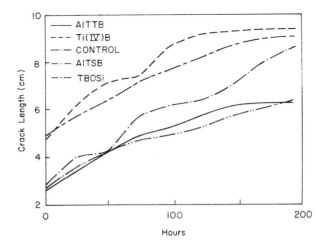

Figure 3. Crack length as a function of time for mild steel primed at 18% RH and bonded with PES and placed in a 100°C water bath.

The crack length in the systems primed at 34% RH and bonded with PES was periodically measured at room temperature for samples aged 1000 h in a desiccator after the wedge was introduced into the bondline. The results are plotted in Fig. 4. The initial crack length for the systems primed with both AlTSB and AlTTB was 2.5 cm while the crack length for the TBOSi and Ti(IV)B primed and control samples was 3.0 cm. The crack propagation curves follow the same pattern seen in Figs 1 and 2. As before, the performance of the alkoxides fell into two groups, with the AlTSB and AlTTB primed samples showing the better durability. With these aluminum alkoxide primers, the crack propagated 1.5 cm and stopped.

The crack propagation in the system primed under 34% RH bonded with FM 300U and immersed in DI water at 100°C was also monitored. Results of this study are reported in Fig. 5. The initial crack length for both the AlTSB and AlTTB primed systems was 2.5 cm. The initial crack length for the control

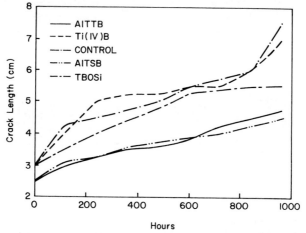

Figure 4. Crack length as a function of time for mild steel primed at 34% RH and bonded with PES and placed in a desiccator at room temperature.

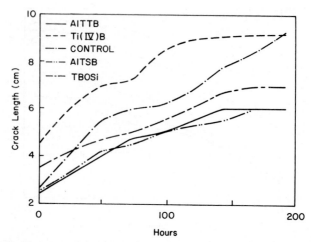

Figure 5. Crack length as a function of time for mild steel primed at 34% RH and bonded with FM 300U and placed in a 100°C water bath.

FM 300U system was also 2.5 cm, while the initial crack length for the TBOSi and Ti(IV)B primed systems was 3.5 and 4.5 cm, respectively. The crack propagation in the AlTSB and AlTTB primed systems was slow and the maximum total crack length attained was 5.8 cm. The Ti(IV)B primed samples and control showed rapid crack propagation and the joints completely delaminated during the test while the TBOSi gave an intermediate response.

Based on the results for the control wedge specimens in Figs 1 and 5, it was concluded that the FM 300U–unprimed steel interface is more durable than the PES–steel interface. Use of the aluminum alkoxide primers eliminated this difference, with both adhesives giving essentially the same response. The effect of humidity on the performance of the aluminum primers was negligible, with the initial crack length and propagation being essentially the same in all cases. The TBOSi primer varied considerably in initial crack length and had the longest crack propagation at the lowest humidity. The titanium primer generally had the longest initial crack length at any humidity; the longest propagation occurred at the lowest humidity. The wedge test results are summarized in Table 1.

Table 1.
Crack length propagation

Alkoxide	Relative humidity (%)	Adhesive	Initial crack length (cm)	Final crack length (cm)	ΔL (cm)
AlTTB	18	PES	2.5	6.0	3.5
	34	PES	2.5	6.0	3.5
	34	PES[a]	2.5	4.0	1.5
	51	PES	2.5	5.5	3.0
	34	FM 300U	2.5	5.5	3.0
AlTSB	18	PES	2.5	6.0	3.5
	34	PES	2.5	6.0	3.5
	34	PES[a]	2.5	4.2	1.7
	51	PES	2.5	5.2	2.7
	34	FM 300U	2.5	5.5	3.0
TBOSi	18	PES	5.0	9.1	4.1
	34	PES	3.5	5.3	1.8
	34	PES[a]	3.0	5.0	2.0
	51	PES	4.5	8.5	4.0
	34	FM 300U	3.5	6.8	3.3
Ti(IV)B	18	PES	4.5	9.5	5.0
	34	PES	4.0	9.0	5.0
	34	PES[a]	3.0	6.8	3.8
	51	PES	4.3	9.0	4.7
	34	FM 300U	4.5	9.0	4.5

[a] Tested at room temperature for 1000 h. All other samples were immersed in DI water at 100°C for 200 h.

3.2 Failure surface analysis

The failure surfaces were analyzed using XPS. The results from the XPS analysis of the samples primed with each of the four alkoxides at 34% RH and the control sample, each bonded with PES and immersed in DI water at 100°C are listed in

Table 2.
XPS analysis of failure surfaces of wedge samples primed at 34% RH and bonded with PES and placed in a 100°C water bath

Alkoxide	Side[a]	Atomic concentration (%)						
		C 1s	O 1s	Fe 2p	S 2p	Al 2p	Ti 2p	Si 2p
AlTTB	A	60	23	1	3	13	—	—
	B	58	22	1	5	14	—	—
AlTSB	A	58	26	1	1	14	—	—
	B	57	27	2	1	13	—	—
Control	A	52	41	5	2	—	—	—
	B	47	45	5	3	—	—	—
Ti(IV)B	A	68	26	2	1	—	3	—
	B	73	24	0.5	0.5	—	2	—
TBOSi	A	70	23	1	2	—	—	4
	B	69	25	2	—	—	—	4

[a] A and B designate the two surfaces resulting from failure of the wedge sample.

Table 2. The two failure surfaces are denoted as 'A' and 'B'. In the case of the AlTSB and AlTTB primed samples, the atomic concentration of carbon was about 60% on both sides and is possibly due to carbon originating from partially unreacted alkoxide or residual adhesive. The concentration of oxygen was about 25%. Very little iron was detected on either side, which indicates that failure did not occur close to the steel substrate. The atomic concentration of sulfur for the AlTSB case is not high enough to conclude that failure occurred primarily within the adhesive. The atomic concentration of aluminum was about 14% on each of the surfaces of both samples. The XPS results support the assignment that failure occurred mainly within the alkoxide layer. The Ti(IV)B primed and TBOSi primed samples also failed mainly within the alkoxide layer.

Based on the concentration of iron and sulfur in the control samples, it may be concluded that failure has occurred primarily at the interface between the steel substrate and the adhesive.

The XPS analysis of the samples primed at 34% RH indicated that there was a consistent failure of the wedge samples which occurred mainly within the alkoxide layer in all systems. Partial hydrolysis may have resulted in the formaton of a weak hydrated oxide layer and was the zone through which the crack propagated to debond the samples. Based on the relative humidity in the chamber during the priming process and the failure surface analysis results, it was concluded that this level of 34% RH was not sufficient to complete the hydrolysis of the alkoxides and produce a stabilized oxide structure. As noted above, however, the wedge crack results did not indicate any instability.

The XPS analysis of the failure surfaces of the wedge samples primed at 51% RH and bonded with PES and immersed in DI water at 100°C are reported in Table 3. The concentration of sulfur in all of the primed samples bonded with PES was about 5% and suggests that failure occurred primarily within the adhesive. The atomic concentrations of aluminum, silicon, and titanium were below 0.5%, precluding assignment of failure within the alkoxide layer. The control sample failed at the interface between the steel and the adhesive.

Table 3.
XPS analysis of failure surfaces of wedge samples primed at 51% RH and bonded with PES and placed in a 100°C water bath

Alkoxide	Side	Atomic concentration (%)						
		C 1s	O 1s	Fe 2p	S 2p	Al 2p	Ti 2p	Si 2p
AlTTB	A	68	26	1	5	—	—	—
	B	70	24	2	4	—	—	—
AlTSB	A	69	25	—	6	—	—	—
	B	71	24	1	4	—	—	—
Control	A	68	24	3	5	—	—	—
	B	71	23	3	3	—	—	—
Ti(IV)B	A	65	29	2	4	—	—	—
	B	67	28	1	4	—	—	—
TBOSi	A	70	23	2	5	—	—	—
	B	70	25	—	5	—	—	—

Therefore, in general, the samples that were primed at 51% RH exhibited failure primarily in the adhesive. Since failure occurred in the adhesive layer in this case, it may be concluded that the alkoxides form a more stable hydrated oxide layer and therefore once the crack was initiated, it propagated through the weaker adhesive. A possible reason why the crack did not propagate close to the steel substrate was the formation of a strong steel–alkoxide (hydroxide) interface.

The XPS analysis of the samples primed at 18% RH and bonded with PES and immersed in DI water at 100°C are reported in Table 4. The atomic concentration of iron in these samples was about 13%. About 1–2% of sulfur was detected on these surfaces, which indicated that some adhesive was retained on these surfaces. Since iron was seen in such significant amounts on both failure surfaces, it was concluded that failure had occurred within the steel (oxide) layer.

Table 4.
XPS analysis of failure surfaces of wedge samples primed at 18% RH and bonded with PES and placed in a 100°C water bath

Alkoxide	Side	Atomic concentration (%)						
		C 1s	O 1s	Fe 2p	S 2p	Al 2p	Ti 2p	Si 2p
AlTTB	A	61	24	14	1	—	—	—
	B	62	23	14	1	—	—	—
AlTSB	A	61	24	13	2	—	—	—
	B	62	23	13	2	—	—	—
Control	A	68	23	15	—	—	—	—
	B	60	25	15	—	—	—	—
Ti(IV)B	A	62	23	14	1	—	—	—
	B	63	24	11	2	—	—	—
TBOSi	A	62	23	13	2	—	—	—
	B	61	25	13	1	—	—	—

4. CONCLUSIONS

The two aluminum alkoxide primers on mild steel showed improved adhesion and better resistance to crack propagation with both the thermoplastic polyether-sulfone (PES) and the FM 300U thermoset epoxy adhesive. The titanium alkoxide exhibited the poorest resistance to crack growth, while tetrabutyl orthosilicate showed intermediate behavior between the two aluminum alkoxides and the titanium alkoxide.

There is an apparent optimum relative humidity level required to achieve good adhesion and durability. Priming the steel adherends at 18% RH caused failure in the wedge samples within the steel (oxide) layer. Adherends primed at 34% RH failed within the alkoxide primer layer, whereas at 51% RH failure occurred primarily within the adhesive layer. This change in locus of failure with humidity was not evident using the wedge crack test when the adherends were primed with aluminum alkoxides. A peel-type test would probably be more sensitive in detecting these shifts in failure mode.

Acknowledgements

We acknowledge the Commonwealth of Virginia for financial support of the work. We thank American Cyanamid for supplying the FM 300U film adhesive and ICI for providing polyethersulfone. The technical assistance of Mr. Frank Cromer in the Surface Analysis Laboratory at Virginia Tech is appreciated.

REFERENCES

1. E. P. Plueddemann, *Silane Coupling Agents*, 2nd edn. Plenum Press, New York (1990).
2. B. E. Yoldas, *Ceram. Bull.* **54**, 289 (1975).
3. D. C. Bradley, R. C. Mehrotra and D. P. Gaur, *Metal Alkoxides*. Academic Press, London (1978).
4. B. E. Yoldas, *J. Am. Ceram. Soc.* **65**, 387 (1982).
5. B. E. Yoldas, *J. Mater. Sci.* **21**, 1087 (1982).
6. D. G. Altenpohl, *Corrosion* **18**, 143 (1962).
7. R. A. Pike, *Proc. 17th Nat SAMPE Techn. Conf.*, p. 448 (Oct. 1985).
8. R. A. Pike, *Int. J. Adhesion Adhesives* **5**, 3 (1985).
9. R. A. Pike, *Int J. Adhesion Adhesives* **6**, 21 (1986).
10. R. A. Pike and F. P. Lamm, in: *Adhesives, Sealants and Coatings for Space and Harsh Environments*, L. H. Lee (Ed.), pp. 141–152. Plenum Press, New York (1988).
11. R. A. Pike and F. P. Lamm, *J. Adhesion* **26**, 171 (1988).
12. J. A. Filbey and J. P. Wightman, *J. Adhesion* **28**, 23 (1989).
13. S. A. Hardwick, J. S. Ahearn and J. D. Venables, *J. Mater. Sci.* **19**, 223 (1984).
14. R. A. Pike, US Patent 4,623,591, 1986.

Dave Armstrong

Cardinal Newman
Q & A in Theology, Church
History, and Conversion

[I]t will be easy to gather together from friends and strangers a vast number of [my] letters on subjects mostly theological, of which a sufficient number may be found interesting to fill a volume.

--- Blessed John Henry Cardinal Newman, "Memorandum on Future Biography," 15 November 1872

[L]etters always have the charm of reality. I have before now given this as the reason why I like the early Fathers more than the Medieval Saints viz: because we have the letters of the former. I seem to know St. Chrysostom or St. Jerome in a way in which I never can know St. Thomas Aquinas.

--- Cardinal Newman to Mrs. Sconce, 15 October 1865